Metallic Nanoparticles

HANDBOOK OF METAL PHYSICS

SERIES EDITOR

Prasanta Misra

*Department of Physics, University of Houston,
Houston, TX 77204, USA*

Metallic Nanoparticles

Volume Editor

JOHN A. BLACKMAN

*Department of Physics, University of Reading, Whiteknights,
Reading RG6 6AF, UK*

and

*Department of Physics and Astronomy, University of Leicester, University Road,
Leicester LE1 7RH, UK*

ELSEVIER

AMSTERDAM • BOSTON • HEIDELBERG • LONDON • NEW YORK • OXFORD
PARIS • SAN DIEGO • SAN FRANCISCO • SINGAPORE • SYDNEY • TOKYO

Elsevier
Radarweg 29, PO Box 211, 1000 AE Amsterdam, The Netherlands
Linacre House, Jordan Hill, Oxford OX2 8DP, UK

First edition 2009

Copyright © 2009 Elsevier B.V. All rights reserved

No part of this publication may be reproduced, stored in a retrieval system
or transmitted in any form or by any means electronic, mechanical, photocopying,
recording or otherwise without the prior written permission of the publisher

Permissions may be sought directly from Elsevier's Science & Technology Rights
Department in Oxford, UK: phone (+44) (0) 1865 843830; fax (+44) (0) 1865 853333;
email: permissions@elsevier.com. Alternatively you can submit your request online by
visiting the Elsevier web site at http://www.elsevier.com/locate/permissions, and selecting
Obtaining permission to use Elsevier material

Notice
No responsibility is assumed by the publisher for any injury and/or damage to persons
or property as a matter of products liability, negligence or otherwise, or from any use
or operation of any methods, products, instructions or ideas contained in the material
herein. Because of rapid advances in the medical sciences, in particular, independent
verification of diagnoses and drug dosages should be made

Library of Congress Cataloging-in-Publication Data
A catalog record for this book is available from the Library of Congress

British Library Cataloguing in Publication Data
A catalogue record for this book is available from the British Library

ISBN: 978-0-444-51240-6
ISSN: 1570-002X

For information on all Elsevier publications
visit our website at www.elsevierdirect.com

Printed and bound in Hungary

09 10 11 12 13   10 9 8 7 6 5 4 3 2 1

Working together to grow
libraries in developing countries

www.elsevier.com | www.bookaid.org | www.sabre.org

ELSEVIER   BOOK AID International   Sabre Foundation

The Book Series 'Handbook of Metal Physics' is dedicated to my wife

Swayamprava

and to our children

Debasis, Mimi and Sandeep

Preface

Metal Physics is an interdisciplinary area covering Physics, Chemistry, Materials Science and Engineering. Due to the variety of exciting topics and the wide range of technological applications, this field is growing very rapidly. It encompasses a variety of fundamental properties of metals such as Electronic Structure, Magnetism, Superconductivity, as well as the properties of Semimetals, Defects and Alloys, and Surface Physics of Metals. Metal Physics also includes the properties of exotic materials such as High-Tc Superconductors, Heavy-Fermion Systems, Quasicrystals, Metallic Nanoparticles, Metallic Multilayers, Metallic Wires/Chains of Metals, Novel-Doped Semimetals, Photonic Crystals, Low-Dimensional Metals and Mesoscopic Systems. This is by no means an exhaustive list and more books in other areas will be published. I have taken a broader view and other topics, which are widely used to study the various properties of metals, will be included in the Book Series. During the past 25 years, there has been extensive theoretical and experimental research in each of the areas mentioned above. Each volume of this Book Series, which is self-contained and independent of the other volumes, is an attempt to highlight the significant work in that field. Therefore the order in which the different volumes will be published has no significance and depends only on the timeline in which the manuscripts are received.

The Book Series "Handbook of Metal Physics" is designed to facilitate the research of Ph.D. students, faculty and other researchers in a specific area in Metal Physics. The books will be published by Elsevier in hard cover copy. Each book will be written by either one or two authors who are experts and active researchers in that specific area covered by the book or multiple authors with a volume editor who will co-ordinate the progress of the book and edit it before submission for final editing. This choice has been made according to the complexity of the topic covered in a volume as well as the time that the experts in the respective fields were willing to commit. Each volume is essentially a summary as well as a critical review of the theoretical and experimental work in the topics covered by the book. There are extensive references after the end of each chapter to facilitate researchers in this rapidly growing interdisciplinary field. Since research in various sub-fields in Metal Physics is a rapidly growing area, it is planned that each book will be updated periodically to include the results of the latest research. Even though these books are primarily designed as reference books, some of these books can be used as advance graduate-level textbooks.

The outstanding features of this Book Series are the extensive research references at the end of each chapter, comprehensive review of the significant theoretical work, a summary of all important experiments, illustrations wherever necessary, and discussion of possible technological applications. This would spare the active researcher in a field to do extensive search of the literature before she or he would start planning to work on a new research topic or in writing a research paper on a piece of work already completed. The availability of the Book Series in hard copy would make this job much simpler.

Since each volume will have an introductory chapter written either by the author(s) or the volume editor, it is not my intention to write an introduction for each topic (except for the book being written by me). In fact, they are much better experts than me to write such introductory remarks.

Finally, I invite all students, faculty and other researchers, who would be reading the book(s) to communicate to me their comments. I would, particularly, welcome suggestions for improvement as well as any errors in references and printing.

Acknowledgements

I am grateful to all the eminent scientists who have agreed to contribute to the Book Series. All of them are active researchers and obviously extremely busy in teaching, supervising graduate students, publishing research papers, writing grant proposals and serving on committees. It is indeed gratifying that they have accepted my request to be either an author or volume editor of a book in the Series. The success of this Series lies in their hands and I am confident that each one of them will do a great job. In fact, I have been greatly impressed by the quality of the book "Metallic Nanoparticles" edited by Professor John Blackman of the University of Reading and the University of Leicester. He is one of the leading experts in the field of Theoretical Condensed Matter Physics and has made significant contributions to the research in the area of Metallic Nanoparticles. In addition to writing several chapters himself, he has assembled a team of experts from Oxford University, the University of Leicester, the University of Strathclyde in Glasgow and AstraZeneca International (all in the U.K.) to write the other chapters of the book.

The idea of editing a Book Series on Metal Physics was conceived during a meeting with Dr. Charon Duermeijer, publisher of Elsevier (she was Physics Editor at that time). After several rounds of discussions (via e-mail), the Book Series took shape in another meeting where she met some of the prospective authors/volume editors. She has been a constant source of encouragement, inspiration and great support while I was identifying and contacting various experts in the different areas covered by this extensive field of Metal Physics. It is indeed not easy to persuade active researchers (scattered around the globe) to write or even edit an advance research-level book. She had enough patience to wait for me to finalize a list of authors and volume editors. I am indeed grateful to her for her confidence in me.

I am also grateful to Dr. Anita Koch, Manager, Editorial Services, Books of Elsevier, who has helped me whenever I have requested her, i.e. in arranging to write new contracts, postponing submission deadlines, as well as making many helpful suggestions.

She has been very gracious and prompt in her replies to my numerous questions. I have profited from conversations with my friends who have helped me in identifying potential authors as well as suitable topics in my endeavor to edit such an ambitious Book Series. I am particularly grateful to Professor Larry Pinsky (Chair) and Professor Gemunu Gunaratne (Associate Chair) of the Department of Physics of University of Houston for their hospitality, encouragement and continuing help.

Finally, I express my gratitude to my wife and children who have loved me all these years even though I have spent most of my time in the physics department(s) learning physics, doing research, supervising graduate students, publishing research papers and writing grant proposals. There is no way I can compensate for the lost time except to dedicate this Book Series to them. I am thankful to my daughter-in-law Roopa who has tried her best to make me computer literate and in the process has helped me a lot in my present endeavor. My fondest dream is that when my grandchildren Annika and Millan attend college in 2021 and Kishen and Nirvaan in 2024, this Book Series would have grown both in quantity and quality (obviously with a new Series Editor in place) and at least one of them would be attracted to study the subject after reading a few of these books.

Prasanta Misra
Department of Physics, University of Houston,
Houston, TX, USA

Volume Preface

Nanoscience is generally defined as the study of phenomena on the scale 1–100 nm. Metallic nanoparticles in that size range are clusters containing a few to about 10^7 atoms. Metallic nanoparticles display fascinating properties that are quite different from those of individual atoms or bulk materials. At the lower end of the size range (say <100 atoms), their properties are affected by the discreteness of their electronic energy levels, which contrasts with the continuum of energy states found in bulk materials. The existence of a surface has a major influence. Atoms at the surface are in a different environment from those in bulk and this will modify the overall electronic, chemical and magnetic properties of the cluster (even for clusters of 2000 atoms, about 20% of the atoms lie on the surface). Throughout the size range, the surface is defining an entity that is smaller than the wavelength of light, and this results in optical properties that are different from those of the bulk material.

Understanding the novel behaviour of these systems provides a challenge to the experimental and theoretical techniques of fundamental science, but coupled with this is their huge potential in nanotechnology. Applications, or potential applications, are diverse. They include, for example, catalysis, chemical and biological sensors, systems for nanoelectronics and nanostructured magnetism (e.g. data storage devices) and, in medicine, there is interest in their potential as agents for drug delivery.

This book describes the production and detection of metallic nanoparticles, their optical, chemical and magnetic properties, theoretical models for describing their structure and properties, and a number of their important applications. A vital first step in understanding the properties of nanoparticles is to study their behaviour as free particles, but in most situations they are deposited on a substrate or embedded in a matrix of another material and it is necessary to study the influence of the environment. In many applications, the attachment of chemical or biological molecules to the nanoparticles is of major interest. These various aspects are discussed.

The book will provide a reference work for researchers who are active in the field, but it is also aimed at newcomers such as research students entering a Ph.D. programme in nanoscience or workers in fields other than physics or chemistry who are users of nanoparticles and wish to gain a deeper understanding of the fundamental science. With the newcomer in mind, we have included an overview of the background to a number of the topics so that these topics should be accessible without the need to consult a more basic text. We have also provided references to books and review articles that may be useful to those readers who wish to pursue particular aspects in more detail.

Nanoscience and nanotechnology are inherently multidisciplinary. The fundamental understanding about the nature of metallic nanoparticles has largely come from the physics and chemistry communities, although interest in nanoparticles and their exploitation spreads across many branches of science. The authors of this volume are physicists and chemists, both experimentalists and theorists. Paul Mulheran and John Blackman are both theoretical/computational physicists, although Mulheran is now based in a Department of Chemical and Process Engineering. Chris Binns and Edman Tsang are both experimentalists, Binns in physics and Tsang in chemistry. Two of Tsang's co-authors (Oduro and Yu) are members of his group and his other co-author (Tam) is based at AstraZeneca, an international pharmaceutical company.

We have attempted to give the book coherence so that it does not read as independent and disconnected chapters. If we have succeeded in this objective, it has been facilitated largely because of existing research collaborations between the contributors.

The authors would like to thank their families for their support and patience while this book was being assembled. Inevitably time spent in writing is time sacrificed elsewhere.

Finally, as editor of this volume, I would like to acknowledge the support of Prasanta Misra (the series editor of the *Handbook of Metal Physics*) and Anita Koch (of *Elsevier*) for their encouragement in bringing this book to fruition.

John A. Blackman
Volume Editor

Contents

Preface	vii
Volume Preface	xi

Chapter 1. Introduction
J.A. Blackman and C. Binns

1.1. Nanoscience and nanotechnology	1
1.2. A note on etymology, neologisms and terminology	3
1.3. Metallic nanoparticles	4
References	13

Chapter 2. Shell Models of Isolated Clusters
J.A. Blackman

2.1. Introduction	17
2.2. Electronic shell structure	19
2.3. Geometric shell structure	35
References	46

Chapter 3. Production of Nanoparticles on Supports Using Gas-Phase Deposition and MBE
C. Binns

3.1. Introduction	50
3.2. Gas-phase nanoparticle production	50
3.3. Deposition of gas-phase nanoparticles	62
3.4. Self-assembly of nanoscale islands on surfaces	67
References	69

Chapter 4. Theory of Cluster Growth on Surfaces
P.A. Mulheran

4.1. Introduction	73
4.2. Monte Carlo simulations	75
4.3. Mean-field rate equations and diffusion-reaction calculations	78
4.4. Scaling analyses of the mean-field rate equations	81
4.5. Confrontation with experiments and the island size distributions	85

4.6. Beyond mean-field theory for island nucleation and growth	89
4.7. Island ripening	97
4.8. Atomistic modelling	101
4.9. Other growth methods	107
References	110

Chapter 5. Chemical Methods for Preparation of Nanoparticles in Solution

C.-H. Yu, Kin Tam, and Edman S.C. Tsang

5.1. Introduction	114
5.2. General synthesis and properties of nanoparticles	114
5.3. Synthesis of monodisperse nanoparticles by chemical methods	124
5.4. Surface chemical modification of nanoparticles	126
5.5. Tailoring the size of nanoparticles	129
5.6. Synthesis of binary metallic nanoparticles	131
5.7. Synthesis of core-shell structures	131
References	139

Chapter 6. Structure of Isolated Clusters

J.A. Blackman

6.1. Introduction	143
6.2. Theoretical methodology	144
6.3. Overview of specific materials	152
6.4. Very large clusters	165
References	168

Chapter 7. Photoexcitation and Optical Absorption

J.A. Blackman

7.1. Introduction	176
7.2. Electron excitation techniques	177
7.3. Optical spectroscopy	196
References	223

Chapter 8. Magnetism in Isolated Clusters

C. Binns and J.A. Blackman

8.1. Introduction	231
8.2. Background	233
8.3. Two experimental techniques	244
8.4. Magnetism in free clusters	254
References	271

Chapter 9. Magnetism in Supported and Embedded Clusters

C. Binns and J.A. Blackman

9.1. Introduction	278
9.2. Magnetic characterisation techniques	280
9.3. Very small supported clusters ($N<10$) and 2D nanostructures	301
9.4. Supported clusters	330
9.5. Clusters embedded in a matrix	333
9.6. Exchange bias	346
References	353

Chapter 10. Some Applications of Nanoparticles

C.-H. Yu, W. Oduro, Kin Tam, and Edman S.C. Tsang

10.1. Introduction	365
10.2. Functionalisation of nanoparticles with biomolecules	366
10.3. Magnetic nanoparticles	370
10.4. Fundamentals in heterogeneous nanocatalysis	374
References	378

Index   **381**

Chapter 1

Introduction

J.A. Blackman[1,2] and C. Binns[2]
[1]*Department of Physics, University of Reading, Whiteknights, Reading RG6 6AF, UK*
[2]*Department of Physics and Astronomy, University Road, University of Leicester, Leicester LE1 7RH, UK*

1.1.	Nanoscience and nanotechnology	1
1.2.	A note on etymology, neologisms and terminology	3
1.3.	Metallic nanoparticles	4
	1.3.1. Size and surface/volume ratio	4
	1.3.2. Geometric and electronic structure	6
	1.3.3. The production of nanoparticles	8
	1.3.4. Photoexcitation and optical absorption	11
	1.3.5. Magnetic nanoparticles	12
	1.3.6. Applications of metallic nanoparticles	13
	References	13

1.1. Nanoscience and nanotechnology

Since the 1980s, there has been a rapid expansion in the field now known as nanoscience [1]. The research area encompasses physics, chemistry, biology, engineering, medicine and materials science, and impacts on other disciplines as well. Nanoscience addresses a large number of important issues, with many of these having the potential for novel technical applications. When the focus moves from the basic science towards the applications, the term nanotechnology is more commonly used.

Nanoscience is generally defined as the study of phenomena on the scale of 1–100 nm, although it is often convenient to extend the range a little at either end. Nanotechnology is the ability to create, control and manipulate objects on this scale with the aim of producing novel materials that have specific properties (functionalised materials).

As defined above, nanoscience is nothing new. Faraday [2] studied colloidal gold particles and lectured on his investigations in 1857. The motivation for his study was the red colour of the gold particles, a striking contrast to the familiar yellow appearance of gold in its bulk form. By a similar argument, the use of small metal particles (smaller than 100 nm) to produce decorative effects in stained glass for church windows or in pottery glazes would date nanotechnology back to mediaeval times or even earlier [3,4].

However, it is the recent invention of a variety of tools for studying systems at the atomic level, coupled with the development of techniques for producing nanoparticles, that has led to the emergence of nanoscience as a new field of study. Of primary importance are the scanning probe microscopes, that make it possible not just to "see" individual atoms and molecules on the surfaces of materials but to move them on the nanoscale as well. The scanning tunnelling microscope (STM), the first scanning probe microscope, appeared in 1982 [5,6] and, in 1986, Gerd Binnig and Heinrich Rohrer were awarded the Nobel Prize in physics for its design. Important also is the ability to study the properties of isolated nanoclusters. Even if the technological interest is in clusters on a surface or embedded in a material, an investigation in their isolated state unencumbered by the background material is an essential first step for understanding their properties. New sources to produce clusters in the gas phase were developed during the 1960s and 1970s [7], but it was in the 1980s that Knight's group [8] first produced clusters of alkali metals with up to about 100 atoms and systematically studied their properties.

Nano-objects have a size that is intermediate between atoms (or molecules) and bulk matter. Even for clusters of 1000 atoms, more than one-quarter of the atoms lie on the surface, resulting in properties that are very different from those of atoms or bulk. The properties vary strongly with the size, shape and composition of the nanoparticle. This has been exploited, certainly since the 1960s, in heterogeneous catalysis [9]. Catalysts comprise metallic nanoparticles dispersed on a porous material, with optimisation for the best reaction rate and selectivity largely by trial and error. However, the catalysts can now be fully characterised using the available nanoscience techniques.

More recently, there have been developments towards biological and medical uses of nanoparticles. The entrapment of anticancer drugs in nanoparticles, and the decoration of the particles with molecular ligands for the targeting of cancerous cells, offers the prospect of more effective cancer therapy with reduced side-effects [10]. An alternative strategy for the use of nanoparticles in cancer treatment is photothermal tumour ablation [11]. Nanoparticles absorb laser light at a characteristic frequency and the associated heating can be used to destroy solid tumours. The aim would be to tune the particle to a frequency for which the absorption by tissue is low.

The binding of biological molecules (DNA) to metallic nanoparticles provides the basis for a number of possible applications. DNA is highly programmable and this characteristic can, in principle, be exploited to self-assemble functionalised nanoparticles into structures with more complex architecture [12]. One possible use is as sensors of biological or chemical molecules.

There is considerable potential for exploiting the magnetic properties of nanoscale structures in spintronics (also known as magnetoelectronics) [13].

High-density data storage is one of the goals, and possible systems for quantum information devices [14] are being explored that are based on the quantum tunnelling of the magnetisation of a cluster through a magnetic anisotropy barrier. The most successful spintronics device to date is the spin valve. This is based on two magnetic layers, one hard and one soft magnetically. An external magnetic field can switch the direction of the magnetisation of the soft layer while leaving that of the other layer unchanged. The switch is accompanied by a sharp increase in the electrical resistance due to spin-dependent scattering of the electrons.

1.2. A note on etymology, neologisms and terminology

The prefix nano- is variously said to derive from the Greek word νάνος or the Latin word nannus, both meaning dwarf. It was adopted as an official SI prefix, meaning 10^{-9} of an SI base unit, at the 11th Conférence Générale des Poids et Mesures (CGPM) in 1960 (Comptes Rendus de la CGPM, 87, Rés. 12) [15,16], although it had informal status before that.

The Oxford English Dictionary (OED) [17] credits Taniguchi [18] with the first use of the word "nanotechnology" in 1974, followed sometime after by Drexler [19] in a book entitled *Engines of Creation*. Interestingly, the OED does not see fit to include "nanoscience" in its compilation of nano- words, although the term was certainly in use by the early 1990s [20,21].

Taniguchi, a precision engineer, had in mind technologies operating to tolerances of less than 100 nm, whereas Drexler's idea of nanotechnology concerned the manipulation and assembly of structures at the molecular scale, a concept already discussed in 1959 by Feynman in a lecture entitled *There's Plenty of Room at the Bottom* [22]. One of the legacies of Drexler's imaginative work is the term "grey goo", which describes a nightmare scenario where nanorobots run out of control and destroy everything while replicating themselves. Unfortunately the term can be (and has been) used to spice up articles about nanoscience in the sensational press and give it a negative slant, inhibiting sensible discussion.

Although nanoscience and nanotechnology are generally the preferred words used to describe national research programmes [23], the terms are so broad that the umbrella covers many topics that have little in common. As a consequence, the prefix is increasingly appearing in front of traditional disciplines as in nanophysics, nanomaterials or nanomedicine. Doubtless the use of the word nano in product names [24], such as the iPod and the car from Tata Motors, provides further confusion for those who are not part of the scientific community.

We shall be concerned in this book with metallic nanoparticles. The scope of material covered is summarised in the rest of this chapter. A brief note on terminology concludes this section. Two main approaches are used in creating nanostructures: "bottom-up" and "top-down". The top-down method involves starting with a larger piece of material and forming a nanostructure from it by removing material through etching or machining. The various lithography techniques (e.g. electron beam and focused ion beam (FIB) lithography) use etching, and are examples of top-down techniques. In the bottom-up approach, molecules or nanoparticles are produced by chemical synthesis or other means, followed by self-assembly into ordered structures by physical or chemical interactions between the units. Feynman was anticipating the bottom-up approach.

1.3. Metallic nanoparticles

Nanoscale particles or clusters (we use the terms interchangeably) can be formed from most elements of the periodic table, and they can be classified as metallic, semiconductor, ionic, rare gas or molecular according to their constituents. Carbon nanoclusters form a special class that includes the famous buckyball (C_{60}), related fullerenes and the carbon nanotubes. Clusters are characterised as homogeneous if they contain a single type of atom, or heterogeneous if they comprise more than one constituent. They may be neutral or charged (anions or cations).

1.3.1. Size and surface/volume ratio

This book is concerned specifically with metallic nanoparticles, their properties, means of production, experimental techniques for studying them, models for describing them and some of their applications. It is instructive first to give a rough estimate of the size of these objects. The simplest description is the liquid drop model (LDM). In the LDM, we ignore the internal structure of the cluster and simply represent it as a sphere of radius R, the size of which is related to the number of atoms N. The relation can be written in terms of the Wigner–Seitz radius r_s. The Wigner–Seitz radius comes from solid state physics [25] and is defined as the radius of a sphere whose volume, v, is equal to the volume occupied by one atom in the bulk material. Strictly, it is the volume per valence electron, but the distinction is immaterial for monovalent metals. Then, setting the volume of the cluster equal to Nv, it immediately follows that

$$R = N^{1/3} r_s. \tag{1.1}$$

Wigner–Seitz radii are usually expressed in atomic units (au). The au of length is the Bohr radius: 1 au = 0.05292 nm. The diameters ($2R$) of Cu and Au clusters as a function of the number of atoms are shown in Table 1.1. We have included $N = 10$ in the table, but it should be noted that there are large departures from the simple picture at small sizes. This is particularly true for Au, which tends to form a planar cluster even at $N = 10$.

The surface of a cluster is the most obvious feature that distinguishes it from bulk material. The fraction of atoms that are on the surface is a measure of how much the cluster differs from the bulk. To get an estimate of this, we can cut out a regularly shaped object from a face-centred cubic (fcc) lattice and count the number of surface atoms for different size clusters. Atom-centred clusters thus formed have six square and eight triangular faces and look like the illustration in Figure 1.1.

1.3. Metallic nanoparticles

Table 1.1. Diameter (2R) of Cu and Au clusters as a function of N, based on Eq. (1.1) and quoted Wigner–Seitz radii of the bulk materials (Cu: 2.669 au; Au: 3.002 au).

N	2R (nm)	
	Cu	Au
10	0.6	0.7
10^2	1.3	1.5
10^3	2.8	3.2
10^4	6.1	6.8
10^5	13.1	14.7
10^6	28.2	31.8
10^7	60.9	68.5
10^8	131.1	147.5

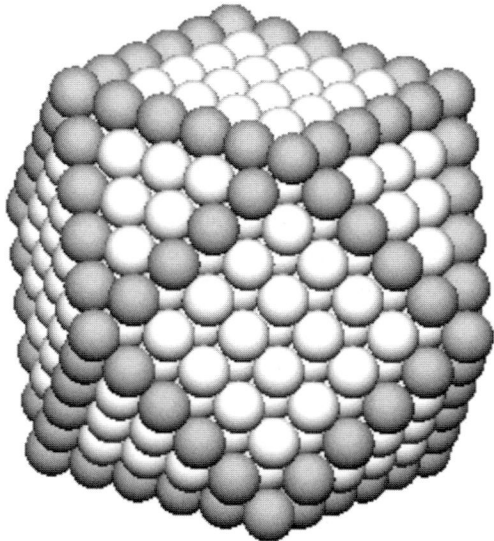

Figure 1.1. A 561-atom cuboctahedral cluster cut from a bulk fcc crystal. Edge atoms are shaded to make the shape clearer.

The fraction of atoms that are on the surfaces of the clusters is shown in Figure 1.2. At $N \approx 300$, half of the atoms are on the surface, and the fraction does not fall below 10% until $N \approx 80000$.

Clusters represent a state of matter that is intermediate between atoms and the solid or liquid state, with properties that depend strongly on the size, shape and material of the particle and also on its environment. At the simplest level, we might expect that an arbitrary property (let us call it X) of a particle of size N could be expressed as

$$X = aN + bN^{2/3}, \tag{1.2}$$

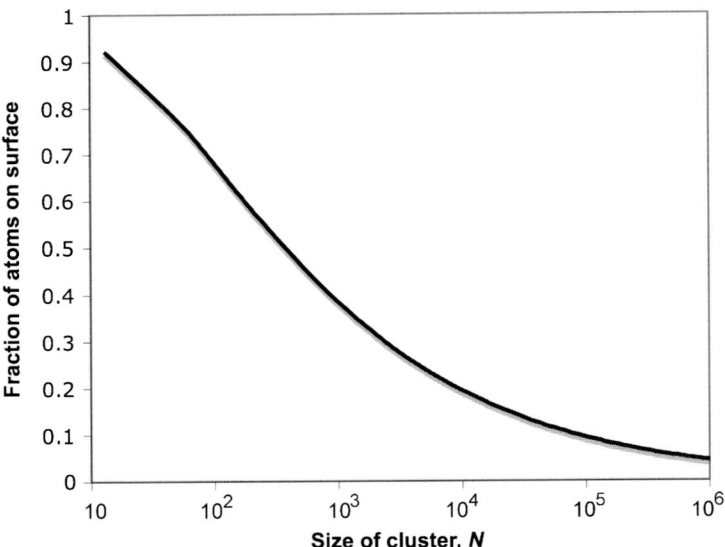

Figure 1.2. Fraction of atoms on the surface of cuboctahedral clusters of N atoms formed from a section of an fcc lattice.

where the first term is a "bulk" contribution that scales with the volume of the particle and the second term, a "surface" contribution, scales with the surface area and represents the deviation from bulk behaviour. On a per atom basis, with $x(N) = X(N)/N$, this can be written as

$$x(N) = a + bN^{-1/3}, \tag{1.3}$$

with convergence to the bulk value, a, as $N \to \infty$.

1.3.2. Geometric and electronic structure

Although Eq. (1.3) often describes a general trend, the behaviour is usually far from monotonic and, particularly for small clusters, oscillations with cluster size are observed. These have a quantum mechanical origin. An early quantum theory of metallic clusters is the electronic shell model. In this model, the valence electrons are treated as free particles confined in a spherical box of radius R (as defined in Eq. (1.1)). This is analogous to the free electron theory of solids [23], except that the solid box has macroscopic dimensions with a continuum of energy levels, whereas the cluster box is a nanoscale entity and the energy levels are discrete.

The energy levels for a spherical box are ordered as in Figure 1.3. Most are degenerate and the numbers of electrons required to fill the states up to each level are indicated in the figure. Clusters with completely filled energy levels are expected to be particularly stable and, for monovalent metals like Na, this simple picture predicts "magic numbers" at $N = 2, 8, 18, 20, 34, \ldots$ Peaks and troughs in the mass spectra of free clusters produced in the gas phase show that certain size particles are

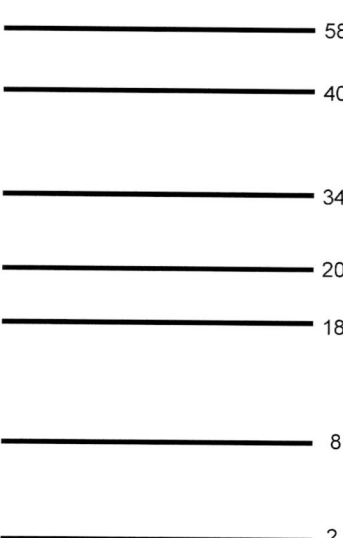

Figure 1.3. First few energy levels for non-interacting electrons in a spherical box. The numbers of electrons required for complete filling of the shells are shown to the right.

generated in greater numbers than others, and this spectral structure can be related to the electronic shell model. Oscillations in properties such as the ionisation potential (the energy required to ionise a cluster) as a function of cluster size can also be related to the electronic shell model. With various refinements, the model has been rather successful in interpreting experimental results for clusters of the simple metals with $N<100$.

An alternative approach suggested by Figure 1.1 is the geometric shell model. The rare gases (except helium) and many metals crystallise in bulk with the fcc structure. This is a close-packed structure that maximises the number (12) of nearest neighbours and also the binding energy. The high coordination is retained in the central part of an fcc cluster but is obviously reduced for the surface atoms. We can see that atoms in the triangular faces in Figure 1.1 have the maximum number of neighbours within the surface layer (6) and nine neighbours in total, but those in the square faces have four surface layer neighbours and only eight in total. It is found that with small deviations from crystallinity, a structure can be formed with close-packing in all its surfaces. This is based on the icosahedron, an object with fivefold symmetry, and is a motif that occurs very frequently in cluster geometries. Of course, fivefold symmetry does not occur in crystalline materials and so the question arises as to the size of cluster at which the structure switches to that of the bulk.

The electronic and geometric shell models have been used with considerable success to describe, respectively, small and large clusters, since the mid-1980s when the first experiments on free clusters were performed. Of course, as the temperature is raised, a point is reached at which any geometric structure is lost. The melting temperature of a cluster is less than that of the bulk and depends on the size of the particle. The shell models and the melting phenomenon are described in Chapter 2.

Most recent theoretical studies of nanoparticle structure are based on density functional calculations where the details of the individual atoms are included explicitly, or by other methods appropriate to the size of the cluster. Besides providing a more accurate description than the shell models, various interesting phenomena can be explored such as the metal–insulator transition. In divalent atoms such as magnesium or mercury, the highest occupied quantum mechanical energy level is an s-state. This is also the case with monovalent elements like sodium but, while the Na s-state contains just one electron, the s-level in the divalent elements is fully occupied with a spin up and a spin down electron. The next highest level is an empty p-state. As the atoms are brought together to form a solid, the s- and p-levels form bands and overlap so that the hybridised s–p band is only partially occupied and the divalent elements are metallic in bulk. Without this overlap, the divalent metals would be insulators. One of the questions for nanoparticle research is the size at which this transition from atomic to bulk-like behaviour occurs. Chapter 6 provides an overview of various theoretical methodologies, a survey of the nanoparticle structures predicted for metallic elements from different parts of the periodic table, and includes a discussion on the metal–insulator transition.

We conclude this section with a note on terminology that could cause confusion. In atoms and molecules, the valence electrons are those in the highest occupied level above a closed electronic shell. For the alkali and noble metals, these would be the s-electrons. In condensed matter physics, one refers to the valence and conduction electron bands. In going from atoms to solids, the valence electrons of the atoms correspond to the electrons in the conduction band rather than the valence band. The use of the terms in the case of nanoparticles should be clear from the context.

1.3.3. *The production of nanoparticles*

Experimentally, one wishes to produce metal nanoparticles with a controlled size so that size-dependent phenomena can be studied experimentally. The four basic generic methods for doing this are summarised in the tableau in Figure 1.4.

(i) Production of pre-formed nanoparticles in the gas phase (Figure 1.4(a))
  There are various methods to produce gas-phase nanoparticles but all involve the production of a super-saturated metal vapour that condenses into particles. These can then be mass-filtered and accumulated in an ion trap or deposited onto a surface or co-deposited with an atomic vapour of another material to produce matrix-isolated nanoparticles [26]. This is the most flexible synthesis technique and allows the production of tightly mass-selected nanoparticles [27] of virtually any material or alloy in environments ranging from free particles in vacuum to embedded nanoparticles in solid matrices.

(ii) Deposition and self-assembly on surfaces (Figure 1.4(b))
  In the simplest case, this can be Volmer–Weber growth on low surface energy materials such as graphite where a degree of size control can be achieved by optimising the coverage, substrate temperature and deposition rate. A more powerful method is to deposit onto surfaces with a natural patterning, such as

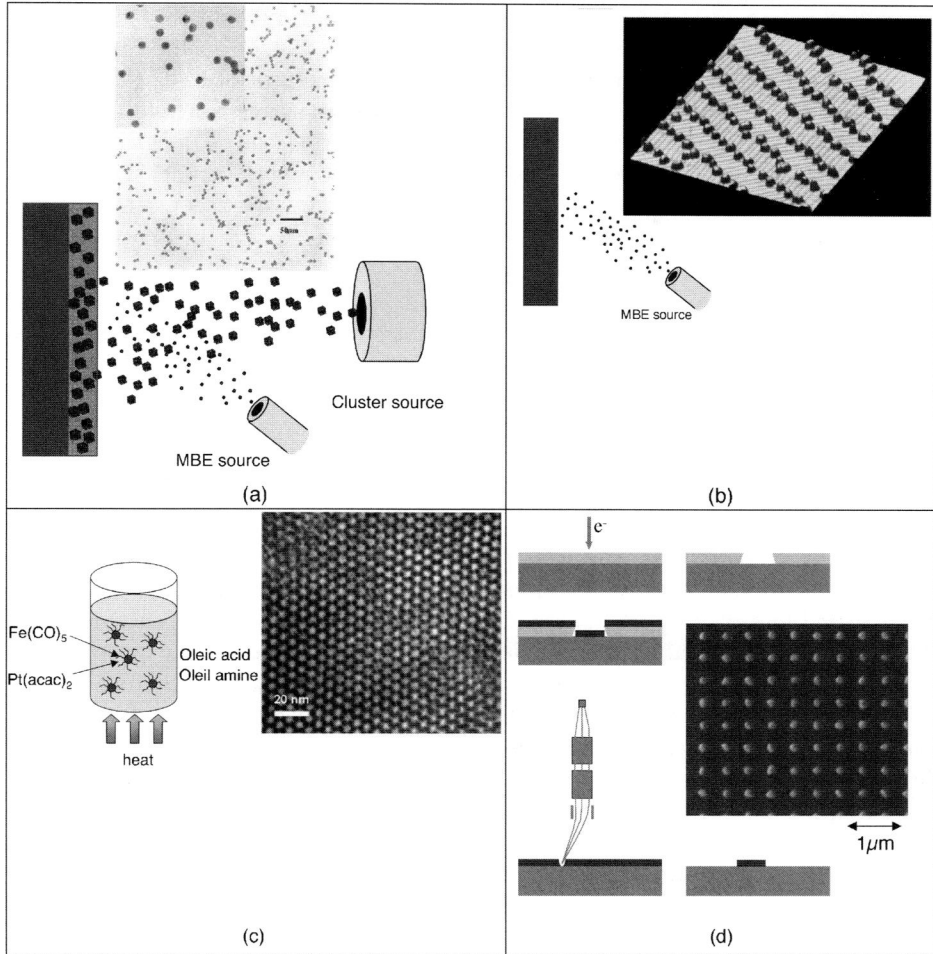

Figure 1.4. Generic methods for producing nanoparticles. (a) Production of gas-phase particles where a supersaturated metal vapour condenses into particles, which can then be size-selected and either studied as free particles in the gas phase or deposited onto a surface. They can also be prepared as matrix-isolated particles by co-depositing a film from a conventional evaporator. The image is from deposited Au nanoparticles [27]. (b) Formation of nanoscale metal islands by depositing an atomic vapour onto a suitable substrate. The image is from Co islands deposited onto a Au(1 1 1) surface with the "herringbone" reconstruction [29]. (c) Wet chemical methods that precipitate nanoparticles into suspension that can then be deposited as self-assembled arrays on surfaces. The image is from an ordered array of 3 nm FePt particles [31]. (d) Top-down methods that cut nanoparticles from larger structures. The image is from 100 nm Co nanoparticles on Si produced by FIB milling. Images reproduced with the permission of (a) the Institute of Physics from Eastham et al. [27], and (b) Elsevier Science from Bansmann et al. [29] and the American Chemical Society from Liu et al. [31].

the "herringbone reconstruction" on Au(1 1 1) surfaces on which metal islands tend to nucleate at specific points on the reconstruction [28]. The metal islands thus formed are then in ordered arrays, which is an advantage for some experimental techniques. Electro-deposited films are also included under this generic heading.

(iii) Wet chemical methods (Figure 1.4(c))

There are a number of methods to reduce metal salts and obtain aggregation into metallic nanoparticles as a suspension in a solvent. A drop of this suspension can then be pipetted onto a surface and with carefully controlled evaporation conditions, the suspension can condense into an ordered flat monolayer on a surface. Alternatively techniques based on Langmuir–Blodgett deposition can be used to obtain the ordered monolayer. Chemical methods can produce a wide variety of metal nanoparticles comprising one or more elements. One of the most widely reported techniques is the so-called polyol method [30]. For example, monodisperse assemblies of FePt alloy nanoparticles with sizes in the range 3–10 nm are produced via the reduction of $Pt(acac)_2$ (acac: acetylacetonate) combined with the thermal decomposition of $Fe(CO)_5$ in the presence of oleic acid and oleyl amine.

(iv) Top-down methods (Figure 1.4(d))

The three generic techniques above are bottom-up methods, where atoms are assembled into nanoparticles. As the name implies, a top-down technique is where nanoparticles are cut from a larger structure, and the two main methods are electron beam lithography (EBL) and FIB milling. In EBL, the required substrate is first spun-coated with an electron resist material, most commonly polymethyl methacrylate (PMMA), which is then exposed to an electron beam focused to a spot ~5 nm across that is rastered to write the desired pattern across the resist. PMMA is a positive resist, that is, exposure to the electron beam makes it impervious to a solvent that is then used to dissolve away all unexposed areas. It is also possible to use negative resists in which only the areas exposed to the electron beam are dissolved. The patterned resist is then used as a mask through which nanoscale metal islands can be deposited onto a substrate. Then after deposition, the rest of the resist material with its metal layer can be rinsed off in a suitable solvent leaving just the nanoscale metal features on the surface. In FIB processing, the metal layer required to be cut into nanoparticles is first deposited onto the required substrate and then the unwanted parts are milled away using a beam of Ga^- ions with a typical energy of 30 kV; the Ga^- ion beam is focused to a spot ~5 nm at the surface. Both techniques have a limiting resolution of ~20 nm at the present state-of-the-art, so they are unable to produce particles as small as the bottom-up approaches. They are however more flexible in the shape of the resulting nanoparticles and can produce square, circular, triangular and elliptical particles as well as nanorings, antidots, etc.

We focus in this book on bottom-up approaches, and discuss the details of methods (i) and (ii) in Chapter 3 and of method (iii) in Chapter 5. The essence of method (ii) is the deposition of atoms on a surface and their diffusion and aggregation to initiate the growth of clusters. The cluster size distribution can be controlled by varying the diffusion rate (via temperature) and the rate of deposition. The theory behind the diffusion, aggregation and growth processes is outlined in Chapter 4.

1.3.4. Photoexcitation and optical absorption

Shining light onto nanoparticles has a number of interesting effects. The photons can be scattered or absorbed. The absorbed photons can create an internal electronic excitation in the particle (optical absorption) or, if they have sufficient energy, can excite an electron out of the particle (photoionisation or photoemission spectroscopy).

The factor determining the optical absorption characteristics is the surface plasmon resonance, which can be excited in particles with a diameter D less than the wavelength λ of light, but not in bulk material. With $D \ll \lambda$, the electric field due to the light beam is virtually uniform within the particle. The electrons in the metal are set into oscillation with respect to the positive background, with a restoring force due to the surface polarisation of the particle. The resonant behaviour, with a peak in the absorption spectrum, typically occurs in the visible part of the spectrum. The theory for the behaviour was first developed by Mie in 1908 [32] using classical dielectric formalism.

For spherical particles, as long as $D \ll \lambda$, there is a single resonance frequency that is independent of the particle diameter. The absorption spectrum is more complex if the particle is not spherical, and there are departures from the classical Mie treatment for small particles (smaller than ~ 100 atoms). In practical situations, metallic nanoparticles are embedded in some medium and, in these cases, the optical absorption is modified by the dielectric function of the embedding material.

Much of the research on small particles has focused on relating the departures from Mie theory to structural models of the particles. The optical absorption phenomenon is also the basis of many applications ranging from the aesthetic one using gold particles mentioned in Section 1.1 to surface enhanced Raman spectroscopy, tumour therapy and the use as sensors of biological or chemical molecules.

To ionise a nanoparticle, by ejecting an electron, requires the energy of photons in the laser beam to exceed a certain threshold. In bulk, this is known as the work function. Ionisation may simply be a necessary first stage to mass analysing clusters in a mass spectrometer to determine the relative probability of forming particles of various sizes (abundance spectra). Alternatively one might focus on the threshold itself and use a tunable laser to measure the threshold (ionisation potential) as a function of cluster size. Peaks in plots of the ionisation potential against cluster size point to clusters of particular stability and this can be related to models of particle structure.

A different experimental strategy is to use a fixed laser frequency (above the threshold) and measure the kinetic energy of the ejected electrons (the terms photoemission or photoelectron spectroscopy, PES, are used). In this process, electrons are being excited from a range of occupied electronic levels of the nanoparticle and the kinetic energy spectrum of the emitted electrons is a map of the particle's energy level density of states. This is a particularly powerful technique because it allows one to monitor the evolution of the quantum mechanical states of the nanoparticle as a function of cluster size, from atomic or molecule like for the very smallest to bulk like for the larger ones.

Photoionisation and PES are discussed in Chapter 7. Photons can also be used to excite electrons from core states of a cluster to unoccupied valence band states. This requires X-ray photons; a powerful technique based on this process is introduced in Section 1.3.5.

1.3.5. *Magnetic nanoparticles*

In bulk materials, magnetism occurs in a limited range of the periodic table. Iron, cobalt and nickel exhibit ferromagnetism due to the unfilled 3d electron bands, while their neighbours chromium and manganese display antiferromagnetism of various degrees of complexity. The rare earths, with unfilled 4f shells, exhibit complex magnetic order. At the atomic level, on the other hand, the majority of elements show a non-zero magnetic moment in the ground state.

The contrast between atomic and bulk behaviour is much stronger in the 3d transition metals than in the rare earths because the 3d orbitals are less localised than the 4f orbitals and there is a much stronger overlap of the former between orbitals on neighbouring atoms when they are brought together to form a solid. The Hund's rule exchange on an individual atom tends to maximise the total spin, but that is competed against by the hybridisation of orbitals on neighbouring atoms and the consequent delocalisation of the electrons.

Nanoparticles are intermediate between atoms and bulk and an enhanced magnetic moment is observed for Fe, Co and Ni clusters up to several hundred atoms in size. It was anticipated that some metals that are non-magnetic in bulk would exhibit a magnetic moment when in the form of small clusters but, apart from rhodium, for which a moment has been observed in clusters of fewer than 100 atoms, the outcome has been negative. The focus remains therefore on metals that are magnetic in bulk.

An essential feature in a description of magnetic behaviour is the anisotropy energy. This arises from dipolar interactions or the spin–orbit interaction and leads to a preferred axis (or axes) along which the moment tends to align (easy axis). Below the Curie temperature, the strong exchange interaction between spins on different atoms aligns them parallel (in the case of ferromagnetism). If a magnetic field is applied to reverse the spin direction, the anisotropy barrier has to be overcome; the strength of field required to do this is known as the coercivity and is a factor in determining the shape of the hysteresis loop.

A magnetic domain structure forms in bulk but clusters smaller than about 10 nm are single domain particles. The size of the anisotropy barrier scales with the volume of the particle, and as the cluster size is reduced there comes a point at which the barrier is insufficient to maintain the stability of the moment against thermal excitations. The temperature at which this instability sets in depends on the particle size and is known as the blocking temperature. We can regard the particle as having a giant moment (the net moment of the particle) and above the blocking temperature this giant moment is able to reorient freely exhibiting what is known as superparamagnetism.

Magnetic grains are used for data storage and the ability to reduce grain size enables an increase in the density at which data can be stored. Data storage media need to operate at a temperature of ~ 350 K and this sets a lower limit to the blocking temperature and consequently size at which magnetic nanoparticles are useful for this application. The search for materials with enhanced anisotropy is a particular focus of research in this field.

We split the discussion of magnetic systems into two sections. Chapter 8 deals with free clusters and describes both theoretical and experimental work on their magnetic structure. The background of atomic and bulk magnetism is summarised, and two experimental techniques (one measures the magnetic moment; the other is a method for inferring atomic structure) are discussed.

Chapter 9 deals with nanoparticles deposited on a surface or embedded in a matrix of another material. The nature of the surface influences the behaviour of the nanoparticles. A basic understanding is obtained from a consideration of clusters of just a few atoms on a variety of different surfaces. We then consider larger particles both supported and embedded. Even at quite modest particle densities, interactions between the particles can be important. The possibility of enhancing the anisotropy through interfacing the nanoparticle with another material is discussed. We also discuss a range of experimental techniques for magnetic characterisation. Most of these are outlined briefly but one, X-ray magnetic circular dichroism (XMCD), is described in some detail. This technique is based on the excitation of electrons from core states of a cluster to unoccupied valence band states and requires use of synchrotron radiation. XMCD is a powerful method because not only does it allow one to probe both the spin and orbital contributions to the magnetic moment, but by tuning the incident X-rays to a particular core level it can also monitor the contributions from a particular element. This is, of course, very valuable for analysing systems comprising more than one element.

1.3.6. *Applications of metallic nanoparticles*

We have alluded to a number of applications of magnetic nanoparticles and some of these are described at appropriate points throughout the chapters. Chapter 10 is devoted entirely to applications with particular focus on clusters produced by chemical preparation methods and their use in biotechnology and heterogeneous catalysis.

References

[1] H.S. Nalwa (Ed.), Encyclopedia of Nanoscience and Nanotechnology, American Scientific Publishers, New York, 2004.
[2] M. Faraday, Philos. Trans. R. Soc. Lond. **147** (1857) 145.
[3] D.J. Barber and I.C. Freestone, Archaeometry **32** (1990) 33.
[4] I. Borgia, B. Brunetti, I. Mariani, A. Sgamellotti, F. Cariati, P. Fermo, M. Mellini, C. Viti and G. Padeletti, Appl. Surf. Sci. **185** (2002) 206.
[5] G. Binnig and H. Rohrer, Helv. Phys. Acta **55** (1982) 128.
[6] G. Binnig, H. Rohrer, C. Gerber and E. Weibel, Phys. Rev. Lett. **49** (1982) 57.
[7] E.J. Robbins, R.E. Leckenby and P. Willis, Adv. Phys. **16** (1967) 739.
[8] W.D. Knight, K. Clemenger, W.A. de Heer, W.A. Saunders, M.Y. Chou and M.L. Cohen, Phys. Rev. Lett. **52** (1984) 2141; **53** (1984) 510(E).
[9] C.R. Henry, Surf. Sci. Rep. **31** (1998) 235.

[10] I. Brigger, C. Dubernet and P. Couvreur, Adv. Drug Deliv. Rev. **54** (2002) 631.
[11] L.R. Hirsch, R.J. Stafford, J.A. Bankson, S.R. Sershen, B. Rivera, R.E. Price, J.D. Hazle, N.J. Halas and J.L. West, Proc. Natl. Acad. Sci. U.S.A. **22** (2003) 47.
[12] J.J. Storhoff and C.A. Mirkin, Chem. Rev. **99** (1999) 1849.
[13] I. Žutić, J. Fabian and S. Das Sarma, Rev. Mod. Phys. **76** (2004) 323.
[14] M.N. Leuenberger and D. Loss, Nature **410** (2001) 789.
[15] B.N. Taylor and A. Thompson (Eds.), The International System of Units (SI), NIST Special Publication 330, 2008 Edition, National Institute of Standards and Technology, Gaithersburg, MD (http://physics.nist.gov/Pubs/SP330/SP330.pdf).
[16] http://www.bipm.org/fr/CGPM/db/11/12
[17] http://dictionary.oed.com (subscriptions).
[18] N. Taniguchi, Proceedings of the International Conference on Production Engineering **T 2** (1974) 18, The Japan Society for Precision Engineering, Tokyo.
[19] K.E. Drexler, Engines of Creation: The Coming Era of Nanotechnology, Anchor Books, New York, 1986.
[20] P.A. Bianconi, J. Lin and A.R. Strzelecki, Nature **349** (1991) 315.
[21] F.A. Buot, Phys. Rep. **234** (1993) 73.
[22] R.P. Feynman, There's Plenty of Room at the Bottom, *Engineering and Science magazine*, Vol. XXIII. No. 5, February 1960, California Institute of Technology, Pasadena (http://www.its.caltech.edu/~feynman/plenty.html).
[23] For example, the US National Nanotechnology Initiative, with five Nanoscale Science Research Centers.
[24] Finlo Rohrer discusses the sprawling use of the word 'nano' in an article entitled 'The joy of nano' (BBC News Magazine, January 2008; http://news.bbc.co.uk/1/hi/magazine/7183085.stm). The most exotic 'nano' product is in the world of haircare: it is the Philips TRESemmé HP4669/17 Salon Shine Nano Diamond Digital Slim Straightener, and it comes with nano-diamond ultragliss ceramic plates. The Tata Nano is 3.1×10^9 nm long.
[25] C. Kittel, Introduction to Solid State Physics, Wiley, New York, 1971.
[26] A. Perez, P. Mélinon, V. Dupuis, P. Jensen, B. Prevel, J. Tuaillon, L. Bardotti, C. Martet, M. Treilleux, M. Broyer, M. Pellarin, J.L. Vaille, B. Palpant and J. Lermé, J. Phys. D **30** (1997) 709.
[27] D.A. Eastham, B. Hamilton and P.M. Denby, Nanotechnology **13** (2002) 51.
[28] S. Padovani, I. Chado, F. Scheurer and J.P. Bucher, Phys. Rev. B **59** (1999) 11887.
[29] J. Bansmann, S.H. Baker, C. Binns, J.A. Blackman, J.-P. Bucher, J. Dorantes-Dávila, V. Dupuis, L. Favre, D. Kechrakos, A. Kleibert, K.-H. Meiwes-Broer, G.M. Pastor, A. Perez, O. Toulemonde, K.N. Trohidou, J. Tuaillon and Y. Xie, Surf. Sci. Rep. **56** (2005) 189.
[30] C. Burda, X. Chen, R. Narayanan and M.A. El-Sayed, Chem. Rev. **105** (2005) 1025.
[31] C. Liu, X. Wu, T. Klemmer, N. Shukla, A.G. Roy, M. Tanase and D. Laughlin, J. Phys. Chem. B **108** (2004) 6121.
[32] G. Mie, Ann. Phys. (Leipzig) **25** (1908) 377.

Further reading

The object of this book is to provide a broad introduction to the subject of metallic nanoparticles. Inevitably some of the topics chosen for discussion will reflect the authors' research interests, but the intention has been to maintain overall a broad scope without confinement by individual predilections.

Some readers will wish to explore various aspects in more detail and a selection from the many available books and review articles is listed below. The ordering in the list implies no particular preferences. However, the subject matter is more general in the upper part of the list and more specific in the lower part, as will be apparent from the titles.

[i] R.L. Johnston, Atomic and Molecular Clusters, Taylor and Francis, London, 2002.
[ii] H. Haberland (Ed.), Clusters of Atoms and Molecules, Vols. I and II, Springer, Berlin, 1994.
[iii] C. Guet, P. Hobza, F. Spiegelman and F. David (Eds.), Atomic Clusters and Nanoparticles, Proceedings of the Les Houches Summer School Session LXXIII, Springer, Berlin, 2001.
[iv] S.N. Khanna and A.W. Castleman (Eds.), Quantum Phenomena in Clusters and Nanostructures, Springer Series in Cluster Physics, Springer, Berlin, 2002.
[v] W. Ekardt (Ed.), Metal Clusters, Wiley, Chichester, 1999.
[vi] W.A. de Heer, The physics of simple metal clusters: Experimental aspects and simple models, Rev. Mod. Phys. **65** (1993) 611.
[vii] M. Brack, The physics of simple metal clusters: Self-consistent jellium model and semiclassical approaches, Rev. Mod. Phys. **65** (1993) 677.
[viii] T.P. Martin, Shells of atoms, Phys. Rep. **273** (1996) 199.
[ix] K.-H. Meiwes-Broer (Ed.), Metal Clusters at Surfaces, Springer Series in Cluster Physics, Springer, Berlin, 1999.
[x] L.D. Marks, Experimental studies of small particle structures, Rep. Prog. Phys. **57** (1994) 603.
[xi] V. Bonačić-Koutecký, P. Fantucci and J. Koutecký, Quantum chemistry of small clusters of elements of groups Ia, Ib, IIa: Fundamental concepts, predictions, and interpretation of experiments, Chem. Rev. **91** (1991) 1035.
[xii] F. Baletto and R. Ferrando, Structural properties of nanoclusters: Energetic, thermodynamic, and kinetic effects, Rev. Mod. Phys. **77** (2005) 371.
[xiii] D.J. Wales, Energy Landscapes with Applications to Clusters, Biomolecules and Glasses, Cambridge University Press, Cambridge, 2003.
[xiv] W. Eberhardt, Clusters as new materials, Surf. Sci. **500** (2002) 242.
[xv] U. Kreibig and M. Vollmer, Optical Properties of Metal Clusters, Springer-Verlag, Berlin, 1995.
[xvi] S. Link and M.A. El-Sayed, Optical properties and ultrafast dynamics of metallic nanocrystals, Annu. Rev. Phys. Chem. **54** (2003) 54.
[xvii] A. Moores and F. Goettmann, The plasmon band in noble metal nanoparticles: An introduction to theory and applications, New J. Chem. **30** (2006) 1121.
[xviii] J.A. Alonso, Electronic and atomic structure, and magnetism of transition-metal clusters, Chem. Rev. **100** (2000) 637.
[xix] C. Binns, Nanoclusters deposited on surfaces, Surf. Sci. Rep. **44** (2001) 1.
[xx] J. Bansmann, S.H. Baker, C. Binns, J.A. Blackman, J.-P. Bucher, J. Dorantes-Dávila, V. Dupuis, L. Favre, D. Kechrakos, A. Kleibert, K.-H. Meiwes-Broer, G.M. Pastor, A. Perez, O. Toulemonde, K.N. Trohidou, J. Tuaillon and Y. Xie, Magnetic and structural properties of isolated and assembled clusters, Surf. Sci. Rep. **56** (2005) 189.
[xxi] C. Binns, K.N. Trohidou, J. Bansmann, S.H. Baker, J.A. Blackman, J.-P. Bucher, D. Kechrakos, A. Kleibert, S. Louch, K.-H. Meiwes-Broer, G.M. Pastor, A. Perez and Y. Xie, The behaviour of nanostructured magnetic materials produced by depositing gas-phase nanoparticles, J. Phys. D: Appl. Phys. **38** (2005) R357.
[xxii] J. Nogués, J. Sort, V. Langlais, V. Skumryev, S. Suriñach, J.S. Muñoz and M.D. Baró, Exchange bias in nanostructures, Phys. Rep. **422** (2005) 65.
[xxiii] U. Hafeli, W. Schütt, J. Teller and M. Zborowski (Eds.), Scientific and Clinical Applications of Magnetic Materials, Plenum, New York, 1997.
[xxiv] M.B. Knickelbein, Reactions of transition metal clusters with small molecules, Annu. Rev. Phys. Chem. **50** (1999) 79.
[xxv] B. von Issendorff and O. Cheshnovsky, Metal to insulator transitions in clusters, Annu. Rev. Phys. Chem. **56** (2005) 549.

[xxvi] C.R. Henry, Surfaces studies of supported model catalysts, Surf. Sci. Rep. **31** (1998) 231.
[xxvii] C. Burda, X. Chen, R. Narayanan and M.A. El-Sayed, Chemistry and properties of nanocrystals of different shapes, Chem. Rev. **105** (2005) 1025.
[xxviii] C.B. Murray, C.R. Kagan and M.G. Bawendi, Synthesis and characterization of monodisperse nanocrystals and close-packed nanocrystal assemblies, Annu. Rev. Mater. Sci. **30** (2000) 545.
[xxix] J.J. Storhoff and C.A. Mirkin, Programmed materials synthesis with DNA, Chem. Rev. **99** (1999) 1849.
[xxx] I. Brigger, C. Dubernet and P. Couvreur, Nanoparticles in cancer therapy and diagnosis, Adv. Drug Deliv. Rev. **54** (2002) 631.

Chapter 2

Shell Models of Isolated Clusters

J.A. Blackman
*Department of Physics, University of Reading, Whiteknights, Reading RG6 6AF, UK;
Department of Physics and Astronomy, University of Leicester, University Road,
Leicester LE1 7RH, UK*

2.1.	Introduction	17
2.2.	Electronic shell structure	19
	2.2.1. Spherical jellium model (phenomenological)	19
	2.2.2. Spherical jellium model (self-consistent)	22
	2.2.3. Ellipsoidal shell model	26
	2.2.4. Non-alkali clusters	28
	2.2.5. Large clusters	31
	2.2.6. Supershells	33
2.3.	Geometric shell structure	35
	2.3.1. Close-packing	35
	2.3.2. Wulff construction	35
	2.3.3. Polyhedra	36
	2.3.3.1. fcc	36
	2.3.3.2. Mackay icosahedron	38
	2.3.3.3. Truncated decahedron	40
	2.3.3.4. bcc	40
	2.3.4. Filling between complete shells	41
	2.3.5. Cluster melting	42
	References	46

2.1. Introduction

Experiments in the early 1980s by Knight *et al.* [1] on molecular beams of alkali metal clusters represented a major advance in the understanding of the structure of metallic clusters. They produced clusters by the supersonic expansion of a metal/carrier gas mixture. The metal is first vaporised and then seeded in the inert carrier gas (typically argon). The supersonic expansion of the mixture, which takes place adiabatically, results in cooling; the vapour becomes supersaturated and clusters condense. The clusters are then ionised and passed through a mass spectrometer to determine their abundance as a function of cluster size N (where N is the number of atoms in the cluster).

Knight et al. [1] observed marked peaks in the mass spectrum of the clusters indicating high stability at particular sizes. These observations were related to an electronic structure model that had its origins in nuclear physics and the "magic numbers" were associated with complete filling of electronic shells. The results for Na clusters are displayed in Figure 2.1(a) and magic number peaks are seen for cluster sizes $N = 8, 20, 40, 58$ and 92. At about the same time, Ekardt [2,3] predicted

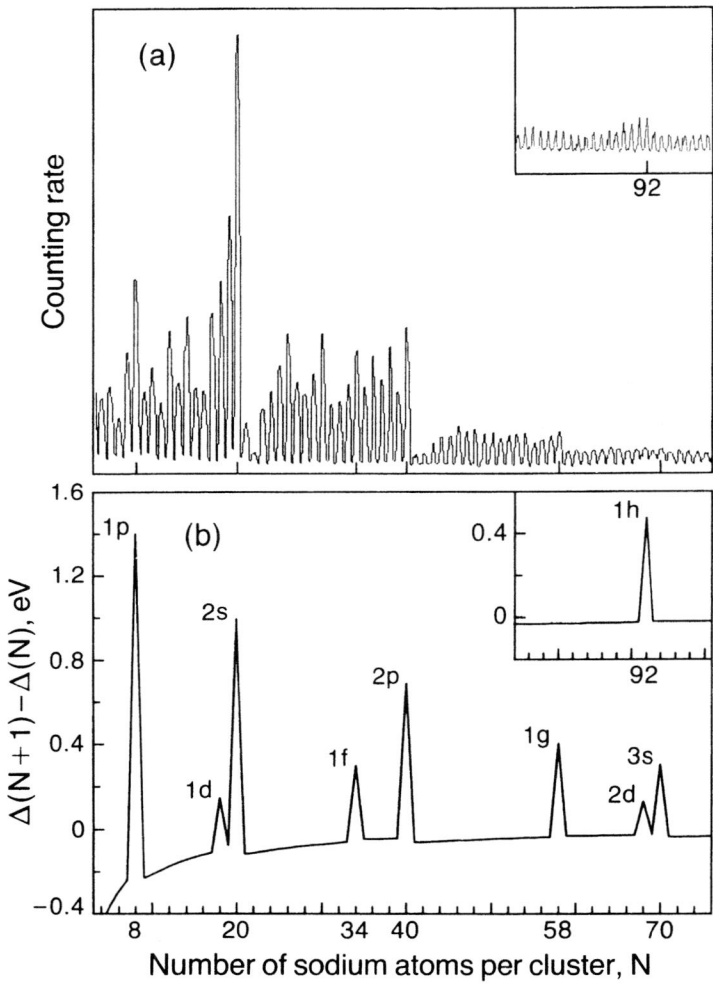

Figure 2.1. (a) The mass spectrum of sodium clusters with $N = 4$–75 and 75–100 (inset); (b) the calculated energy difference between successive clusters as a function of N using the phenomenological electronic shell model (see Section 2.2.1). Peak labelling corresponds to closed-shell orbitals. Reproduced with the permission of the American Physical Society from Knight et al. [1].

this electronic shell structure using a jellium model in which the positive ions are replaced by a uniform charge.

The features in the mass spectra of other alkali metals follow a similar pattern to Na as, to a large extent, do those of the noble metals. For large clusters, the pattern of magic numbers appears different and is more suggestive of geometric shell filling. The geometric shell structure is well known in the rare gas clusters and is characteristic of short-range interactions in which the tendency is for close-packing of the atoms tempered by the need to minimise the surface energy. Packing of atoms to form a cluster with an icosahedral shape is the common form but other polyhedral structures are preferred at very large sizes.

We begin (Section 2.2) with an overview of the main features of the electronic shell model starting with the simplest phenomenological models as they provide a good qualitative account of the main features, and then consider the added sophistication necessary to obtain a more quantitative agreement with experiment. The geometric shell model follows (Section 2.3) with a description of the principal modes of shell filling. We conclude the section with an outline of the melting phenomenon in clusters, which is observed through the loss of the shell signature in the mass spectra with increasing temperature.

The electronic shell model is based on quantum mechanics but ignores the details of individual atoms, whereas the geometric shell model arises from a purely classical energy minimisation based on pairwise interactions between atoms. If there are clear trends in the sequence of magic sizes in a particular system, these models can provide a useful insight into the underlying physics of the clusters, but it is obviously desirable to perform full quantum mechanical calculations that take into account the details of the individual atoms of the cluster. Quantum mechanical calculations will be discussed in Chapter 6.

There are a number of useful reviews that cover the topics of this chapter in some detail. De Heer [4] and Brack [5] give a comprehensive survey of the experimental and theoretical work on metal clusters to the early 1990s, with particular emphasis on interpretations based on the electronic shell model. Martin [6] discusses the growth modes of the geometric shell model and the relevance for large clusters in particular.

2.2. Electronic shell structure

2.2.1. *Spherical jellium model (phenomenological)*

For monovalent simple metals, the bulk properties can be understood in terms of a homogeneous electron gas. The band structure can be described with weak pseudopotentials [7] and there is nearly free electron behaviour. The Fermi surface is approximately spherical and fits in the first Brillouin zone.

The counterpart for a cluster is the spherical jellium model. The details of the ionic cores are ignored completely; they are replaced by a uniform positive charge background of radius R. The electrons move in a mean field that can be calculated self-consistently or, alternatively, they can be considered as independent particles moving in a parameterised phenomenological potential. The latter approach captures the essential quantum mechanics and provides a remarkably good account of the observed behaviour. The basic parameter of the model is the Wigner–Seitz radius, r_s, the radius of a sphere containing one electron. The radius of an N atom (or valence electron) cluster is given by

$$R = r_s(N)^{1/3}. \tag{2.1}$$

Before discussing calculations based on realistic forms for the effective potential, it is useful to consider the electronic structure arising with some simple empirical potentials. For spherically symmetric potentials, the wave function is separable into a radial and an angular part

$$\psi_{n,l,m}(r, \theta, \phi) = R_{nl}(r) Y_{lm}(\theta, \phi), \tag{2.2}$$

with the familiar quantum numbers and $2(2l+1)$-fold degeneracy (including spin) with respect to m. Three potentials, familiar from models in nuclear physics and useful for the study of cluster physics, are the harmonic oscillator, square well and Woods–Saxon potentials shown in Figure 2.2.

The model potentials $V(r)$ leading to the greatest numerical simplicity are the harmonic oscillator potential

$$V(r) = \frac{1}{2} m \omega_0 r^2, \tag{2.3}$$

and the spherical square well potential (constant for $r < R$ and infinity otherwise). The energy levels for the harmonic oscillator potential are equally spaced and highly degenerate and are labelled by the quantum number v as shown in the left-hand side of Figure 2.2

$$E_v = \left(\frac{3}{2} + v\right) \hbar \omega_0. \tag{2.4}$$

The radial wave function $R_{nl}(r)$ for the spherical square well potential is written in terms of spherical Bessel functions $j_l(\kappa_{nl}R)$, where $\kappa_{nl}^2 = 2m|E|/\hbar^2$. The energy levels are determined from the boundary condition $j_l(\kappa_{nl}R) = 0$ and, for each l, the first zero of j_l is given the quantum number $n=1$, the second $n=2$ and so on. The order in which the energy levels occur (1s, 1p, 1d, 2s, 1f, 2p, 1g, 2d, etc.) is shown to the left of panel (b) in Figure 2.2. There is no restriction on the values of l. The ordering rule is: if two solutions have the same number of radial nodes, the one with higher l has the higher energy. Note that the labelling of states originates from nuclear physics and is different from that used in atomic physics. The principal quantum number in atomic physics is equal to $n+l$ in the above notation.

Obviously quantum numbers (n, l) can be used for any spherically symmetric potential. For the harmonic oscillator potential, all orbitals with the same value of $(2n+l)$ are degenerate and the energies are written in terms of the single quantum number $(v = 2n+l-2)$, as can be seen in Figure 2.2.

2.2. Electronic shell structure

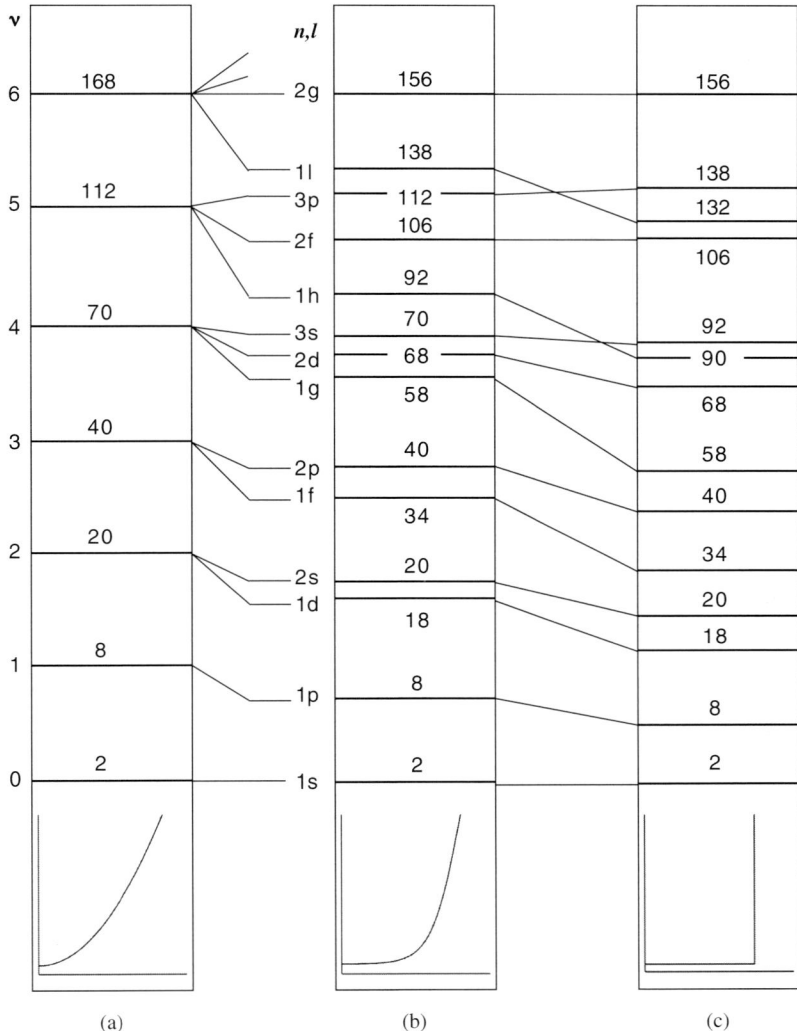

Figure 2.2. Ordering of energy levels for spherical three-dimensional harmonic (a), Woods–Saxon (b) and square well (c) potentials. The cumulative totals of electrons are indicated above the energy levels. See text for level notation. A sketch of the potentials as a function of r is shown at the bottom of the figures.

The potential energy due to the background charge is

$$V(r) = \begin{cases} \dfrac{3e^2 N}{8\pi\varepsilon_0 R^3}\left(\dfrac{r^2}{3} - R^2\right) & \text{if } r<R, \\ \dfrac{e^2 Z}{e^2 N} & \\ \dfrac{e^2 N}{4\pi\varepsilon_0 r} & \text{if } r>R. \end{cases} \qquad (2.5)$$

So the harmonic oscillator potential, at least inside the cluster, mimics the potential felt by the electrons if the electron–electron interaction is ignored completely.

Including the electrostatic potential of the electrons and the exchange–correlation contribution, the interior of the jellium cluster will be electrically neutral and the effective potential will be nearly constant, which to some extent is represented by the spherical square well potential. A better phenomenological representation is a potential that is flat in the middle of the cluster and rounded at the edges and a popular approximation is the Woods–Saxon potential [8]

$$U(r) = -\frac{U_0}{\exp[(r-R)/\varepsilon]+1}. \tag{2.6}$$

The degeneracy of the states is similar to that of the square well potential but the ordering of energy levels is different as can be seen in Figure 2.2. Knight et al. [1] used this potential in the analysis of the mass spectra shown in Figure 2.1. They took U_0 as the sum of the Fermi energy (3.23 eV) and the work function (2.7 eV) of bulk sodium. R was determined by Eq. (2.3) with r_s, the Wigner–Seitz radius of the bulk (3.93 a.u.). The parameter, ε, was taken as 1.5 a.u. to match the variation in potential at the surface obtained in a calculation [9] on a planar surface.

The calculated total energy, $E(N)$, is obtained by summing the eigenvalues of the occupied states. The energy per atom shows minima at the values of N where shell filling occurs. A quantity that reflects the peaks in the mass spectra more closely is the following, which is basically the second derivative of the energy with respect to N:

$$\Delta_2(N) = E(N+1) + E(N-1) - 2E(N). \tag{2.7}$$

Actually, Knight et al. [1] used the first derivative which displays a similar structure and is shown in Figure 2.1. It can be seen that the main magic number peaks in the abundance spectra are reproduced rather nicely and correspond to the stability associated with clusters with filled electronic shells.

2.2.2. Spherical jellium model (self-consistent)

Density functional theory (DFT) is used for a proper treatment of the interacting electron gas. The basic assumption of DFT is that the total energy of the system is a functional of the electron density, $\rho(\bar{r})$. Formally this is based on the Hohenberg–Kohn theorem [10], which states that the exact ground state energy of a correlated electron system is a functional of the density and that it has its minimum for the exact ground state density.

The density, $\rho_I(r)$, of the smeared out positive charge of the ions is written as

$$\rho_I(r) = \rho_0 \theta(r-R), \tag{2.8}$$

where R is the cluster radius given by Eq. (2.1) and ρ_0 the constant bulk density of the metal

$$\rho_0 = \left[\frac{4\pi r_s^3}{3}\right]^{-1}. \tag{2.9}$$

2.2. Electronic shell structure

It is usual to write the total energy as the sum of three parts

$$E[\rho] = E_{\text{kin}}[\rho] + E_{\text{es}}[\rho] + E_{\text{xc}}[\rho]. \tag{2.10}$$

The first term corresponds to the kinetic energy of a system of independent particles with density ρ. The second term is the electrostatic energy

$$E_{\text{es}}[\rho] = \iint d\vec{r}' d\vec{r} \frac{[\rho(\vec{r}) - \rho_1(\vec{r})][\rho(\vec{r}') - \rho_1(\vec{r}')]}{|\vec{r} - \vec{r}'|}, \tag{2.11}$$

and $E_{\text{xc}}[\rho]$ is the exchange–correlation term. The task is to find the variational minimum of $E[\rho]$ subject to the constraint

$$\int \rho(\vec{r}) d\vec{r} = N, \tag{2.12}$$

where N is the total number of valence electrons of the cluster. The units used are atomic units: 1 bohr $= \hbar^2/me^2$ is the unit of length; the energy unit, defined as the Coulomb repulsion between two electrons separated by 1 bohr, is 1 hartree $= me^4/\hbar^2$.

Kohn and Sham [11] proposed writing the electron density in terms of normalised single particle wave functions

$$\rho(\vec{r}) = \sum_{i=1}^{N} |\psi_i(\vec{r})|^2. \tag{2.13}$$

The variation of the energy functional can now be done through the wave functions rather than the density and this leads to the Kohn–Sham equation

$$\frac{1}{2}\nabla^2 \psi_i(\vec{r}) + V_{\text{KS}}(\vec{r})\psi_i(\vec{r}) = \varepsilon_i \psi_i(\vec{r}). \tag{2.14}$$

It is just Schrödinger's equation with an effective potential $V_{\text{KS}}(\vec{r})$, where

$$V_{\text{KS}}(\vec{r}) = V_{\text{H}}(\vec{r}) + V_{\text{xc}}(\vec{r}). \tag{2.15}$$

$V_{\text{H}}(\vec{r})$ is the Hartree term describing the Coulomb interaction of the electrons and background positive charge

$$V_{\text{H}}(\vec{r}) = 2 \int d\vec{r}' \frac{\rho(\vec{r}') - \rho_1(\vec{r}')}{|\vec{r} - \vec{r}'|}. \tag{2.16}$$

The local density approximation (LDA) assumes that $\rho(\vec{r})$ is slowly varying and, in which case, one can write the exchange–correlation part of the total energy as

$$E_{\text{xc}}[\rho] = \int d\vec{r} \rho(\vec{r}) \varepsilon_{\text{xc}}(\rho(\vec{r})) \tag{2.17}$$

and

$$V_{\text{xc}}(\vec{r}) = \frac{d}{d\rho}[\rho \varepsilon_{\text{xc}}(\rho)], \tag{2.18}$$

where $\varepsilon_{xc}(\rho)$ is the exchange and correlation energy of a uniform electron gas of density ρ.

A number of functional forms are used for ε_{xc}. Ekardt [3], for example, used

$$\varepsilon_{xc}(\rho(\vec{r})) = -\frac{0.916}{r_s(\vec{r})} - 0.0666 G(x(\vec{r})), \tag{2.19}$$

where the first term is the exchange part derived by Dirac [12] and $r_s(\vec{r}) = [3/4\pi\rho(\vec{r})]^{1/3}$. The second is the correlation energy functional of Gunnarson and Lundqvist [13]

$$G(x) = (1 + x^3) \ln\left[1 + \frac{1}{x}\right] - x^2 + \frac{x}{2} - \frac{1}{3}, \tag{2.20}$$

where $x(\vec{r}) = r_s(\vec{r})/11.4$.

The procedure then is to obtain the wave functions self-consistently from the Kohn–Sham equation and hence the total energy. Again the wave functions can be split into a radial and an angular part as in Eq. (2.2) and the radial equation is solved numerically. The kinetic contribution to the total energy is evaluated either from

$$E_{kin}[\rho] = \frac{-\hbar^2}{2m} \int d\vec{r} \sum_{i=1}^{N} |\nabla \psi_i(\vec{r})|^2 \tag{2.21}$$

or from

$$E_{kin}[\rho] = \sum_{i=1}^{N} \varepsilon_i - \int d\vec{r} \rho(\vec{r}) V_{KS}(\vec{r}). \tag{2.22}$$

Chou et al. [14] used a local density functional scheme to calculate the energies of Li, Na and K clusters. The Kohn–Sham potentials and energy levels are shown in Figure 2.3 for $N=40$. It can be seen that the potentials are fairly flat in the centre of the cluster reminiscent of the Woods–Saxon potential discussed in the previous section. They obtain an energy derivative that is similar to that shown in Figure 2.1(b) apart from two main differences. One was the absence of peaks at $N=68$ and 70, which is due to the closeness of the 2d, 3s and 1h levels in the self-consistent calculation and is in agreement with experimental observations. The other is a stronger peak at $N=34$ than at $N=40$, which is contrary to experiment and is due to a too large splitting of the 1f and 2p states.

It should be noted that the wave functions $\psi_i(\vec{r})$ used in the density functional calculations are just variational tools used to obtain the approximate ground state density and one should not attach the same physical meaning to them as one would in, say, Hartree–Fock theory. Likewise, the ε_i do not have the meaning of single particle energies. However, an exception can be made [5] for the two energy levels on either side of the Fermi energy (ε_N and ε_{N+1}), which are equivalent to the highest occupied molecular orbital (HOMO) and lowest unoccupied molecular orbital (LUMO) of molecular orbital theory and can be used to estimate the ionisation potential (IP) and electron affinity (EA). This is essentially Koopman's theorem, which is exact in DFT but can be unreliable because of the use of approximate

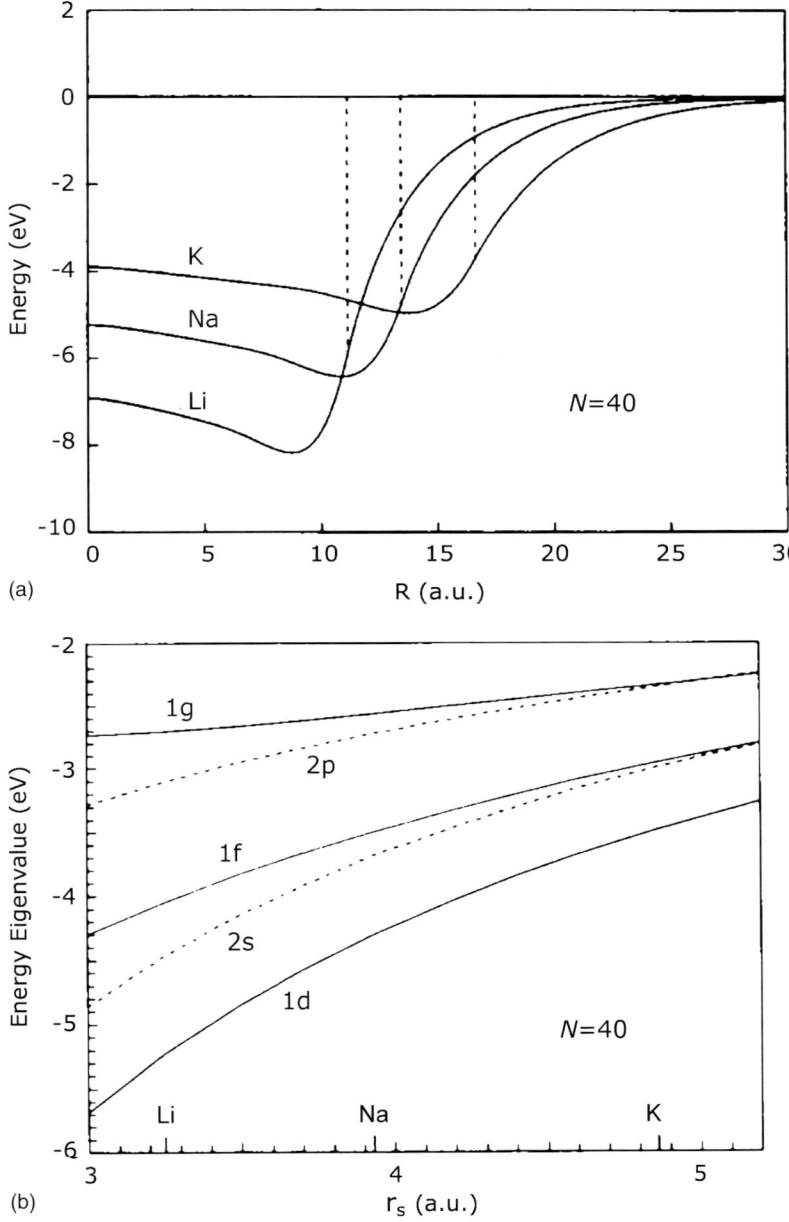

Figure 2.3. (a) Kohn–Sham effective radial electronic potentials of a jellium sphere with 40 atoms of Li, Na and K. Vertical broken lines indicate the radii of the jellium spheres. (b) Electron eigenvalues of a jellium sphere with 40 atoms plotted against Wigner–Seitz radius. The 2p level is just filled for $N=40$. Reproduced with the permission of Elsevier Science from Chou et al. [14].

energy functionals like the LDA [6]

$$IP \approx -\varepsilon_{HOMO} = -\varepsilon_N \tag{2.23}$$

$$EA \approx -\varepsilon_{LUMO} = -\varepsilon_{N+1} \tag{2.24}$$

There are a number of shortcomings of the simple jellium model because of its absence of ionic structure. This was first realised in studies of planar surfaces. Generally it gave good results for alkali metals but failed in higher density materials where, in aluminium, for example, it gave a negative surface energy. The incorrect prediction of the relative stability of the $N=34$ and 40 clusters has been noted above. These difficulties have been overcome by the introduction of ion pseudopotentials by Perdew et al. [15] through a modified exchange–correlation functional in their "stabilised jellium model", which was subsequently applied to clusters [16].

2.2.3. Ellipsoidal shell model

Although the spherical electronic shell model accounts for major features in the mass spectrum of sodium associated with shell filling, there is a lot of structure that is not accounted for. The use of spheres is a justified approximation only for closed-shell structures. For open-shell structures, energy is lowered by a Jahn–Teller distortion [17]. Clemenger [18] borrowed the Nilsson model from nuclear physics [19] that includes ellipsoidal distortions. The model is based on the harmonic oscillator Hamiltonian

$$H = \frac{p^2}{2m} + \frac{1}{2}m\omega_0^2[\Omega_\perp^2(x^2+y^2)+\Omega_z^2 z^2] - U\hbar\omega_0[l^2 - \langle l^2 \rangle_n], \tag{2.25}$$

with two equal axes (x and y) and one unequal axis (z) (an ellipsoid with two equal axes is called a spheroid). Constant volume is maintained by imposing the condition $\Omega_\perp^2 \Omega_z = 1$, and the distortion is expressed by a parameter δ

$$\delta = \frac{2(\Omega_\perp - \Omega_z)}{\Omega_\perp + \Omega_z}. \tag{2.26}$$

The final term is an empirical addition which splits states of different angular momenta and gives the same ordering as the Woods–Saxon potential. It is straightforward to show that $\langle l^2 \rangle_n = (\frac{1}{2})n(n+3)$. The energy of a cluster of size N is minimised with respect to the distortion parameter, δ, and the results are shown in Figure 2.4.

When δ is positive, the potential is enhanced in the x and y directions compared to the z direction and the cluster is elongated along the z-axis (prolate distortion). Oblate distortion and expansion in the x–y plane occurs with negative δ.

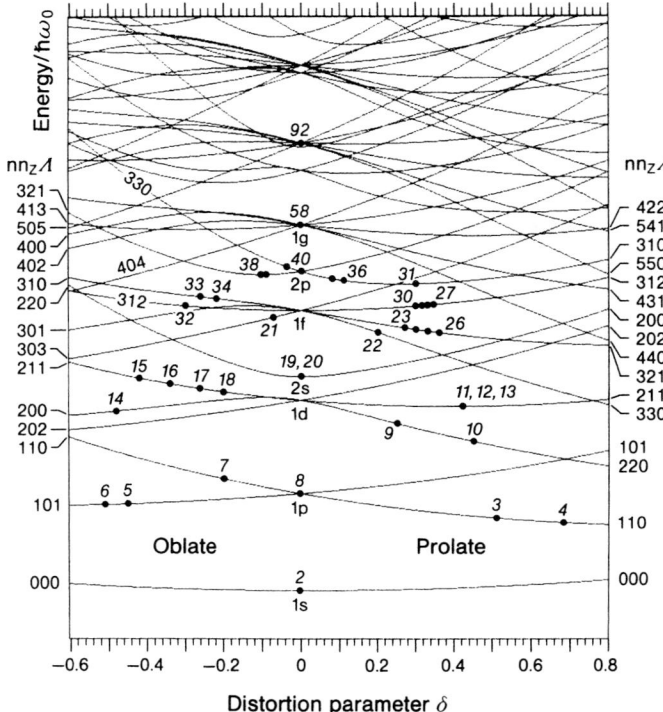

Figure 2.4. Cluster energy levels as a function of the distortion parameter δ for U=0.04. The energy scale is dimensionless. The points on the figure indicate the highest occupied level for a cluster of that size at the value of δ corresponding to the ground state energy. Reproduced with the permission of the American Physical Society from Clemenger [18].

Ignoring the anharmonic terms in Eq. (2.13)

$$E(n_1, n_2, n_3) = \hbar\omega_0 \left[\Omega_\perp(n_1 + n_2 + 1) + \Omega_z \left(n_3 + \frac{1}{2}\right) \right]. \quad (2.27)$$

The shape of the plots in Figure 2.4 can be seen from an expansion of the above equation in powers of δ

$$E(n_1, n_2, n_3) = \hbar\omega_0 \left[n + \frac{3}{2} + \frac{\delta}{3}(n - 3n_3) + \frac{\delta^2}{18}(n + 3n_3 + 3) \right], \quad (2.28)$$

where $n = n_1+n_2+n_3$. The wave functions are characterised by the quantum numbers (n, n_3, Λ), where Λ is the component of l along z. Shell filling occurs with a prolate distortion until half filling and then reverts to oblate. Closed-shell clusters are spherical, but interestingly there are sub-shell closings that are ellipsoidal in the model. $N = 34$ is an example.

The calculated energy differences for sodium as defined by Eq. (2.7) are shown in Figure 2.5. The oscillations in peak height as a function of N that are seen in the

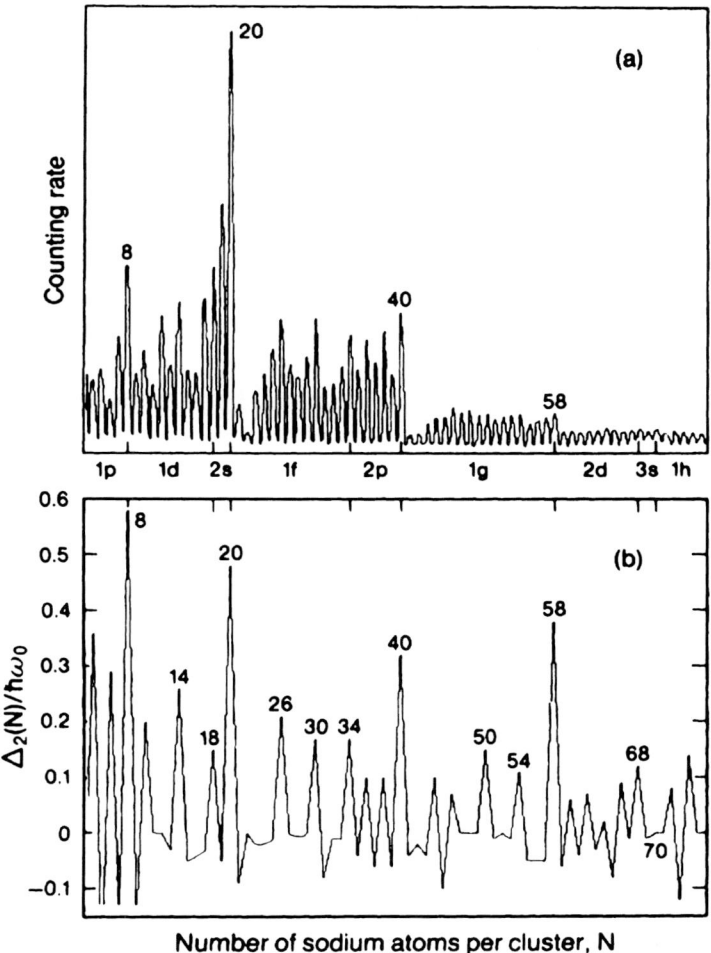

Figure 2.5. Experimental mass spectrum of sodium clusters (as in Figure 2.1) in the upper panel (a), and the second difference of energy (Eq. (2.7)) from the ellipsoidal shell model in the lower panel (b). Reproduced with the permission of the American Physical Society from Clemenger [18].

experimental data are now reflected in the calculations, and clearly the distortions from spherical geometry capture an essential part of the physics of clusters. The even–odd effects are due to a lifting of the orbital degeneracy. Clusters with a doubly occupied HOMO level are more stable than clusters with a singly occupied level and this results in peaks in the abundance spectra and the ionisation potential for even N clusters.

2.2.4. *Non-alkali clusters*

Our discussion so far has focused on alkali clusters as the paradigm of simple metals; they are characterised by electrons in s-atomic orbitals that are non-localised and are essentially free to wander throughout the cluster. The noble metals (Cu, Ag and Au) lie at the end of the 3d, 4d and 5d periods, respectively, with a filled d-shell of 10 electrons and a single valence s-electron. In bulk, the d-band falls well below the Fermi level, and so we might expect that the valence electrons would behave reasonably independently of the d-electrons and lead to behaviour that is similar to that observed in the alkali metal clusters.

The noble metals show clear electronic shell structure in the mass spectra as illustrated for copper and silver in Figure 2.6. In the experiments of Katakuse et al. [20], the clusters were created as positively charged ions, so the number of electrons

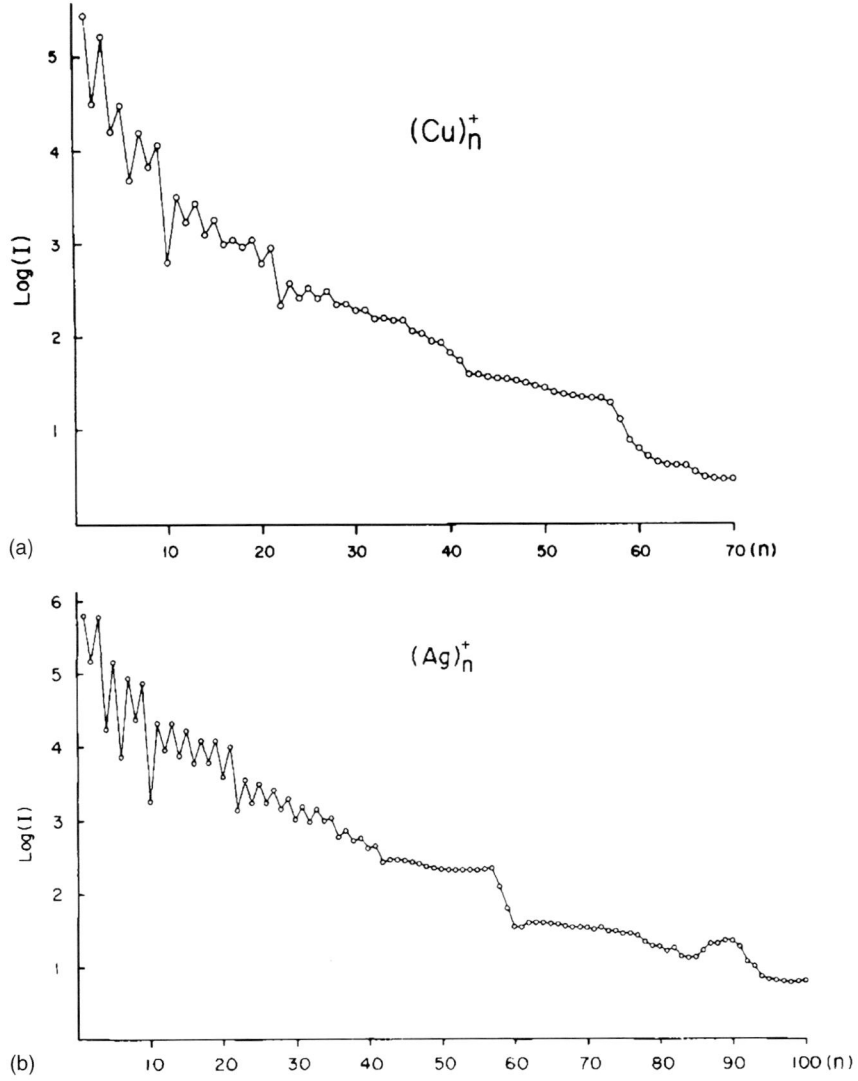

Figure 2.6. Experimental mass spectrum of (a) copper and (b) silver cluster ions. Reproduced with the permission of the Elsevier Science from Katakuse et al. [20].

in a Cu_N^+ or Ag_N^+ cluster is $N-1$. Features at $N-1 = 8$, 20 and 40 corresponding to electronic shell closings are apparent. The one expected at 34 is absent (but see comment earlier). The magic number at 58 which should lead to a shell step at $N = 59$ is in fact smeared out over several clusters. However, the agreement with the electronic shell model is very satisfactory. Similar behaviour is also observed in the negatively charged cluster ions where $N+1$ corresponds to the magic numbers.

If the electronic shell model applies to divalent metals, one would expect that $2N$ would correspond to the shell filling numbers shown in Figure 2.2 leading to magic numbers at $N = 4$, 9, 10, 17, 20, 29, 34, 35, 46, 53, 56, 69, ..., or at least some of them since as we have seen close spaced energy levels can mask the effect of shell filling. Katakuse et al. [21] performed similar experiments on zinc and cadmium to those on the noble metals with the results for the mass spectra shown in Figure 2.7.

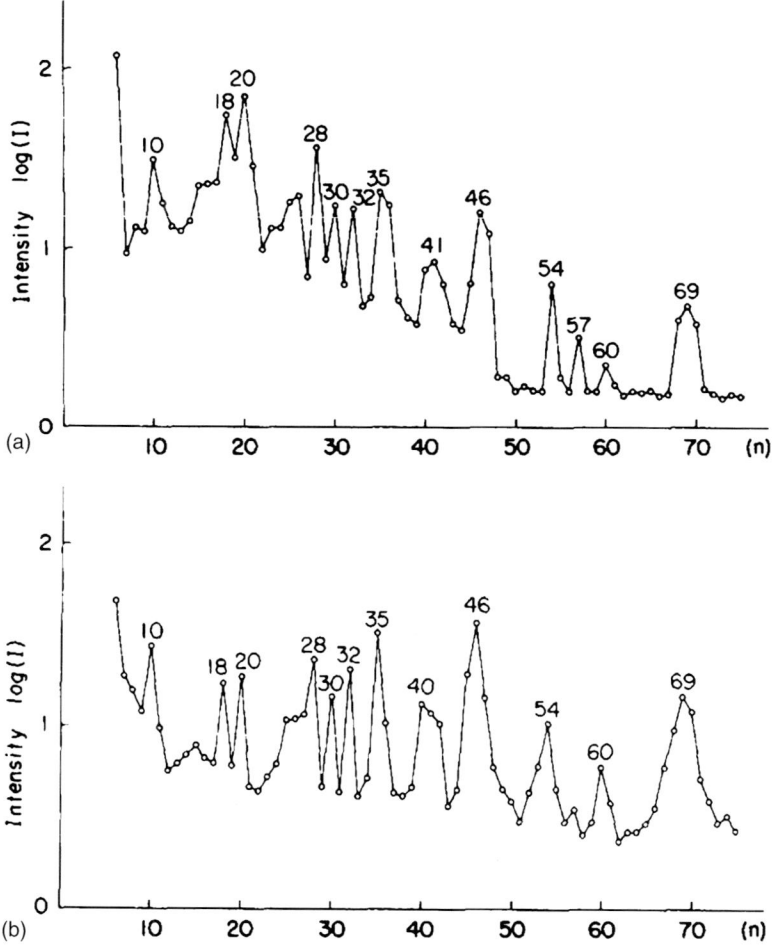

Figure 2.7. Experimental mass spectrum of (a) zinc and (b) cadmium cluster ions. Numbers correspond to spherical shell closings. Reproduced with the permission of the Elsevier Science from Katakuse et al. [21].

The two spectra are very similar and, although there is not a perfect match in the structure with the predictions of the electronic shell model, there are many correspondences, giving a strong indication that the electronic shell model plays a major role. However, hybridisation of the s and p bands is important in the bulk material and the dependence of the degree of hybridisation on cluster size is likely to have an influence on the behaviour.

The development of a clear pattern in the mass spectra based on the electronic shell model is less apparent in small clusters of other metals. In fact, barium clusters show features that have more in common with geometric shell model predictions [4]. Features characteristic of the geometric shell model appear for quite a wide range of materials at larger cluster sizes.

2.2.5. Large clusters

The mass spectra of sodium clusters have been recorded up to sizes of about 25000 atoms with again magic numbers being evident. Portions of spectra using two different laser-light wavelengths for ionisation from the work of Martin et al. [22] are shown in Figure 2.8. Light very close to the ionisation threshold is used to ionise the clusters for mass analysis. Closed-shell clusters have relatively larger ionisation potentials than open-shell clusters with the result that electronic shell filling manifests itself as minima in the mass spectra. The resolution of the minima can be enhanced by varying the wavelength of light.

For these larger clusters, the electronic energy levels tend to bunch together in groups so that we have a series of approximately degenerate levels (shells). From the even spacing of the minima in Figure 2.8, it is apparent that these shells fill on a scale that is proportional to $N^{1/3}$. This assertion can be verified numerically from calculations using the Woods–Saxon potential. However, useful insight is gained by considering the electronic shells that occur for potentials for which exact degeneracy occurs and for which there is an analytic solution.

Consider first the harmonic oscillator potential and the energy level scheme on the left of Figure 2.2. We define a shell index, $K (= v$ in Figure 2.2). Using (n, l) to label states, all states with $K = 2n+l$ are degenerate, as noted in Section 2.2. The number of atoms in a cluster with complete level filling up to and including shell K ($K = 0, 1, 2, \ldots$) is

$$N(K) = \frac{1}{3}(K+1)(K+2)(K+3) \xrightarrow[\text{Large} K]{} \frac{1}{3} K^3. \qquad (2.29)$$

A familiar example of an exactly degenerate set of electronic shells is the $1/r$ potential (i.e. the hydrogen atom). In this case, all states $K = n+l$ are degenerate. Note that the principal quantum number (n_{at}) used in atomic physics is related to the radial quantum number used here by $n_{at} = n+l$. The number of states for a complete filling up to shell K ($K = 1, 2, 3, \ldots$) is

$$N(K) = \frac{2}{3}K\left(K+\frac{1}{2}\right)(K+1) \xrightarrow[\text{Large} K]{} \frac{2}{3} K^3. \qquad (2.30)$$

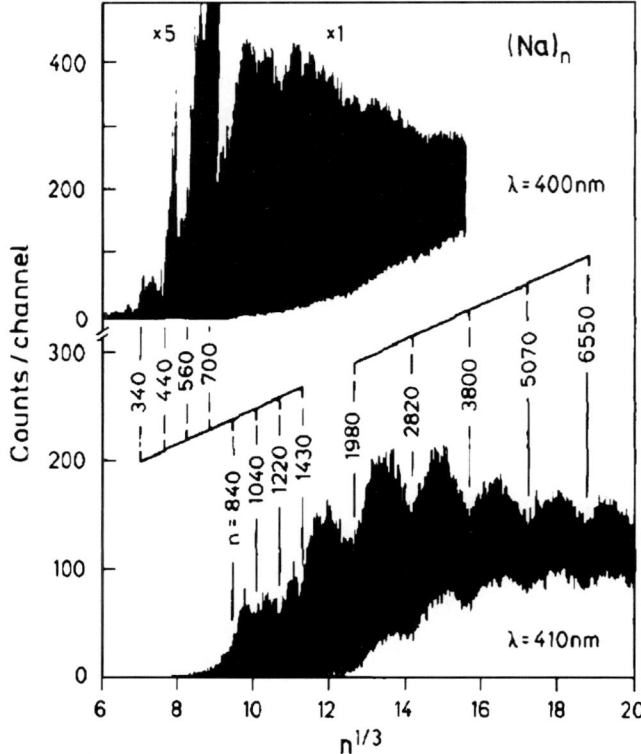

Figure 2.8. Mass spectra of Na$_N$ clusters photoionised with 400 and 410 nm wavelength light. Two sequences of structures are observed at equally spaced intervals on a $N^{1/3}$ scale. Reproduced with the permission of Elsevier Science from Martin et al. [22].

Martin et al. [22] observe that, with the Woods–Saxon potential, approximate degeneracy occurs with $K = 3n+l$ and a numerical estimate of the large K limit of the cluster size for shell filling is

$$N(K) \approx 0.21 K^3. \tag{2.31}$$

It is apparent then that generally shell filling scales as $K \propto N^{1/3}$, which explains the scaling of the magic numbers with $N^{1/3}$ seen in Figure 2.8.

The mass spectra shown in Figure 2.9 extend the size range up to 25000 sodium atoms. The shell index associated with the magic numbers is plotted against $N^{1/3}$ in Figure 2.10. There is a very clear break in the behaviour at about $N = 1500$. Below this size, clusters conform well to the electronic shell analysis. For larger clusters, the geometric shell model appears to give a better account of the stability (see Section 2.3.3). A similar $K \propto N^{1/3}$ dependence arises in the geometric shell model but with a different constant of proportionality, which enables one very easily to distinguish the two regimes.

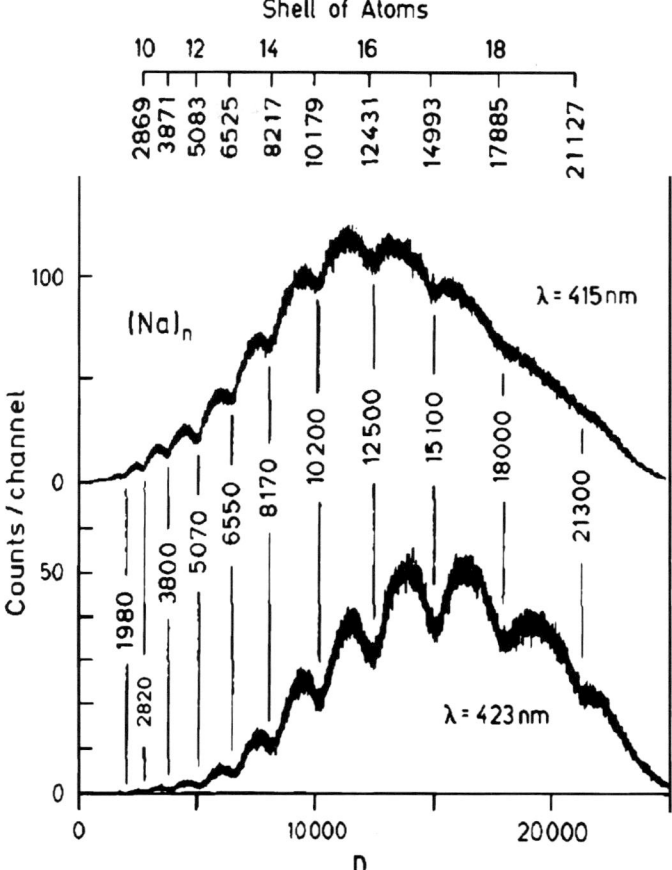

Figure 2.9. Mass spectra of Na$_N$ clusters photoionised with 415 and 423 nm light up to $N=25000$ atoms. Reference is to geometric shells. Reproduced with the permission of Elsevier Science from Martin et al. [22].

2.2.6. Supershells

Nishioka et al. [23,24] performed calculations of shell energies using spherical potential mean field models (including the Woods–Saxon potential) for clusters of up to 4000 valence electrons. Their shell energies were defined as follows:

$$E(N) = \sum_{i=1}^{N} E_i = E_{\text{av}}(N) + E_{\text{shell}}(N), \qquad (2.32)$$

where $E(N)$ is just the sum of the single particle energies within the model and $E_{\text{av}}(N)$ a smoothed average part comprising a volume ($\propto N$) and a surface ($\propto N^{2/3}$) contribution

$$E_{\text{av}}(N) = -aN + bN^{2/3}. \qquad (2.33)$$

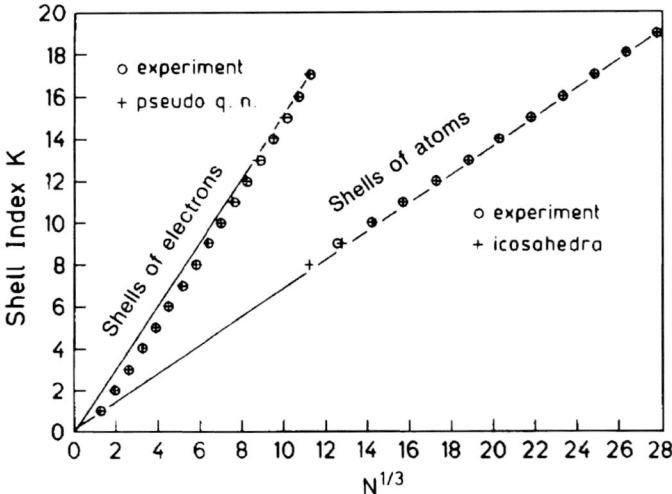

Figure 2.10. Shell index K plotted against $N^{1/3}$ for Na$_N$ clusters. Reproduced with the permission of Elsevier Science from Martin [6].

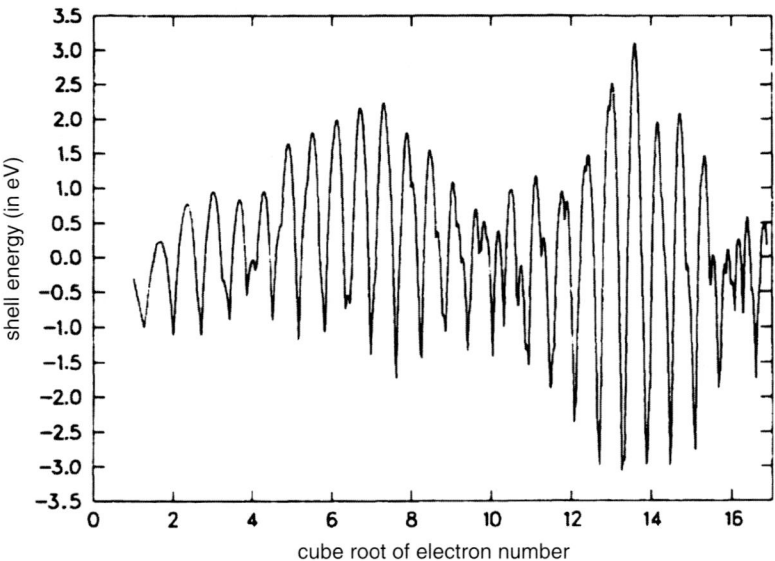

Figure 2.11. The shell part, $E_{shell}(N)$, of the electronic binding energy plotted against $N^{1/3}$. Reproduced with the permission of Springer Science+Business Media from Nishioka [24].

The two parameters were found by a numerical fit. The deviation from average behaviour, $E_{shell}(N)$, represents the contribution from the shell structure. A plot of $E_{shell}(N)$ is shown in Figure 2.11. The minima represent the most stable structures with an even spacing between minima when plotted against $N^{1/3}$, but in addition there is a long period beat pattern with minima in the envelope for N near 500 and

4000 and maxima near 500 and 2500. This supershell structure was explained by an appeal to a semiclassical theory of Balian and Bloch [25].

The analysis is based on the correspondence between classical and quantum physics in the limit of large quantum numbers. The motion of a particle in a quantum state is related to classical motion in a closed orbit. The classical counterparts of the electronic states of a particle moving in a $1/r$ potential are elliptic orbits, for example. The closed orbits of a particle moving in a spherical cavity (square well potential) are polygons (free particle motion with bouncing off the walls). The Woods–Saxon potential is flat in the centre and we can think of it as a cavity with soft walls, so a classical orbit in such a potential will be roughly a polygon with rounded corners. Nishioka et al. [23,24] find that there is interference between the classical triangular and square orbits which accounts for the beat pattern. Experimental evidence of the supershell effect has been claimed in sodium [26], lithium [27] and gallium[28].

2.3. Geometric shell structure

2.3.1. Close-packing

It is clear from Figure 2.10 that a new criterion of stability sets in for sodium clusters larger than about 1500 atoms. This is based on the close-packing of atoms (the fcc or hcp structures in bulk) suitably modified by the presence of a surface (the geometric shell model). For other materials, the geometric shell model is usually the starting point for studies of stability and associated magic numbers and is the dominant mechanism even for quite small clusters. For metals like nickel, say, the bonding is dominated by the d–d interaction which is short-ranged and there is a strong tendency toward a configuration that maximises the number of nearest neighbours. There is a caveat with the transition metals, of course, in that those with unfilled d-shells show directional dependence in their bonding as witnessed by the occurrence of bcc and hcp ordering in bulk iron and cobalt.

The prototype materials for close-packing are the rare gases [29], for which the bonding is via the non-directional van der Waals interaction. They adopt an icosahedral geometry for sizes up to many thousands of atoms. This is a common motif for close-packed structures although, for metals, there is generally a transition to structures other than icosahedral at a much smaller size than for rare gas clusters. We consider now some of the geometries based on close-packing.

2.3.2. Wulff construction

We begin by recalling the macroscopic approach to the stability of metallic particles based on the Wulff construction [30]. The equilibrium morphology is a polyhedron constructed so that the perpendicular distance from the centre of the particle to a face of the polyhedron is proportional to the specific surface energy of the face. Using this procedure for an fcc metal produces the polyhedron comprising eight

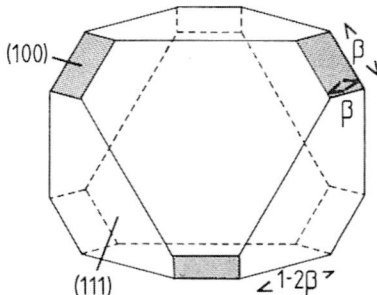

Figure 2.12. Polyhedral structure based on the Wulff construction. Reproduced with the permission of Elsevier Science from Marks [31].

(1 1 1) and six (1 0 0) faces as shown in Figure 2.12. Defining the perpendicular distance of the two faces as p_{111} and p_{100}, it is fairly easy to show that the lengths of the sides of the faces are β and $(1-2\beta)$ in units of $\sqrt{6}p_{111}$, as shown in the diagram. The scale factor, β, is given by $\beta = 1 - p_{100}/\sqrt{3}p_{111}$, and so

$$\beta = 1 - \frac{\gamma_{100}}{\sqrt{3}\gamma_{111}}, \tag{2.34}$$

where γ_{100} and γ_{111} are the specific surface energies for the respective faces.

The simplest model for the surface energy assumes pairwise interactions (denoted by ϕ) between nearest neighbours (broken bond model). The specific surface energy is given by $\gamma = n_d\phi/A$, where n_d is the number of neighbours a surface atom is deficient in compared with bulk and A the surface area per atom. An atom on the (1 1 1) surface has nine neighbours and so is deficient by three compared with an atom in the bulk and so $\gamma_{111} = 2\sqrt{3}\phi/d^2$, where d is the nearest neighbour distance. Similarly $\gamma_{100} = 4\phi/d^2$ since an atom on the (1 0 0) surface has a deficiency of four neighbours. Consequently, $\beta = 1/3$ and the lengths of the sides of the square and hexagonal faces are equal. This is illustrated in Figure 2.13 for a 586-atom polyhedron, although the Wulff criterion itself strictly only applies for large sizes when edge and vertex effects become negligible.

2.3.3. Polyhedra

2.3.3.1. fcc

There are a large number of ways in which a symmetric cluster can be constructed from a fcc close-packed lattice. A basic shape is the octahedron (Figure 2.14) with eight triangular faces which are close-packed (1 1 1) planes with low surface energy. The total number of atoms in a cluster containing K octahedral shells is

$$N = \frac{1}{3}(2K^3 + K). \tag{2.35}$$

A shell number, K, is defined as the number of atoms along the edge of a face. A cluster with an odd value of K has a single atom at its centre, while one with an even value of K is built round an elementary octahedron of six atoms.

2.3. *Geometric shell structure*　　37

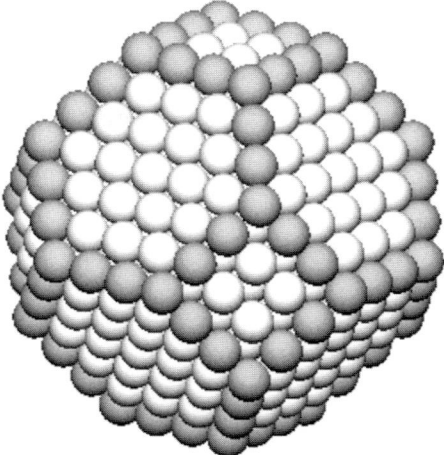

Figure 2.13. A 586-atom truncated octahedron with square and hexagonal faces. Edge atoms here and in subsequent figures are shaded to make the shape clearer.

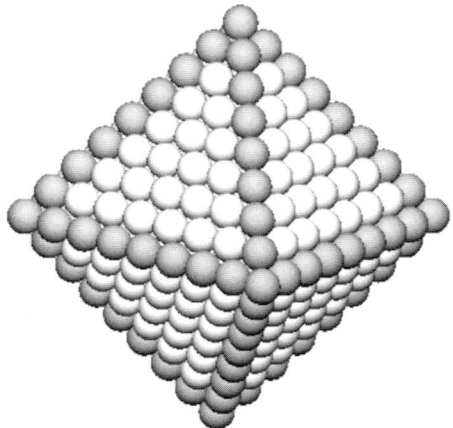

Figure 2.14. A 489-atom octahedron.

The octahedron has a large surface area and this can be reduced by truncation, albeit with the introduction of planes of higher specific energy than (1 1 1). A cuboctahedron, an octahedron truncated by a cube, can be produced in two ways. Perhaps the simplest is that shown in Figure 2.15, which has six square (1 0 0) faces and eight triangular (1 1 1) ones.

The total number of atoms in a cluster comprising K shells is

$$N = \frac{1}{3}(10K^3 - 15K^2 + 11K - 3), \tag{2.36}$$

where again K is the number of atoms along the edge of a face. This polyhedron has a central atom irrespective of the size of the cluster and we can consider it built of

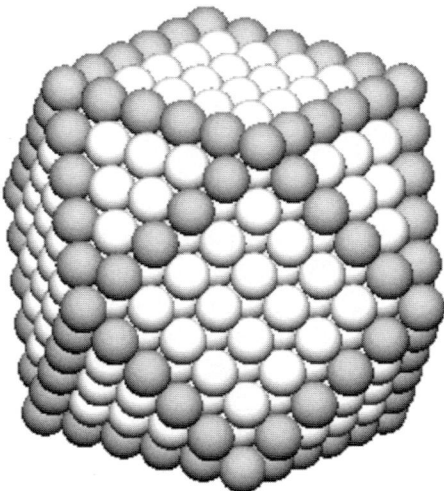

Figure 2.15. A 561-atom cuboctahedron.

successive shells covering interior shells. The Kth shell contains $(10K^2+2)$ atoms, which leads to Eq. (2.36) for the total number of atoms in a cluster comprising K shells. The magic numbers associated with complete geometric shell filling are therefore 1, 13, 55, 147, 309, 561, 923, 1415, ... A cuboctahedron with K_{cub} shells can be cut from an octahedron with K_{oct} shells, where $K_{oct} = 2K_{cub} - 1$.

An alternative truncation is the one with hexagonal and square faces already shown in Figure 2.13. Like the octahedron itself, the cluster is built round a central atom or a central octahedron of atoms depending on whether the shell number is odd or even. Unlike the polyhedron in Figure 2.15, we cannot generate a cluster of K shells simply by covering a cluster of $(K-1)$ shells with a complete new layer of atoms over all the faces. To avoid ambiguity, it is best to stay with the definition that K is the number of atoms along the edge of a face. The number of atoms in this cluster is given by

$$N = 16K^3 - 33K^2 + 24K - 6, \qquad (2.37)$$

and the magic numbers are therefore 1, 38, 201, 586, 1289, ... A K_{toct}-shell figure can be cut from an octahedron with $K_{oct} = 3K_{toct} - 2$.

2.3.3.2. Mackay icosahedron

The total surface energy can be minimised if all the faces of the polyhedron are (1 1 1) planes and the surface area can be kept at a minimum. This can be accomplished with an icosahedral structure with 20 triangular faces [32,33], as in Figure 2.16(a). The icosahedron is one of the five regular Platonic solids [34]; the octahedron, already mentioned, is another. The Mackay icosahedron can be regarded as being made up of 20 tetrahedra with slight distortion as shown in Figure 2.16(b). The icosahedra can be constructed by arranging 12 neighbours of a central atom at the corners of an icosahedron. Successively larger clusters can be

2.3. Geometric shell structure

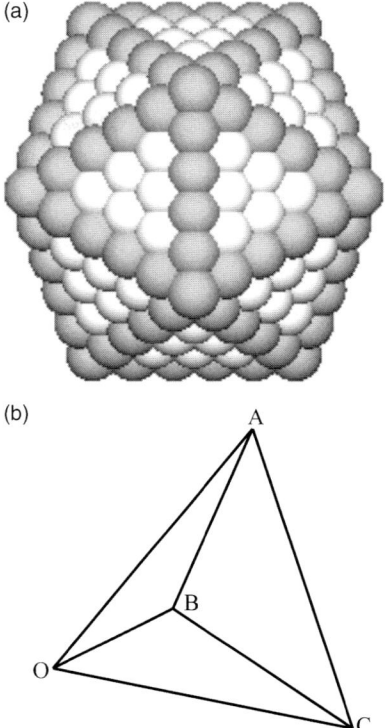

Figure 2.16. (a) An icosahedron of 561 atoms; (b) the Mackay icosahedron can be considered as a packing of 20 distorted tetrahedra. The vertex O is at the centre. If the three edges meeting at the centre (OA, OB and OC) are of unit length, the three sides of the equilateral triangle ABC are extended to 1.05146. The angles subtended at O are 63°26′.

built around this 13-atom core by covering it with a second layer of 42 atoms, a third of 92 atoms and so on. The total number of atoms in an icosahedron of K shells is identical to that of the cuboctahedron with triangular faces (Eq. (2.36)).

Because of the distortion, the packing density of the icosahedron is lower (0.6882 in the large cluster limit [32]) than that for close-packing (0.7495). However, it has been shown [33] that, compared with the close-packed cuboctahedral structure, the reduction in surface energy achieved by the icosahedron more than compensates for the extra strain energy resulting from the distortion, and even up to quite large clusters the icosahedral packing is energetically favourable. The icosahedron, like the cuboctahedron with triangular faces, has 12 vertices. Mackay [32] has shown how the cuboctahedron can be transformed into the icosahedron by a continuous rotation of its triangular faces.

The reason for the assignment of magic numbers to the geometric shell structure for large sodium clusters is clear. The positions of the minima in the mass spectra in Figure 2.9 fit rather well to the values from Eq. (2.36) with $K = 10$–19 as indicated in the diagram, implying an icosahedral (or cuboctahedral) configuration. Furthermore, in the K against $N^{1/3}$ plots of Figure 2.10, a good fit to Eq. (2.36)

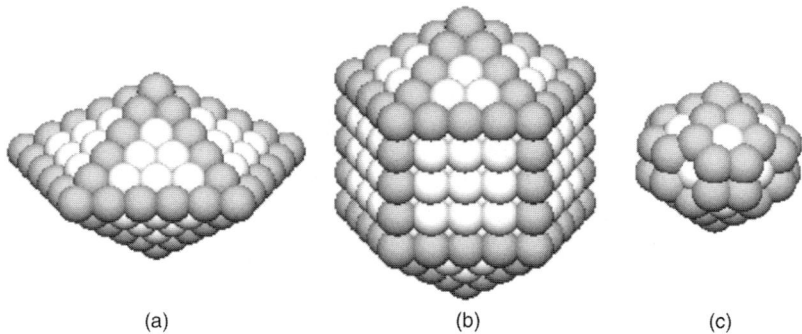

Figure 2.17. (a) A decahedron of 181 atoms; (b) a truncated decahedron of 309 atoms; (c) a Marks decahedron of 75 atoms.

(in the large K limit) is obtained for $N > 1500$, whereas Eq. (2.31) works well for the smaller clusters.

2.3.3.3. Truncated decahedron

The icosahedron has a fivefold symmetry axis. An alternative arrangement with this symmetry is the decahedron constructed out of five slightly distorted tetrahedra with a common edge (Figure 2.17(a)). The five edges in the x–y plane are longer than the other tetrahedron edges by a factor of 1.018. The number of atoms in a decahedron of K shells is

$$N = \frac{1}{6}(5K^3 + K). \tag{2.38}$$

Decahedra with odd values of K are atom-centred and those with even K have a seven-atom decahedron at the centre.

Like the icosahedron, the surfaces are close-packed, but in this case the total surface area is very large. This can be reduced [33] by truncation (Figure 2.17(b)), but this exposes high energy (1 0 0) faces. The optimum configuration can be obtained by the Wulff construction [30]. The total number of atoms in a truncated decahedron with square (1 0 0) faces is given by the same expression as for the icosahedron and cuboctahedron with triangular faces (Eq. (2.36)). A truncated decahedron is atom centred, and one with K_{tdec} shells can be cut from a decahedron with K_{dec} shells, where $K_{\text{dec}} = 2K_{\text{tdec}} - 1$.

Further truncation can be performed to expose re-entrant (1 1 1) faces in addition to the (1 0 0) faces [31,35] as is shown in Figure 2.17(c).

2.3.3.4. bcc

The rhombic dodecahedron is the primary cluster shape based on the bcc lattice, as shown in Figure 2.18. It is made up of 12 identical faces whose diagonals have lengths with the ratio $1 : \sqrt{2}$, and is the dual of the cuboctahedron. The number of atoms for a cluster of K shells is given by

$$N = 4K^3 - 6K^2 + 4K - 1. \tag{2.39}$$

2.3. Geometric shell structure

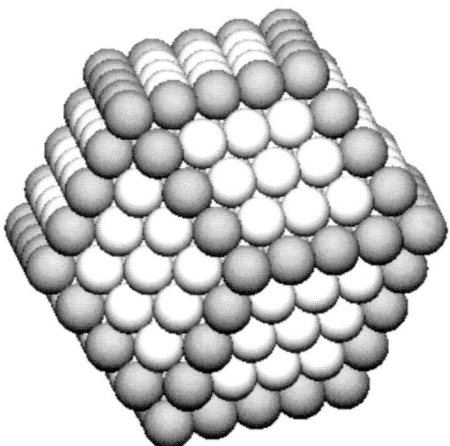

Figure 2.18. The 369-atom rhombic dodecahedron.

The cluster sizes at successive shell fillings are therefore 1, 15, 65, 175, 369, 671, 1105, ...

2.3.4. Filling between complete shells

The filling of sub-shells in between the completion of the main shells can help to determine the cluster structure. We focus here on the Mackay icosahedra for which two different ways of filling the geometric shells have been proposed.

The first model was proposed by Northby [36]. In it, we start from a complete icosahedron and add atoms. They preferentially sit on the threefold hollow formed by three atoms on the (1 1 1) face, thereby maximising their coordination number (anti-Mackay overlayer). However, continuation of this process does not lead to the formation of the next shell of the icosahedron, since the next shell has atoms sitting above pairs of atoms on the original icosahedron. To reconcile this fact, the atoms rearrange themselves to the Mackay packing arrangement when the shell is half filled. This model was originally used in a calculation based on a Lennard–Jones potential and applied to rare gas clusters, and successfully described certain features in the mass spectra.

The other approach is the so-called "umbrella" model [37]. In this case, atoms are added to icosahedral sites on the surface of the core icosahedron. The covering of a complete face achieves enhanced stability and this is further enhanced when a vertex and its five surrounding faces are covered to form an umbrella shape. The filling sequence is different from the Northby model and this allows one to determine which process is operating and, further, to distinguish between the icosahedral structure and the cuboctahedral one (with triangular faces) for which the magic numbers are the same. Such an analysis has been carried out for transition metal clusters by Pellarin *et al.* [38].

2.3.5. Cluster melting

The plots in Figure 2.10 indicated that, for sodium clusters larger than about 2000 atoms, the magic numbers fitted better to the geometric shell model than to the electronic shell structure. The temperature dependence of the shell structure has been investigated by Martin [6] from measurements of the mass spectra. The wavelength of the ionising light was chosen so that the photon energy corresponds to the ionisation threshold of the sodium cluster, and the difference mass spectra were analysed. The results are displayed in Figure 2.19. At low temperatures, an oscillatory structure is observed in the difference spectra with minima that

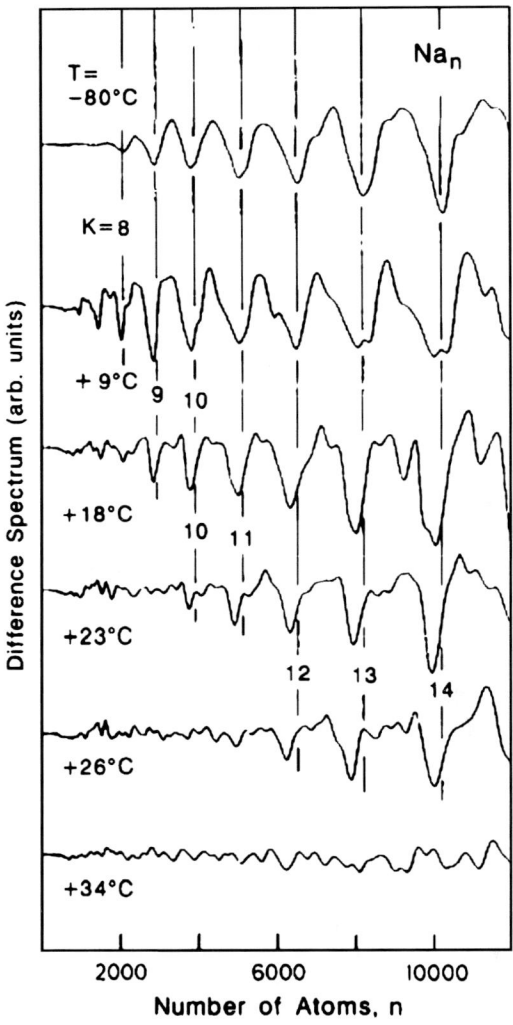

Figure 2.19. Difference mass spectra of sodium clusters using near-threshold photoionisation. Vertical lines indicate geometric shell closings; numbers refer to the shell index, K, from Eq. (2.36). Reproduced with the permission of Elsevier Science from Martin [6].

correspond to the number of atoms needed to complete icosahedral (or cuboctahedral) shells. As the temperature is raised, the structure associated with the shell filling gradually disappears, first with the smaller clusters until at 34°C it has vanished completely. The temperature at which a minimum disappears is identified as the melting temperature, T_m, for a cluster of that size.

The melting temperature of clusters had been predicted [39] to deviate from the bulk value in inverse proportion to the cluster radius, R, as in the following equation where A is a constant:

$$\frac{T_m}{T_{bulk}} = 1 - \frac{A}{R}. \tag{2.40}$$

One of the earliest attempts to verify this prediction was made [40] with gold particles deposited on an amorphous carbon substrate, using scanning electron microscopy to detect melting through the change in the diffraction pattern. The results and fit to Eq. (2.40) are shown in Figure 2.20. The melting temperature falls from its bulk value of 1336 K to as low as 300 K for particles of 20 Å diameter.

The change in the diffraction pattern enables the melting phenomenon to be studied in relatively large clusters on a surface, with good agreement with the size dependence predicted by Eq. (2.40). However, in experiments using calorimetric methods, quite irregular variations in the melting point of smaller free clusters (sodium atoms in the size range 50–350 atoms) have been observed [41–44].

Calorimetric measurements yield the internal energy, U, of the clusters as a function of temperature and its derivative, the heat capacity. The results of measurements [41] on sodium anions of a particular size are shown in Figure 2.21. Unlike in the bulk (as indicated in the diagram), the transition does not take place

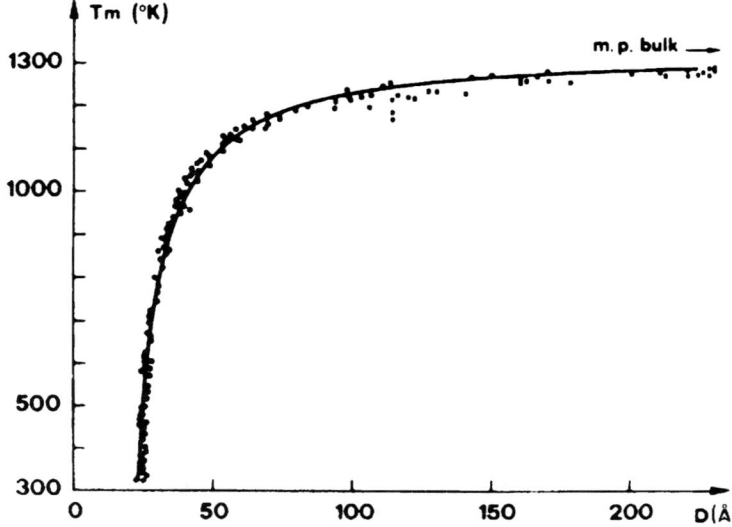

Figure 2.20. Melting temperature of gold clusters. D is the particle diameter. Data points are experimental results and the line is a fit to the phenomenological expression. Reproduced with the permission of the American Physical Society from Buffat and Borel [40].

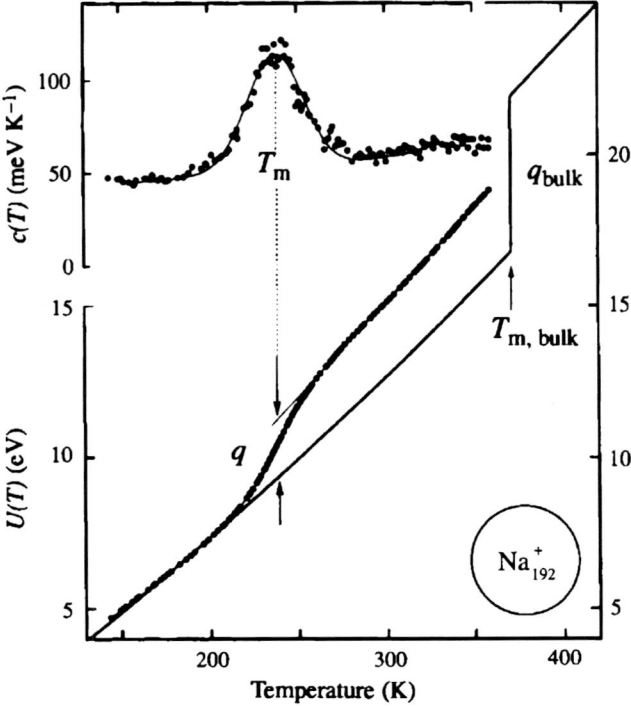

Figure 2.21. The heat capacity, $c(T)$, and caloric curve $U(T)$ of the Na_{192}^+ anion. The peak position of $c(T)$ determines the melting temperature and q is the latent heat of fusion. The melting temperature and latent heat of the bulk are shown. Reproduced with the permission of Macmillan Publishers Ltd. (*Nature*) from Schmidt et al. [41].

sharply at a definite temperature but occurs smoothly over a finite temperature range. Also the latent heat, q, is smaller than in the bulk and is extracted from the data by extrapolation of the straight line regions as shown. The free energy is given by $F = U - TS$, where S is the entropy, and matching the solid and liquid free energy at the melting temperature gives the relation

$$T_m = \frac{U_{\text{liq}} - U_{\text{sol}}}{S_{\text{liq}} - S_{\text{sol}}} = \frac{q}{\Delta s}, \tag{2.41}$$

where Δs is the specific entropy difference between the two states at the melting temperature. T_m and q are obtained directly and Δs is deduced from Eq. (2.41) [42–44].

The principle of the cluster calorimetry of Haberland [43] is as follows. A beam of cluster ions is produced and thermalised in vacuum in a heat bath at temperature T_1. The clusters are then irradiated by photons from a laser. Energy is absorbed from a single photon ($\delta U = h\nu$) and the temperature is raised to T_2 (to be determined) at which the clusters do not lose atoms in the 100 μs time-scale of the experiment. The absorption of more photons from the same laser pulse raises the temperature above that required for the evaporation of atoms from the cluster.

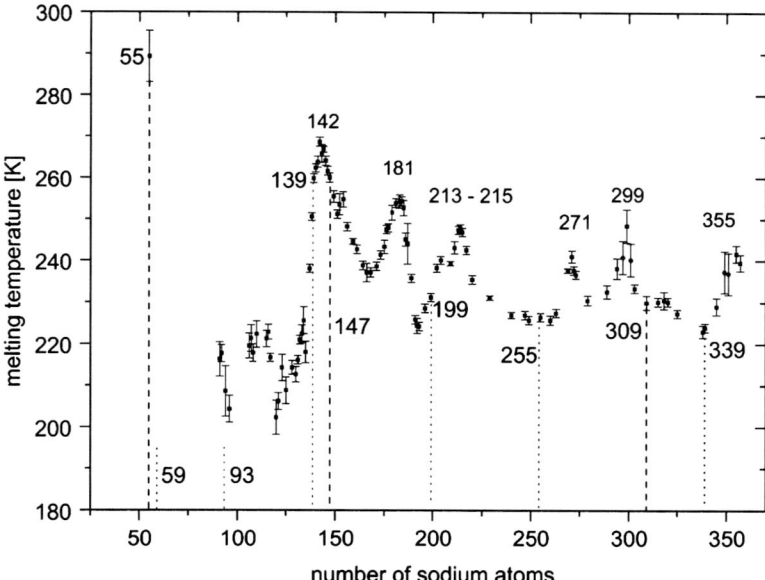

Figure 2.22. The melting temperature T_m of Na_N^+ clusters plotted against the number of atoms N. Vertical lines indicate electronic (dotted) and geometric (dashed) shell closings. The bulk melting temperature is 371 K. Reproduced with the permission of Elsevier Science from Schmidt and Haberland [44].

The size distribution of the remaining cluster ions is measured. The temperature T_2 is obtained by repeating the procedure with a raised heat bath temperature until the photofragmentation behaviour matches that occurring in the laser-heated clusters. Of course, the number of photons absorbed can vary, but different numbers of absorbed photons produce clearly separated groups of fragments with an approximately Gaussian distribution. The distance between the maxima of the Gaussians corresponds exactly to one photon energy. Thus, δU is determined and $\delta T = T_2 - T_1$ and, for small enough δT, the ratio $\delta U/\delta T$ gives the heat capacity $c(T)$.

The results for the melting temperature are shown in Figure 2.22. There are size-dependent fluctuations which bear little relation to the smooth behaviour predicted by Eq. (2.40), but rather indicate some sort of shell structure. Over the size range $135 < N < 350$, the melting temperature is reduced from the bulk value of 371 K by roughly 35%. Δs is reduced by a similar amount, and q drops by about 58% [42–44]. Haberland et al. [42,43] also employ photoelectron spectroscopy (PES) in their analysis since it can give an indication of those clusters that have a high symmetry.

The fluctuations seen in Figure 2.22 do not appear to correlate with the electronic shell closings (indicated by dotted lines in the figure) that dominate most of the known properties of the alkali metal clusters. Some of the peaks in Figure 2.22 can be associated with the geometric shell structure. The Na_{55}^+ cluster has the highest melting point and is associated with the stable icosahedral structure. The next closed-shell icosahedron is at $N = 147$ and a peak in the melting temperature occurs just below this point. The corresponding peaks in the entropy and latent heat do occur at $N = 147$ [42,44]. PES measurements confirm closed-shell icosahedra at 55 and 147, and also give a clear pointer to a closed-shell icosahedron at 309, but the lack of structure in the melting properties at this size remains unexplained. However, it is suggested [42] that a solid–solid transition may occur at about 40 K below T_m and that the cluster, which has an icosahedral shape at low temperatures, may have a different structure at T_m. Other peaks are tentatively attributed to icosahedra with caps added or subtracted, but there are no obvious structural assignments to many of the features. It is suggested [42] that some peaks may result from the entropy associated with mobile atoms in the outer layers of the cluster, a "premelting" feature that has been observed in many computer simulations.

References

[1] W.D. Knight, K. Clemenger, W.A. de Heer, W.A. Saunders, M.Y. Chou and M.L. Cohen, Phys. Rev. Lett. **52** (1984) 2141.
[2] W. Ekardt, Phys. Rev. Lett. **52** (1984) 1925.
[3] W. Ekardt, Phys. Rev. B **29** (1984) 1558.
[4] W.A. De Heer, Rev. Mod. Phys. **65** (1993) 611.
[5] M. Brack, Rev. Mod. Phys. **65** (1993) 677.
[6] T.P. Martin, Phys. Rep. **273** (1996) 199.
[7] N.W. Ashcroft and N.D. Mermin, Solid State Physics, Holt, Rinehart and Winston, New York, 1976.
[8] R.D. Woods and D.S. Saxon, Phys. Rev. **95** (1954) 577.
[9] N.D. Lang and W. Kohn, Phys. Rev. B **1** (1970) 4555.
[10] P. Hohenberg and W. Kohn, Phys. Rev. **136** (1964) B864.
[11] W. Kohn and L.J. Sham, Phys. Rev. **140** (1965) A1133.
[12] P.A.M. Dirac, Proc. Cambridge Philos. Soc. **26** (1930) 376.
[13] O. Gunnarson and B.I. Lundqvist, Phys. Rev. B **13** (1976) 4274.
[14] M.Y. Chou, A. Cleland and M.L. Cohen, Solid State Commun. **52** (1984) 645.
[15] J.P. Perdew, H.Q. Tran and E.D. Smith, Phys. Rev. B **42** (1990) 11627.
[16] P. Ziesche, J.P. Perdew and C. Fiolhais, Phys. Rev. B **49** (1994) 7916.
[17] H.A. Jahn and E. Teller, Proc. R. Soc. Lond. Ser. A **161** (1937) 220.
[18] K. Clemenger, Phys. Rev. B **32** (1985) 1359.
[19] S.G. Nilsson, K. Dan Vidensk. Selsk. Mat.-Fys. Medd. **29**(16), (1955) 1.
[20] I.T. Katakuse, Y. Ichihara, Y. Fujita, T. Matsuo, T. Sakurai and H. Matsuda, Int. J. Mass Spectrom. Ion Processes **67** (1985) 229.
[21] I.T. Katakuse, Y. Ichihara, Y. Fujita, T. Matsuo, T. Sakurai and H. Matsuda, Int. J. Mass Spectrom. Ion Processes **69** (1986) 109.
[22] T.P. Martin, T. Bergmann, H. Göhlich and T. Lange, Chem. Phys. Lett. **172** (1990) 209.
[23] H. Nishioka, K. Hansen and B.T. Mottelson, Phys. Rev. B **42** (1990) 9377.
[24] H. Nishioka, Z. Phys. D **19** (1991) 19.
[25] R. Balian and C. Bloch, Ann. Phys. (N.Y.) **69** (1971) 76.
[26] J. Pedersen, S. Bjørnolm, J. Borggreen, K. Hansen, T.P. Martin and H.D. Rasmussen, Nature **353** (1991) 733.
[27] T.P. Martin, S. Bjørnolm, J. Borggreen, C. Bréchnignac, Ph. Cazuzac, K. Hansen and J. Pedersen, Chem. Phys. Lett. **186** (1991) 53.
[28] J. Lermé, M. Pellarin, J.L. Vialle, R. Baguenard and M. Broyer, Phys. Rev. Lett. **68** (1992) 2818.
[29] H. Haberland, Rare gas clusters in H. Haberland (Ed.), Clusters of Atoms and Molecules, Springer Series in Chemical Physics, Vol. 52, Springer, Berlin, 1994, pp. 379–395.

[30] G. Wulff, Z. Krist. **34** (1901) 449.
[31] L.D. Marks, J. Cryst. Growth **61** (1984) 556.
[32] A.L. Mackay, Acta Crystallogr. **15** (1962) 916.
[33] S. Ino, J. Phys. Soc. Jpn. **27** (1969) 941.
[34] Plato, Timaeus and Critias (trans. D. Lee), Penguin Classics, London, 1977.
[35] L.D. Marks, Philos. Mag. A **49** (1984) 81.
[36] J.A. Northby, J. Chem. Phys. **87** (1987) 6166.
[37] T.P. Martin, T. Bergmann, H. Göhlich and T. Lange, J. Chem. Phys. **176** (1991) 343.
[38] M. Pellarin, R. Baguenard, J.L. Vialle, J. Lermé, M. Broyer and A. Perez, Chem. Phys. Lett. **217** (1994) 349.
[39] P. Pawlow, Z. Phys. Chem. **65** (1909) 54.
[40] Ph. Buffat and J.P. Borel, Phys. Rev. A **13** (1976) 2287.
[41] M. Schmidt, R. Kusche, B. von Issendorff and H. Haberland, Nature (London) **393** (1998) 238.
[42] H. Haberland, T. Hippler, J. Donges, O. Kostko, M. Schmidt and B. von Issendorff, Phys. Rev. Lett. **94** (2005) 035701.
[43] H. Haberland, Melting of clusters in C. Guet, P. Hobza, F. Spiegelman and F. David (Eds.) Atomic Clusters and Nanoparticles, Proceedings of the Les Houches Summer School Session LXXIII, Springer, Heidelberg, 2001, pp. 29–56.
[44] M. Schmidt and H. Haberland, C. R. Phys. **3** (2002) 327.

Chapter 3

Production of Nanoparticles on Supports Using Gas-Phase Deposition and MBE

C. Binns
Department of Physics and Astronomy, University Road, University of Leicester, Leicester LE1 7RH, UK

3.1.	Introduction	50
3.2.	Gas-phase nanoparticle production	50
	3.2.1. Cluster beam sources	51
	3.2.2. Generic types of cluster beam source	53
	3.2.2.1. Seeded supersonic nozzle source	53
	3.2.2.2. Thermal gas aggregation source	54
	3.2.2.3. Sputter gas aggregation source	55
	3.2.2.4. Laser ablation source	55
	3.2.2.5. Pulsed arc cluster ion source	55
	3.2.2.6. Pulsed microplasma cluster source	56
	3.2.3. Aerodynamic lensing and mass selection of neutral clusters	56
	3.2.4. Mass selection of charged clusters	58
	3.2.4.1. RF quadrupole mass filter	59
	3.2.4.2. Wien filter	61
	3.2.4.3. Time-of-flight mass spectrometer	61
	3.2.4.4. Pulsed field mass selector	62
3.3.	Deposition of gas-phase nanoparticles	62
	3.3.1. Deposition of clusters onto surfaces	63
	3.3.2. Embedding nanoparticles in matrices	66
3.4.	Self-assembly of nanoscale islands on surfaces	67
	3.4.1. UHV environment	67
	3.4.2. Evaporation sources	67
	References	69

HANDBOOK OF METAL PHYSICS
ISSN 1570-002X/DOI 10.1016/S1570-002X(08)00203-6

© 2009 ELSEVIER B.V.
ALL RIGHTS RESERVED

3.1. Introduction

This chapter presents methods currently used to produce metal nanoparticles with a controlled size deposited on substrates or incorporated as three-dimensional assemblies in matrices. Pre-forming and then depositing clusters from the gas-phase is a highly controllable and flexible method enabling production out of just about any material in the size range 1–100 nm with good size control. Methods have also been developed to produce gas-phase alloy and core–shell particles, again with complete flexibility of choice of core and shell materials. Thus, despite the technical demands and the expense of gas-phase cluster methods, they remain a favourite for producing supported nanoparticle assemblies for research. The field was reviewed recently by Wegner et al. [1].

Gas-phase production comprises a host of related but distinct methods, each specialised to different groups of materials. This chapter presents the various kinds of nanoparticle source that have been constructed, methods to filter the size of the particles in the gas-phase and methods to deposit them on substrates or incorporate them into matrices.

Due to its experimental simplicity and inherent cleanliness, direct deposition of the metal vapour in ultra-high vacuum (UHV) onto suitable substrates, where it can condense into nanoscale islands, is a commonly used technique. Achieving control over the film properties such as island size, shape and density relies on a good understanding of the film growth kinetics, which for a number of metal/substrate combinations has been achieved; the theory of growth kinetics is discussed in detail in Chapter 4. Some of the practical considerations for producing nanoscale assemblies this way are presented at the end of this chapter (see Section 3.4).

3.2. Gas-phase nanoparticle production

Sources capable of producing free beams of nanoscale metal clusters were first reported nearly 30 years ago [2,3], based on a large body of existing work on rare-gas clusters formed in supersonic beams (see Ref. [4] for example). Since then the technology has developed to enable the production of clusters out of almost any metal, including refractory materials and binary clusters [5,6]. A consistent pressure on the development is the demand for ever higher flux as the usefulness of cluster-assembled films in producing new materials and devices is realised [1]. The field has benefited recently from interdisciplinarity with the aerosol community, leading to a better understanding of the cluster formation process and a greater ability to control it [7,8].

Gas-phase nanoparticle production encompasses a variety of techniques but can be generally divided into methods that operate at atmospheric pressure, including plasma [9], flame [10] and spark [11] sources and those that generate supersonic beams of clusters at low pressure [1]. The outputs from the former group are normally referred to as aerosols and from the latter group as cluster beams. In this chapter, the focus will be entirely on cluster beam sources, as these are required to produce pristine metal particles suitable for the study of fundamental size-dependent properties. A detailed description of cluster beam generation is available in a book by Milani and Ianotta [12]. In order to scale up production of nanoparticles of a particularly useful element and size, however, the technology developed by the aerosol community is required and modern cluster beam deposition systems incorporate features used in aerosol sources, such as aerodynamic lensing [1] (see Section 3.2.3).

3.2.1. Cluster beam sources

At the heart of most cluster beam sources is a region in which a supersaturated vapour of the material to be studied is generated by ejecting an atomic metal plume into a flow of a cooled inert gas. The metal vapour can be generated by laser ablation [5,13,14], sputtering [15–17], a pulsed [18,19] or continuous [20] arc, or thermally [2,21]. The classical theory of the condensation of supersaturated vapours into clusters is highly developed, and although it is formulated in terms of quantities such as surface tension that are not clearly defined in small clusters, it still distinguishes the growth mechanism. If the metal vapour in the supersaturated region has a vapour pressure P_0 at the temperature within the bath gas and P is its actual pressure, then the critical diameter, d_{crit}, for a cluster to grow is given by the Gibbs–Kelvin equation:

$$d_{crit} = \frac{4\gamma V}{k_B T \ln(S)}. \tag{3.1}$$

Here, γ is the surface tension of the cluster, V its volume and S the supersaturation ratio P/P_0. If a cluster is smaller than d_{crit}, it will evaporate atoms faster than absorbing them from the surrounding vapour and will disappear. If it is larger than d_{crit}, the incident flux onto the particle is greater than the evaporation rate, and it will grow. The equation distinguishes two types of growth. If d_{crit} is smaller than an atom, every cluster is stable and growth occurs by coagulation due to Brownian motion. If it is larger than an atom, then initially seeds must form by homogenous nucleation followed by further growth that can either be by coagulation of clusters or surface growth where the incident atomic vapour forms new layers on the existing clusters. Generally, cluster beam sources operate in the latter regime and the initial formation is by homogenous nucleation, which is a bottleneck for the formation of clusters.

Going beyond the basic insights provided by classical models requires a realistic model, such as a Monte Carlo model that builds in realistic atomistic processes. A test of any model is whether it can reproduce the experimentally observed log-normal distribution of particle sizes that appears to be produced by all cluster sources [22–26]. This is given by:

$$F(n) = \frac{1}{\sqrt{2\pi} \ln \sigma} \exp - \left(\frac{\ln n - \ln \bar{n}}{\sqrt{2\pi} \ln \sigma} \right)^2, \tag{3.2}$$

where n is the number of atoms per cluster and σ the variance. A Monte Carlo simulation of Cu clusters produced in an Ar bath gas [27], which included elastic and inelastic collisions between Cu–Cu and Cu–Ar atoms, dimer formation, cluster growth and cluster decay through atom evaporation, reproduced this size distribution. The work showed that the size distribution depends on the dwell time in the source, that is the time till the mixture of bath gas and clusters exits the aperture into the high vacuum region, at which point clustering essentially stops. Figure 3.1 shows the evolution of the Cu cluster size distribution as a function of time for the initial conditions: Ar pressure = 1 mbar, Cu pressure = 0.1 mbar. The simulation reveals that the average size of Cu clusters increases with time and settles into a stable distribution, closely following a log-normal shape (Eq. (3.2)) after about 5 ms.

The simulation also shows that the dominant cooling mechanism for the growing clusters is by evaporation of atoms. This is approximately balanced by the heating due to the adsorption of new atoms (latent heat) and the cooling by the background gas, which, although inefficient, provides the *extra* cooling required to allow the clusters to grow. The simulation also revealed that the initial formation of dimers is

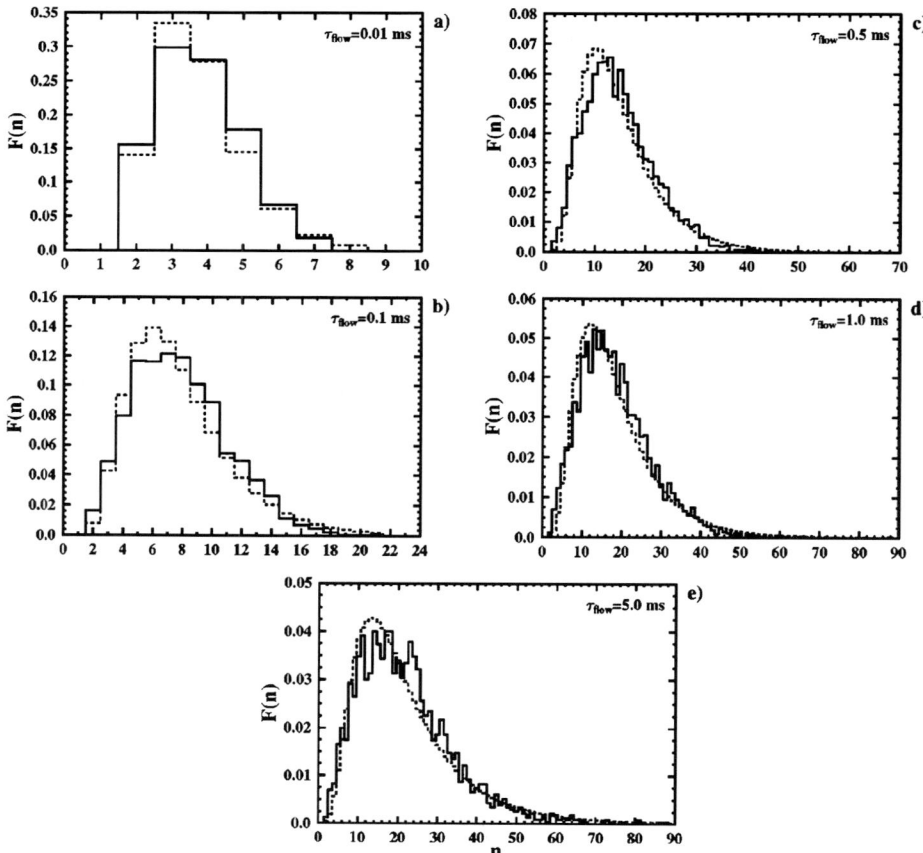

Figure 3.1. Dependence of normalised size distribution $F(n)$ of Cu$_n$ clusters on the dwell time (t_{flow}) of gas in the aggregation volume [27]. The solid lines show the full simulation and the dashed lines are a fit to a log-normal distribution (Eq. (3.2)). Reused with permission from Boris Briehl and Herbert M. Urbassek [27].

a critical bottleneck to the clustering process. Thus, in sources that rely on sputtering (see Section 3.2.2.3) in which the metal vapour plume is rich in dimers, the clustering is very efficient.

Simulations have grown more sophisticated to include a realistic description of the atomic interaction based on the embedded atom method [28]. This has enabled the exploration of the atomic structures of gas-phase Fe clusters and has revealed that in the early stage of growth, icosahedral structures are preferred. These evolve into close-packed fcc or hcp clusters with increasing size.

3.2.2. Generic types of cluster beam source

The region in which the metal vapour initially mixes with the rare gas is in the pressure range of a few millibar to a few bar depending on the type of source, and at the end of this region, there is an aperture with a diameter in the range of a few micrometres to a few millimetres. The expansion from this aperture accelerates the clusters, and in the limit of a high-pressure differential between the two regions separated by the aperture, the clusters can acquire the full thermal distribution of the bath gas atom velocities. In this case, the clusters all have the same velocity distribution irrespective of their size, which enables mass filtering by a simple pair of charged plates (see Section 3.2.4). In weaker expansions, there is a velocity slip between the carrier gas and the metal clusters. Below are given brief descriptions of the operation of various sources based on gas condensation (Figure 3.2). For more comprehensive descriptions of these and other types, the reader is referred to existing reviews [1,12,29]. The generic types of cluster source operating in vacuum and producing a broad range of cluster sizes are shown schematically in Figure 3.2 and described briefly in the following sections.

3.2.2.1. Seeded supersonic nozzle source

The seeded supersonic nozzle sources (SSNSs) are designed to produce high fluxes of low melting point (usually alkali) metals (Figure 3.2(a)). The furnace containing the melt is heated to a sufficiently high temperature to yield a metal vapour pressure in the region of 10–100 mbar and this vapour is mixed with (seeded into) a rare gas introduced at a pressure of several atmospheres. The hot mixture expands adiabatically into vacuum through a small aperture and the rapid cooling occurring close to the nozzle condenses the metal into clusters. The clustering continues till the mean free path becomes too long to allow significant interactions between the condensed particles. The technical difficulty of the containment of the melt in a relatively large furnace has restricted the temperatures achievable to below 1600 K. The source is thus confined to the study of high vapour pressure materials. It is, however, capable of producing a flux in excess of 10^{18} atoms/s of clustered material. The very high rate of material consumption has encouraged the development of *in situ* refilling devices [30].

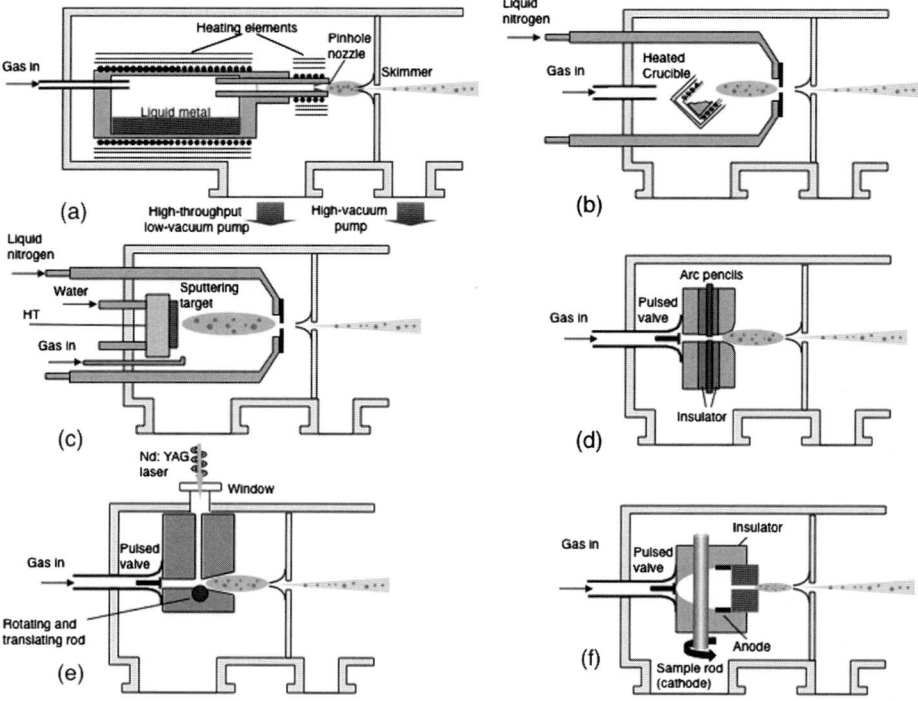

Figure 3.2. Basic layout of sources using rare gases to produce supersaturated vapours. (a) Seeded supersonic nozzle source. (b) Thermal gas aggregation source. (c) Sputter gas aggregation source. (d) Pulsed arc cluster ion source. (e) Laser ablation source. (f) Pulsed microplasma cluster source.

3.2.2.2. Thermal gas aggregation source

The thermal gas aggregation source (TGAS) was the first type of metal cluster source reported [2], and a common design is illustrated in Figure 3.2(b). The metal vapour is produced by a Knudsen cell similar to that used for molecular beam epitaxy (MBE) (see Section 3.4). The vapour pressure of metal required in a cluster source, however, is typically 10^{-3} mbar and is three orders of magnitude higher than used in MBE film growth, requiring correspondingly higher temperatures. Improvements since 1980 include the *in situ* positioning of the crucible assembly within the gas flow to optimise the output and an adjustable gap between the first aperture and the skimmer [21]. Generally, there is a larger aperture to the high vacuum region and the free-jet expansion is weaker than in the SSNS. Because only the crucible is heated, with a careful design, materials can be contained at temperatures above 2000 K, and thermal sources have been built that are capable of producing beams of magnetic transition metal clusters [21]. With careful outgassing, the source is particularly suited to producing very clean cluster beams and to UHV-compatible operation [21,31] with reported vapour pressures of contaminant gases other than the Ar or He bath gas in the 10^{-11} mbar region [21].

3.2.2.3. Sputter gas aggregation source

The inert gas pressure within the clustering region in a typical gas aggregation source is compatible with magnetron sputtering, and the thermal source can be replaced by a sputtered target (Figure 3.2(c)) [15,16]. This type of source was originally developed by the Haberland group in Freiburg and has been adopted by a number of groups because it brings several advantages, including the ability to produce clusters of virtually any solid including refractory metals. In addition, with AC sputtering, it can be adapted to produce clusters of insulating materials. Further developments include the use of a hollow cathode target geometry with which very high fluxes can be attained [32]. Clustering is highly efficient as the sputtered vapour is rich in dimers overcoming the initial bottleneck for condensation, as discussed in Section 3.2.1. In addition, the emerging cluster beam is ionised with a greater proportion (up to 50%) of ions than achievable with a conventional ioniser. This is an important factor if the clusters are to be mass-selected because most mass analysers filter charged particles.

3.2.2.4. Laser ablation source

Light pulses from a Nd-YAG laser focused onto a suitable target can vaporise even refractory materials, and if the laser pulse coincides with a gas burst across the target produced by a pulsed valve, suitable conditions for clustering can be achieved (Figure 3.2(e)). The first report of this type of source [3] was quickly followed by improvements, including a mechanism for driving the target rod in a screw motion so that a fresh region is exposed to each laser pulse [14]. Clustering occurs within the nozzle as the metal vapour encounters the rare gas and continues in the strong expansion as the mixture is ejected. There is a strong adiabatic cooling despite the relatively large nozzle diameters because the instantaneous carrier gas pressure is very high – up to several atmospheres. The processes are thus a combination of those found in the SSNS and the sputter gas aggregation source (SGAS) and TGAS and, as in the former type, clustering is complete within a nozzle diameter of the aperture. Despite the short duty cycle of the laser, the peak output is so high that the average flux from the source competes with the other types described above. A useful variable is the phase of the laser pulse relative to the valve opening that can be used to control the size distribution of clusters by altering the average gas pressure during the pulse [33]. Phasing becomes particularly powerful in sources employing two targets to produce binary clusters [6,34] because it can be used to control the distribution of elements within the cluster. The pulsed output couples efficiently to time–of–flight analysers (see Section 3.2.4.3).

3.2.2.5. Pulsed arc cluster ion source

The pulsed arc cluster ion source (PACIS) is related to the laser ablation source (LAS) (Figure 3.2(e)) but, in this case, the evaporated plume is produced by a pulsed arc coinciding with the gas burst [18]. The clustering process is similar to that found in the LAS. As with the gas aggregation source employing sputtering, the cluster output contains a high proportion of ions ($\sim 10\%$) and is particularly suited to charged-particle mass analysers. More recently, a continuous arc source (ACIS) has been developed in which a continuous arc is driven around a hollow cathode by a magnetic field [20]. This is capable of very high fluxes of clusters.

3.2.2.6. Pulsed microplasma cluster source

Recently, an improved understanding of the operation of cluster sources has been achieved by applying fluid dynamics simulations to the gas flows within them. This has led to designs that actively control the velocity field of the bath gas and established the conditions that optimise the cluster flux. An example presented in Section 3.2.3 is aerodynamic lensing, but another is the development of the pulsed microplasma cluster source (PMCS) [35] (Figure 3.2(f)), which combines design characteristics of the SGAS, LAS and PACIS sources. Like the LAS and PACIS sources, the solenoid valve that introduces the inert carrier gas is backed by a pressure of up to 50 bar and, while it is open (typically 0.5 ms), there is a strong expansion that directs a gas jet against the target cathode. The target cathode is made of the material to be formed into clusters and, as in the SGAS source, the vapour plume is produced by sputtering. The electric discharge is also applied as a pulse (typically 50 µs) during the introduction of the gas jet, and the simulations show that with the correct design of expansion chamber, there is a high gas pressure confined close to the rod in the ablation discharge that increases the sputtering yield. The high pressure also promotes direct sputtering of clusters that seed further cluster growth. With a repetition rate of 5 Hz, the source has produced carbon nanoparticle films at a rate of 100 µm h^{-1}, making it suitable for coating and device applications [1].

3.2.3. Aerodynamic lensing and mass selection of neutral clusters

Manipulating gas flows is a familiar exercise to researchers working on aerosols and the cluster community has benefited from interacting with them. An example is the application of aerodynamic focusing. A series of axial restrictions in the gas flow (Figure 3.3(a)) produces a set of vortices [36], and if there are nanoparticles within the gas flow, simulations show that they become increasingly axially confined as they traverse through the series of apertures [37] (Figure 3.3(b)). Using this method, deposited features as small as 50 µm were demonstrated [37]. The effect can be used to boost the flux from cluster sources by inserting the aerodynamic lenses in the high-pressure region of the source just after the clustering is complete and focusing the cluster beam into the skimmer at the start of the high vacuum region. In addition, the focusing effect is size-dependent and, as demonstrated in Figure 3.3(c), the flow tends to remove the largest and smallest particles from the central beam. Thus, the focused beam that emerges from the end of the lens system has a narrowed size distribution relative to the newly formed cluster population before the lens.

The tendency of a small particle to follow an underlying gas flow through an obstacle rather than impact a wall due to its inertia is given by the dimensionless

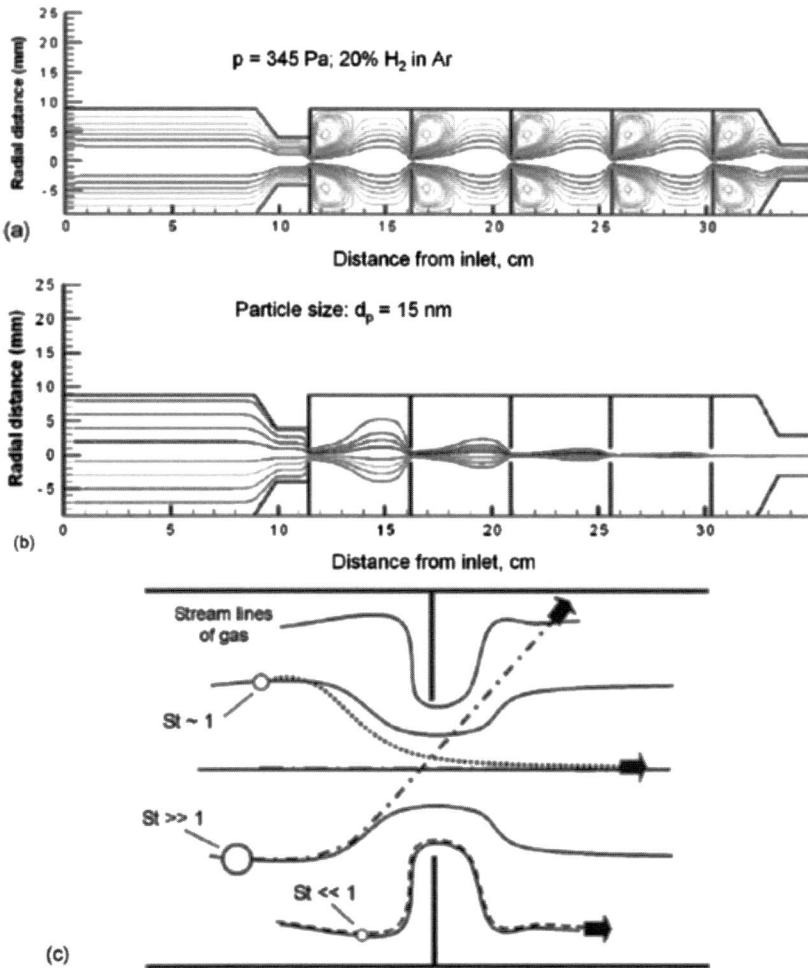

Figure 3.3. Illustration of the principle of aerodynamic lensing and the resulting [36,37] partial mass selection of particles. (a) A series of axial restrictions in the flow introduces a set of vortices. (b) These tend to confine particles in the central beam and after a series of restrictions; particle beam widths of less than 50 μm can be achieved. (c) The process depends on particle size and for particles with Stokes numbers $St \gg 1$, their inertia prevents them from joining the central gas flow and they are deposited on a wall. For $St \ll 1$, the particles closely follow the gas flow and become trapped in the vortices so that only particles with $St \sim 1$ join the central beam. Reproduced with permission of the Institute of Physics from Wegner et al. [1].

Stokes number (St) defined by:

$$St = \frac{\tau v_F}{d}, \tag{3.3}$$

where t is the relaxation time of the particle, v_F the fluid velocity well away from the disturbance produced by the obstacle and d the characteristic size of the obstacle. Figure 3.3(c) illustrates the fate of a particle with $St \gg 1$, that is it is deflected from its path by the orthogonal gas flow near the constriction but its inertia prevents it from rejoining the axial flow and it is deposited on the wall. On the other hand, particles with $St \ll 1$ follow the gas flow closely and do not end up in the central beam, thus getting trapped in the vortices and eventually ending up deposited on a wall. It is only particles with $St \sim 1$ that are focused into the central beam. By altering the d value, the lens can be designed to pass different characteristic sizes and can thus be used as a low-resolution mass filter.

3.2.4. Mass selection of charged clusters

Generally, the requirement for mass selection in the gas phase falls into two different regimes. Very tight mass selection, able to resolve the number of atoms in the clusters, can be used in gas-phase experiments, for example Stern–Gerlach deflection measurements, in which the detector counts the number of incident clusters and very low particle fluxes are measurable. On the other hand, in deposition experiments in which measurements are made on some property of the cluster film, for example the magnetisation, large numbers of clusters are required and very high resolution filters would produce too small an output flux. In this case, a lower resolution filter operating at $\Delta M/M$ ratios in the range 1–10% is required. Ideally one would use a filter in which $\Delta M/M$ is easily adjustable so that flux and resolution can be interchanged. Some sources employ two mass filters; a high resolution instrument such as a time-of-flight analyser to measure the mass spectrum in the free beam and a lower resolution high-throughput device to narrow the native mass distribution prior to deposition [24]. In many experiments involving magnetometry, where at least 1 μg of cluster material is required, the films are produced using the full unfiltered output of the source with control over the average size achieved by altering the source conditions such as bath gas pressure and temperature or, as discussed in Section 3.2.3, modifying the parameters of aerodynamic lenses within the high-pressure region of the source.

Most mass separators require the clusters to be charged and, with the exception of sources that rely on sputtering to produce the metal plume, which inherently produce a large proportion of charged clusters, a separate ioniser is required. The simplest design is the electron impact ioniser, which typically ionises $\sim 5\%$ of the flux.

In the case of a hard free-jet expansion in which the clusters all acquire the same velocity distribution (i.e. the same velocity distribution as the bath gas atoms), mass filtering of the cluster ions can be achieved by a simple parallel-plate electrostatic deflector as shown in Figure 3.4(a). If the clusters enter the plates with a velocity v_x, then particles deflected by an angle θ have a mass, M, given by:

$$M = \frac{qEL^2}{2v_x^3 \tan \theta}, \qquad (3.4)$$

where q is the charge on the cluster, L the length of the plates and E the electric field in V m^{-1} across the plates. Figure 3.4(b) demonstrates in a highly visual way this mode of filtering using an ACIS source, which naturally produces a high proportion

3.2. Gas-phase nanoparticle production

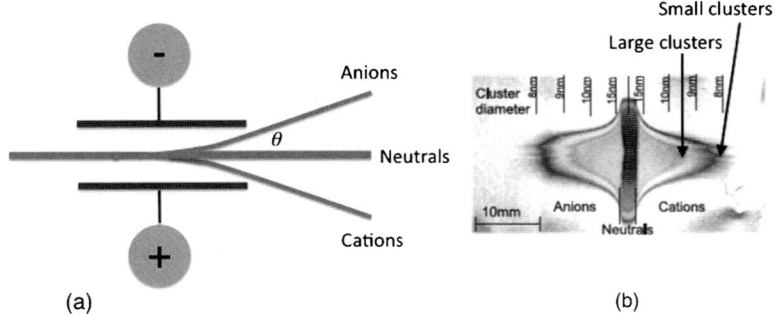

Figure 3.4. (a) Mass separation by a simple parallel-plate deflector in the case of a hard free-jet expansion where the clusters all have the same velocity distribution. (b) Illustration of the method using Fe clusters deposited onto an Al foil from an ACIS source, which produces positive and negative cluster ions as well as neutrals. The clusters have been separated into visible anion and cation deposits either side of the neutral beam and the charged particles have been separated according to size. Reproduced from Methling et al. [20] with permission of *The European Physical Journal* (EPJ), courtesy K.-H. Meiwes-Broer.

of positive and negative cluster ions as well as neutral clusters. Iron clusters passed through a parallel-plate deflector and then deposited onto an Al foil have been separated into visible cation and anion deposits either side of the neutral beam and the charged particles show a spread due to the range of cluster sizes [20]. In practice, mass filtering using this method is usually achieved using a static quadrupole rather than a parallel-plate deflector. The mass resolution using this method is limited by the velocity distribution of the bath gas.

If a higher resolution is required or there is significant velocity slip between the bath gas and the clusters in the free jet expansion, then the mass selector has to be able to determine the mass independently of the cluster ion velocity. The various types of mass filters developed to do this are described below.

3.2.4.1. RF quadrupole mass filter

This instrument, illustrated in Figure 3.5(a), operates using electric fields only and was first described by Paul et al. [38,39]. Four cylindrical poles in the geometry shown have an RF modulation on a DC offset ($U+V\cos\omega t$) applied to one pair of opposite poles and the negative of the same potential to the other pair. In order to be transmitted by the filter, the oscillatory motion of an ion as it moves along the poles must have a stable amplitude and this can only occur, irrespective of its mass, if the ratio U/V is less than 0.168. For ratios lower than this, the instrument passes specific masses given by the absolute value of V. For a quadrupole operating at the stability limit, the range of masses passed depends on the energy of the ions entering the filter, the frequency, ω, the length of the poles and the mechanical accuracy of the pole alignment. Operating with U/V ratios below the stability limit produces a band-pass filter with a top-hat amplitude function. One of the attractive features of the instrument is that a simple electronic adjustment of U/V allows a trade off between flux and resolution. Other advantages are that the filter is light

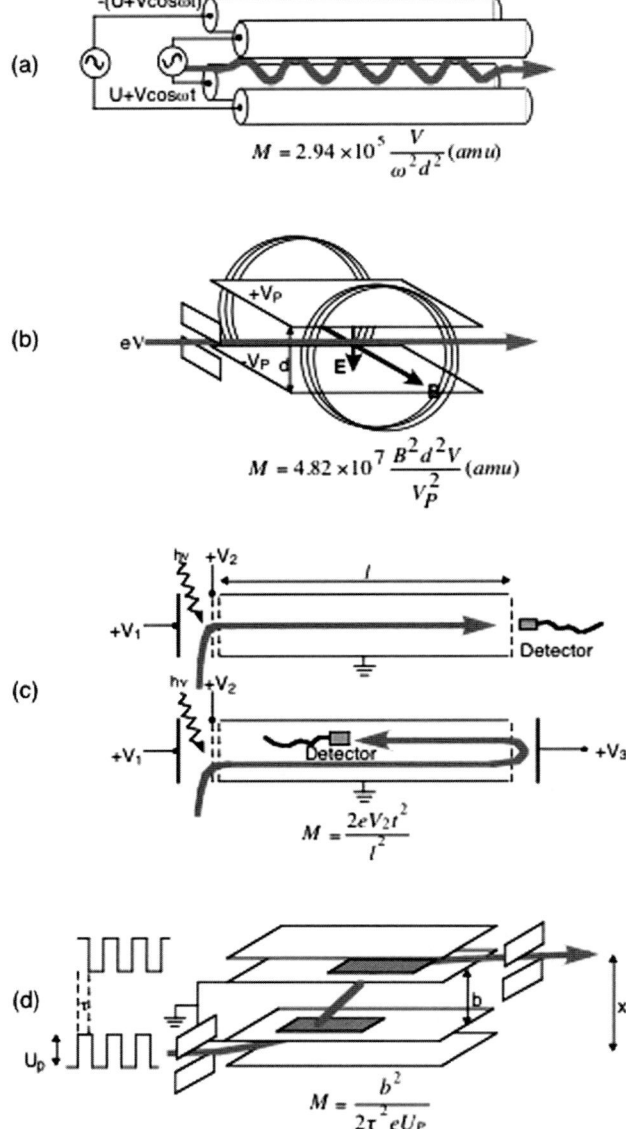

Figure 3.5. Mass separators for charged clusters: (a) RF quadrupole; (b) Wien filter; (c) TOF mass spectrometer with direct and reflectron configurations; (d) pulsed field mass selector.

and compact, and it operates axially, which is convenient in a deposition source. Technically, the maximum resolving power ($M/\Delta M$) obtainable from this type of mass spectrometer is about 4000. Therefore, for a resolution of 1 amu (atomic mass unit) or better, the mass range is restricted to clusters smaller than about 100 atoms. In cluster sources, this type of mass filter is usually treated as a band-pass filter with a variable bandwidth. If obtaining the ultimate resolution is not a priority, lowering the frequency makes it possible to pass arbitrarily large masses. For a quadrupole operating at the stability (high resolution) limit ($U/V = 0.168$), the central mass passed is given by:

$$7.45 \times 10^7 \frac{V}{f^2 d^2} \quad (\text{amu}) \tag{3.5}$$

and the resolution is

$$\Delta M = 3.92 \times 10^9 \frac{E_z}{f^2 L^2} \quad (\text{amu}), \tag{3.6}$$

where E_z is the ion energy entering the filter (in V), f the RF frequency ($= \omega/2\pi$), d the diameter and L the length of the rods. The maximum mass range reported for a quadrupole filter [21] is 60000 amu with a resolution of about 200 amu (resolving power $= 300$). In the case of Fe clusters for example, this corresponds to particles containing 1000 ± 4 atoms, or in terms of size, particles with a diameter of 2.82 ± 0.008 nm.

3.2.4.2. Wien filter

This type of mass filter, illustrated in Figure 3.5(b), applies orthogonal electric and magnetic fields to a charged particle with velocity v. The anti-parallel forces on the ion balance when:

$$v = \frac{2V_p}{Bd}, \tag{3.7}$$

where $2V_p$ is the voltage applied across the electrostatic plates and d the plate spacing. The device is really a velocity filter and it is operated as a mass filter by initially accelerating the ions through a fixed potential V, so each mass has a different velocity. The mass passed by the filter is then given by:

$$4.82 \times 10^7 \frac{B^2 d^2 V}{V_p^2}. \tag{3.8}$$

The resolution is determined by the collimating slits and the velocity spread in the ion beam prior to acceleration. The effect of the latter can be minimised by using a high accelerating potential, but this puts a heavy demand on the magnetic field strength required to operate up to a high mass. A thorough analysis of the performance of a commercial Wien filter for the mass separation of clusters emerging from a PACIS source was carried out by Wrenger and Meiwes-Broer [40]. They showed that the instrument was capable of distinguishing clusters with specific numbers of atoms, N, up to about $N = 20$. The device could be used as a band-pass filter for masses up to $N \approx 1000$.

3.2.4.3. Time-of-flight mass spectrometer

In this scheme (Figure 3.5(c)), a well-defined pulse of ions is created by a pulsed ultraviolet laser or pulsed electron impact ioniser. The packet of ions is accelerated into a field-free drift tube and the arrival time at the end of the tube is recorded by an ion detector. The resolution of the simple instrument is limited mainly by the timing accuracy and the pre-ionisation spread in position and velocity. These latter two effects can be minimised by the reflectron configuration [41–43] shown in the figure, which uses a simple diode reflector at the end of the drift tube. Slower ions of a given mass take a shorter path in the reflector than faster ones, so ions of the same mass will be bunched closer together on reflection. This type of mass spectrometer has the highest resolution of those described here and resolving powers of 10^5 have been reported. The analyser is not so convenient to use as a filter for cluster deposition but filtering can be achieved by adding an electrostatic kicker mechanism that pulses a set of steering plates a set time after each ionising pulse to expel ions of a given mass from the drift tube. This type of mass analyser is most efficiently matched to a pulsed cluster source such as the PACIS or LAS that typically operate at around 100 Hz. The ionising laser in the case of the ablation source is operated at the same frequency as the evaporation laser, so the ion packets are separated by around 10 ms while typical flight times are of the order of 0.1 ms. Operating at much higher frequencies can lead to a problem of harmonics, that is simultaneously selecting slow heavy ions from a previous pulse and lighter ions from the present pulse.

3.2.4.4. Pulsed field mass selector

This is a simple and ingenious mass filter that uses an electric field pulse to displace an ion beam sideways into a field-free region (Figure 3.5(d)) [44]. As discussed in Section 3.2.2, if the ions all have the same velocity, they can be separated in mass by a simple pair of electrostatic plates. The pulsed field mass selector exploits the fact that orthogonal to the beam the ions all do have the same velocity, that is close to zero. For a given strength and timing of the pulse, the lateral velocity given to the ions in the field-free region depends on their mass. A given time later, a second decelerating pulse deflects the ions onto a path parallel to the original one and through an aperture. The timing between the lateral accelerating and decelerating pulses determines the selected mass. The instrument can filter masses up to an arbitrarily high limit with a transmission >50% and with a mass resolution given by:

$$\frac{M}{\Delta M} = \frac{x}{\Delta x}, \tag{3.9}$$

where x is the ion beam lateral displacement and Δx the exit slit width. In practice, $M/\Delta M$ is in the range 20–50, which is adequate for most deposition experiments. The compact inexpensive nature of the filter makes it an attractive option.

3.3. Deposition of gas-phase nanoparticles

The previous section described methods to produce beams of gas-phase nanoparticles and mass select them. One can do various experiments on the clusters in the gas phase, for example photoemission (see Chapter 7) or Stern–Gerlach deflection measurements to determine the magnetic moment per atom (see Chapter 8). In addition, it is possible to collect gas-phase clusters in ion traps such as the Penning trap [45] to increase the density, enabling experiments such as electron diffraction. In order to make cluster assemblies accessible to a full range of structural, electronic and magnetic measurements, and certainly for any materials or device applications, the particles have to be deposited on a surface or embedded in a matrix. Gas-phase measurements are described in detail in Chapters 7 and 8, but here the preparation of supported cluster assemblies is described. Methods have been developed to assemble gas-phase nanoparticles in environments ranging from sparse assemblies adsorbed on a surface in UHV to dense assemblies in solid and liquid matrices.

3.3.1. Deposition of clusters onto surfaces

The most important parameter determining the morphology of deposited clusters and the films they form is the kinetic energy per atom, K_i, when the cluster impacts the surface. Generally, three distinct regimes are distinguished, which are low energy or soft landing ($K_i \sim 0.1$ eV), medium energy (1 eV $< K_i <$ 10 eV) and high energy ($K_i >$ 10 eV). The minimum energy a cluster can have in the gas phase is the kinetic energy it acquires in the free-jet expansion, which in the limiting case of a high-pressure differential across the first aperture (achieved in many of the sources described above) is the Maxwellian speed distribution function:

$$f(v) \propto \left(\frac{m}{2\pi kT}\right)^{3/2} v^2 \exp\left(\frac{-(1/2)mv^2}{kT}\right). \tag{3.10}$$

For example, for Ar at 100 K, the velocity at the peak of the distribution is about 200 m s^{-1}, so the impact energy of a Au cluster would be about 0.04 eV/atom irrespective of size, which is well inside the soft-landing regime. This can be considered a best-case scenario, whereas, if the bath gas was He at 300 K, the velocity at the peak of the distribution would be 1200 m s^{-1}, giving Au clusters an impact energy of 1.5 eV in the medium energy regime. With this kind of expansion only carbon clusters would be soft-landed. If the clusters are charged, it is possible to decelerate them electrostatically, but in practice, because of the width of the Maxwell distribution, it is only possible to achieve a significant decrease in kinetic energy if a large proportion of the clusters are repelled from the surface and lost. In cases where the kinetic energy is too high, soft-landing can also be achieved by depositing the clusters onto a layer of rare gas adsorbed onto a low-temperature substrate [46].

Charged clusters can also be accelerated so that they impact at medium or high energies in order to achieve a required film morphology. In the medium energy regime, although the clusters are significantly flattened, they remain intact. This deposition regime can be useful to prevent cluster diffusion on surfaces, which causes clusters to agglomerate on substrates such as carbon. The impact energy creates a surface defect at the landing site that pins the cluster as demonstrated by Carroll et al. for Ag clusters deposited on highly oriented pyrolytic graphite (HOPG) [47]. Deposition at variable energies in the medium energy regime can be used as a method to control the cluster morphology and its behaviour. For example, in the case of magnetic clusters, the magnetic anisotropy can be controlled by varying the deposition energy.

In the high-energy deposition regime, the cluster integrity is lost but it has been shown that depositing fast clusters can produce exceptionally good coatings [15,48]. Fragmentation followed by the diffusion and rapid annealing of the constituent atoms tends to remove hills and fills valleys in the film, producing a very smooth coating. Highly energetic clusters can also be used to metallise small (micron size) holes with a high aspect ratio (depth/diameter) in dielectric coatings [16].

Because this book focuses on the fundamental properties of the clusters, it is the soft-landing regime that is of most interest, as this best preserves the innate cluster properties. For deposition at low energies, the morphology of the cluster-assembled film is strongly dependent on the substrate. Three main types of film growth can be identified and these are illustrated in Figure 3.6, which shows scanning tunnelling microscopy (STM) and atomic force microscopy (AFM) images of Au clusters deposited on atomically flat Au(1 1 1), Ag(1 1 1), HOPG and mica substrates [49]. This is an important study as it compares directly the growth modes of cluster films, under identical conditions, on four different substrates, and because they are all atomically flat (on the scale of the images), it isolates the effect of the substrate material.

The Au and Ag substrates were grown *in situ* on mica at elevated temperatures, and following annealing, films containing flat terraces with dimensions up to $500 \times 500\,\text{nm}^2$ could be produced. The growth regime of the Au clusters on these two substrates is similar (Figure 3.6(a), (b)) and consists of a random paving of the substrate. The bright features seen in the images correspond to individual or very small groups of clusters. There is no evidence for diffusion of the clusters or preferential binding to steps and defects on the substrate. There is evidence for epitaxy between the clusters and the substrates provided by the observation of abundance peaks for cluster heights corresponding to integer multiples of a Au(1 1 1) monolayer. Earlier observations by Bardotti *et al.* on Au clusters deposited on Au(1 1 1) [50] also found the random paving growth mode and in general small diffusion coefficients are expected in epitaxial systems [51]. The height-to-width ratio of the clusters after correcting for the tip radius was ~ 0.3, showing a significant flattening that may be due to a partial wetting of the surface. A recent AFM study of Au clusters deposited on a dithiol self-assembled monolayer also showed a random paving of isolated Au clusters due to the strong S–Au bond [52].

Figure 3.6(c) is an image of Au clusters from the same source grown on HOPG and illustrates the growth of ramified islands formed when the clusters diffuse on the surface but do not coalesce when they come into contact. This mode has also been observed previously in Sb clusters on HOPG [53] and analysed using the deposition diffusion aggregation (DDA) model [54]. Finally, at the opposite extreme to the random paving mode, Figure 3.6(d) shows that Au clusters on mica not only diffuse but coalesce on contact as observed previously for sufficiently small Sb clusters on HOPG [54].

For the low-energy deposition of noble and simple metals, the substrate thus plays a pivotal role in determining the cluster growth mode and the properties of the cluster-assembled film. In the case of transition metals or noble metals on metal substrates, the most likely growth mode is the random paving mode in which

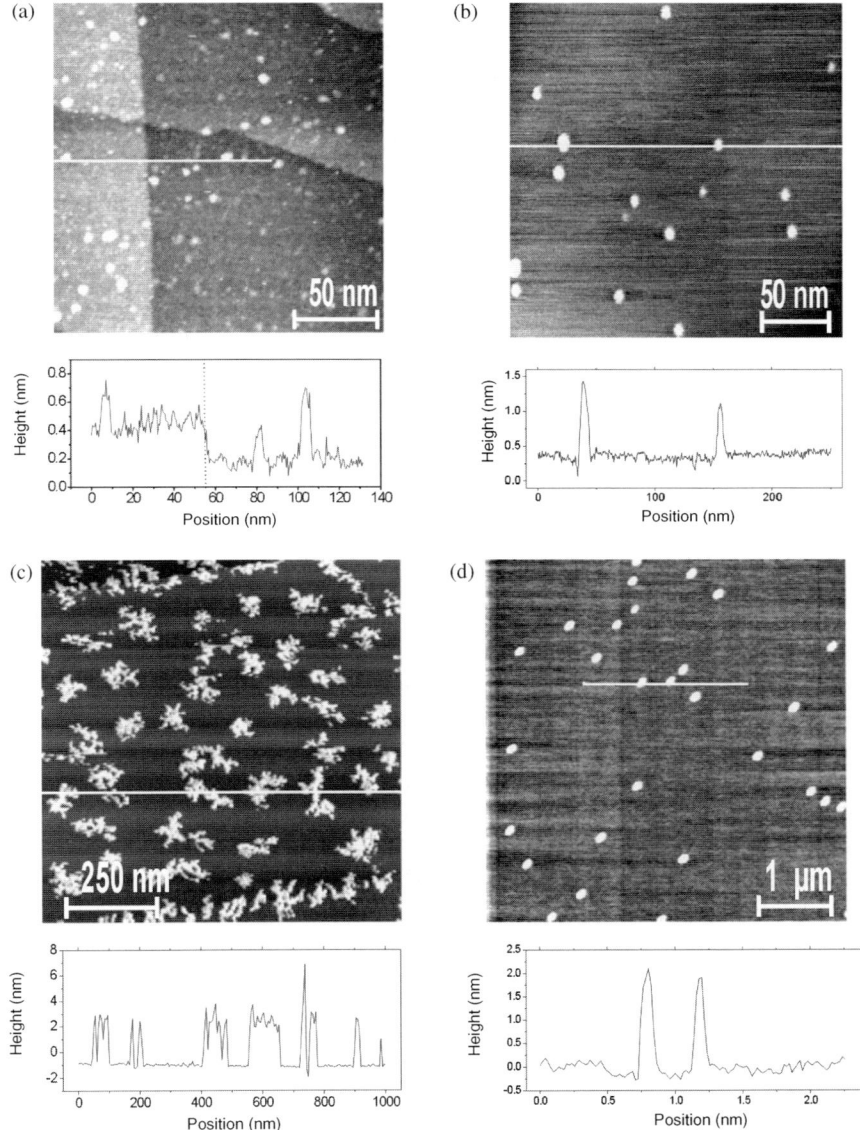

Figure 3.6. (a) STM image of Au$_N$ clusters (N = 100–300) on Au(1 1 1) showing random paving. (b) STM image of Au$_N$ clusters on Ag(1 1 1) showing random paving. (c) STM image of Au$_N$ clusters on HOPG showing ramified islands. (d) AFM image of Au$_N$ clusters on mica showing cluster diffusion and coalescence. Reproduced with permission of the Institute of Physics from Vandamme et al. [49].

deposition beyond a cluster monolayer coverage results in a porous film. This was verified by Palasantzas et al. who analysed the observed power law increase in the surface roughness of films of Cu nanoclusters deposited on Si(100); moreover, they estimated the density of the film to be 50% of the Cu bulk value [55]. The nanoporous nature of cluster-assembled films is a useful technological attribute for applications requiring materials with a very large effective surface area per unit mass, which in a cluster-assembled film can reach $1000\,\mathrm{m^2\,g^{-1}}$.

3.3.2. Embedding nanoparticles in matrices

One can also assemble three-dimensional ensembles of nanoparticles by co-depositing the clusters emerging from a cluster source with an atomic vapour from an MBE source (see Section 3.4) onto a common substrate so that the clusters are continuously buried by the matrix material (Figure 3.7). The method was originally developed by Perez and co-workers [5] and has since been adopted by most cluster groups. The clusters can be embedded into any material that is UHV-compatible at room temperature and has a sufficiently high vapour pressure ($\sim 10^{-5}$ mbar) at an achievable temperature. Using a combination of Knudsen cell and electron beam evaporators (Section 3.4), this includes all metallic elements. As described in the previous section, the impact energy of the clusters influences the morphology of the film, but in this case, even in the soft-landing regime, diffusion of clusters is limited and the random paving morphology is observed irrespective of the cluster and matrix materials.

The production method shown in Figure 3.7 is a uniquely flexible way of making granular thin films in which there is independent control over the grain size and volume fraction. The cluster size can be chosen by any of the various methods described in Section 3.2 and the volume fraction can vary simply by controlling the relative deposition rate of the clusters and matrix. The metallurgy of cluster-deposited films would thus appear trivial, but this is superficial as the reality is a rich complexity of structure. For example, the atomic structure remains to some extent an open question even for the particles in the gas phase (see Chapter 6). When they are embedded in matrices, the structure and morphology is modified and becomes dependent on the matrix material as shown by a series of structural studies of embedded clusters using extended X-ray absorption fine structure (EXAFS) [56,57]. The dominant factor determining the cluster structure, at least for small (<5 nm) particles, is epitaxy with the matrix. Thus, the most stable cluster atomic structures

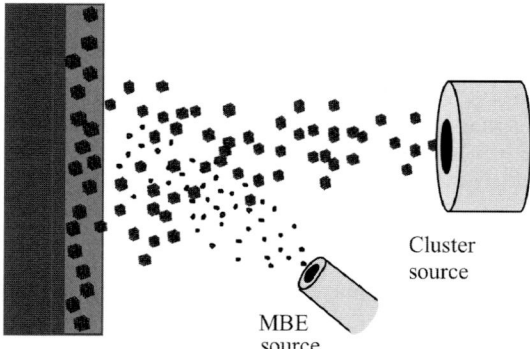

Figure 3.7. Formation of a cluster-assembled film by co-deposition of gas-phase nanoclusters and an atomic vapour from a conventional deposition source.

are those for which there is a good lattice match between the cluster and the matrix. For example, Fe clusters in a Cu matrix form the fcc structure because of the good match between Fe fcc and Cu fcc crystal lattices, whereas Fe clusters in Ag matrices adopt the bulk Fe bcc structure. Similarly, embedded Co clusters can be formed with the hcp or bcc structure by embedding them in Ag or Fe matrices, respectively. A comprehensive review of the behaviour of gas-phase nanoparticles embedded in matrices is given in Ref. [58].

3.4. Self-assembly of nanoscale islands on surfaces

3.4.1. UHV environment

A simpler method to form nanoparticles on surfaces than depositing pre-formed clusters is to impinge an atomic beam onto a non-wetting surface and to allow the subsequent kinetics to form nanoscale islands. The growth of films due to atoms impinging on a surface is discussed in detail in Chapter 4, so here we focus on the practical requirements to produce the vapour and conditions for growth. The most basic requirement for growth is UHV because the process is trying to produce pristine films with a typical thickness of a monolayer. In a closed vessel at pressure P (mbar) and temperature T (K), the rate of impingement of molecules, v, on any surface in the vessel is given by:

$$v = \frac{2.68 \times 10^{22}}{\sqrt{MT}} P \text{ molecules cm}^{-2}\text{s}^{-1}, \quad (3.11)$$

where M is the molecular weight of the gas. Typically, a surface contains $\sim 10^{15}$ atoms/cm^2, so in order to maintain the contamination level at say 1% of a monolayer per hour, the pressure, assuming the residual gas is N$_2$ ($M = 28$), needs to be 10^{-11} mbar. This is routinely achievable in bakeable stainless steel vessels fitted with turbomolecular or ion pumps.

3.4.2. Evaporation sources

The atomic vapour is normally produced by a heated Knudsen effusion cell or an electron beam evaporator illustrated in Figure 3.8. The Knudsen cell (Figure 3.8(a)) is a heated container made of a material that does not react with the material to be evaporated with a hole at one end. Normally, in order to maintain UHV clean conditions during evaporation, the crucible assembly is placed within a cooled jacket to trap the gas given off by the heated components.

The rate of effusion of atoms from the cell is given by the Knudsen equation:

$$\frac{dN_e}{dt} = \frac{A_e}{\sqrt{2\pi m k_B T}} P \quad (3.12)$$

where A_e is the area of the aperture, m the mass of the atoms and P the vapour pressure of the evaporant material at temperature T. A typical deposition rate

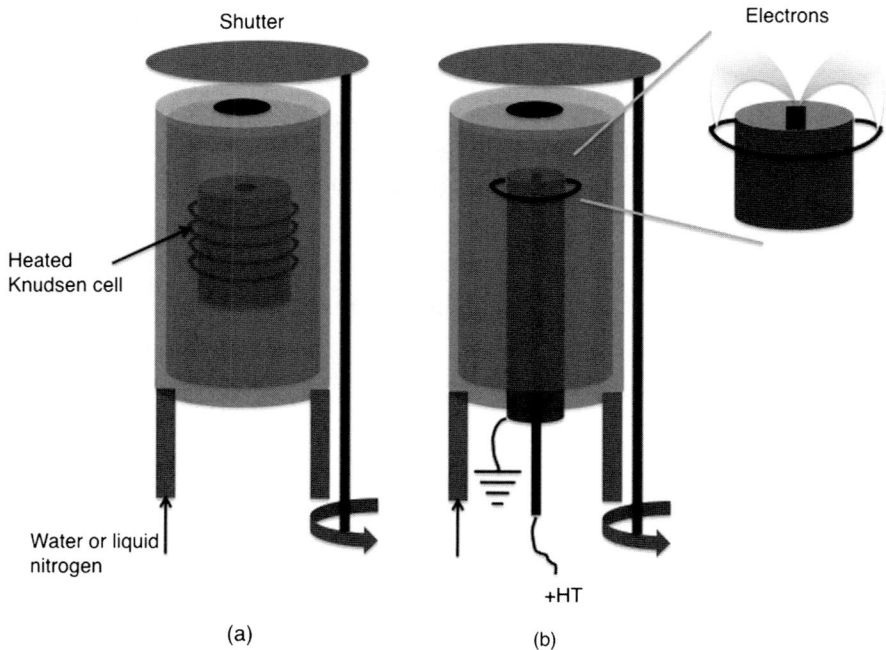

Figure 3.8. (a) Knudsen cell evaporator consisting of a heated crucible contained within a water or liquid nitrogen-cooled shroud. (b) Electron beam evaporator that focuses an electron beam onto the end of a target rod (inset) that is heated by electron bombardment. As in (a), this is normally contained within a cooled jacket.

would be 0.1 monolayers per second or $dN_e/dt = 10^{14}\,\text{s}^{-1}$. For an orifice diameter of 5 mm and assuming that all the evaporant material is incident on the substrate, this gives a required vapour pressure of $\sim 10^{-6}$ mbar. The temperature required to reach this pressure is given for a range of elements in Table 3.1.

The maximum temperature achievable in practice in a Knudsen cell is about 1800 K, above which it becomes impossible to find materials that will contain the melt without reacting with it. Crucible materials normally used are alumina, boron nitride and graphite depending on which element is contained.

Several elements in Table 3.1 require higher temperatures than achievable with a Knudsen cell and for these, the normal UHV-compatible method is electron beam evaporation (Figure 3.8(b)). With this method, a rod of the evaporant material, raised to a high positive potential (typically 1–2 kV), is placed so that it is flushed with the edge of an earthed can. A heated filament just below this level acts as a thermionic emitter and the electron beam is focused onto the end of the rod. Typically, emission currents of tens of milliamperes are used so that up to 100 W of power can be applied to a small spot, sufficient to reach well over 2000 K. Because there are no containment problems, the temperature achieved can be much higher than a Knudsen cell. The evaporator acts as an isotropic point source and is not as efficient as a Knudsen cell, whose output is much more directional, so much more material is lost. One needs to compensate by running the evaporator at a higher vapour pressure than a Knudsen source, but because the vapour pressure is a fast

Table 3.1. Temperature required for a Knudsen cell effusion rate of about 0.1 monolayers per second.

Element	Temperature required (K) for about 0.1 monolayers per second
Ag	958
Al	1085
Au	1220
C	214*
Co	1340
Cr	1250
Cu	1125
Fe	1305
Mg	519
Mn	884
Mo	2095*
Nb	2200*
Ni	1345
Pb	702
Pd	1265
Pt	1765*
Re	2490*
Rh	1745*
Si	1420
Ti	1500
V	1605
W	2680*

*Denote temperatures unreachable by Knudsen cells. These materials must be evaporated using electron beam heating.

function of temperature, the required increase in temperature is usually ~ 100 K. In the most sophisticated designs, the evaporant rod can be fed into the evaporator as the material is used up.

References

[1] K. Wegner, P. Piseri, H. Vahedi Tafreshi and P. Milani, J. Phys. D: Appl. Phys. **39** (2006) R439.
[2] K. Sattler, J. Mhlback and E. Recknagel, Phys. Rev. Lett. **45** (1980) 821.
[3] T.G. Dietz, M.A. Duncan, D.E. Powers and R.E. Smalley, J. Chem. Phys. **74** (1981) 6511.
[4] O. Echt, K. Sattler and E. Recknagel, Phys. Rev. Lett. **11** (1981) 1121.
[5] A. Perez, P. Mélinon, V. Dupuis, P. Jensen, B. Prevel, J. Tuaillon, L. Bardotti, C. Martet, M. Treilleux, M. Broyer, M. Pellarin, J.L. Vaille, B. Palpant and J. Lerme, J. Phys. D: Appl. Phys. **30** (1997) 709.
[6] W. Bouwen, P. Thoen, F. Vanhoutte, S. Bouckaert, F. Despa, H. Weidele, R.E. Silverans and P. Lievens, Rev. Sci. Instrum. **71** (2000) 54.
[7] P. Piseri, H. Vahedi Tafreshi and P. Milani, Curr. Opin. Solid State Mater. Sci. **8** (2004) 192.
[8] X. Wang, F.E. Kruis and P.H. McMurry, Aerosol Sci. Technol. **39** (2005) 611.

[9] S.L. Girshick, C.P. Chiu, R. Muno, Y.U. Wu, L. Yang, S.K. Singh and P.H. McMurry, J. Aerosol Sci. **24** (1993) 367.
[10] S.E. Pratsinis, Prog. Energy Combust. Sci. **24** (1998) 197.
[11] R.P. Camala, H.A. Atwater, K.J. Vahala and R.C. Flagan, Appl. Phys. Lett. **68** (1996) 3162.
[12] P. Milani and S. Iannotta, Cluster Beam Synthesis of Nanostructured Materials, Springer, Berlin, 1999.
[13] S. Maruyama, L.R. Anderson and R.E. Smalley, Rev. Sci. Instrum. **61** (1990) 3686.
[14] P. Milani and W.A. de Heer, Rev. Sci. Instrum. **61** (1990) 1835.
[15] H. Haberland, M. Karrais, M. Mall and Y. Thurner, J. Vac. Sci. Technol. A **10** (1992) 3266.
[16] H. Haberland, M. Mall, M. Moseler, Y. Qiang, T. Reiners and Y. Thurner, J. Vac. Sci. Technol. A **12** (1994) 2925.
[17] E. Barborini, P. Piseri and P. Milani, J. Phys. D: Appl. Phys. **32** (1999) L105.
[18] H.R. Siekmann, Ch. Luder, J. Faehrmann, H.O. Lutz and K.H. Meiwes-Broer, Z. Phys. D: At., Mol. Clusters **20** (1991) 417.
[19] G. Ganteför, H.R. Siekmann, H.O. Lutz and K.H. Meiwes-Broer, Chem. Phys. Lett. **165** (1990) 293.
[20] R.P. Methling, V. Senz, D. Klinkenberg, T. Diederich, J. Tiggesbaumker, G. Holzhuter, J. Bansmann and K.H. Meiwes-Broer, Eur. Phys. J. D **16** (2001) 173.
[21] S.H. Baker, S.C. Thornton, K.W. Edmonds, M.J. Maher, C. Norris and C. Binns, Rev. Sci. Instrum. **71** (2000) 3178.
[22] C.G. Granqvist and R.A. Buhrman, J. Appl. Phys. **47** (1976) 2200.
[23] H. Haberland, M. Mall, M. Moseler, Y. Qian, T. Reiners and Y. Thurner, J. Vac. Sci. Technol. A **12** (1994) 2925.
[24] F. Frank, W. Schultze, B. Tesche, J. Urban and B. Winter, Surf. Sci. **156** (1985) 90.
[25] J.G. Pruett, H. Windischmann, M.L. Nicholas and P.S. Lampard, J. Appl. Phys. **64** (1988) 2271.
[26] K.M. McHugh, H.W. Sarkas, J.G. Eaton, C.R. Westgate and K.H. Bowen, Z. Phys. D: At., Mol. Clusters **12** (1989) 3.
[27] B. Briehl and H.M. Urbassek, J. Vac. Sci. Technol. A **17** (1999) 256.
[28] N. Lümmen and T. Kraska, Nanotechnology **15** (2004) 525.
[29] W.A. de Heer, Rev. Mod. Phys. **65** (1993) 611.
[30] L. Bewig, U. Buck, Ch. Mehlmann and M. Winter, Rev. Sci. Instrum. **63** (1992) 3936.
[31] I.M. Goldby, B. von Issendorf, L. Kuipers and R.E. Palmer, Rev. Sci. Instrum. **68** (1997) 3327.
[32] K. Ishii, K. Amano and H. Hamakake, J. Vac. Sci. Technol. A **17** (1999) 310.
[33] R.T. Laaksonen, D.A. Goetsch, D.W. Owens, D.M. Poirer, F. Stepniak and J.H. Weaver, Rev. Sci. Instrum. **65** (1994) 2267.
[34] A. Nakajima, K. Hoshino, T. Naganuma, Y. Sone and K. Kaya, J. Chem. Phys. **95** (1991) 7061.
[35] H.V. Tafreshi, P. Piseri, G. Benedek and P. Milani, J. Nanosci. Nanotechnol. **6** (2006) 1140.
[36] P. Liu, P.J. Ziemann, D.B. Kittelson and P.H. McMurry, Aerosol. Sci. Technol. **22** (1995) 293.
[37] F. Di Fonzo, A. Gidwani, M.H. Fan, D. Neumann, D.I. Iordanoglou, J.V.R. Heberlein, P.H. McMurry and S.L. Girshick, Appl. Phys. Lett. **77** (2000) 910.
[38] W. Paul and H. Steinwedel, Z. Naturforsch **8a** (1953) 448.
[39] W. Paul, H.P. Reinhard and U. von Zahn, Z. Physik **152** (1958) 143.
[40] Bu. Wrenger and K.H. Meiwes-Broer, Rev. Sci. Instrum. **68** (1997) 2027.
[41] V.I. Karataev, B.A. Mamyrin, Y.I. Gaziev, A.D. Melnik and G.I. Petrenko, J. Anal. Chem. **42** (1987) 214.
[42] T. Bergmann and T.P. Martin, Rev. Sci. Instrum. **60** (1989) 347.
[43] T. Bergmann and T.P. Martin, Rev. Sci. Instrum. **60** (1989) 792.
[44] B. von Issendorf and R.E. Palmer, Rev. Sci. Instrum. **70** (1999) 4497.

[45] L. Schweikhard, K. Hansen, A. Herlert, M.D. Herraiz Lablanca, G. Marx and M. Vogel, Int. J. Mass Spectr. **219** (2002) 363.
[46] K. Bromann, C. Felix, H. Brune, W. Harbich, R. Monot, J. Buttet and K. Kern, Science **274** (1996) 956.
[47] S.J. Carroll, P. Weibel, B. von Issendorf, L. Kuipers and R.E. Palmer, J. Phys.: Condens. Matter **8** (1996) L617.
[48] H. Haberland, M. Mall, M. Moseler, Y. Quiang, Th. Reiners and Y. Thurner, Nucl. Instrum. Meth. B **80/81** (1993) 1320.
[49] N. Vandamme, E. Janssens, F. Vanhoutte, P. Lievens and C. Van Haesendonck, J. Phys.: Condens. Matter **15** (2003) S2983.
[50] L. Bardotti, B. Prevel, M. Treilleux, P. Melinon and A. Perez, Appl. Surf. Sci. **164** (2000) 52.
[51] J.-M. Wen, S.-L. Chang, J.W. Burnett, J.W. Evans and P.A. Thiel, Phys. Rev. Lett. **73** (1994) 2591.
[52] N. Vandamme, J. Snauwaert, E. Janssens, E. Vandeweert, P. Lievens and C. Van Haesendonck, Surf. Sci. **558** (2004) 57.
[53] B. Yoon, V.M. Akulin, Ph. Cahuzac, F. Carlier, M. De Frutos, A. Masson, C. Mory, C. Colliex and C. Bréchignac, Surf. Sci. **443** (1999) 76.
[54] L. Bardotti, P. Jensen, A. Horeau, M. Treilleux and B. Cabaud, Phys. Rev. Lett. **74** (1995) 4694.
[55] G. Palasantzas, S.A. Koch and Th.M. De Hosson, Appl. Phys. Lett. **81** (2002) 1089.
[56] S.H. Baker, M. Roy, S.J. Gurman, S. Louch, A. Bleloch and C. Binns, J. Phys.: Condens. Matter **16** (2004) 7813.
[57] S.H. Baker, M. Roy, S. Louch and C. Binns, J. Phys.: Condens. Matter **18** (2006) 2385.
[58] C. Binns, K.N. Trohidou, J. Bansmann, S.H. Baker, J.A. Blackman, J.-P. Bucher, D. Kechrakos, A. Kleibert, S. Louch, K.-H. Meiwes-Broer, G.M. Pastor, A. Perez and Y. Xie, J. Phys. D **38** (2005) R357.

Chapter 4

Theory of Cluster Growth on Surfaces

P.A. Mulheran

Department of Chemical and Process Engineering, University of Strathclyde, James Weir Building, 75 Montrose Street, Glasgow G1 1XJ, UK

4.1.	Introduction	73
4.2.	Monte Carlo simulations	75
4.3.	Mean-field rate equations and diffusion-reaction calculations	78
4.4.	Scaling analyses of the mean-field rate equations	81
4.5.	Confrontation with experiments and the island size distributions	85
4.6.	Beyond mean-field theory for island nucleation and growth	89
4.7.	Island ripening	97
4.8.	Atomistic modelling	101
4.9.	Other growth methods	107
	References	110

4.1. Introduction

In this chapter, theoretical modelling of cluster growth at surfaces will be reviewed. The predominant scenario that we will discuss involves the growth of two-dimensional islands, as encountered during the earliest stages of thin film growth by vapour deposition. This is an ideal model system since it allows confrontation with experiment through comparisons with epitaxial growth in ultra-high vacuum (UHV) conditions, which provides the most controlled situation for experimentation. The islands themselves can be considered the building blocks for three-dimensional clusters (and other nanoscale structures), and in any case the theoretical methodology we describe is rather versatile and can be adapted for other types of cluster growth. We will touch on some of these towards the end of the chapter. Further background information on theoretical approaches and experimental results can be found in reviews by Brune [1] and Evans et al. [2].

The most widely used theoretical approach to cluster nucleation and growth employs rate equations that are implicitly mean-field in nature. In other words, the equations describe the growth rates of clusters that depend only on time and their size, but not on their immediate surroundings. These equations describe average behaviour extremely well, and are often amenable to further theoretical analysis. The analyses that we describe often use the language of scaling theory. This is due to the scale-invariant properties often observed during non-equilibrium growth situations. For example, we will often find that although the average cluster size changes during growth, the statistical distribution of cluster sizes relative to the average does not. Consequently, much effort has gone into understanding the scaling properties observed in cluster growth at surfaces, as we will see below when we look beyond the traditional mean-field approach.

Scaling theory draws heavily on the concept of power-law relationships between (say) cluster density and deposition rate. The classic argument for these relationships is that because the statistical properties of the growing clusters are scale-invariant, there is only one significant length scale and that is the average cluster size (which is allowed to change over time). Any theoretical expressions for this scale length cannot introduce another reference length, and power-laws fit the bill nicely. Therefore, experimental data for cluster density versus deposition rate are often presented on log–log axes whose gradients reveal the exponents in the power-law relationships, and these exponents provide an easy point of comparison to theory alongside the cluster size distributions. Different models yield different growth exponents and size distributions, so this approach is highly successful in drawing out details of the atomistic growth mechanisms and their rates.

A complementary approach often used to test and improve the theoretical methods is to employ Monte Carlo simulation. These simulations are often intuitively simple, employing straightforward local rules for deposition and growth, yet they can yield realistic statistical behaviour. Indeed such is the simplicity and economy of the approach that one might ask why bother developing theory at all? However, Monte Carlo simulation can be regarded as a form of numerical experiment, albeit at the extremes of controllability. It is ideal to test the consequences for the emerging properties of the growing clusters caused by simple changes to the growth rules. However, in itself it does not *explain* how the growth might depend on temperature or deposition rate; only theoretical analysis can do that. Furthermore, the accessible system size is limited by memory and execution time, and scaling analyses are required to extend results beyond the directly observed ranges to others that might well be of experimental relevance.

Whilst on the topic of simulation, another obvious question occurs: why not simply use atomistic modelling for the growth of clusters? Even if *ab initio* methods are too slow to simulate the growth of clusters, surely empirical potentials such as the Embedded Atom Method for metals can be employed in a molecular dynamics simulation? The answer to this lies in the timescales involved. Experimental growth of clusters occurs over seconds and minutes, so that the deposited material has time to arrange itself into stable structures, whereas traditional molecular dynamics simulations can only simulate up to tens or hundreds of nanoseconds depending on system size and interaction potentials. The important events in the life-cycle of a cluster often occur on the millisecond timescale, way beyond the scope of direct molecular dynamics. However, recent developments using so-called accelerated dynamics methods do now allow one to access such timescales in unbiased ways, revealing the relevant atomistic dynamics of small clusters and providing insight into the growth processes.

Interest in the evolution of clusters at surfaces is not of course restricted to nucleation and growth, but also to what happens to the clusters afterwards. The phenomenon of coarsening, whereby large clusters grow larger still at the expense of smaller neighbours, thereby reducing the interfacial energy of a system, is well known. In the context of clusters supported on a surface, such ripening has been observed experimentally to induce scale-invariant spatial and size distributions. Once again, the mean-field approach is found to fail in this regard and efforts to develop a theoretical framework that takes local surroundings into account are needed.

In this chapter, we will review the work mentioned above in the following order. In Section 4.2, we will review some key Monte Carlo simulation results to set the context for what follows. We will present the theory of island nucleation and growth during epitaxy using the bread-and-butter workhorse of mean-field rate equations in Section 4.3. Scaling analyses that reveal the growth exponents for various models are reviewed in Section 4.4, and in Section 4.5 we will consider the island size distribution (ISD) and its comparison with experiment. The failing of the mean-field approach to successfully describe the scale-invariance of the ISD is also discussed here. In Section 4.6, we review the theoretical methodology to look beyond the mean-field approach and show how this resolves the problem. In Section 4.7, we will review the theory for island growth (ripening) post-deposition, looking both at the mean-field approach and beyond. In Section 4.8, we will describe atomistic modelling approaches to describe the island nucleation processes. Finally in Section 4.9, we will provide some context with other growth methods at surfaces.

4.2. Monte Carlo simulations

Perhaps the simplest way to introduce the key ideas that theorists seek to explain is through Monte Carlo simulation of cluster (or island) nucleation and growth during vapour deposition. The first comprehensive (but simplified) set of simulations is from Bartelt and Evans in the early 1990s [3]. They studied growth using a lattice, initially empty, to represent the substrate. Monomers are deposited at random into this lattice at a constant deposition rate. The monomers can diffuse by random nearest neighbour hops, and when they collide they nucleate new 'point' islands. The immobile islands themselves 'grow' whenever new monomers diffuse onto them; however, they remain as single points on the lattice whilst their recorded size increments by the number of adsorbed monomers.

The density of islands nucleated in the simulation depends on the ratio (often called R) of the monomer diffusion rate to the monolayer deposition rate. This ratio alone dictates the Monte Carlo simulation procedure, by changing the probabilities of the next simulation step being a deposition event or a monomer diffusion event. Typically $R \sim 10^5$–10^{10} in experimental growth conditions. The higher the value of R, the slower the deposition rate relative to the monomer diffusion rate, and the further a deposited monomer can travel to become incorporated into an existing island. For this reason, simulated island density decreases with increasing R. We expect exactly the same consideration to apply to real experiments, so determining exactly how the island density will change as the deposition rate or growth temperature changes is an important issue. Of course, island density directly affects the average island size and spatial separation at any given coverage θ. Bartelt and Evans [3] showed that we expect the island density at fixed monolayer deposition to vary as

$$N \sim \theta^{1/3} R^{-1/3} \qquad (4.1)$$

for the growth of point islands, and their simulation results support this. We shall see the type of arguments used to find this result below.

As computers became increasingly powerful, researchers lifted the restriction on the physical point size of the islands used by Bartelt and Evans. A common strategy was to simulate epitaxy, so that the lattice representing the substrate becomes the crystal lattice exposed on a crystal facet. For example, fcc(1 0 0) surfaces have square symmetry just like the most obvious simulation lattice. Amar et al. [4] simulated the growth of fractal islands by allowing the monomers to freeze in place whenever they diffuse to nearest neighbour sites of existing islands or other monomers. Note again that the islands are immobile, and furthermore are stable since monomers cannot dissociate from an island in this model. The critical island size i is defined as the size above which the islands are stable. In this and the previous point island model, $i = 1$.

Results from the Amar et al. model are shown in Figure 4.1. We see how the density of islands has essentially reached a plateau by a coverage of $\theta = 0.1$. The authors divide the growth into four regimes: low-coverage (L) where the monomer

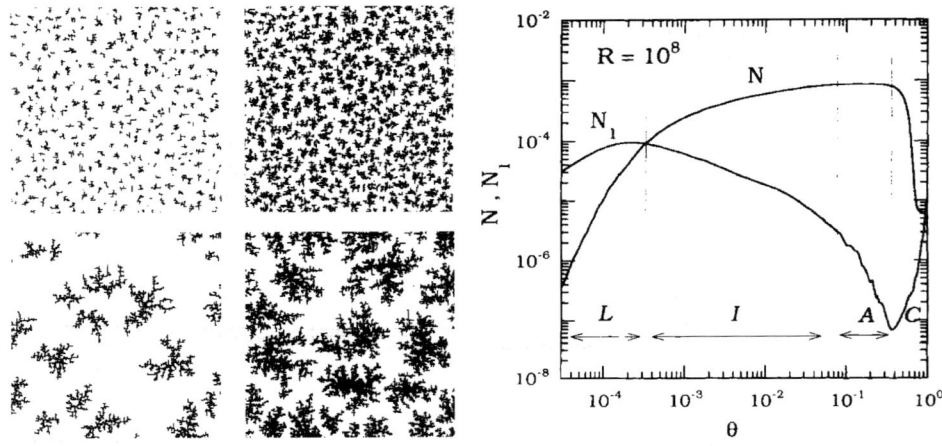

Figure 4.1. The left panels show the fractal islands nucleated and grown during the simulations of Amar et al. [4]. The top left and right images are for $R = 10^5$ at coverage $\theta = 0.1$ and 0.3, respectively, and the lower images for $R = 10^7$. The right-hand panel shows the variation of island (N) and monomer (N_1) density against coverage at $R = 10^8$. Reproduced with the permission of the American Physical Society from Amar et al. [4].

density is larger than that of the islands; intermediate (I) where the situation is reversed by significant nucleation of new islands; aggregation (A) where the island density has saturated but islands continue to grow by capturing new monomers; and coalescence (C) where islands grow into one another to cause the lattice to percolate. We are usually interested in the later stages I and A before significant island coalescence occurs.

Ratsch et al. [5] also simulated a simple solid-on-solid model using a square mesh to represent the substrate. Again the simulation allows only the movement of single monomers, but now any monomer might be able to hop to a vacant nearest neighbour site. However, the probability to hop now depends on the number of nearest neighbour 'bonds' n the monomer would have to break in order to move. The activation energy to be overcome has the form $E_n = E_S + nE_N$, where the substrate binding energy is $E_S = 1.3\,\text{eV}$, and the lateral bond energy E_N can be changed along with the simulated temperature. Note that this simulation does allow for the dissociation of islands, so the critical island size i now depends on temperature and the strength of E_N.

Typical results are shown in Figure 4.2. As can be seen, the morphology of the islands changes with the strength of the lateral bond (the simulation temperature for these images is 800 K). For strongly bound monomers ($E_N = 1.0\,\text{eV}$), fractal islands tend to grow since edge-diffusion is suppressed by the algorithm. For lower lateral

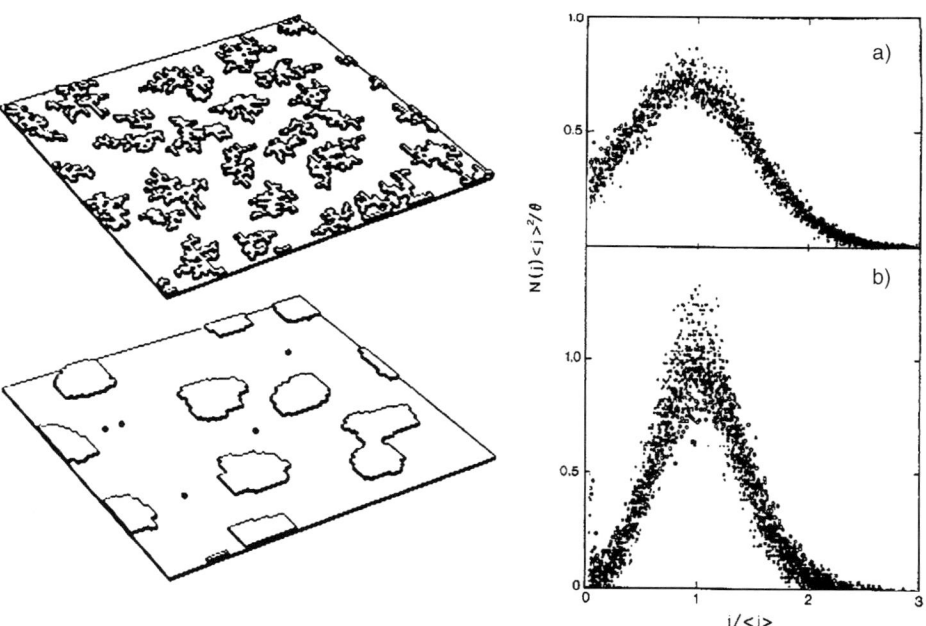

Figure 4.2. Simulated morphology and corresponding island size distributions for different values of the lateral binding energy E_N in a Monte Carlo simulation of homoepitaxy: $E_N = 1.0\,\text{eV}$ (top left) and $0.3\,\text{eV}$ (bottom left). Reproduced with the permission of the American Physical Society from Ratsch et al. [5]. The right panels show the scaled island size distributions at various times and deposition rates for the following two cases: (a) $E_N = 1.0\,\text{eV}$ and (b) $E_N = 0.3\,\text{eV}$. Reproduced with the permission of Elsevier Science from Zinke-Allmang [6].

interaction strength ($E_N = 0.3 \text{ eV}$), more compact islands reflecting the square symmetry of the lattice are created. At the same time, the ISD has a different shape. The reason for this is that the critical island size $i = 1$ for high E_N, so that a dimer island is immobile and stable. However, with lower E_N, the dimer islands can dissociate before new monomers attach themselves, and the critical island size $i = 3$ meaning that a square of four monomers, each with two nearest neighbour lateral bonds, is needed to create a stable island. This dependence of ISD on the critical island size became a major theme in the theory of island nucleation and growth as we shall see below.

The most intriguing aspect brought into sharp relief by the Monte Carlo simulations is the scale-invariance of the ISD. The data in the right-hand panels of Figure 4.2 are taken at six different values of coverage (and therefore times) between 7.5 and 25%. The data are seen to 'collapse' to a (i-dependent) universal curve when size j is plotted relative to the average $\langle j \rangle$ at the time of recording. This invariance to change of scale, here defined by the average size, is a common feature of growth simulations.

Let us summarise this brief review of Monte Carlo simulation of island nucleation and growth during vapour deposition. Suitable algorithms can easily be created and, on modern computers, simulations are usually quick to execute. The simulations throw up many questions relevant to experiment, such as: how does the island density depend on growth rate and temperature; how does the critical island size i change with these conditions; how does the ISD depend on this critical island size; and why does it display scale-invariance? These theoretical challenges come from the most straightforward of growth simulations, and one can easily extend the list by including other plausible growth mechanisms such as: evaporation of monomers from the substrate; the mobility of small islands; the effects of strain in hetero-epitaxy or non-epitaxial growth, compounded by the effects of intermixing of the species. However, it is often adequate only to address the initial list of questions which we shall do in the following sections. It is worth emphasising again that the analyses we present are rather general in nature and not specific to the epitaxial nature of the simulations, not withstanding the physical effects neglected in our model systems.

4.3. Mean-field rate equations and diffusion-reaction calculations

The most widely used theoretical framework for investigating island nucleation and growth during the earliest stages of vapour deposition utilises rate equations for the evolution of the density of islands and monomers. The key processes in the nucleation and growth of the islands must be decided upon. As in the Monte Carlo simulations discussed above, typical assumptions used are that: only monomers are mobile; islands of size $s \geq 2$ grow by capturing the diffusing monomers; monomers are produced on the substrate by random deposition from vapour or by release from an existing island. Assuming that we are interested in the early stages of nucleation and growth (regime I in Figure 4.2), so that direct impingement from vapour onto existing monomers and islands can be neglected, we obtain for the time evolution of monomer density N_1 and density N_s of islands of size s:

$$\frac{dN_1}{dt} = F - 2D\sigma_1 N_1^2 - DN_1 \sum_{s \geq 2} \sigma_s N_s + 2\frac{N_2}{\tau_2} + \sum_{s \geq 3} \frac{N_s}{\tau_s}, \quad (4.2)$$

$$\frac{dN_s}{dt} = DN_1(\sigma_{s-1} N_{s-1} - \sigma_s N_s) + \frac{N_{s+1}}{\tau_{s+1}} - \frac{N_s}{\tau_s}. \quad (4.3)$$

In the above equations, F is the monomer deposition rate (monolayers per second), D the monomer diffusion rate, σ_s the so-called 'capture number' of monomers by an island of size s and $1/\tau_s$ the dissociation rate (rate of monomer release) of an island size s.

The form of the rate equations is fairly intuitive and looks rather straightforward. A point worth emphasising is the mean-field nature of the equations. All islands of a given size, and at a given time, experience the same growth probabilities. There is no variation in growth rate due to different neighbourhoods that the islands might have. For example, a small island might be 'shadowed' by a large near-neighbour and therefore grow more slowly than if it had no near-neighbours at all. We will return to this point in following sections, but for now we shall see what can be learnt from the mean-field approach.

Provided one has expressions for the capture numbers and dissociation rates, there is no obstacle to (at least) numerically integrating the equations up to some finite time and some maximum possible island size at that time. The question of course is how to create realistic expressions for these quantities. One simplification often made is to assume that above a critical size i, islands no longer dissociate, so that $1/\tau_s = 0$, $s > i$. This is used in the following review of the diffusion-reaction calculations of the capture numbers and the subsequent scaling analyses for their approximate forms.

The theory for the capture numbers was developed in the late 1960s, and here we follow Venables' treatment [7]. We consider the evolution of the monomer density $n_1(r, t)$ in the vicinity of a circular island of radius $r_s = \sqrt{s/\pi}$ at the origin, utilising cylindrical symmetry. The monomer density is zero at the adsorbing island edge and rises to the global average value N_1 a long way from the island. The capture number then comes from the diffusive flux into the island:

$$DN_1\sigma_s = D2\pi r_s \frac{\partial n_1}{\partial r}\bigg|_{r_s}$$

To evaluate the monomer density field, we solve the diffusion equation. Consider first a case where the monomers can evaporate from the substrate at the rate $1/\tau_a$ and that $N_1 \to F\tau_a$ a long way from the island (this implies the density of islands is sufficiently low). Then the diffusion equation reads

$$\frac{\partial n_1(r, t)}{\partial t} = D\nabla^2 n_1(r, t) - \frac{n_1(r, t)}{\tau_a} + F.$$

From the long time $t > \tau_a$ steady-state solution, the capture number is found in terms of Bessel functions K_1 and K_0 to be

$$\sigma_s = 2\pi X_s \frac{K_1(X_s)}{K_0(X_s)}, \quad X_s = \frac{r_s}{\sqrt{D\tau_a}}. \tag{4.4}$$

To extend this analysis, Venables makes a 'uniform depletion approximation' whereby the stay time τ of the monomer is reduced by capture by stable islands. If we assume the islands around our central one are randomly distributed and all have the same average size $\langle s \rangle$ and thus the same capture number σ_x, we can replace τ_a in Eq. (4.4) with τ such that

$$\frac{1}{\tau} = \frac{1}{\tau_a} + D\sigma_x N, \tag{4.5}$$

where $N = \sum_{s>i} N_s$ is the total density of stable islands. The evaporation of the monomers can now be switched off ($1/\tau_a = 0$) in Eq. (4.5) to model the commonly encountered 'complete condensation' regime of film growth.

Consider now the average capture number σ_x for the stable islands, obtained by evaluating Eq. (4.4) with the average island radius. Since the substrate coverage $\theta = N\langle s \rangle$, from Eq. (4.5), we find that σ_x is a function of the product $\sigma_x \theta$ and so must be solved self-consistently. Thereafter, τ in Eq. (4.5) is also known and so all σ_s can be found from Eq. (4.4).

The results are presented in Figure 4.3 as a function of $Z_s = \pi N r_s^2 = Ns = \theta s/\langle s \rangle$. Also shown are results from a complementary calculation for the 'lattice

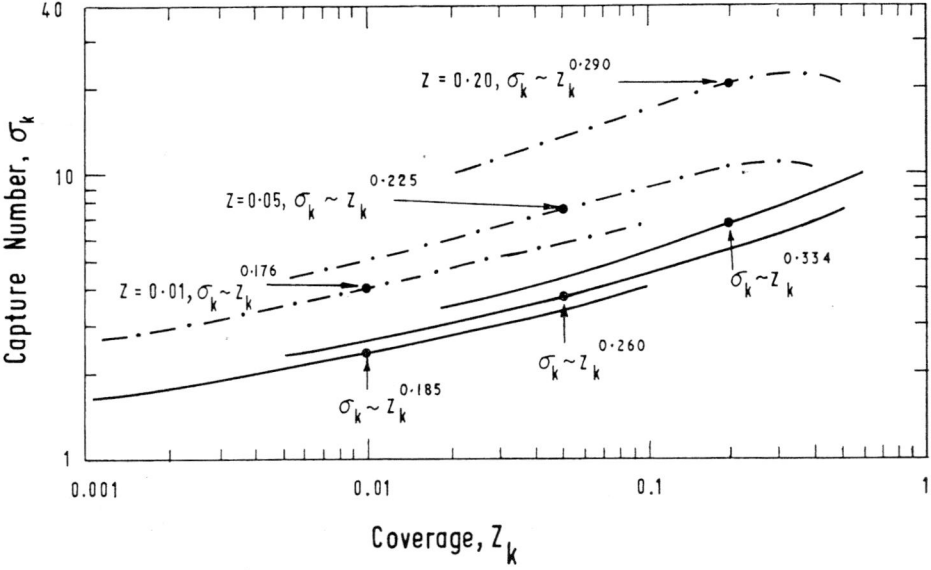

Figure 4.3. Calculated capture numbers: the solid lines are from the uniform depletion approximation discussed in the text, and the dot-dash lines from the lattice approximation. The three curves show how the capture number varies with size at different substrate coverage of $\theta = 0.01$, 0.05 and 0.20. Reproduced with the permission of Taylor & Francis from Venables [7].

approximation' where the island traps are regularly rather than randomly placed. We will not consider this case here, only noting that it represents the opposite extreme to the randomly placed island traps of the 'uniform depletion' approximation. Whilst island nucleation is stochastic, it does induce a degree of spatial order that is somewhere between these two extremes as we discuss below. However, for the purposes of our discussion, it is sufficient to concentrate only on the uniform depletion case.

Figure 4.3 shows that the capture numbers can usefully be represented through power-law relationships:

$$\sigma_s \approx (Ns)^p, \tag{4.6}$$

where p ranges from 0.185 to 0.334 as the coverage increases from 1 to 20%. We will consider this form further in the following section.

Before leaving this section on the diffusion-reaction calculations, we review the work of Bales and Chrzan who showed how successful this approach can be [8]. They presented results from kinetic Monte Carlo simulations and showed how a self-consistent solution for all σ_s (not just the average as in Venables [7]) can be used in conjunction with Eqs. (4.2) and (4.3) to reproduce the evolution of the monomer and total island density. In this work, Eq. (4.5) for the average capture number is replaced by

$$\frac{1}{\tau} = 2D\sigma_1 N_1 + \sum_{s \geq 2} D\sigma_s N_s + F\kappa_1. \tag{4.7}$$

This allows for monomer trapping by other monomers and direct hits from vapour deposition onto existing monomers as well as capture by islands. The results shown in Figure 4.4 are extremely impressive and convincingly show that the mean-field approach implicit to the rate equations is entirely appropriate for calculating average quantities. However, we also see that the ISDs are not reproduced well at all, which is the greatest failing of the approach.

4.4. Scaling analyses of the mean-field rate equations

The above section for the self-consistent calculation of capture numbers hints at the way a scaling analysis of the rate equations (4.2) and (4.3) can be performed. The key idea is to represent capture numbers through simple power-law relationships, and use these in the rate equations to deduce corresponding growth exponents such as those of Eq. (4.1) for average size and cluster density. The latter quantities then play a pivotal role in analysing experimental and Monte Carlo simulation data, relating observed dynamical behaviour to fundamental growth mechanisms.

To proceed with this analysis, we must first assume that the solutions to the rate equations will be scale-invariant in the intermediate and aggregation regimes of interest (recall Figure 4.1). As mentioned in Section 4.2, this idea builds upon earlier studies of model growth systems, and was brought to the fore in the field of island nucleation and growth during vapour deposition by Bartelt and Evans in their study of point island models [3]. Scaling results from simulations of Bales and

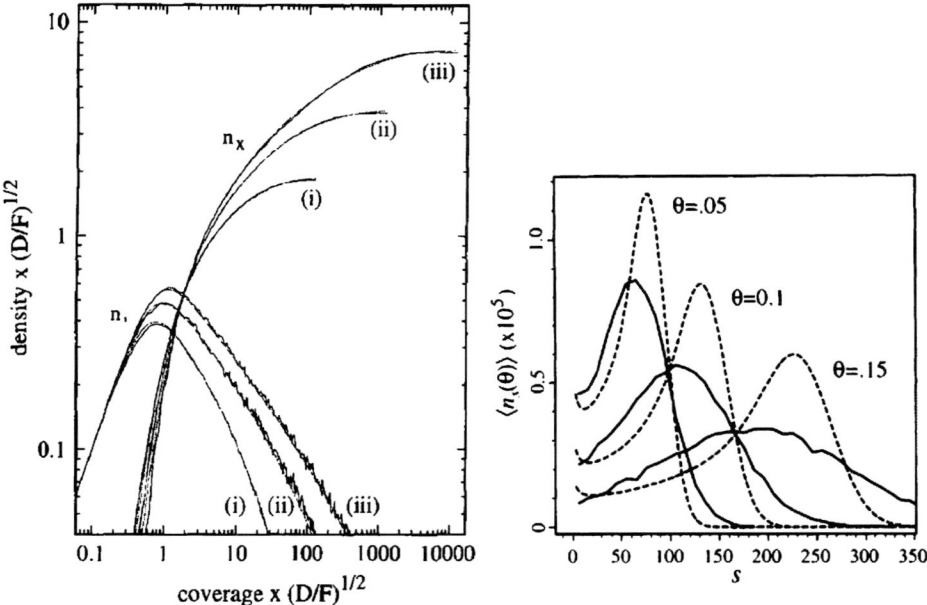

Figure 4.4. The left panel shows results of the self-consistent mean-field rate equations (dashed lines) for monomer and island density as growth proceeds, compared to Monte Carlo simulations (solid lines). The curves are virtually superposed and very difficult to distinguish. Reproduced with the permission of Elsevier Science from Brune [1]. In the right panel, the island size distributions measured in the simulations (solid lines, $R = 10^8$) are shown alongside the rate equation calculations. Reproduced with the permission of the American Physical Society from Bales and Chrzan [8].

Chrzan [8] mentioned above are shown in Figure 4.5. The data collapse onto a single curve found with appropriate scaling is compelling.

We shall follow the notation of Ratsch *et al.* [5], who analysed the scaling properties of results found in their own Monte Carlo growth model. There is only one significant size scale that is set by the average size $\langle s \rangle$. The normalised distribution of island sizes $g(u)$ will always look the same when the data are plotted as size divided by the average at that particular time, $u = s/\langle s \rangle$. In terms of the unscaled ISD appearing in Eqs. (4.2) and (4.3), we can write

$$N_s(\theta) = N(\theta)g(u)\frac{du}{ds} = \frac{\theta}{\langle s \rangle^2} g\left(\frac{s}{\langle s \rangle}\right). \tag{4.8}$$

This is also the scaling form used by Amar *et al.* [4]. As in the previous section, we assume the total density of stable islands $N \approx \theta/\langle s \rangle$ in the intermediate and aggregation regimes where the substrate coverage $\theta = Ft$.

The scaling analysis of Blackman and Wilding [9] proceeds by assuming the monomer density decays, and the average island size grows, as power-laws over time (coverage):

$$N_1 \sim \theta^{-r}, \quad \langle s \rangle \sim \theta^z \Rightarrow N \sim \theta^{1-z}. \tag{4.9}$$

4.4. *Scaling analyses of the mean-field rate equations*

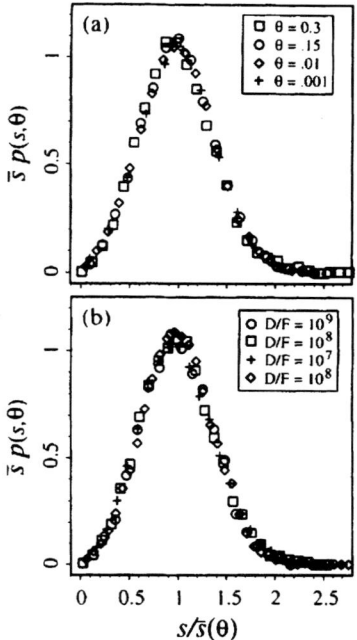

Figure 4.5. The scaled island size distributions found by for extended, ramified islands: (a) different coverages with $R = 10^8$; (b) $\theta \approx 0.2$ where results for compact square islands at $R = 10^8$ are also shown. Reproduced with the permission of the American Physical Society from Bales and Chrzan [8].

In addition, they assume a power-law form of the capture number

$$\sigma_s \sim s^p, \tag{4.10}$$

where the exponent p might come from consideration of the island diameter on the substrate. For compact two-dimensional islands, $p = 1/2$, but would be larger for fractal objects.

Consider the nucleation of stable islands implied by Eq. (4.3). At large enough times, all islands of size $s \leq i$ will decay faster than the grow, and the nucleation rate of stable islands is therefore given by

$$\frac{dN}{dt} \approx DN_1 N_1^i \Rightarrow \frac{dN}{d\theta} \sim R\theta^{-r(1+i)}. \tag{4.11}$$

Utilising the relationship (4.9) and equating exponents, we find

$$z = r(1 + i). \tag{4.12}$$

The next step is to use moments of the distribution, defining the nth moment as

$$M_n(\theta) = \sum s^n N_s(\theta) = \theta^{z(n-1)+1} \int u^n g(u) du. \tag{4.13}$$

The evolution of this moment can be related to lower moments through Eq. (4.3):

$$\frac{dM_n}{d\theta} \approx RN_1 M_{n+p-1}(\theta). \tag{4.14}$$

This result is obtained by the binomial expansion of the resulting terms on the right of Eq. (4.3), and neglecting lower order moments in the long time limit. Combining Eqs. (4.13) and (4.14), we find $z(1-p) = 1-r$, and from Eq. (4.12)

$$r = \frac{1}{2 + i - p(1+i)}, \quad z = (1+i)r. \tag{4.15}$$

From Eq. (4.9), we see that we require $z \leq 1$, otherwise the total island density would decrease over time which cannot occur in the system of Eqs. (4.2) and (4.3) with our restrictions on island dissociation ($1/\tau_s = 0$, $s > i$). This implies that $p < 1/(1+i)$. For $p = 1/(1+i)$, Eq. (4.11) implies $N \sim \ln|\theta|$ and $z = 1$ so that the island density saturates and the average size $\langle s \rangle \sim \theta$. For larger values of the capture number exponent, we keep the idea of the island density saturating with $z = 1$, but Eq. (4.9) must be amended to read

$$N_1 \sim \theta^{-r}, \quad \langle s \rangle \sim \theta \Rightarrow N \sim C - a\theta^{-q}, \tag{4.16}$$

so that Eq. (4.11) yields $1+q = r(1+i)$. Eqs. (4.13) and (4.14) then yield

$$r = p, \quad q = (i+1)p - 1.$$

This result is worthy of some note. If the capture number exponent is large enough for a given critical island size i, a log–log plot of the monomer density against coverage should yield the value of that exponent. Ratsch et al. [5] tested this idea directly and found that $p \approx 0.5$ for the growth of compact islands, and $p \approx 0.8$ for fractal islands, which accords with the estimate of the fractal dimensions of the islands and their perimeters.

Ratsch et al. [5] extended the scaling analysis of Blackman and Wilding [9] to include dependence on $R = D/F$. Focusing on the regime where $p > 1/(1+i)$, Eqs. (4.16) become

$$N_1 \sim R^{-\omega}\theta^{-r}, \quad \langle s \rangle \sim R^\chi \theta \Rightarrow N \sim R^{-\chi}(C - a\theta^{-q}). \tag{4.17}$$

Eqs. (4.13) and (4.14) then yield $\chi(1-p) = (1-\omega)$, a relationship that is not supported by the Monte Carlo data.

The reason for the failure in this part of the scaling analysis is apparent from the previous section (Eq. (4.6)). When the monomer diffusion fields are established around existing islands, the capture number does not depend solely on size s but also on the coverage or equivalently the island density N. Ratsch et al. included this dependence by replacing Eq. (4.10) with

$$\sigma_s \sim N^Q s^p. \tag{4.18}$$

Note however that this form is only valid for the stable islands. Therefore, in the high R scaling regime with low total island density, the first equation in Eq. (4.11) for the nucleation rate still holds, yielding $\chi = \omega(1+i) - 1$. With Eq. (4.18),

Eq. (4.14) now yields $\chi(1+Q-p) = 1-\omega$, so that we find

$$\omega = \frac{2+Q-p}{i(1+Q-p)+2+Q-p}, \quad \chi = \frac{i}{i(1+Q-p)+2+Q-p}. \tag{4.19}$$

The Monte Carlo data now support the notion that $Q \approx p$, as does Eq. (4.6), resulting in scaling exponents for R that are independent of p. The simplified exponents of $\omega = 2/(2+i)$ and $\chi = i/(2+i)$ are found in a variety of simulation data including that for point and extended islands [3,4]. We shall also see how the result holds up against experiment below. Overall we can say that the scaling analyses of Blackman and Wilding [9] and Ratsch et al. [5] are remarkably successful.

4.5. Confrontation with experiments and the island size distributions

The scaling analyses and the Monte Carlo simulations between them provide ample opportunity for comparison to experiment. It was recognised early on that successful application of the theory would help to identify and indeed quantify the atomistic processes at play during island nucleation and growth, and this methodology has been successfully used many times (see, for example, Refs. [1,2,10]). A classic comparison of this kind was performed by Stroscio and Pierce [11] who performed scanning tunnelling microscopy on Fe/Fe(0 0 1) islands after growth at different temperatures. The results shown in Figure 4.6 show excellent agreement with the scaling analysis. They find that at low temperature, the data conform to critical island size $i = 1$, and at high temperature to $i = 3$, in the scaling relation $\chi = i/(2+i)$ for the variation of island density with monomer diffusion rate in Eq. (4.17). Furthermore, the ISD displays good data collapse to different forms in the two regimes. In Figure 4.6, the point ISDs are shown, but referring back to the distributions of Ratsch et al. for extended islands in Figure 4.2 suggests an even more impressive comparison to Monte Carlo data.

The usefulness of the ISD and its scale-invariance as a means of identifying the dominant growth processes through the critical island size has inspired much theoretical work. Amar and Family [13] investigated the growth of extended islands on substrate lattices with both square and triangular symmetries, the latter being found on fcc(1 1 1) surfaces. They showed the critical island size at high temperature (or low lateral bond strength) is $i = 2$ for the triangular lattice as might be anticipated for a stable triangular island. Furthermore, this group created a useful family of curves that fit both their simulated ISD data and the experimental data of Stroscio and Pierce shown in Figure 4.7. The functional form Amar and Family proposed reads

$$f_i(u) = C_i u^i \exp(-ia_i u^{1/a_i}), \tag{4.20}$$

where C_i is a normalisation factor and the parameter $(ia_i)^{a_i} = (\Gamma([i+2]a_i)/((\Gamma([i+1]a_i))$. However, this form does not follow from theory, although it does appear to work extremely well for $i = 1$–3.

Amar and Family also presented the ISDs for spontaneous nucleation ($i = 0$), where a diffusing monomer has a small probability of creating a fixed, stable island

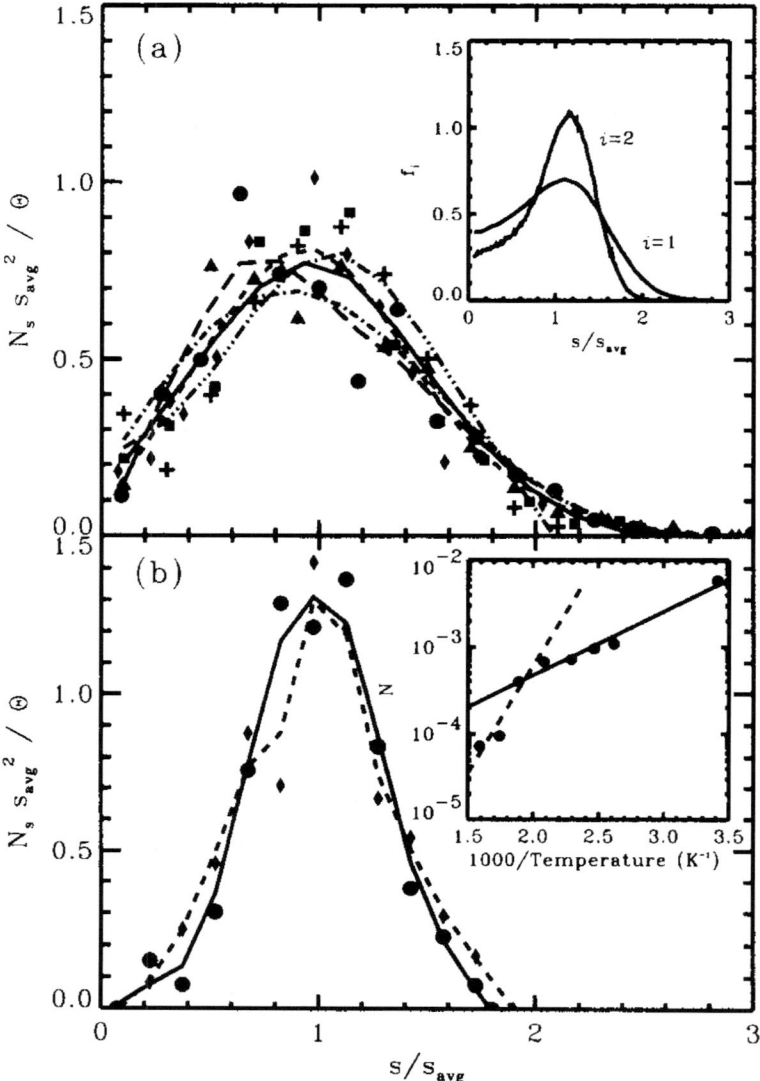

Figure 4.6. The scaled island size distributions measured for Fe/Fe(100) homoepitaxy at low temperatures (a) and high temperatures (b). The top inset shows island size distributions from point island Monte Carlo studies [12] and the lower inset shows the variation of island density with inverse temperature, revealing the transition between $i = 1$ and 3 behaviour. Reproduced with the permission of the American Physical Society from Stroscio and Pierce [11].

without interacting with any other monomers. This is believed pertinent to the experiments of Chambliss and Johnson [15] who studied Fe deposition onto Cu(100), where the island formation is explained by the thermally activated incorporation of diffusing Fe adatoms into the Cu surface plane. The simulation again captures the form of the distribution observed experimentally, and once again good scale-invariance is found in the simulated data. However, Amar and Family were not able to offer a functional form for the $i = 0$ ISD.

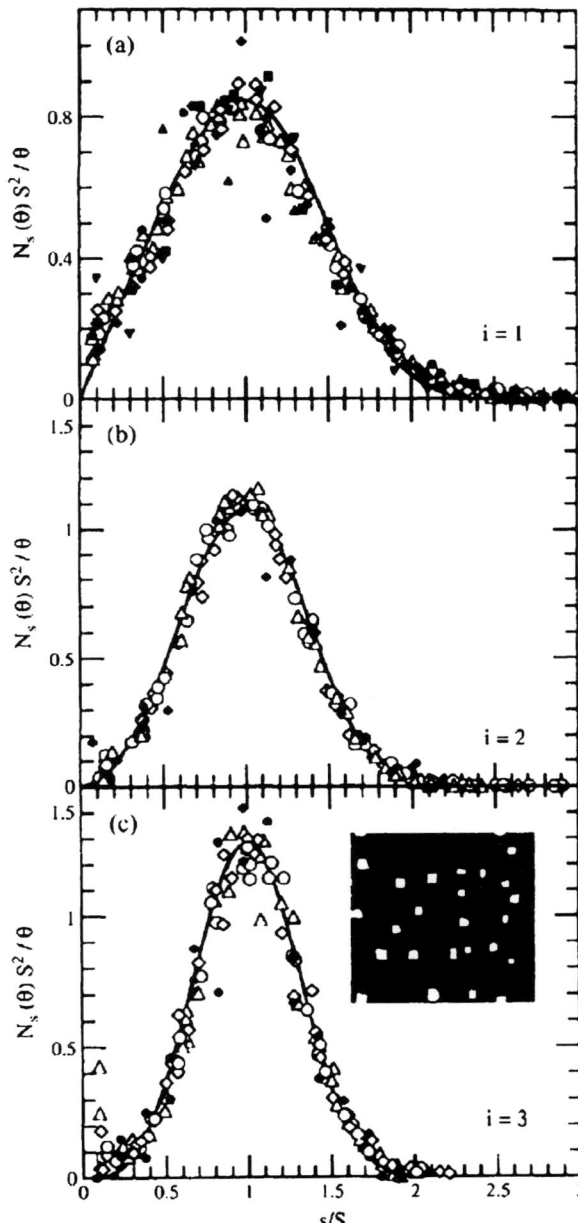

Figure 4.7. Scaled island size distributions using the empirical form of Eq. (4.20) compared to Monte Carlo simulation data (open symbols). Fe/Fe(1 0 0) experimental data (closed symbols [11,14]) are also shown at increasing temperature where the critical island size is believed to change from $i = 1$ to 2 to 3. Reproduced with the permission of Elsevier Science from Evans et al. [2].

So what of theoretical predictions for the ISD using the rate equations (4.2) and (4.3)? We have already remarked on the poor relationship to simulation data found by Bales and Chrzan [8] using their self-consistent solution for the capture numbers (see Figure 4.4), and there have been no reports of consistent success using other intuitive forms for the capture numbers such as those discussed above in Section 4.4. Bartelt and Evans [16] provided a relationship between the capture numbers and ISD in the rate equation approach, and showed what exactly is going wrong for this aspect of the problem.

The analysis starts by assuming the scaling form of the ISD given in Eq. (4.8). A corresponding scaling ansatz is made for the distribution of capture numbers, namely that

$$\sigma_s = \sigma_{av} c(u), \qquad (4.21)$$

where as before $u = s/\langle s \rangle$ and σ_{av} is the average capture number. In the intermediate/aggregation regime, the deposition flux F is matched by the monomer adsorption by islands, so that

$$F = DN_1 \sigma_{av} N. \qquad (4.22)$$

For $\langle s \rangle \gg 1$, which happens for large values of R, we can take a 'hydrodynamic' limit and treat u as a continuous variable. Using the expression of Eq. (4.9) for $N \sim \theta^{1-z}$, we then find

$$\frac{dN_s}{d\theta} = \frac{1}{\langle s \rangle^2}\left[(1-2z)g(u) - zu\frac{dg(u)}{du}\right]. \qquad (4.23)$$

Similarly the right-hand side of Eq. (4.3), with $1/\tau_s = 0$ for stable islands, can be expressed in terms of the derivative of $g(u)$ and $c(u)$:

$$\frac{D}{F}N_1(\sigma_{s-1}N_{s-1} - \sigma_s N_s) = \frac{-1}{\langle s \rangle^2}\frac{d[c(u)g(u)]}{du}, \qquad (4.24)$$

where we have used the relation (4.22). Equating Eqs. (4.23) and (4.24) and integrating, we find the key result

$$g(u) = g(0)\exp\left\{\int_0^u \frac{2z - 1 - c'(x)}{c(x) - zx}dx\right\}. \qquad (4.25)$$

This relates the presumed scale-invariant distribution of capture numbers to the corresponding distribution of island sizes.

Bartelt and Evans applied this result to their point island model with $i = 1$ [16]. This provides a particularly useful test case, because the capture numbers here will be unrelated to island size in either of the approaches discussed in Sections 4.3 and 4.4, implying $c(u) = 1$. Furthermore, it is known that $z = 2/3$ for this model. Thus, the ISD from Eq. (4.25) has a particularly simple solution that reads

$$g(u) = \begin{cases} 2(1-zu)^{(1-2z)/z}, & u < \dfrac{1}{z} \\ 0, & u > \dfrac{1}{z} \end{cases}. \qquad (4.26)$$

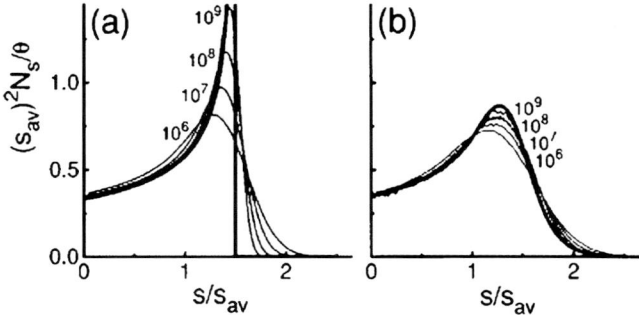

Figure 4.8. Island size distribution for the point island model. (a) The numerical solutions to the mean-field rate equations with constant capture number for all sizes, and the corresponding asymptotic form from Eq. (4.26) is in bold. (b) The distributions measured in Monte Carlo simulations for various values of R at coverage 0.2 ML. Reproduced with the permission of the American Physical Society from Bartelt and Evans [16].

In particular, the divergence at $u = z^{-1} = 3/2$ is caused by the denominator in the integral of Eq. (4.25) approaching zero as the numerator remains positive. This divergence will happen for all forms of $c(u)$ when these conditions apply.

Eq. (4.26) is shown in the left panel of Figure 4.8 alongside numerical results of the mean-field rate equations with constant capture number. It is apparent that the analysis works extremely well for the asymptotic limit of large R. However, it is also apparent that the Monte Carlo simulation data in the right-hand panel do not agree with the mean-field predictions for the ISD. The mean-field approach is neglecting a crucial ingredient for the growth rates of islands as we shall see in the following section.

In summary, it has been found that the mean-field rate equation approach with its corresponding scaling analyses has proved very successful in explaining a wide range of experimental and simulation data concerning island densities and its dependence on critical island size, deposition rate and temperature. However, confrontations with ISDs consistently fail, and furthermore the asymptotic divergence of the mean-field ISD identified by Bartelt and Evans is not observed in reality. The excellent scale-invariance of the ISDs therefore remains unexplained, despite its central role in the scaling analyses. The empirical form of the ISDs, whilst convenient, is no substitute for detailed theory.

4.6. Beyond mean-field theory for island nucleation and growth

To move beyond the mean-field description, it is convenient to consider a simple heterogeneous island nucleation and growth model [17]. In this scenario, we suppose islands can only nucleate and grow from seed sites on the substrate (these might be point or other defects in reality) that are pre-determined before the deposition proceeds. Then the deposited monomers will diffuse around until they encounter a bare seed site or an island that has already nucleated on a seed site. Rapidly all the seed sites will be occupied by islands, which then compete with one another for the diffusing monomers. Assuming isotropic diffusion, a deposited monomer is more likely to first encounter its geometrically closest island than any other. This means that the growth rates of each island depend on how large its *capture zone* is; in other words, how much substrate is closer to it than to any other island. Therefore, the Voronoi tessellation[1] of the substrate using the pre-determined seed sites will reveal the unchanging growth rates of the islands, and the distribution of island sizes at all coverage prior to the onset of coalescence will mirror that of the Voronoi cell areas. A scale-invariant ISD thus naturally emerges from the spatial arrangements of the seeds and nucleated islands. Simulations by Mulheran and Blackman [17] demonstrated the utility of this idea. Note that in contrast to the mean-field model, which would predict that all islands would have the same size, we end up with a range of sizes dictated by the stochastic geometry of Voronoi tessellations.

Mulheran and Blackman extended this idea to homogeneous island nucleation and growth simulations [18]. Here islands are nucleated not from defect sites but whenever $i+1$ monomers come together on an otherwise bare substrate site. In Figure 4.9, we show the simulated arrays of dendritic islands and compact circular ones for $i = 1$ along with their corresponding capture zones. For the case of the dendritic islands, where monomers remain at the point they hit the island, it is clear how the islands grow to fill their capture zones, due to the incoming flux of diffusing monomers from the bare substrate. In this case, it is adequate to use the Voronoi construction for the cell centres to construct the capture zones, due to the suppression of late nucleation events by the extended nature of the dendritic islands. Some refinement of the Voronoi tessellation is required for the growth of compact islands that are often found in room temperature molecular beam epitaxy experiments. In Ref. [18], circular islands were grown for computation convenience, but the conclusions drawn seem fairly general. Due to the ongoing nucleation of islands throughout the $i = 1$ simulation, the network of capture zones is continually being modified by the creation of new islands. At later stages, the finite size of the near-neighbours needs to be accounted for in the estimation of the capture zone of the new island. For this reason, rather than using the Voronoi tessellation for island centres, a modified construction using island edges is employed. As shown in Figure 4.9, these capture zones provide reliable calculations of the actual ISD, but one has to follow the evolution of the capture zone areas through the growth process to get accurate results.

A crucial aspect that emerged from this study of capture zones is the excellent scale-invariance that they display in the growth simulations, also shown in Figure 4.9. In fact, the capture zone areas can be well represented by a semi-empirical functional form derived for random Voronoi networks (RVN) [17–20]

$$F(y) = \frac{\beta^\beta}{\Gamma(\beta)} y^{\beta-1} \exp(-\beta y), \tag{4.27}$$

---

[1] Voronoi tessellation: the partitioning of a plane containing n points (seed sites in the current context) into regions (capture zones); each region contains just one point and any position within a region is closer to the generating point of that region than to any other point. The regions are space-filling convex polygons.

4.6. Beyond mean-field theory for island nucleation and growth

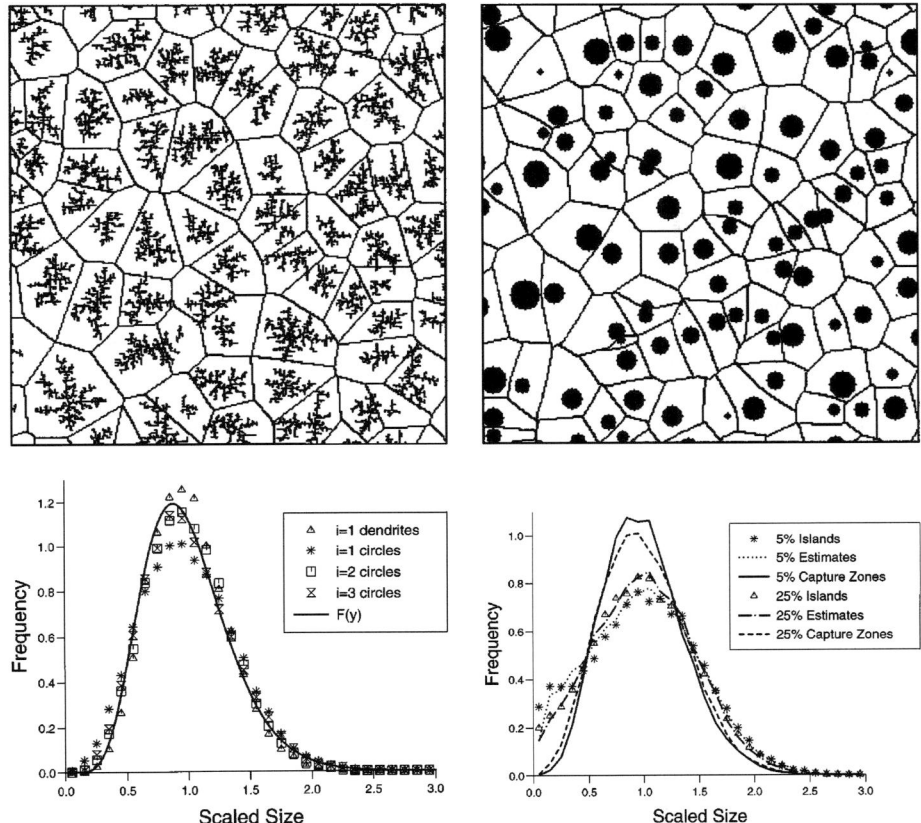

Figure 4.9. Capture zones superimposed over simulated island arrays for homogeneous nucleation and growth. The top two panels show islands at 15% substrate coverage with $i = 1$ for fractal and compact (circular) island growth, respectively. The lower left panel shows the invariance of the distribution of capture zone areas found in different Monte Carlo models. The lower right panel shows how the $i = 1$ island size distribution (compact islands) is broader than that of the capture zones, but nevertheless shows approximate scale-invariance. Also shown are the corresponding island size distribution estimates, drawn from the capture zones (see Ref. [18]).

where y is the scaled cell area and β a convenient parameter. It is found that $\beta \approx 3.61$ for the RVN with a random distribution of points across a substrate. However, it is apparent that the circular islands shown in Figure 4.9 are not randomly placed, because the island nucleation occurs where the average monomer density is highest, that is between existing islands. It emerges that the capture zone area distribution can be described well by Eq. (4.27) with $\beta \approx 8$, as can a RVN with nearest neighbour exclusion of 30% of the original seeds [18]. The distribution of capture zone areas is thus consistent with the underlying picture of the nucleation process.

The capture zone concept directly allows further insight into the functional forms of the ISDs. We see in Figure 4.9 that for $i = 1$, the ISD is broader than that of the capture zones themselves, because in this case there is considerable ongoing nucleation of islands throughout the intermediate and aggregation regimes of

Figure 4.1; indeed from the scaling analysis of Section 4.4, the nucleation rate varies inversely with coverage (or time). This continuing disruption to an island's capture zone size in a stochastic manner, plus the creation of new islands with small capture zones, leads to the broadening of the ISD. In contrast, for $i = 2$ simulations also studied in Ref. [18], the nucleation rate falls off rapidly and the island density saturates in the aggregation regime. Then the ISD follows more closely that of the capture zones which are established early on in the intermediate-coverage regime (Figure 4.1) and remain largely intact throughout the majority of the deposition. Indeed the ISD for $i = 3$ simulations faithfully followed the function $F(y)$ of Eq. (4.27) shown in Figure 4.9, so that the extreme of static capture zones reminiscent of the heterogeneous nucleation scenario naturally emerges in these homogeneous simulations [18]. Finally, the ISDs found in the simulations of Ref. [18] were compared to the empirical form of Eq. (4.20) showing how the behaviour observed in other experiments and simulations correlates well with these findings.

The capture zone concept had been considered before Mulheran and Blackman by Venables and Ball [21], who calculated the Voronoi capture zones for xenon crystals growing on a graphite surface and found excellent correlation with the observed growth rates. However, the idea lay dormant until its 'rediscovery' as the underpinning concept for the scaling properties of the ISD. The concept has also been developed by other modellers, notably Bartelt and Evans who used it to understand why their mean-field theory failed for their point island simulations [16].

In subsequent work, Bartelt et al. [22] studied the capture zones for Ag islands growing on Ag(1 0 0) substrates. They showed that the growth rates of the islands accurately followed those predicted from solutions of the diffusion equation, and furthermore were well represented by using Voronoi-type 'edge cells' as used by Mulheran and Blackman [18]. In Figure 4.10, we show the scaled capture number distribution observed in experiment, along with the measured edge cell area distribution. The comparison between the two distributions is very compelling.

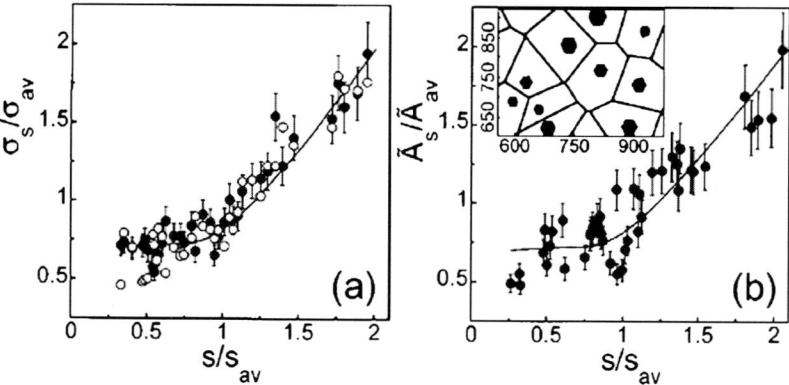

Figure 4.10. Distributions of the capture numbers and Voronoi-type edge cell areas as a function of scaled size. The lines come from Monte Carlo simulation and the data from Cu/Co/Ru(0 0 0 1) experimental observation. Reproduced with the permission of the American Physical Society from Bartelt et al. [23].

Beyond providing an intuitive picture for the evolution of the island sizes during growth, the question arises as to how the concept can be used in a quantitative way to calculate the precise form of the island and capture zone size distributions. An early success was achieved by Blackman and Mulheran who provided a rather thorough examination of an $i = 1$ model for the growth of point islands on a one-dimensional substrate [24]. It is straightforward to solve for the monomer diffusion equation in one dimension, allowing one to predict nucleation rates and the distribution of inter-island gap sizes. From these, under the assumption that the gaps on either side of an island are uncorrelated in the long term due to the significant degree of nucleation in this model, a random convolution of the gap distribution yields very accurate capture zone (and consequently island size) distributions as shown in Figure 4.11. In addition, reconciliation with the self-consistent rate equation approach of Bales and Chrzan [8] is achieved. A similar reconciliation with the rate equation approach is much more challenging for two-dimensional substrates, though the approach of Amar et al. [25,26] seems to enjoy success in the ISD at least.

Mulheran and Robbie [27] aimed to create a suitable theoretical framework that could encapsulate the capture zone concept. Since we are interested not only in the size of an island, but also its capture zone area, they focus not just on the ISD N_s as in the mean-field rate equations (4.2) and (4.3), but on a joint probability distribution (JPD) $f(a, s, t)$. This represents the number of islands with capture zone area a and size s at time t.

To make a concrete model, they introduce the following rules for its evolution during the growth:

1. Islands grow continuously at a rate proportional to their capture zone area a.
2. Island nucleation occurs at the rate dictated by mean-field theory.
3. Each nucleation fragments an existing capture zone area, taking the proportion λ for the new island and leaving the rest for the parent island.
4. The probability for the fragmentation occurring inside a capture zone area a is $\propto a^{2(i+1)}$.

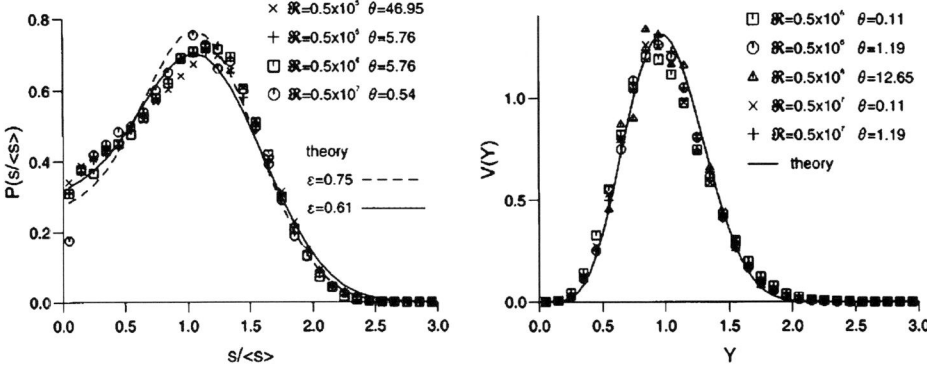

Figure 4.11. The scaled distribution of island sizes (left panel) and the distribution of capture zone lengths (right panel) from Monte Carlo simulation compared to theory for the growth of point islands on a one-dimensional substrate. Figures courtesy of Blackman (see Ref. [24]).

The final rule comes from a consideration of the average lifetime of a monomer in a capture zone of area a, the probability of a monomer being deposited into the capture zone, and the fact that $i+1$ are needed to create a new island. The fragmentation rule 3 is an obvious simplification of the complicated geometrical rearrangements that occur in the Voronoi-type network of capture zones when a new island nucleates, but nevertheless it proves adequate to capture the essence of the capture zone area redistribution. Clearly there is a free parameter in λ, and it is found that $\lambda \approx 0.4$ works well.

Expressing the difference equations as continuous differentials, these rules are embodied in the following equations for the evolution of the JPD:

$$\frac{\partial f(a,s,t)}{\partial t} = -\frac{\partial}{\partial s}\left[\frac{Ra}{\Omega} f(a,s,t)\right]$$

$$+ \dot{N}(t) \frac{a^{2i+2}}{m_{2i+2}(t)}\left[-f(a,s,t) + \frac{1}{(1-\lambda)^{2i+3}} f\left(\frac{a}{1-\lambda}, s, t\right)\right]. \qquad (4.28)$$

Here the first term on the right describes the smooth growth of islands inside their capture zone areas a, where Ω is the total substrate area and R the deposition rate. The second terms describe the fragmentation processes which occur at the total island nucleation rate, wherein $m_{2i+2}(t)$ is the $2i+2$ moment of the distribution with respect to a, which normalises the fragmentation probability. Note that islands with size s and area a can be created from fragmenting suitably large areas containing islands of size s. A similar equation describes the evolution of the islands of size $i+1$ created on nucleation [27].

The scaling behaviour of Eq. (4.28) is investigated by introducing the scaled variables for size and capture zone area:

$$y = \frac{a}{\bar{a}(t)}, \quad z = \frac{s}{\bar{s}(t)}, \quad \bar{a}(t) = \frac{\Omega}{N(t)}, \quad \bar{s}(t) = \frac{Rt}{N(t)}.$$

In the limit of large average size $\bar{s}(t)$ and large average capture zone area $\bar{a}(t)$, these scaled variables can be considered as continuous.

Under a scaling ansatz, the number distribution then becomes, in terms of a scaled time-invariant JPD $F(y,z)$:

$$f(a,s,t) = \frac{N^3(t)}{\Omega Rt} F(y,z).$$

Eq. (4.28) can now be re-written:

$$(3p-1)F + py\frac{\partial F}{\partial y} + (p-1)z\frac{\partial F}{\partial z}$$

$$= -y\frac{\partial F}{\partial z} + p\frac{y^{2i+2}}{M_{2i+2}}\left[-F + \frac{1}{(1-\lambda)^{2i+3}} F\left(\frac{y}{1-\lambda}, z\right)\right]. \qquad (4.29)$$

Here $p = t\dot{N}(t)/N(t)$ is the only time-dependence remaining in the above equation. Furthermore, for $i=0$, we have $N \sim t^{1/3}$ so that $p = 1/3$ independent of time, and the scaling solution to Eq. (4.28) is indeed truly time-independent. The scale-invariance observed in simulations and experiments is then explicable. For $i=1$,

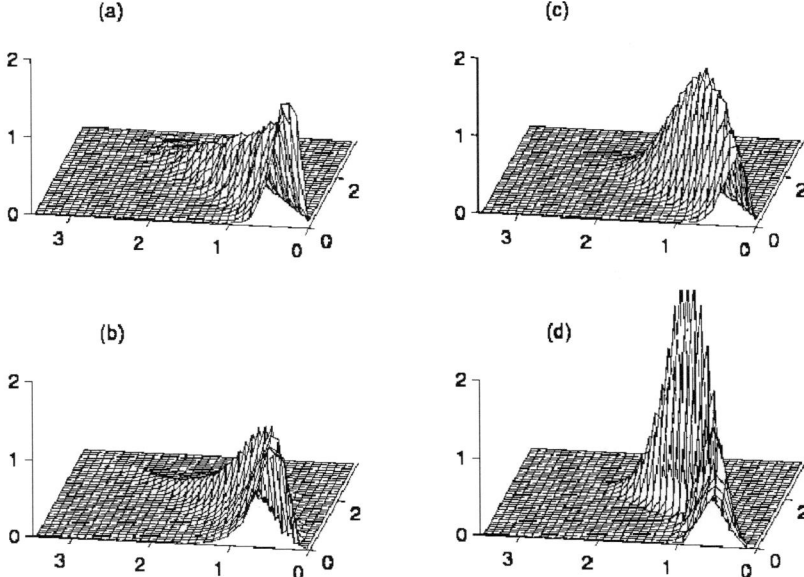

Figure 4.12. Joint probability distributions for scaled island size (into page) and scaled capture zone area (right-to-left). (a) $i = 0$ Monte Carlo simulation and (b) $i = 0$ theory; (c) $i = 1$ simulation and (d) $i = 1$ theory. See Ref. [27].

$N(t) \sim \ln|t|$ so that the time-dependence of p is slowly varying for large times; in fact, $p \sim 0.16$ at 10% dropping to $p \sim 0.14$ at 30% coverage. Again an approximately time-independent solution to Eq. (4.28) is expected, and the scale-invariance observed in simulation and experiment explained.

Mulheran and Robbie [27] used a relaxation method to numerically solve Eq. (4.29) which contains non-local terms that are difficult to handle in other ways. The resulting JPDs for $i = 0$ and 1 are shown in Figure 4.12, alongside results from Monte Carlo simulations. The results show how the theoretical model reproduces many of the essential features observed in the simulation data. In particular, the JPDs are aligned along the $y = z$ diagonal as might be expected given the identification of capture zone area as island growth rate, but the spread of capture zone areas for each island size embodies the changing environment of the capture zones caused by continuing nucleation. For $i = 0$, this spread is most pronounced, since the greater degree of ongoing nucleation throughout growth, represented by the higher value of p in Eq. (4.29), disturbs the capture zone network the most. For $i = 1$, there is less nucleation at later times (p is lower in Eq. (4.29)) and the spread in capture zone area is less pronounced.

It is interesting to ask if the trend in reduced spread of the capture zones continues as higher values of i are used in the Monte Carlo simulation. In Figure 4.13, we show the JPDs measured for $i = 2$ and 3 [28], and indeed we find that the JPD is now tightly aligned along the diagonal. This is entirely expected since at high i, the nucleation events occur early during the deposition process once the monomer density has grown sufficiently high. However, once islands are formed, they act as sinks for the monomers and quickly lower the monomer

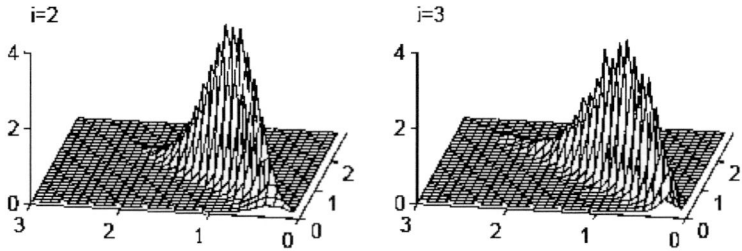

Figure 4.13. The JPDs measured in Monte Carlo simulations for $i = 2$ and 3. See Ref. [28].

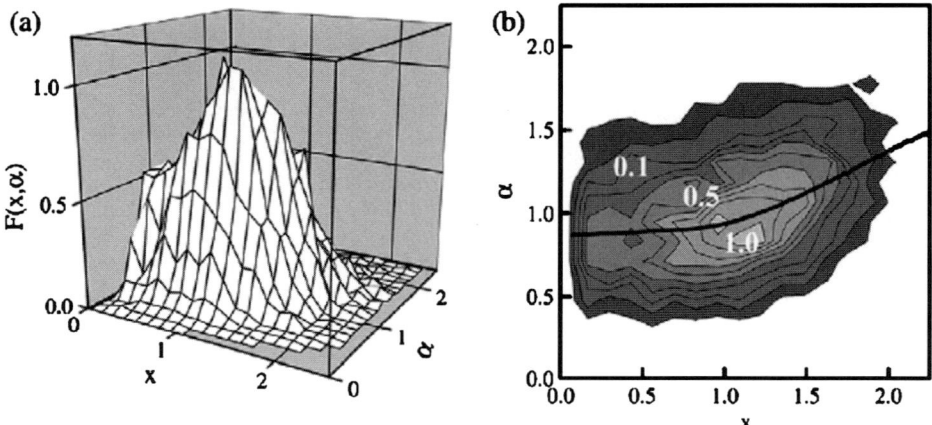

Figure 4.14. The scaled JPD for the point island model. Reproduced with the permission of Elsevier Science from Evans et al. [2].

density so that further nucleation becomes increasingly rare given that $i+1$ monomers must collide to create a stable island.

Theoretical work on the capture zone model that utilises the JPD approach is ongoing. In Figure 4.14, we show the JPD measured by Evans and Bartelt [29] for their point island model. This group remains particularly active, analysing in detail the evolutionary equations and the relationship with the mean-field rate equation formalism, and employing the ideas of stochastic geometry to find improvements over the original set of rules used above to create a tangible model [29,30]. Other groups are also investigating the geometrical consequences of fluctuation-driven nucleation events [31]. Mulheran has adopted an approach that transcends the need for mean-field approach for the dynamics, albeit using the original fragmentation idea [32].

Elements of the capture zone idea are also utilised by Amar et al. [25,26], who have achieved excellent results for the ISD from improved rate equation calculations. It is intriguing that such good agreement results without recourse to the full JPD, since it implies average capture size is sufficient to predict island sizes rather than full inclusion of the fluctuations caused by the stochastic nature of the nucleation.

On a broader scale, the capture zone model has proved to be a valuable concept in analysing experimental systems such as the growth of GaAlAs quantum dots [33]. It appears to be a useful tool to have alongside the rate equations and scaling theory in the armoury available to investigate the fundamental aspects of nucleation and growth at surfaces.

4.7. Island ripening

So far we have been considering the evolution of islands during vapour deposition. What happens once the flux is turned off? Then the islands can continue to grow through ripening (coarsening). From an equilibrium thermodynamics viewpoint, smaller islands have a higher vapour pressure of substrate-supported monomers than larger ones due to the Gibbs–Thompson effect. Then material can diffuse across the substrate from a smaller to a larger island, thereby increasing the size of the latter at the expense of the smaller. Over long timescales, the island density will decrease and the average size will increase.

A mean-field theory for the coarsening of domains in three-dimensional matrices was famously developed by Lifschitz and Slyozov [34] and by Wagner [35]. Applications to the ripening of islands supported on a surface can readily be made [6]. However, in practice, the predicted ISDs deviate from the mean-field predictions. In addition, it has been discovered that a spatial scale-invariance simultaneously emerges with the size scaling in ripening systems [36] as illustrated in Figure 4.15. This shows that the island arrays display a strong degree of spatial correlation as a consequence of the ripening. The situation is reminiscent of the scaling observed during growth, and as we shall see, a similar approach to it utilising Voronoi tessellation proves a productive line of enquiry.

Island ripening requires the thermal activation of the release of monomers from the islands, a process that can be much slower than the surface diffusion of the monomers. One can thus also envisage ripening where the release of the monomers from the islands is the rate-limiting step, rather than a diffusional flux of material from regions of high density to low. From dimensional arguments, the signature for the rate-limiting step lies in the growth exponent of the average island size over time. If the average area grows linearly in time (for two-dimensional islands), the release rate of the monomers limits the growth, whereas a growth exponent of 2/3 implies diffusion is the limiting process. The former has indeed been observed and makes a simple yet interesting case for study.

We consider the case where the release of monomers from supported clusters is a random process that can occur independently of cluster size. This might be expected if the number of kink sites from which monomer release can occur does not change with size. Would the array of islands still ripen? Yes – primarily because larger islands will have a greater capture cross-section for the diffusing monomers than smaller ones. Thus, larger islands would tend to grow and the smaller shrink. For the case of two-dimensional compact islands, the capture rate is proportional to the square-root of size. A mean-field expression for the island growth rate can

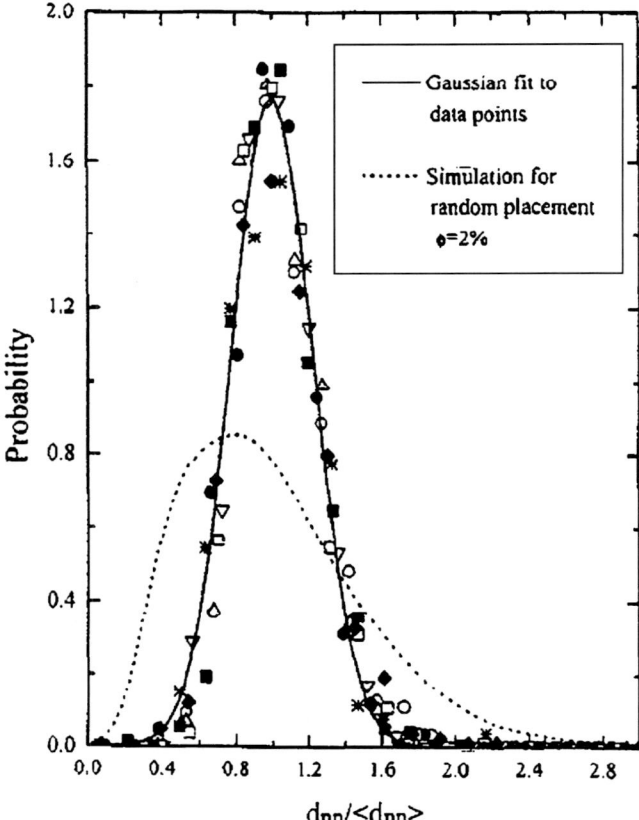

Figure 4.15. Results from the experimental observation of Sn/Si(1 1 1) island ripening, showing the scaled distribution of nearest neighbour distances between the clusters at various final substrate coverage. Reproduced with the permission of Elsevier Science from Zinke-Allmang [6].

then be formed [37,38]:

$$\frac{ds}{dt} = \frac{r}{\bar{r}} - 1. \qquad (4.30)$$

Here s is the island size (in monomers), r the island radius (so that $s = \pi r^2$ for circular islands) and the average radius $\bar{r} \sim t^{1/2}$ on dimensional grounds. The form of Eq. (4.30) is easily understood in terms of the constant release rate of monomers from all islands plus the growth rate due to monomer adsorption in competition with all other islands in the system. The scale-invariant distribution of island sizes follows from Eq. (4.30) when used in the continuity equation; it is in fact the same solution found by Hillert [39] for two-dimensional grain growth following Wagner [35]. It is plotted in Figure 4.16 alongside simulation data. It is clear that the 'normal' simulation data do not agree terribly well with this distribution.

To understand why the size distribution is not following the mean-field prediction, we need only consider the above-mentioned spatial ordering (see the upper left panel in Figure 4.16) that emerges in ripening systems when they reach

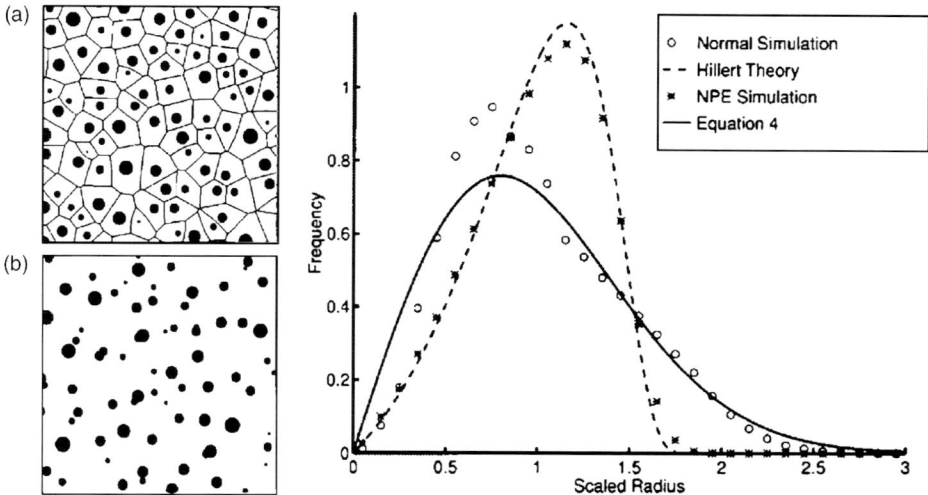

Figure 4.16. The left panels show the results of ripening circular, two-dimensional islands using monomer release as the rate-limiting step. (a) The islands are ripened using the 'normal' rule whereby any island instantly adsorbs a diffusing monomer that reaches its perimeter, increasing its size whilst maintaining its circular shape. A Voronoi-type edge-cell tessellation is imposed on this image. (b) The islands have been ripened using an artificial NPE rule where islands cannot adsorb the monomers they release themselves. In the right panel, the scaled island size distribution (as a function of scaled radius) is shown for both simulations, alongside the Hillert solution to Eq. (4.30), and the modified solution discussed in the text (labelled Equation 4 in this figure). See Ref. [37].

the scale-invariant regime. The ripened islands are not randomly distributed and a mean-field environment regardless of size is not an appropriate model for growth. To affect the ISD, the emergent spatial ordering must include a correlation between size and spatial environment. The origin of this lies in the concept of 'local ripening' that was recognised some time ago in the literature [6,40].

Consider a pair of closely spaced islands in isolation from the rest of the system. Under our ripening rule, they can exchange monomers with each other at random. However, the larger of the pair does have an advantage, since it can shadow its smaller neighbour from those monomers released on its further side more effectively than the smaller one shadows the larger. In other words, the larger island is more effective at consuming its own offspring (the paedophagous effect), and so wins-out over its smaller neighbour in the ripening process. This consideration applies to all closely spaced pairs of islands, which will tend to ripen more rapidly than well-separated pairs, with the larger growing at the expense of the smaller one. In this manner, a degree of spatial ordering is imposed on the system; only well-separated islands enjoy a long-life.

To verify this interpretation, Tarr and Mulheran [37] performed artificial 'non-paedophagous effect' (NPE) simulations where islands cannot adsorb their own offspring. The results of this simulation are shown in Figure 4.16, where it can be seen that the ripened islands no longer display the high degree of spatial disorder. Furthermore, the ISD now follows the mean-field prediction much more faithfully.

To utilise this insight and improve on the mean-field growth law, Tarr and Mulheran [37] used the Voronoi edge cells of the islands to identify how many near-neighbours each island has, following Theis et al. [38,41]. The ripening mechanism tends to preserve local material density, because monomers are unlikely to be able to diffuse beyond the first shell of near-neighbours, which results in a strong correlation between cell area and the size of the occupant island. Furthermore, this also means that the shape of the cell is correlated with island radius, implying that larger islands have more near-neighbours than smaller ones in proportion to radius. As a result, the propensity of an island to grow by capturing monomers depends on the number of near-neighbours releasing monomers and the radius of the island, yielding a modified growth law:

$$\frac{ds}{dt} = \frac{s}{\bar{s}} - 1. \qquad (4.31)$$

The above equation yields a different scale-invariant size distribution [37] also shown in Figure 4.16. The agreement with the 'normal' simulation data is much improved.

The concept of using the Voronoi edge cells has been taken further to explain why the spatial organisation emerges concomitantly with the size invariance during the ripening [42]. The approach has many similarities with the JPD idea of the previous section, but here the variables are island size and cell shape rather than cell area. We extend the growth law of Eq. (4.30) to explicitly contain the cell shape:

$$v(n, s, t) = \frac{ds}{dt} = \frac{nr}{\langle nr \rangle} - 1, \qquad (4.32)$$

which is the growth rate of an island of radius r and cell shape n at time t.

The growth law (Eq. (4.32)) can be used with a modified continuity equation for the distribution $f(n, s, t)$, which is the distribution of islands with size s and cell shape n at time t:

$$\frac{\partial}{\partial t} f(n, s, t) + \frac{\partial}{\partial s}[v(n, s, t) f(n, s, t)] = S(n, s, t). \qquad (4.33)$$

The term on the right-hand side of the above equation describes the redistribution of cell shapes when islands disappear. The cells form a network of triple junctions, similar to that observed in soap froths trapped between two glass sheets, for which the topological rearrangement rules are well studied [43]. However, in the case of surface-tension-driven growth, only cells of shape $n = 3, 4$ or 5 can shrink. In island ripening, larger cells can disappear if the occupant island happens to be much smaller than the average size. The topological rules therefore need extending for the ripening model, and one natural possibility is shown in Figure 4.17. This figure illustrates how many neighbouring cells lose, gain or retain the same number of sides when an n-sided cell disappears, with the overall balance maintaining the average cell shape at 6 as required by Euler's Law for a network of triple junctions. In addition, the connectivity of the cells needs careful consideration. Cells with a large number of sides n tend to lose a side when a neighbour shrinks because of the high average internal angles in large-n cells. Conversely,

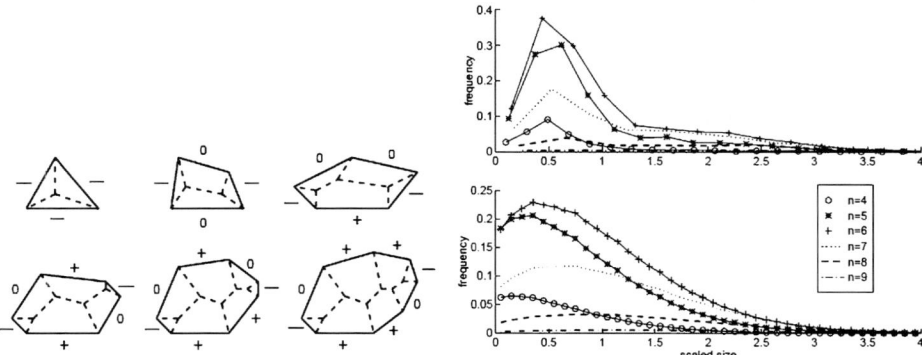

Figure 4.17. The left panel shows the topological rearrangements used to identify the changing number of near-neighbours during island ripening when an island shrinks to size zero. The right panel shows the calculated (top) and Monte Carlo simulation (bottom) island size distributions in the scale-invariant state partitioned by edge-cell shape. See Ref. [42].

low n-sided cells will tend to gain a side when a neighbour disappears. These considerations must be included in the formulation of $S(n, s, t)$ [42].

The systems of Eqs. (4.32) and (4.33) are easily solved numerically and show that scale-invariant solutions naturally emerge regardless of starting conditions. These are shown in Figure 4.17 alongside those observed in the Monte Carlo simulations; the comparison is rather good. The existence of these scale-invariant solutions shows why the system evolves into a scaling state; of course, it encapsulates the overall size distribution, the long-time unchanging cell shape distribution and the correlation between the two. In particular, this model captures the essence of why the system evolves into a spatially invariant state at the same time as the size distribution displays scale-invariance. The explanation has naturally emerged in terms of edge-cell shape rather than the nearest neighbour distributions used by Carlow and Zinke-Allmang [36]; however, the fact that the simulations obey exactly the same distribution of nearest neighbour separations [37] provides one with confidence that the phenomenon is well described by this approach.

4.8. Atomistic modelling

Metallic systems can be modelled using empirical potentials with reasonably high fidelity. As described in Section 4.1, in a typical MD simulation lasting 1 ns, the atoms will be seen only to vibrate around their equilibrium positions. Interesting events that might cause the cluster's centre of mass to move are overwhelmingly unlikely to occur in 1 ns, let alone the dissociation of small clusters that would determine the critical size in experimental deposition conditions. Furthermore, these rare events might be difficult to predict without the insight simulation can provide.

As an example of the sort of dynamical process of interest, in Figure 4.18 we show some EAM potential calculations for the movements of tetramer islands on a fcc(0 0 1) surface [44]. It is seen that the island can migrate through the initial step of one pair of atoms sliding past the other, since this has a lower activation energy

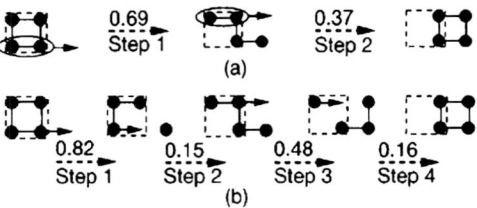

Figure 4.18. The proposed diffusion mechanism for a tetramer island on fcc(1 0 0) surfaces. The dimer shearing mechanism and activation energies (in eV) for the moves are shown in (a), and the corresponding singe atom moves are shown in (b). Reproduced with the permission of the American Physical Society from Shi et al. [44].

(0.69 eV in this calculation) than the more intuitive single atom hop away from the parent island (0.82 eV). From the sheared state, the other half of the island can catch up with modest activation energy (0.37 eV), or alternatively an exposed atom can break away from the parent in a dissociation event (not shown in this figure).

Voter et al. have proposed several approaches to accelerate the MD simulations so that rare events such as these can be observed with increased frequency [45]. Perhaps the most intuitive and useful method for thin film growth is temperature accelerated dynamics (TAD) [46]. The idea of TAD is that by running a simulation at high temperature, one can speed up the occurrence of rare events, thereby showing what can happen at a target lower temperature of interest. However, the problem is that different transitions might have different frequency pre-factors, so that they might occur in a different order at high and low temperatures. Thus, a high-temperature MD simulation alone is not sufficient to reliably show the evolution of island dynamics at a lower growth temperature. The key point in TAD is to not accept blindly the high-temperature evolution of the system, but to stop it each time a basin-hopping transition is detected and ascertain whether it is acceptable for the target low-temperature simulation.

How this can be achieved is illustrated in Figure 4.19. This diagram builds on a typical Arrhenius plot where the logarithm of the frequency of a rare event is plotted against inverse temperature, yielding a straight line whose gradient is the negative of the activation energy of the event. Now imagine performing a tradition MD simulation at high temperature until a basin-hopping event is detected. The simulated time t_1 has provided an estimate of the frequency of the event, $1/t_1$, the logarithm of which is labelled '1' on the β_{high} line in Figure 4.19. A simple estimate for the simulation time required for this event to be seen at the target low-temperature can now be made; slide down the activation energy gradient to determine the frequency, and hence inverse simulation time, at the point '1' on the β_{low} line.

Consider now repeating this process by resuming the high-temperature simulation without accepting transition '1', in other words by reflecting the system's trajectory back into the starting basin. This then continues until another possible transition '2' is detected. Note that this transition took longer to find at high temperature than '1'; however, if it has a lower activation energy, it might occur faster at lower temperature as illustrated in the figure. How do we know when to stop this process of running high-temperature simulation? Voter uses an estimate

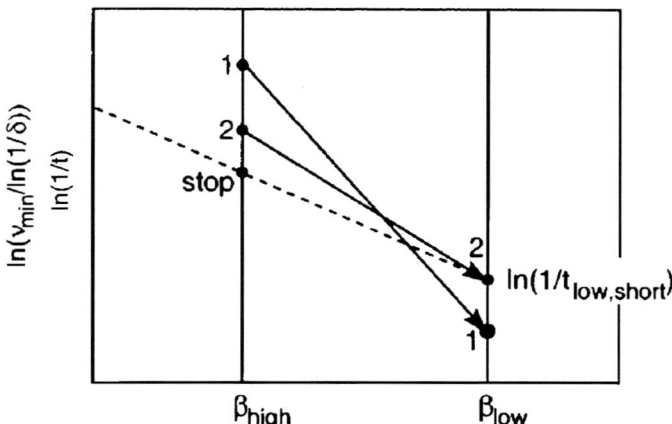

Figure 4.19. Diagram to illustrate the TAD methodology; β is inverse temperature. Reused with permission from Mads R. Sørensen, J. Chem. Phys. **112** (2000) 9599. Copyright 2000, American Institute of Physics.

of the minimum likely frequency pre-factor, combined with a tolerance for errors due to the stochastic nature of the frequency estimates from the high-temperature MD, to create the 'stop' time beyond which any further high-temperature transitions cannot yield shorter times at low temperature. Another possibility has been implemented whereby if one knows the minimum energy barrier possessed by a system, the necessary stopping time can be directly yielded.

To implement TAD, one must be able to detect basin-hopping transitions, which can be done by regularly halting the MD run and checking to see if the instantaneous configuration relaxes back to its starting one. This of course adds a computational overhead to the MD simulation. In addition, the activation energy of the detected transitions must be found (see below) which also costs processor time. Nevertheless, due to the Arrhenius form of activated events, tremendous boosts in simulated time can result. The limiting factor is the lowest activation barrier encountered, which is likely to be repeatedly found in the simulations.

To illustrate the power of TAD, the growth of an Ag/Ag(1 0 0) crystal surface has been simulated [47] as shown in Figure 4.20. Note that the simulations have been performed at very low temperature, so that long-range surface diffusion of adatoms and the nucleation and growth of islands is not favourable. Instead the simulations reveal the evolution of the surface roughness with growth temperature and draw out the key activated processes that determine it. It is remarkable to note that this simulation yields realistic MD trajectories of the evolving and complex surface structures at experimental growth rates. The simulation time boost is many orders of magnitude.

So is TAD the answer to the problems of simulating island nucleation and growth? To date we are not aware of any work that shows this to be the case. The drawbacks of TAD are that it most likely finds low energy transitions, most of which are uninteresting in terms of diffusion or dissociation of small islands. In addition, the larger the system, the more such uninteresting transitions exist, so the direct TAD simulation of island dynamics during growth is going to be very

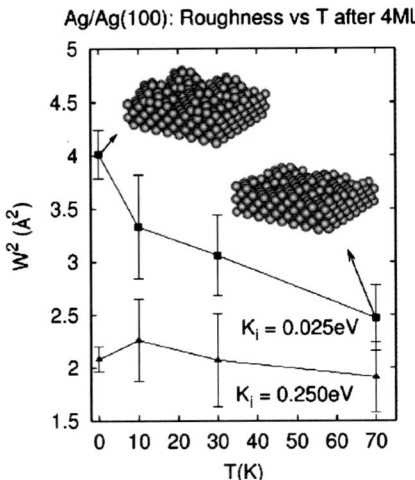

Figure 4.20. Simulated growth of Ag/Ag(1 0 0) using temperature accelerated dynamics. Reproduced with the permission of the American Physical Society from Montalenti et al. [47].

difficult to achieve. However, TAD can be used to reveal some of the dynamics of isolated islands, information that can be used in kinetic Monte Carlo simulation. We shall encounter the same type of dynamical processes discovered by the TAD developers below as we turn to the issue of finding activation energies.

The issue of finding saddle points for the transitions between two potential wells has a long pedigree. The nudged elastic band method provides a good paradigm for some of the issues involved [48]. Imagine the configuration of the system of N atoms as a single point in $3N$-dimensional space. As a first attempt to find the saddle point, join a straight line between potential energy minimum at point A to the one at B. Calculating the potential energy at regular steps (or images) along this straight line will give some information about the barrier to be overcome. Unfortunately there is no guarantee that this barrier will be close to the minimum barrier energy of the saddle point, and indeed often it is much higher and of no use to us. However, now imagine that this straight line is in fact an elastic band which exerts an external force on the images depending on the directions ($3N$ vectors) to the two nearest neighbour images. We now minimise the forces on the images, using only the elastic force parallel to the band and the potential energy surface gradient perpendicular to this. This will 'nudge' the elastic band towards the minimum barrier pathway over the saddle point. This method is reasonably straightforward and works well in TAD. However, to utilise the method, one needs to know the end points of the band, potential minima A and B that are connected by one saddle point. Of course, TAD will provide this in the high-temperature MD simulation.

Consider now a more complicated scenario, where all one knows is the starting minimum energy configuration (state A, for example). If there is a simple way to search in the vicinity of this well for possible saddle points, then we could find the possible transitions and their associated activation energies without needing any MD simulation. Henkelman and Jonsson have provided one such approach which they called the 'Dimer Method' [49]. Here we must think of constructing a pair of

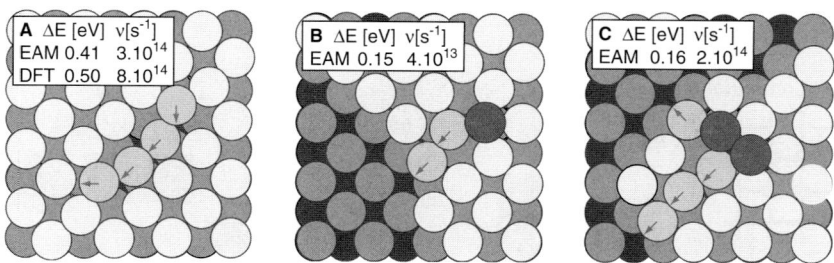

Figure 4.21. Some multi-atom movements seen in simulated growth of Al(1 0 0) layers that cause the adatom to diffuse down from atop an island. Reproduced with the permission of the American Physical Society from Henkelman and Jonsson [50].

images (the dimer) in the $3N$-dimensional space, where one image is slightly displaced from the other in a certain $3N$ direction. In general, each image will feel a slightly different $3N$ force on it, which can be combined to create both a translational force on the dimer and a torque. The latter is key here, because the dimer can be repeatedly rotated (keeping its length fixed) to find the direction of minimum curvature on the potential energy surface. Continuously following this direction is known to lead towards a saddle point. The dimer can now be translated in the right direction by manipulating the parallel and perpendicular forces on the dimer. Iterating the rotation and translation moves the system to the saddle point. Furthermore, by starting from different points in the initial potential well, several different saddles can be found, and if enough searches are done, hopefully all the relevant low-energy saddles will be found.

In Figure 4.21, we show examples of how this Dimer Method for finding saddle points can be used to simulate the growth of thin films similar to those illustrating TAD in Figure 4.20. In this simulation, the authors have developed an 'on-the-fly' kinetic Monte Carlo procedure [50]. In standard kMC, the possible transitions are identified *a priori* for the types of configuration expected in the simulation. Here the authors have circumnavigated this restriction by allowing the simulation to evolve from its latest configuration using transitions discovered by the Dimer Method searches. As more material is deposited, ever more unanticipated moves can occur. Figure 4.21 shows some of the more elaborate movements that occurred in the simulation, which involve chains of co-operating atoms moving together. This type of movement would not routinely be anticipated in a set of pre-defined kMC rules, illustrating the pitfalls of traditional kMC approaches.

This 'on-the-fly', or 'self-learning', kMC concept is well adapted to investigate the kinetics of small islands on a surface. A suitable methodology is illustrated in Figure 4.22, which has been applied to Cu(1 0 0) homoepitaxy as an example [51]. In this approach, a small island is modelled in the kMC as a 'lattice animal'. If a new animal is encountered, the configuration is passed to the atomistic model of the island on a substrate with interactions modelled using EAM potentials. The atomistic model is investigated using the Dimer Method saddle point searches, which discover how the island can change and with what activation energy. Frequency pre-factors are also calculated from the well-known Vineyard expression [45]. Importantly, the atomistic model includes mobile surface layers to allow

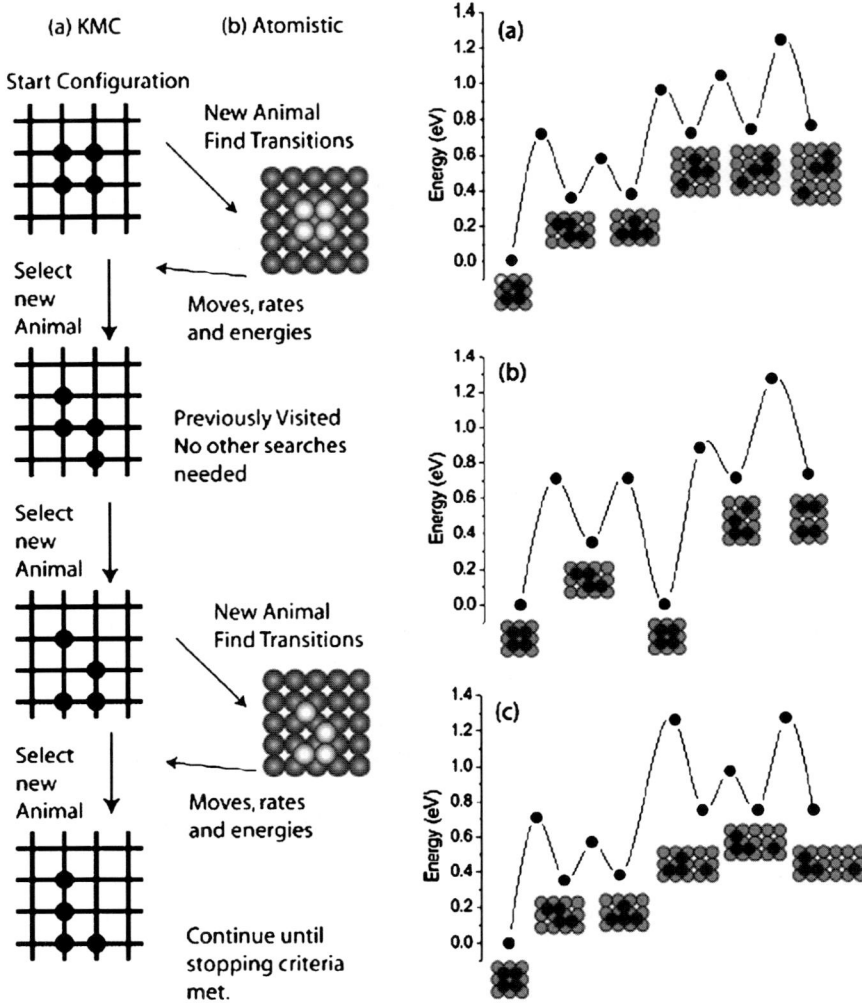

Figure 4.22. The left panels illustrate the self-learning Kinetic Monte Carlo simulations for small Cu/Cu(1 0 0) island dissociation. The right panel shows some of the dissociation pathways discovered for a tetramer island. See Ref. [51].

co-operative movements. When sufficient searches are complete, the information is passed back to the kMC which then selects a possible move for the lattice animal. As the simulation proceeds, more lattice animals are found and thoroughly investigated using the atomistic model. By utilising the translational symmetry of the lattice (other symmetries could be included), information from previous searches can be re-used to greatly speed up the process. In this way, the mobility and dissociation of small islands as a function of temperature can be found without any pre-conceptions about the dominant pathways.

In Figure 4.22, some dissociation pathways for a tetramer island are shown along with the activation energy (~ 1.3 eV). The activation energy and frequency pre-factors for the dissociation of all island sizes of one to eight atoms are similarly found, obtained from Arrhenius plots of the rates at different temperatures. This information can further be used in rate equations of the type (4.2) and (4.3), so that the critical island size and scaling relations can be found for different temperature and deposition rates [52].

A similar methodology to power kMC with atomistic Dimer Method searches has been used to simulate the coalescence of Cu/Cu(1 1 1) islands [53]. This example shows the incredible timescales accessible to this style of simulation; 4×10^7 kMC steps are used to simulate 169.2 s in real time, whilst capturing atomistic realism. These simulation tools hold much promise for illuminating key processes during nanostructure evolution that higher level mathematical models need to capture.

4.9. Other growth methods

In this chapter, an overview of the theory of cluster nucleation and growth at surfaces has been presented. The reader will be aware that such theoretical approaches usually treat idealised systems that might be met during highly controlled experiments on 'well-characterised' surfaces. Even so, the models necessarily make many approximations and neglect many complications of experimental reality, such as strain affects due to lattice mismatch between adsorbate and substrate, and indeed in the substrate surface itself which can lead to reconstructions in the topmost layers. These complications present both a challenge and an opportunity, and so we finish this chapter with a few examples of how these effects can be exploited.

A recurrent theme throughout this chapter has been the importance of capture zones to the growth rates of clusters, both during deposition and in subsequent ripening. Therefore, it is intuitive to imagine that if one has a regular array of nucleation sites on a substrate, deposited monomers can diffuse to these sites and create a regular array of equi-sized clusters. Surface reconstructions provide an obvious template for this scenario. In Figure 4.23, we see how such a pattern of cobalt clusters has been grown on the well-known zigzag (or herringbone, or chevron) reconstruction that self-organises on the Au(1 1 1) surface. The reconstruction consists of narrow 'stripes' of hcp-packed surface atoms separated from wider fcc-packed stripes, and at the elbows of the zigzags the atom is only fivefold coordinated in the plane [54]. These under-coordinated sites then act as the nucleation centres for the diffusing Co, and clusters will grow at these sites below 400 K [55].

Transition metal oxides provide another class of substrates that have technological importance. Rutile titania is one of the most widely studied materials, with the most stable (1 1 0) surface receiving much attention. Upon reduction, this surface is seen to undergo (1×2) reconstructions due to added-rows of Ti_2O_3 [56]. Further reduction induces a regular array of cross-links on the surface which can then be used as nucleation sites for Pd clusters as shown in Figure 4.24. Once again the regularity of the array ensures a narrow size distribution for the clusters nucleated from vapour deposition. High-temperature anneals cause the clusters to ripen, so that once the cluster density has fallen by an order of magnitude or so,

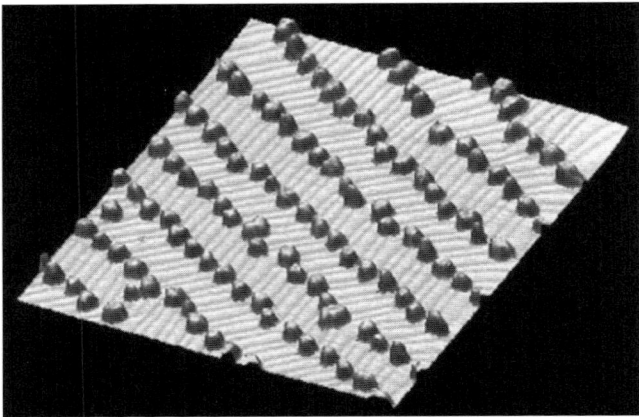

Figure 4.23. Cobalt clusters, each about 300 atoms in size, nucleated on the herringbone reconstruction of the Au(1 1 1) surface. The effective Co dosage is two monolayers. Reproduced with the permission of the Elsevier Science from Bansmann *et al.* [54].

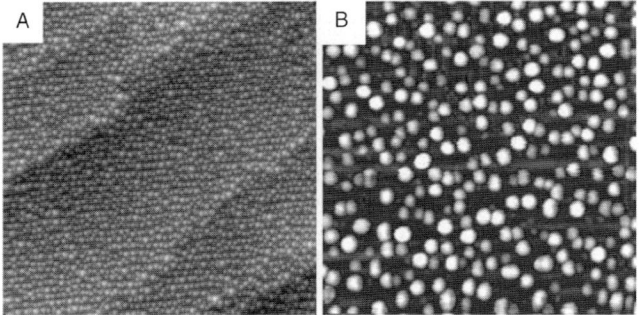

Figure 4.24. Variable temperature scanning tunnelling microscope images of Pd clusters grown on the cross-linked (1 × 2) surface reconstruction of the rutile titania(1 1 0) surface. (A) 1500 Å square image taken at 373 K, showing that the clusters nucleate on the cross-links of the reconstructed surface; (B) 1000 Å square image at 773 K showing that significant ripening occurs at elevated temperatures. Figures courtesy of Bennett (see Ref. [57]).

the initial ordering is lost and the cluster array resembles those found following ripening on homogeneous substrates [57].

Even without reconstructions, many surfaces will present step edges which can either act as nucleation sites or can confine the deposited adatoms to finite regions of the surface. A natural extension of this idea is to create a confined region on a surface by masking. In Figure 4.25, we show some results of gold cluster nucleation and growth from vapour deposition onto a narrow strip created by masking the silicon oxide substrate by PMMA [58]. The metal islands could be used to construct Coulomb blockade devices if their self-organised growth can be steered so that control over size and spatial separation is achieved.

Another effect that is starting to be exploited concerns the strain interactions between neighbouring clusters grown on a substrate. In Figure 4.26, iron clusters

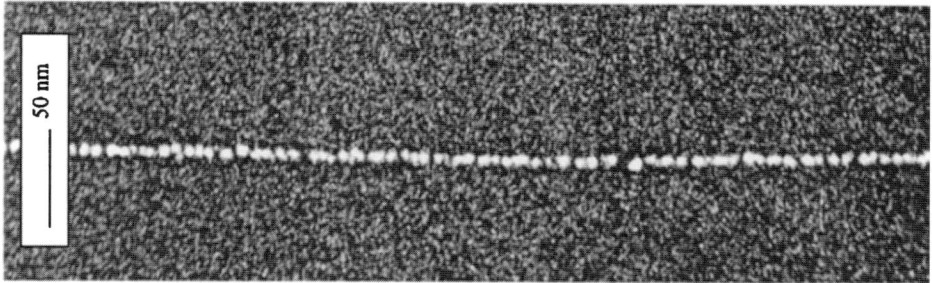

Figure 4.25. A one-dimensional chain of evaporated gold clusters grown in a 7 nm channel on a PMMA-masked silicon oxide substrate [58]. Reproduced with permission from M. Boero, Journal of Applied Physics, **87** (2000) 7261. Copyright 2000, American Institute of Physics.

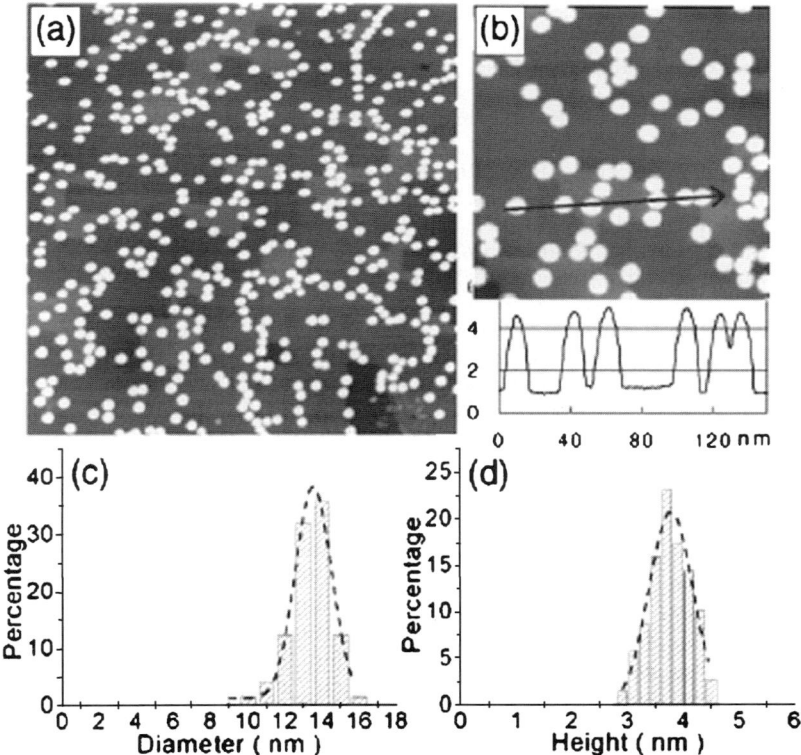

Figure 4.26. (a) 500 nm square non-contact AFM image of Fe clusters grown on NaCl(0 0 1) at 530 K (1.7 ML dose); a line scan (b) reveals uniform cluster heights. The cluster diameter (c) and height (d) size distributions are remarkably narrow. Reproduced with the permission of the American Physical Society from Gai et al. [59].

grown on NaCl(0 0 1) are shown, alongside the narrow diameter and height distributions measured after deposition [59]. The authors believe that the cluster size is selected by strain-mediated dipolar interactions between the clusters. Such strain affects have also been invoked to predict self-organisation through cluster diffusion across the substrate [60]. It will be interesting to see if these effects can be widely applied to control cluster nucleation and growth in the future.

References

[1] H. Brune, Surf. Sci. Rep. **31** (1998) 121.
[2] J.W. Evans, P.A. Thiel and M.C. Bartelt, Surf. Sci. Rep. **61** (2006) 1.
[3] M.C. Bartelt and J.W. Evans, Phys. Rev. B **46** (1992) 12675.
[4] J.G. Amar, F. Family and P.-M. Lam, Phys. Rev. B **50** (1994) 8781.
[5] C. Ratsch, A. Zangwill, P. Smilauer and D.D. Vvedensky, Phys. Rev. Lett. **72** (1994) 3194.
[6] M. Zinke-Allmang, Thin Solid Films **346** (1999) 1.
[7] J.A. Venables, Philos. Mag. **27** (1973) 697.
[8] G.S. Bales and D.C. Chrzan, Phys. Rev. B **50** (1994) 6057.
[9] J.A. Blackman and A. Wilding, Europhys. Lett. **16** (1991) 115.
[10] H. Brune, G.S. Bales, J. Jacobsen, C. Boragno and K. Kern, Phys. Rev. B **60** (1999) 5991.
[11] J.A. Stroscio and D.T. Pierce, Phys. Rev. B **49** (1994) 8522.
[12] M.C. Bartelt and J.W. Evans, Surf. Sci. **298** (1993) 421.
[13] J.G. Amar and F. Family, Phys. Rev. Lett. **74** (1995) 2066.
[14] J.A. Stroscio, D.T. Pierce and R.A. Dragoset, Phys. Rev. Lett. **70** (1993) 3615.
[15] D.D. Chambliss and K.E. Johnson, Phys. Rev. B **50** (1994) 5012.
[16] M.C. Bartelt and J.W. Evans, Phys. Rev. B **54** (1996) R17359.
[17] P.A. Mulheran and J.A. Blackman, Philos. Mag. Lett. **72** (1995) 55.
[18] P.A. Mulheran and J.A. Blackman, Phys. Rev. B **53** (1996) 10261.
[19] D.L. Weaire, J.P. Kermode and J. Wejchert, Philos. Mag. B **53** (1986) L101.
[20] P.A. Mulheran, Philos. Mag. Lett. **66** (1992) 219.
[21] J.A. Venables and D.J. Ball, Proc. R. Soc. Lond. Ser. A **322** (1971) 331.
[22] M.C. Bartelt, C.R. Stoldt, C.J. Jenks, P.A. Thiel and J.W. Evans, Phys. Rev. B **59** (1999) 3125.
[23] M.C. Bartelt, A.K. Schmid, J.W. Evans and R.Q. Hwang, Phys. Rev. Lett. **81** (1998) 1901.
[24] J.A. Blackman and P.A. Mulheran, Phys. Rev. B **54** (1996) 11681.
[25] J.G. Amar, M.N. Popescu and F. Family, Phys. Rev. Lett. **86** (2001) 3092.
[26] M.N. Popescu, J.G. Amar and F. Family, Phys. Rev. B **64** (2001) 205404.
[27] P.A. Mulheran and D.A. Robbie, Europhys. Lett. **49** (2000) 617.
[28] J.A. Blackman and P.A. Mulheran, Comp. Phys. Comm. **137** (2001) 195.
[29] J.W. Evans and M.C. Bartelt, Phys. Rev. B **66** (2002) 235410.
[30] M.C. Bartelt and J.W. Evans, Phys. Rev. B **63** (2001) 235408.
[31] A. Pimpinelli and T.L. Einstein, Phys. Rev. Lett. **99** (2007) 226102.
[32] P.A. Mulheran, Europhys. Lett. **65** (2004) 379.
[33] M. Fanfoni, J. Phys. Condens. Matter **20** (2008) 015222.
[34] I.M. Lifschitz and V.V. Slyozov, J. Phys. Chem. Solids **19** (1961) 63.
[35] C. Wagner, Z. Elektrochem. **65** (1961) 581.
[36] G.R. Carlow and M. Zinke-Allmang, Phys. Rev. Lett. **78** (1997) 4601.
[37] D.M. Tarr and P.A. Mulheran, Phys. Rev. E **68** (2003) 020602(R).
[38] N.C. Bartelt, W. Theis and R.M. Tromp, Phys. Rev. B **54** (1996) 11741.
[39] M. Hillert, Acta Metall. **13** (1965) 227.
[40] K. Morgenstern, G. Rosenfeld and G. Comsa, Surf. Sci. **441** (1999) 289.
[41] W. Theis, N.C. Bartelt and R.M. Tromp, Phys. Rev. Lett. **75** (1995) 3328.
[42] P.A. Mulheran, Europhys. Lett. **71** (2005) 1001.
[43] D.L. Weaire and N. Rivier, Contemp. Phys. **25** (1984) 59.
[44] Z.P. Shi, Z. Zhang, A.K. Swann and J.F. Wendelken, Phys. Rev. Lett. **76** (1996) 4927.

[45] A.F. Voter, F. Montalenti and T. Germann, Annu. Rev. Mater. Res. **32** (2002) 321.
[46] M.R. Sørensen and A.F. Voter, J. Chem. Phys. **112** (2000) 9599.
[47] F. Montalenti, M.R. Sørensen and A.F. Voter, Phys. Rev. Lett. **87** (2001) 126101.
[48] G. Henkelman, B.P. Uberuaga and H. Jonsson, J. Chem. Phys. **113** (2000) 9901.
[49] G. Henkelman and H. Jonsson, J. Chem. Phys. **111** (1999) 7010.
[50] G. Henkelman and H. Jonsson, Phys. Rev. Lett. **90** (2003) 116101.
[51] M. Bahsam, F. Montalenti and P.A. Mulheran, Phys. Rev. B **73** (2006) 045422.
[52] P.A. Mulheran and M. Basham, Phys. Rev. B **77** (2008) 075427.
[53] O. Trushin, A. Karim, A. Kara and T.S. Rahman, Phys. Rev. B **72** (2005) 115401.
[54] J. Bansmann, S.H. Baker, C. Binns, J.A. Blackman, J.-P. Bucher, J. Dorantes-Dávila, V. Dupiuis, L. Favre, D. Kechrakos, A. Kleibert, K.-H. Meiwes-Broer, G.M. Pastor, A. Perez, O. Toulemonde, K.N. Trohidou, J. Tuaillon and Y. Xie, Surf. Sci. Rep. **56** (2005) 189.
[55] S. Padovani, I. Chado, F. Scheurer and J.P. Bucher, Appl. Surf. Sci. **42** (2000) 164.
[56] R.D. Smith, R.A. Bennett and M. Bowker, Phys. Rev. B **66** (2002) 035409.
[57] R.A. Bennett, D.M. Tarr and P.A. Mulheran, J. Phys. Condens. Matter **15** (2003) S3139.
[58] M. Boero, J.K. Vincent, J.C. Inkson, M. Mejias, C. Vieu, H. Launois and P.A. Mulheran, J. Appl. Phys. **87** (2000) 7261.
[59] Z. Gai, B. Whu, J.B. Pierce, G.A. Farnan, D. Shu, M. Wang, Z. Zhang and J. Shen, Phys. Rev. Lett. **89** (2002) 235502.
[60] F. Liu, A.H. Li and M.G. Lagally, Phys. Rev. Lett. **87** (2001) 126103.

Chapter 5

Chemical Methods for Preparation of Nanoparticles in Solution

C.-H. Yu[1], Kin Tam[2] and Edman S.C. Tsang[1]
[1]*Wolfson Catalysis Centre, Inorganic Chemistry Laboratory, University of Oxford, Oxford OX1 3QR, UK*
[2]*AstraZeneca, Mereside, Alderley Park, Macclesfield, Cheshire SK10 4TG, UK*

5.1.	Introduction	114
5.2.	General synthesis and properties of nanoparticles	114
	5.2.1. Introduction: nucleation and nanomaterial growth	114
	5.2.1.1. Nucleation	114
	5.2.1.2. Growth	116
	5.2.2. Micro-emulsion process	117
	5.2.3. Sol-gel process	120
	5.2.4. Polyol process	120
	5.2.5. Nanoparticles in aqueous solution	122
	5.2.6. Nanoparticles on thin films	124
5.3.	Synthesis of monodisperse nanoparticles by chemical methods	124
5.4.	Surface chemical modification of nanoparticles	126
5.5.	Tailoring the size of nanoparticles	129
	5.5.1. Overview	129
	5.5.2. Tailoring the size of silica-encapsulated iron oxide using micro-emulsion	130
	5.5.2.1. Synthesis of silica-encapsulated iron oxide nanoparticles	130
	5.5.2.2. Controlling the size of core-shell silica-coated magnetic nanoparticles	131
5.6.	Synthesis of binary metallic nanoparticles	131
5.7.	Synthesis of core-shell structures	131
	5.7.1. Synthesis of silica-encapsulated FePt–Fe$_3$O$_4$ (SiO$_2$@FePt–Fe$_3$O$_4$)	133
	5.7.2. Synthesis of monodisperse silica-encapsulated iron platinum nanoparticle with external Ti–O–Si coating (TiO$_2$–SiO$_2$@FePt)	135
	5.7.3. Synthesis of polymer-encapsulated iron platinum (polymer@FePt)	137

5.7.4. Formation of glucose-encapsulated iron platinum spheres by
       hydrothermal treatment   138
References   139

5.1. Introduction

As noted in Chapter 1 (Section 1.2), originally the concept of *Nanotechnology* was proposed by Richard P. Feynman as early as 1959 at the annual meeting of the American Physical Society with his often-cited paper 'There's plenty of room at the bottom' [1,2]. He suggested the possibility of manipulating materials on an atomic scale via synthetic processes with the use of high-resolution microscopes. All these ideas were later aided in the development of some new microscopies such as high-resolution transmission electron microscopy (HRTEM), atomic force microscopy (AFM) and scanning tunneling microscopy (STM). Feynman also suggested a new area of nanoscale surgery using robots [2,3]. The technology for synthesis of tailored nanomaterials is one area he envisaged that has caught the attention of scientists recently because of the potential wide-ranging applications, including bio-magnetic separation, drug delivery [4,5], biomedical nanostructured fluid, imaging [4], recyclable catalysts [6] and data storage nanostructure [7]. This chapter will highlight some common chemical methodologies for the synthesis of metal and metal core-shell nanoparticles.

5.2. General synthesis and properties of nanoparticles

5.2.1. Introduction: nucleation and nanomaterial growth

Chemical synthesis of metal nanoparticles generally involves a number of steps taking place in the liquid phase. First, the formation of metal atoms can be accomplished by the reduction of metal precursor(s) by using chemical reductant(s) in solution. The freshly formed metal atoms can then undergo elementary nucleation followed by slow growth processes leading to the formation of nanoparticles. This notion was first described in detail by La Mer (Figure 5.1) [8,9]. It is noted that a good understanding of such processes and experimental parameters can aid the engineering of particle size and shape. In this section, we will give an overview of the nucleation and growth of nanoparticles followed by a brief account of some common experimental procedures.

5.2.1.1. Nucleation

Nucleation takes place because the supersaturated solution is thermodynamically unstable. For the nucleation process to occur, the solution must be supersaturated in order to generate an extremely small size 'sol' particle [9,10]. The overall free energy change, ΔG, is the sum of the free energy due to the formation of a new volume and the free energy due to the new surface created. We consider a spherical

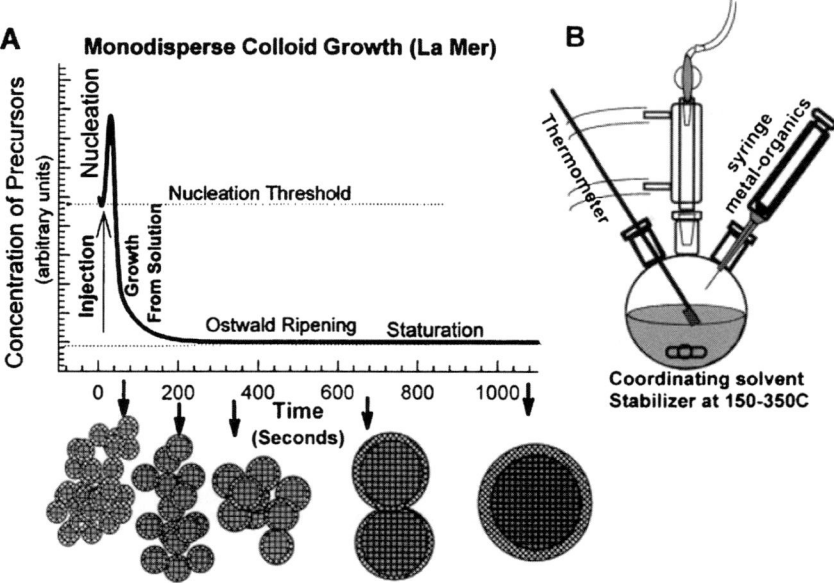

Figure 5.1. The concept of monodisperse colloid growth of La Mer model (A) and typically synthetic apparatus (B). Reproduced with the permission of Annual Reviews from Murray et al. [9].

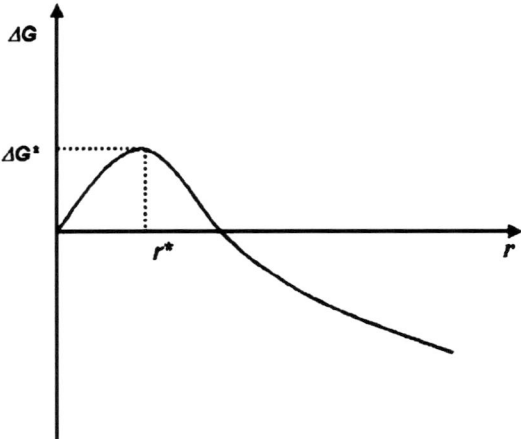

Figure 5.2. Illustration of the overall free energy ΔG as a function of the growth particle size r. Reproduced with the permission of the American Chemical Society from Burda et al. [10].

particle, where V is the molecular volume, r the radius of the nuclei, k_B the Boltzmann constant, S the saturation ratio and γ the surface free energy per unit surface area. When S is greater than 1, G shows a positive maximum at a critical size r^* (Figure 5.2). This maximum free energy is the activation energy for nucleation,

$$\Delta G = -\frac{4}{V}\pi r^3 k_B T \ln(S) + 4\pi r^2 \gamma. \tag{5.1}$$

Nuclei larger than the critical size will further decrease their free energy for growth and form stable nuclei that grow to form particles. The critical nuclei size r^* can be obtained by setting $d\Delta G/dr = 0$,

$$r^* = \frac{2V\gamma}{3k_B T \ln(S)}. \tag{5.2}$$

For a given value of S, all particles with $r > r^*$ will grow and all particles with $r < r^*$ will dissolve. It can be seen from the above equation that a large saturation ratio S implies a small critical nuclei size r^*.

5.2.1.2. Growth

After the nuclei are formed from the solution, they grow via deposition of the soluble species onto the solid surface (molecular addition). Nucleation stops when the concentration falls below the critical level, but the particles continue to grow by the molecular addition process until the equilibrium concentration of the species is reached. There is a differential growth rate for the smaller and larger particles at this stage. Assuming that the larger ones are slightly bigger than the critical size, the free energy driving the growth is larger for the smaller particles and they will grow more rapidly. A nearly monodisperse size distribution can be obtained if the reaction (both nucleation and growth) is stopped quickly. An alternative strategy for a monodisperse distribution is to maintain a saturated condition throughout the reaction by building up the concentration of metal ions through repeated injections of the metal precursor(s).

It should be noted that the relative rates of growth of small and large particles are different when the reactants are depleted due to particle growth. In this case, Ostwald ripening will occur. The saturation ratio (S) is decreasing and the corresponding critical nuclei size (r^*) is increasing [Eq. (5.2). Particles smaller than this new critical size will get smaller and finally dissolve, while the larger ones will continue to grow. An analogous ripening process has already been encountered in the context of island growth on surfaces (see Section 4.7). Stopping the reaction at this point results in a broad size distribution of particles with two regimes (for smaller and larger particles) on either side of the critical size. The only way to get monodisperse particles at this stage is to allow the reaction to continue until the supersaturation is depleted and the smaller nuclei have completely vanished. This is fine if one wants relatively large particles approaching the micrometer-size regime, for example.

An additional factor that has to be considered is secondary growth. This is the growth of particles by aggregation with other particles. Growth by this process is faster than that by molecular addition, and it occurs by stable particles combining with smaller unstable nuclei.

Finally, we should note that nanoparticles are small and are not thermodynamically stable, and it is necessary to stabilise them either by adding surface-protecting reagents, such as organic ligands or inorganic capping materials, or by placing them in an inert environment, such as an inorganic matrix or polymer. A suitable choice of capping material can also provide a barrier to counteract the attraction between nanoparticles due to the van der Waals interaction (or magnetic attraction in the case of magnetic materials). In the subsequent sections, we will discuss some common methods for the synthesis of nanoparticles.

5.2.2. Micro-emulsion process

Micro-emulsions are complex liquids with potentials for current and future application prospects. They consist of oil, water, surfactant and co-surfactant that form a clear solution. The amphiphilic nature of a surfactant, such as cetyltrimethylammonium bromide (CTAB) as described later, makes them miscible with water and oil. Micro-emulsions can be either 'water in oil' (w/o) or 'oil in water' (o/w) stabilised by surfactant molecules with morphologies similar to those of micelles and reverse micelles where the interfacial tension is extremely low between the two phases. With these characteristics, micro-emulsions appear to be an ideal platform to bring together the metal precursor (e.g. water-soluble) and the reactant (e.g. oil-soluble) to enable the reduction of the metal to occur. Generally, the size of micro-emulsions is small and uniform. Hence, they can be regarded as a micro-reactor for the synthesis of nanoparticles. As the water concentration alters, the system can change from a w/o to an o/w micro-emulsion, which is illustrated in Figures 5.3–5.5. These systems offer a wide range of possibility and flexibility to carry out the synthesis.

Winsor has described and reviewed the micro-emulsion phase behaviours [11,12]. The mixtures of water/surfactant–co-surfactant/oil can give rise to different characteristic phases. Typically, a rigid interior region of water exists, which is referred as a 'micro-pool'. In these systems, the surfactant molecules with the polar heads are oriented inwards and the non-polar head tails orient towards the oil phase. Only a small quantity of the surfactant molecules is required at the water–oil interface to decrease the water surface tension. Figure 5.4 shows different types of micro-emulsion structures, the description of which depends on the relative proportion of the three components, namely oil, water and surfactants [11,12]. According to Winsor, these micro-emulsions can be classified as follows:

Winsor I: dispersion of o/w in contact with essentially oil;
Winsor II: dispersion of w/o in contact with essentially water;
Winsor III: both o/w and w/o dispersions are simultaneously present in the same domain in mixed state in separate contacts with both oil and water and
Winsor IV: a homogeneous single phase of dispersion either o/w or w/o not in contact with any other phase [11–13].

Figure 5.6 shows an example of using CTAB as a surfactant in a micro-emulsion for the synthesis of nanoparticles. First, if the concentration of CTAB is above the critical micelle concentration 1 (CMC1), nanoparticles will be formed. If the concentration of CTAB is above critical micelle concentration 2 (CMC2), micelle droplets will be formed. With the addition of water at a CTAB concentration above

118   Chapter 5.   Chemical Methods for Preparation of Nanoparticles in Solution

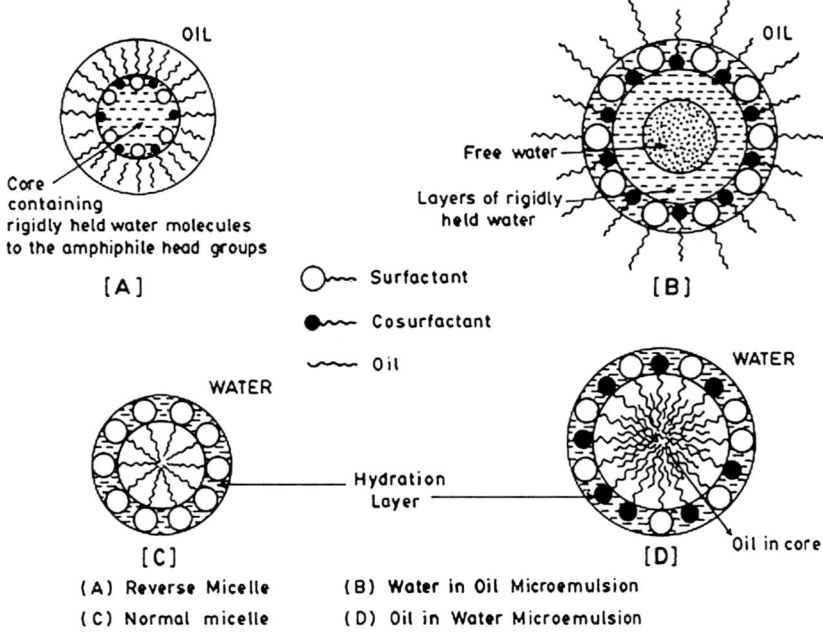

Figure 5.3.   Representations of reverse micelles and micro-emulsions. Reproduced with the permission of Elsevier Science from Moulik and Paul [13].

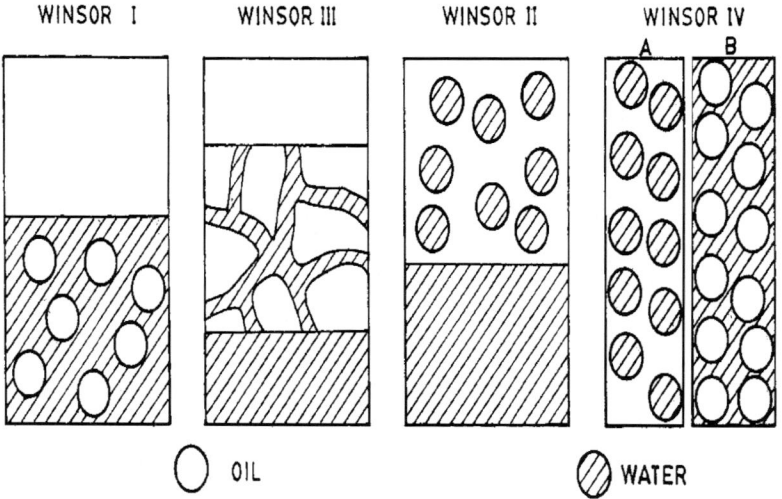

Figure 5.4.   Different phase-forming situations for water–amphiphile–oil mixtures. Reproduced with the permission of Elsevier Science from Moulik and Paul [13].

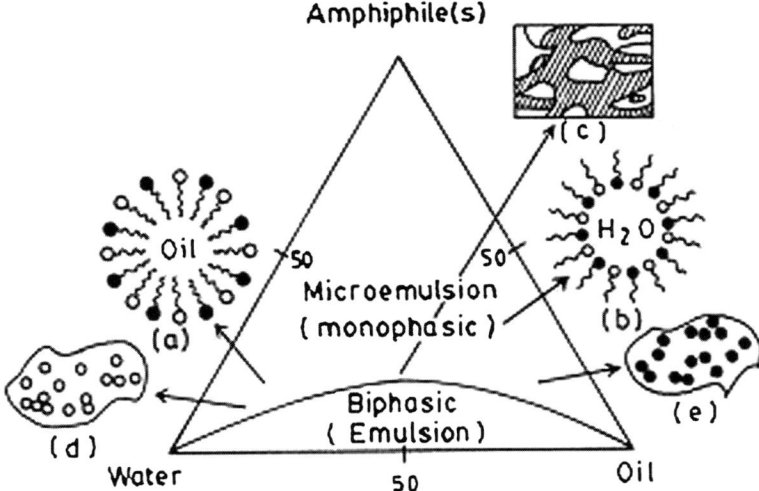

Figure 5.5. A ternary phase diagram showing how various complexes depend on the different surfactants and intrinsic structures: (a) o/w micro-emulsion, (b) w/o micro-emulsion, (c) bicontinuous dispersion, (d) isolated and aggregated o/w dispersion, and (e) isolated and aggregated w/o dispersion. Reproduced with the permission of Elsevier Science from Moulik and Paul [13].

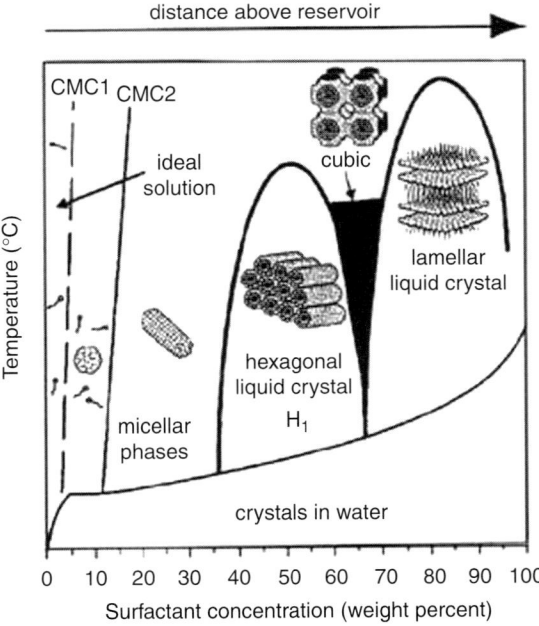

Figure 5.6. Schematic difference of water to surfactant molar ratio [H_2O]/[surfactant] in a microemulsion system for CTAB. Reproduced with the permission of Wiley-VCH from Brinker et al. [15].

CMC2, the nanoparticles will be assembled as a hexagonal liquid crystal, a cubic liquid crystal or a lamellar liquid crystal, depending on the concentration of CTAB. If the temperature falls, these crystals precipitate from the solutions as solids. Thus, this makes it possible to synthesise nanoparticles of different shapes. It has been reported that micro-emulsion systems are well suited for the preparation of nano-sized metals and metal oxides with defined size. The size of aqueous nano-droplets depends on the water to surfactant molar ratio ($Wo = [H_2O]/[surfactant]$) used in the micro-emulsion system. By controlling metal particle growth inside the water droplet, the particle dimension can therefore be manipulated. Thus, the micelle droplets act as 'nano-reactor' templates for the formation of metal clusters upon reductions of their water-soluble precursors [14]. Further examples on using the micro-emulsion technique will be described in Sections 5.3–5.5.

5.2.3. Sol-gel process

The sol-gel process involves the hydrolysis, condensation and thermal decomposition of metal alkoxides or metal precursors in solution [10]. In this process, the metal alkoxides or precursors form a stable solution with all necessary reagents, which is referred to as the *sol*. Then, the *sol* would undergo hydrolysis and condensation to form a networked structure (*gel*), resulting in a marked increase in the viscosity. Scheme 5.1 shows a typical sol-gel process. Water, alcohol, acid or base can be used to control the kinetics of the reactions. The change in precursor concentration, temperature and pH values enables the particle size to be tuned. After the gel formation, an aging step is required to enable the formation of a solid mass. The aging step may take up to several days in which the expulsion of solvent, Ostwald ripening and phase transformation could occur. Finally, the gel would be subjected to a high-temperature treatment to decompose the organic precursors and remove the volatile reagents to yield the nanoparticles. Applications of the sol-gel method to synthesise nanoparticles will be discussed in Sections 5.4 and 5.7.

5.2.4. Polyol process

This synthetic methodology employs high-boiling alcohol as a reductant; for instance, 1,2-hexadecanediol is able to reduce reactive metal precursors at an elevated temperature. In the polyol system, oleic acid and oleylamine are also commonly added as surfactant or stabiliser molecules to control the growth of

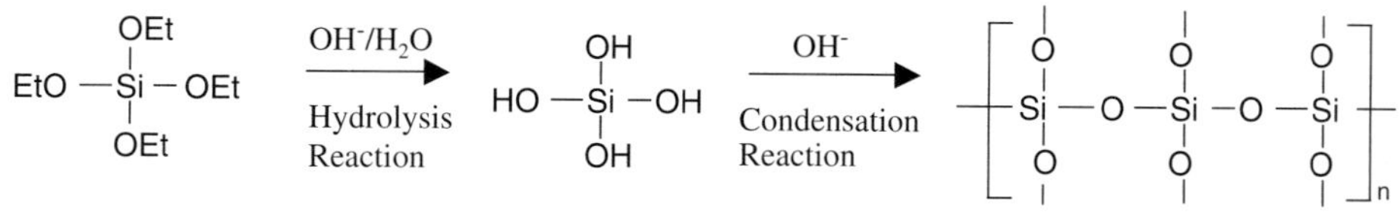

Scheme 5.1. Sol-gel process in hydrolysis and condensation. Reproduced with the permission of the American Chemical Society from Tsang *et al.* [14].

freshly formed metal atoms. A high-boiling solvent such as 1,2-hexadecanediol is essential to allow refluxing the mixture at 286°C [16]. Murray's group at IBM has studied the preparation of monodisperse metallic nanoparticles extensively by this technique. One important class of alloys studied is the iron platinum (FePt) nanoparticle [17]; the paper has been cited more than thousand times until now. In the case of FePt, co-reduction of iron(0) pentacarbonyl [Fe(CO)$_5$] and platinum(II) acetylacetonate [Pt(acac)$_2$] at elevated temperatures gives 3 nm FePt nanoparticles. A similar procedure was used to produce monodisperse magnetic face-centred cubic (*fcc*) (A1 phase) FePt nanoparticles followed by post-synthetic annealing at over 500°C to generate face-centred tetragonal (*fct*) (so-called L1$_0$ phase) FePt crystals (Scheme 5.2) [17]. In addition, Fe$_{30}$Pt$_{70}$ to Fe$_{80}$Pt$_{20}$ could be tuned by adjusting the molar ratio of Fe(CO)$_5$ to Pt(acac)$_2$ and the particle sizes were manipulated by using different precursor concentration, heating rate, heating temperature, aging, etc. [18]. The formation of FePt–Fe$_3$O$_4$ was achieved by passing an air stream during the reduction of Fe(CO)$_5$ in the polyol process [19]. On the other hand, less than 1 ppm of oxygen is required to prepare pure FePt L1$_0$ phase. Very recently, Kang *et al.* used a high-boiling-point solvent and surfactant to produce FePt L1$_0$ phase without any annealing treatment. Hexadecylamine (HDA) was used as a solvent and 1-adamantanecarboxylic acid as a stabiliser and the mixture was allowed to reflux at 360°C to produce partially transformed ordered L1$_0$ phase [20]. Howard *et al.* [21] reported a similar procedure by using tetracosane as solvent and oleylamine as surfactant with refluxing at 390°C [21]. Similar methods to prepare CoPt and FePd nanoparticles have been studied. For example, Co$_2$(CO)$_8$ and platinum hexafluoroacetylacetonate (hfac) were used as metal precursors, oleic acid and oleylamine as stabilisers, with toluene and nonane as solvents, which were mixed under argon to obtain CoPt with an average diameter of 6.27 nm [22]. FePd nanoparticles can be prepared under similar conditions by using oleic acid and oleylamine as stabilisers [23]. Kang *et al.* used a third metal and reduced temperature to form the L1$_0$ phase by adding silver acetylacetonate to the reaction mixture, giving Fe$_{49}$Pt$_{51}$Ag$_{12}$ with a phase transition temperature of only 400°C [24,25]. Liu *et al.* also reported the formation of FePt–Fe$_3$O$_4$ core-shell nanoparticles by a two-step polyol process followed by annealing at 550°C for 30 min to form an L1$_0$ phase [19,26]. Other groups used polymers such as

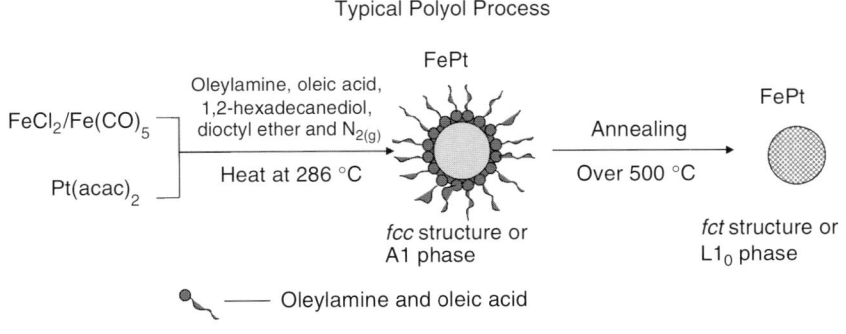

Scheme 5.2. The synthetic procedure for the synthesis of FePt by polyol process.

polyvinylpyrrolidone (PVP) or polyethylenimine (PEI) instead of oleylamine and oleic acid to stabilise FePt from aggregation in the polyol mixture [27]. Examples on the use of polyol process for core-shell nanoparticles synthesis will be discussed in Section 5.7.

5.2.5. *Nanoparticles in aqueous solution*

Colloidal nanoparticles in aqueous solution have recently shown great potential for applications in the biomedical and catalysis areas [14,28–31]. Depending on particular applications, particle agglomeration may either be required to occur in a controllable manner or need to be avoided. This section highlights some considerations in developing these kinds of systems.

To understand particle agglomeration, it is important to understand the surface forces at the interface of the particle and the aqueous portion of the colloid. The zero-charged particles greatly induce aggregation, combining them into large particles, which settle by gravitational force. Figure 5.7 shows zero-charge nanoparticles in the liquid phase, in which it is easy for them to collide and form larger clusters. On the other hand, if each particle carries a 'like' electrical charge, a force of mutual electrostatic repulsion between adjacent particles is produced as in Figure 5.8. Encapsulation of the particles by silica in a pH-7.4 buffer can produce stable and homogenous nanoparticles that do not aggregate because of the lower isoelectric point (IEP) of the silica rendering the overall particle negatively charged in solution. Particles possessing high charges can induce double layers in aqueous environments and are discrete, disperse and in suspension. Figure 5.9 illustrates double layers of colloidal and electrostatic principle. The van der Waals attraction

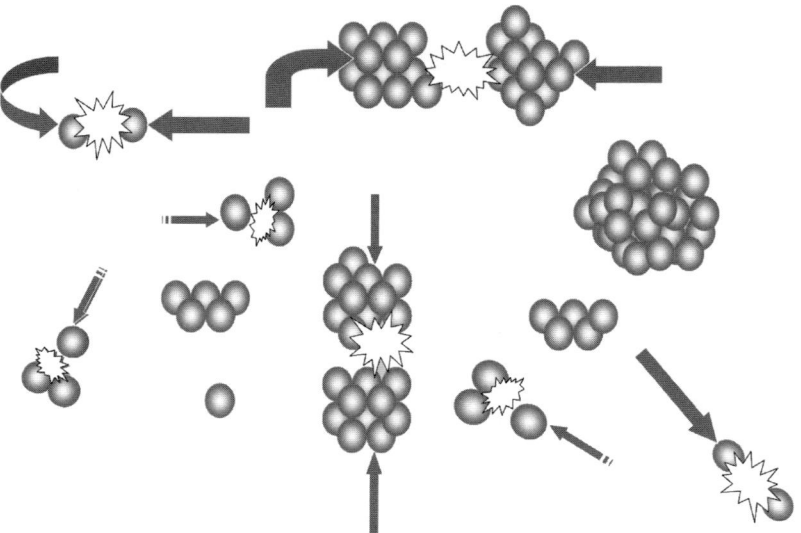

Figure 5.7. It is energetically and kinetically favourable for uncharged nanoparticles to collide and coalescence.

5.2. General synthesis and properties of nanoparticles

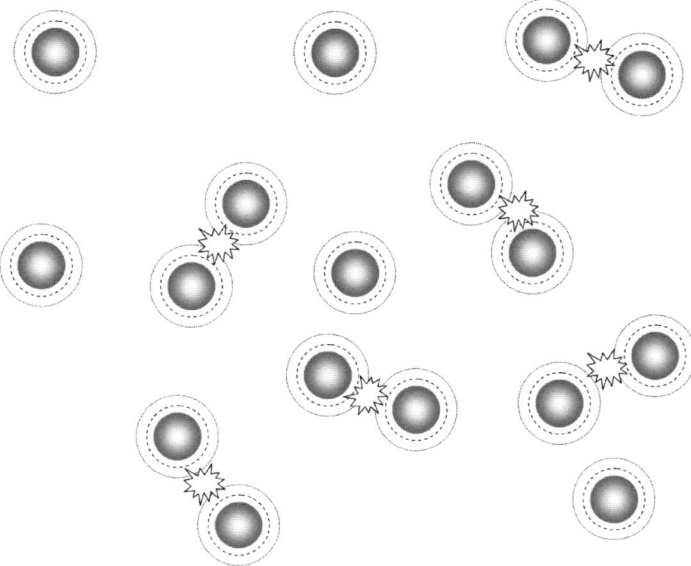

Figure 5.8. Double layers of ions can protect colloid particles against rapid aggregation. They can be kinetically stable in an aqueous medium with identical surface charge.

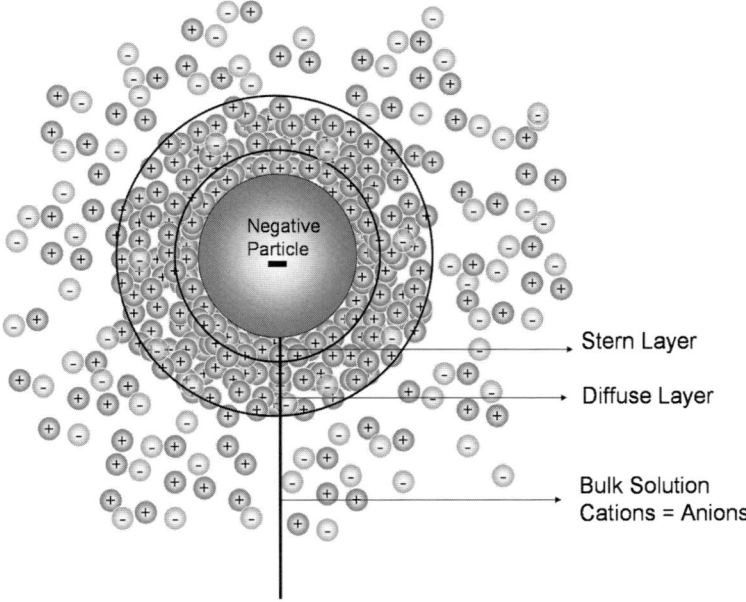

Figure 5.9. A colloidal particle protected by a double layer of ions in an aqueous medium.

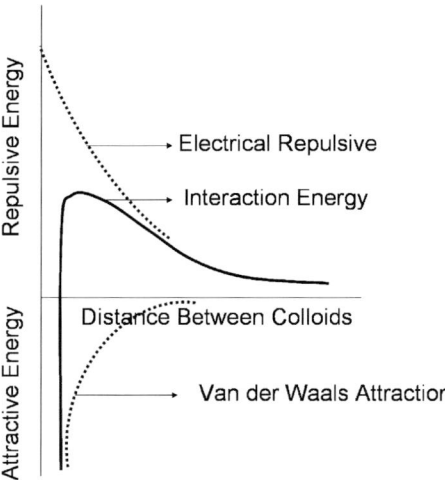

Figure 5.10. The net interaction energy is derived from the resulting combination of repulsive and attractive energy [35].

is another force opposing the repulsive energy. The combination of an attractive force and a repulsive force induces a net interaction energy as shown in Figure 5.10 [32,33]. Thus, particle agglomeration is a complex phenomenon [34].

5.2.6. *Nanoparticles on thin films*

The synthesis of nanoparticles on a thin film has recently received considerable attention [36]. In general, these kind of ordered assemblies and close-packed glassy solids are synthesised by tailoring the different composition, concentration and solvents and concentrating via evaporation using the methodology of so-called evaporation-induced self-assembly (EISA) [37]. While the solvent evaporates, the nanoparticles in solution gradually precipitate as close-packed solids. The nature of the solvent and particle, the evaporation rate and the concentration affect the 'self-assembly' or 'bottom-up' process, and this has a direct influence on the type of ordered structure that arises [9,38,39]. The self-assembly techniques can provide precision in various materials on a thin film. Typically, Murray *et al.* [39] synthesised various crystals of binary nanoparticle superlattices by using oppositely surface-charged nanoparticles to produce ordered structures. Controlling charge and some experimental parameters could provide macroscopic physicochemical properties of nanomaterials suitable for electronic, optical and magnetic applications.

5.3. Synthesis of monodisperse nanoparticles by chemical methods

As mentioned, nanoparticles can be prepared using various synthesis methods such as polyol reduction [16,17,19,36,39], micro-emulsion [14,40–42] and thermal decomposing methods [17,43–46]. However, for most applications, monodisperse nanoparticles will be ideally needed especially regarding to their critical size-dependent physicochemical properties. Chemical synthesis of monodisperse nanoparticles has been studied in the past few years. Recently, Hyeon's group used a simple synthetic method using a high-intensity ultrasound to produce iron colloids from sonic decomposition of $Fe(CO)_5$ in the presence of PVP as stabiliser. Their result showed a very narrow particle distribution at 8 nm when oleic acid was used as a stabiliser [47]. The monodisperse cobalt ferrite was produced by a high-temperature reduction of a metal precursor, oleic acid, lauric acid in dioctyl ether followed by mild air oxidation [57]. Figure 5.11 displays the image of the monodisperse cobalt ferrite sample. Vestal et al. synthesised single-phase $CoCrFeO_4$ using micro-emulsion system. Typically, they showed that monodisperse cobalt can be prepared from the control reduction of a cobalt salt with sodium tetrahydridoborate ($NaBH_4$) in micro-emulsion system [48]. Lu et al. developed the synthesis of carbon-shell-protected monodisperse cobalt nanocrystals as a template. These monodisperse nanomaterials can be stabilised under strongly acidic and basic conditions, and the magnetic function of the cores allows efficient separation of the particles from solution [49].

Controlled synthesis of magnetic nanoparticles of uniform composition, size, shape and crystal structure can also be carried out at a great precision in specific size ratios leading to superlattices with useful properties [51]. For example, iron oxides

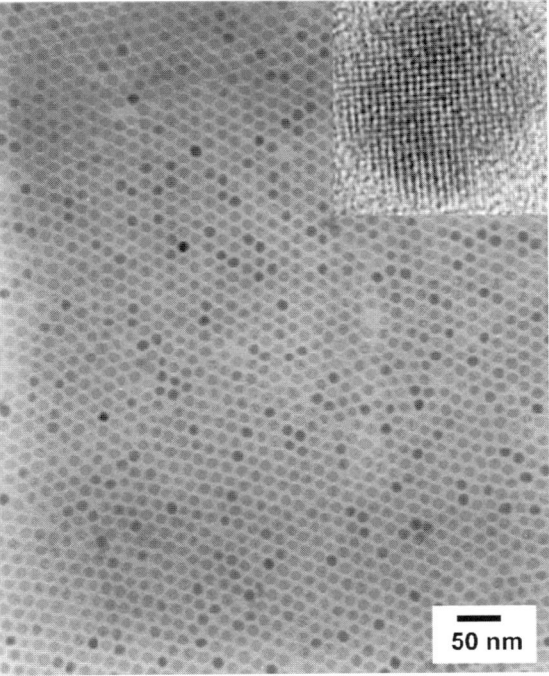

Figure 5.11. TEM image (inset) of 9-nm cobalt ferrite nanocrystals. Reproduced with the permission of the American Chemical Society from Hyeon et al. [50].

have been synthesised from various iron-containing organometallic precursors at high temperature with surfactant as stabiliser in solution. Their structural, magnetic and electronic properties have also been characterised by electron microscopy, X-ray powder diffraction (XRD), Mössbauer spectroscopy and magnetic characterisation. The iron oxide nanoparticles were then deposited onto a patterned wafer to form multilayer for magnetic device fabrication [52].

Murray *et al*. reported the synthesis of nickel and cobalt monodisperse nanoparticles using high-boiling solvent via the polyol route [53–55]. Ho *et al*. reported the preparation of nickel nanoparticles 3.7 nm in size by reduction of precursor nickel acetylacetonate in $NaBH_4$ with HDA that acts as stabiliser and solvent. Further, nanoparticles with a good control in size can be prepared via chemical reduction in solution using borohydride or superhydride in the presence of HDA and trioctylphosphine oxide (TOPO) [56]. Monodisperse nickel nanoparticles in size range 2–7 nm have been used as hydrogenation catalysts [57]. The syntheses of MnP, Co_2P, FeP and Ni_2P nanorods were also conducted by mixing precursor salts with tri-*n*-octylphosphine (TOP) carefully using a syringe pump [45]. Other similar procedures were also reported for the preparation of monodisperse nickel nanoparticles from nickel acetylacetonate using trialkylphosphine oxide or TOPO [58,59].

The formation of monodisperse nanoparticles deviating from the equilibrium spherical shape (minimising the surface energy) is of considerable interest in optical and catalysis areas. Figure 5.12 shows nanocrystals of CdSe and cobalt that can produce anisotropic growth by kinetic shape control to generate rod, arrow or tetrapod shapes. This is achieved because different type of surfactant molecules can selectively adhere to and stabilise particular nanocrystal facets rendering their slower growth rate compared to the other free high-energy crystal facets [60–62]. Figure 5.13 shows that self-assembly of nanorods can produce superstructure with narrow size distribution as well as long-range superlattice with simple hexagonal packing. Also, the results demonstrate excellent luminescent properties, combing high photoluminescence efficiency and giant extinction coefficients [62].

It is noted that a new solvent-less thermolysis chemical synthesis has recently been developed with a capability of controlling particle shape and size [10,64,65]. Figure 5.14 shows that nickel sulphide (NiS) triangular nanoprisms and nanorods can be produced selectively. The resulted α-NiS crystals exhibit anti-ferromagnetic property [63].

5.4. Surface chemical modification of nanoparticles

Physicochemical properties such as the optical, thermal, electrical parameters and magnetic strength [10] of nanoparticles can be altered by the way in which the nanoparticle assemblies are presented. Thus, techniques for the fabrication of nanoparticles as one-, two- or three-dimensional structures are important. During the growth of the nanoparticles in solution, stabilising or capping agents have key importance not only in controlling the particle size, but also in preventing their

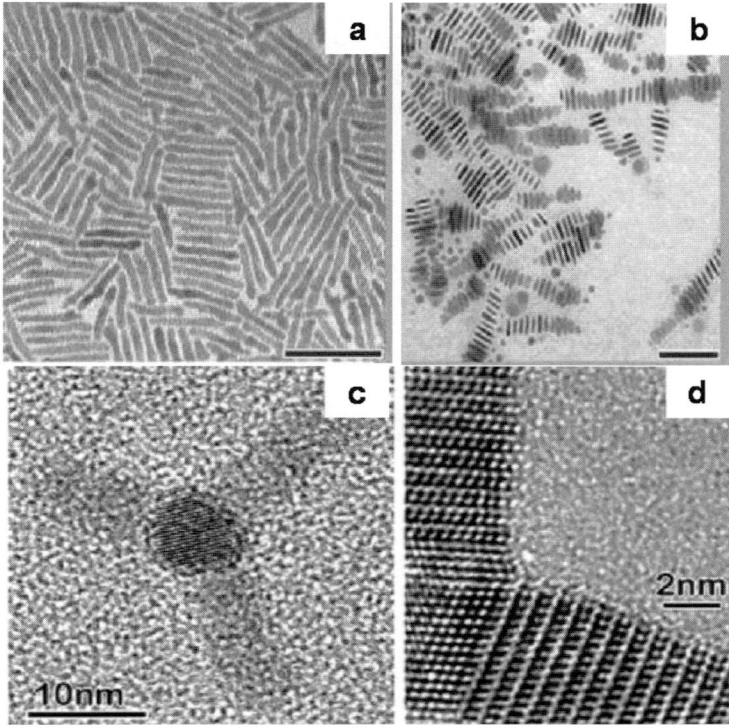

Figure 5.12. (a) TEM images of CdSe nanorods with scale bar, 50 nm. (b) Cobalt nanodisk with scale bar, 100 nm. (c) CdSe–CdS nanotetrapods. (d) HRTEM image of tetrapod fragment between CdSe and CdS. (a) and (b) are reproduced with the permission of Macmillan Publications Ltd. (Nature) from Yin and Alivisatos [61], and (c) and (d) are reproduced with the permission of the American Chemical Society from Talapin et al. [62].

aggregation. Nanoparticles may be assembled as three-dimensional crystals [9], binary nanoparticle superlattices by the slow evaporation technique [38,39], heterodimers [66], core-shell structures [30,67–69], biofunctional nanoparticles for drug delivery and targeting biomolecules, etc. [10,70]. For capping metal or metal oxide, the materials can be organic or inorganic in nature; they are prepared by using chemical preparative methods such as micro-emulsion [40,41,71], sol-gel [36,72] or layer-by-layer techniques [73]. Xu and his co-workers decorated magnetic nanoparticles with bisphosphonate ions to remove uranyl ions from blood through their specific binding affinity [74]. A number of papers have been also published in which decorated nanoparticles with specific terminal groups are used for the detection of various bio-species [75,76]. Other polymer-modified nanoparticles including dendrimers for the affinity of bovine serum albumin (BSA) immobilisation [77], polydipyrrole for DNA hybridisation [69] and specific polymers as anchored species for homogeneous catalyst immobilisation have been developed [78]. In addition, inorganic, silica-modified metal or metal oxide as chemical coats have also been developed [14,29,68,79]. Thus, the chemically modified surfaces play a vital role in nanoparticle fabrication and assembling.

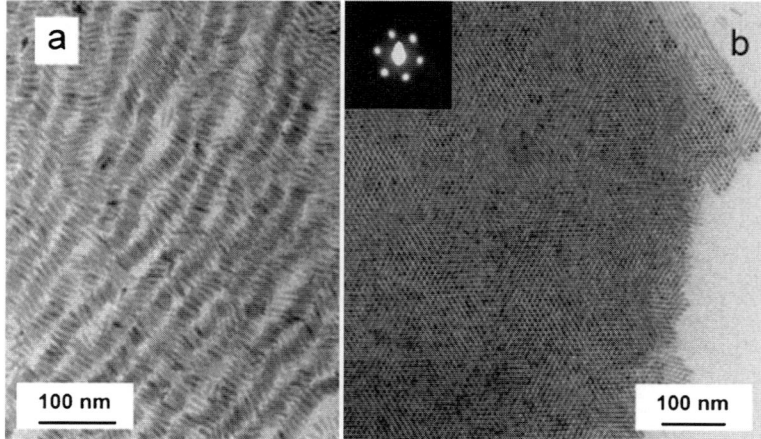

Figure 5.13. Self-assembly of CdSe–CdS nanorods. (a) TEM image of a superstructure with ordering. (b) TEM image of a nanorod superlattice with simple hexagonal packing of nanorods assembled perpendicular to the substrate. Inset shows electron diffraction from a superlattice domain. Reproduced with the permission of the American Chemical Society from Talapin et al. [62].

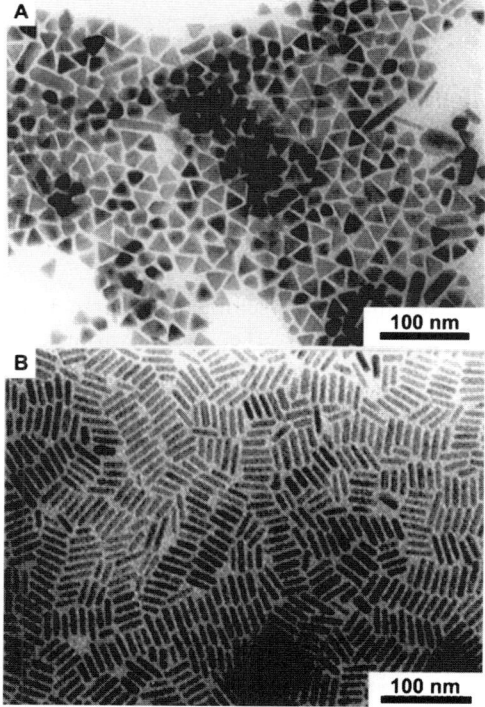

Figure 5.14. TEM images of NiS nanocrystals synthesised with octadecanethiol. The region on the TEM grid shown in (A) is triangular nanoprisms and in (B) is nanorods. Reproduced with the permission of the American Chemical Society from Ghezelbash et al. [63].

5.5. Tailoring the size of nanoparticles

5.5.1. Overview

To control particle size, parameters such as temperature, solvent concentration and terminal ligand are important. There have been a number of chemical methods reported in the literature concerning synthesis of monodisperse iron oxide nanocrystallites with progressive increase in size. Santra et al. [41] used various non-ionic surfactants of different chain lengths to assist the formation of iron oxide nanoparticles via co-precipitation reaction of Fe(II) and Fe(III) salts with base. Ulman also synthesised iron oxide of controlled particle size by the aid of different block co-polymer surfactants [80]. Tartaj and Serna [71] described how particle size can be controlled via a post-synthesis thermal annealing process. Hyeon and his colleagues synthesised their monodisperse iron oxides from 6 to 13 nm by using the 'polyol process' via iron carbonyl decomposition with a seed-mediated growth approach [43]. It was claimed that the particle size could be fine controlled progressively at 1 nm length scale. The process utilised preformed iron oxide nanoparticles as seeds to mediate further particle growth by decomposition of an iron carbonyl precursor such as $Fe(CO)_5$, as shown in Figure 5.15 [43]. Seed-mediated nanoparticles can also be applied to synthesise a larger rod-shaped particle by repeatedly adding $Fe(CO)_5$ in TOP at 320°C [81]. However, the expensive precursor and solvent and cumbersome procedure, in the case of iron oxide, are particularly denoted.

In addition, Boyen controlled gold nanoparticle sizes from 1 to 8 nm, some of which showed a maximum oxidation resistance with SiO_2 and TiO_2 when the gold cluster reached the 'magic numbers' such as Au_{13}, Au_{55} and Au_{147} [82]. Chan demonstrated gold size effects that can apply for diagnostic and therapeutic

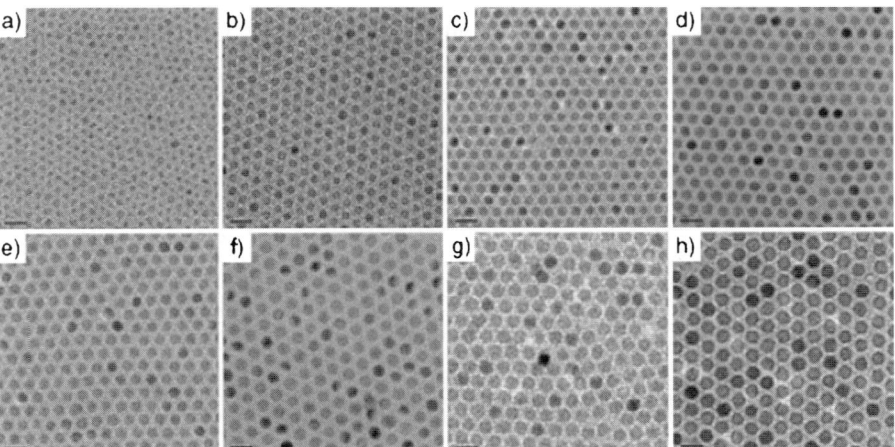

Figure 5.15. TEM images of (a) 6-nm-, (b) 7-nm-, (c) 8-nm-, (d) 9-nm-, (e) 10-nm-, (f) 11-nm-, (g) 12-nm- and (h) 13-nm-sized air-oxidised iron oxide nanoparticles at 1-nm scale. Reproduced with the permission of Wiley-VCH from Park et al. [43].

applications [83]. Wilson also reported that the diameter of the palladium nanoparticles was from 1.3 to 1.9 nm, which presented critical size effects on the rate of hydrogenation of allyl alcohol [84]. In the next section, a study on the effects of some experimental parameters concerning synthesis of monodisperse iron oxide with the application of silica coating using micro-emulsion method is presented.

5.5.2. Tailoring the size of silica-encapsulated iron oxide using micro-emulsion

5.5.2.1. Synthesis of silica-encapsulated iron oxide nanoparticles

In order to protect magnetic nanoparticles from aggregation and introduce surface chemical functionality for biomolecule attachment, Tsang and co-workers have developed a chemical methodology to produce oxide-encapsulated magnetic nanoparticles [14,29,85,86]. The one-pot synthesis procedure can be carried out in two stages: adjusting a range of water to surfactant molar ratios, Wo, to tailor the particle size of the nanomagnetic crystallite core (Fe_3O_4 or FePt) and oxide coating at 23°C under a nitrogen purge. Their typical experimental procedure is carried out as follows: first, a micro-emulsion is created by using an ionic surfactant, toluene and water mixture. The micelle size can be adjusted by using different water to surfactant molar ratios. As a result, magnetic particles can be synthesised inside the micelle by conventional precipitation or chemical reduction. Water-sensitive alkoxide is then added into the microemulsion system, which will undergo hydrolysis and condensation at the water–organic solvent interface on the micelle surface to form the oxide or hydroxyl shells. These reactions are very slow and require about 5–7 days for the whole process to take place. After that, the sample can be placed in a vacuum oven to remove impurities. Typically, the formation of Fe_3O_4 nanoparticle encapsulated in silica can be achieved by the chemical precipitation of Fe(II)/Fe(III) species using ammonium hydroxide inside the micelle, followed by the silica encapsulation using tetraethoxysilane (TEOS) as shown in Scheme 5.3.

$$FeCl_2 + 2FeCl_3 + 8NH_3 \cdot H_2O \rightleftharpoons Fe_3O_4 + 8NH_4Cl + 4H_2O.$$

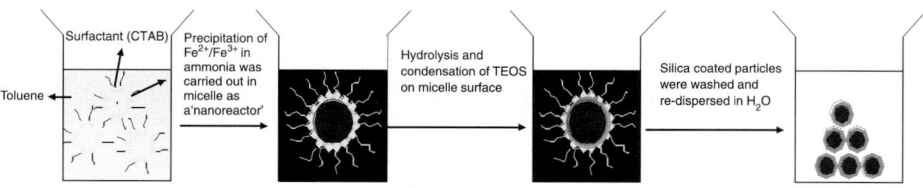

Scheme 5.3. Equation and synthetic procedure for formation of silica-coated iron oxide [14].

5.5.2.2. *Controlling the size of core-shell silica-coated magnetic nanoparticles*

It is evident from the XRD, HRTEM, vibration sample measurement (VSM), and Mössbauer spectroscopy that the sizes of the core magnetic nanoparticles can indeed be successfully controlled from 4.8 to 13.1 nm by adjusting W_o from 10.5 to 75.8 [14].

5.6. Synthesis of binary metallic nanoparticles

According to the procedure of Sun *et al.* [16,17], the synthesis of monodisperse iron platinum–iron oxide (FePt–Fe$_3$O$_4$) nanoparticles can be achieved through the co-reduction of Pt(acac)$_2$ and thermal decomposition of Fe(CO)$_5$ in the presence of oleic acid and oleylamine stabilisers. An 8:1 molar ratio of Fe(CO)$_5$ to Pt(acac)$_2$ gave an FePt–Fe$_3$O$_4$ homogeneously mixed-phase particles [72].

Very recently, Murray's group also reported a generic chemical synthesis of binary nanoparticles using a modified polyol process [39]. More than 15 different binary nanoparticles were demonstrated, using combinations of semiconducting, metallic and magnetic nanoparticles. Figure 5.16 shows various binary nanoparticle superlattice structures with micrometer-scale lattice spacings.

Xu's group prepared FePt–CdS bifunctional hetero-dimers of nanoparticles with sizes of around 7 nm. This procedure also followed the method developed by Sun *et al.* [17] through an initial synthesis of FePt nanoparticles in dioctyl ether without purification and separation. A sulphur-containing compound is then added to the dioctyl ether solution at 100°C to immobilise on the surface of FePt, followed by the subsequent additions of TOPO, 1,2-hexadecanediol and Cd(acac)$_2$ at 100°C. The CdS-encapsulated FePt was then heated to 280°C to form FePt–CdS. Figure 5.17 shows the transmission electron micrographs (TEM) of FePt–CdS nanoparticles of 2.5 nm FePt and 3–4 nm CdS, which demonstrate that they are in crystalline forms [66].

5.7. Synthesis of core-shell structures

In the core-shell nanoparticle structure, a silica coating on the metal particle can protect the core metal from oxidation and corrosion [41]. The sol-gel approach is one of the commonly used methods to synthesise core-shell nanoparticles. Hydrolysis and condensation of added sol-gel precursors, such as silicon alkoxide, will displace the surfactant molecules in the micelle assemblies and produce an ultra-thin porous coating on the core particle at the water–organic interface. This process is based on the hydrolysis of the precursors, such as TEOS, in the presence of water and catalysts, followed by condensation with surface metal hydroxyls. An M–O–Si chemical linkage is established between surface metal atoms and TEOS, followed by the formation of a three-dimensional network of siloxane bonds (Si–O–Si) with increasing TEOS concentration. Once the precursor has been condensed to a gel, the solvent is removed. Very recently, the core-shell

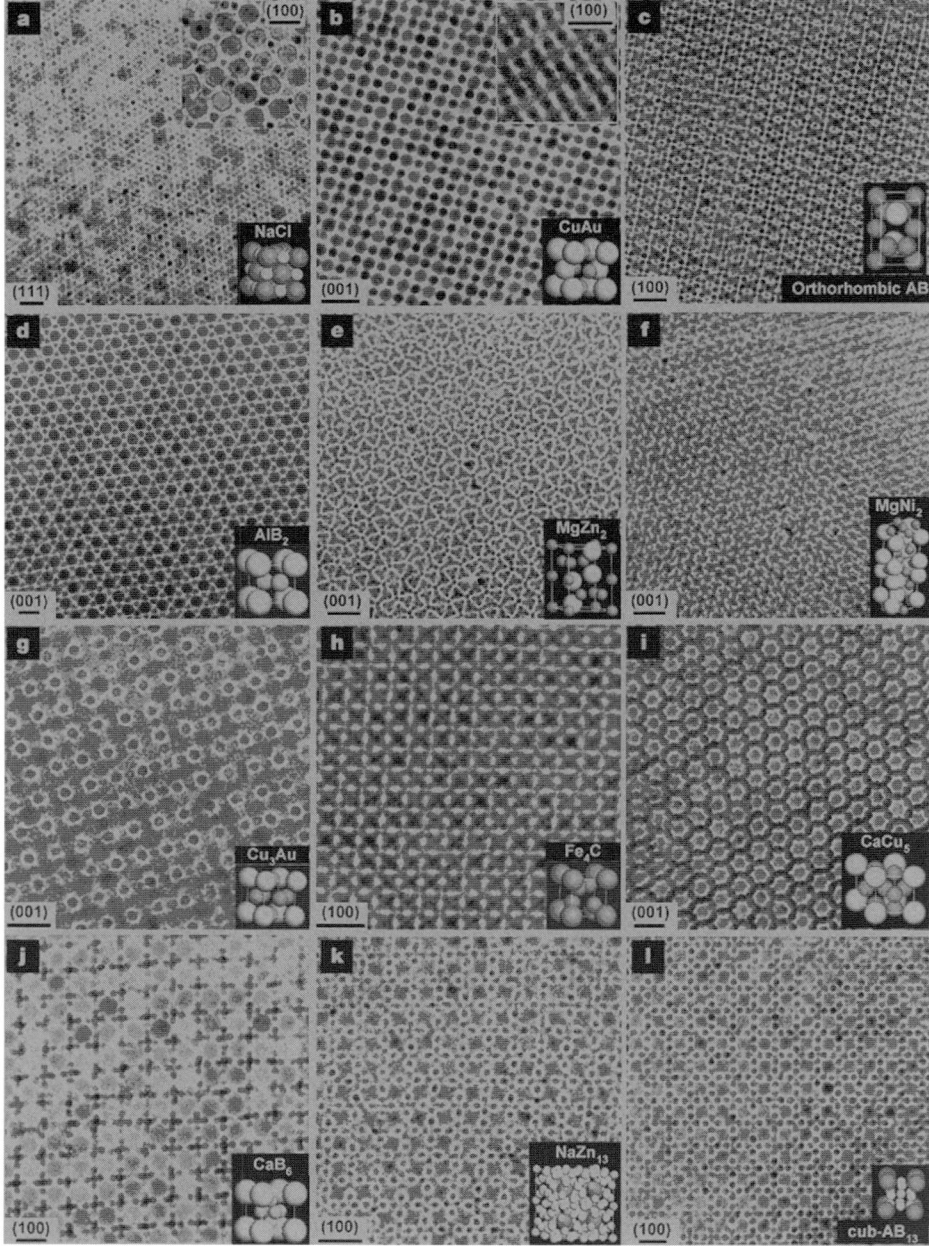

Figure 5.16. TEM images of the characteristic projections of the binary superlattices, self-assembled from different nanoparticles and modelled unit cells of the corresponding three-dimensional structures. The superlattices are assembled from (a) 13.4 nm g-Fe_2O_3 and 5.0 nm Au; (b) 7.6 nm PbSe and 5.0 nm Au; (c) 6.2 nm PbSe and 3.0 nm Pd; (d) 6.7 nm PbS and 3.0 nm Pd; (e) 6.2 nm PbSe and 3.0 nm Pd; (f) 5.8 nm PbSe and 3.0 nm Pd; (g) 7.2 nm PbSe and 4.2 nm Ag; (h) 6.2 nm PbSe and 3.0 nm Pd; (i) 7.2 nm PbSe and 5.0 nm Au; (j) 5.8 nm PbSe and 3.0 nm Pd; (k) 7.2 nm PbSe and 4.2 nm Ag and (l) 6.2 nm PbSe and 3.0 nm Pd nanoparticles. Scale bars: (a–c), (e), (f), (i–l), 20 nm; (d), (g), (h), 10 nm. The lattice projection is labelled in each panel above the scale bar. This figure is adapted from Shevchenko et al. [39] with the permission of Macmillan Publishers Ltd. (Nature).

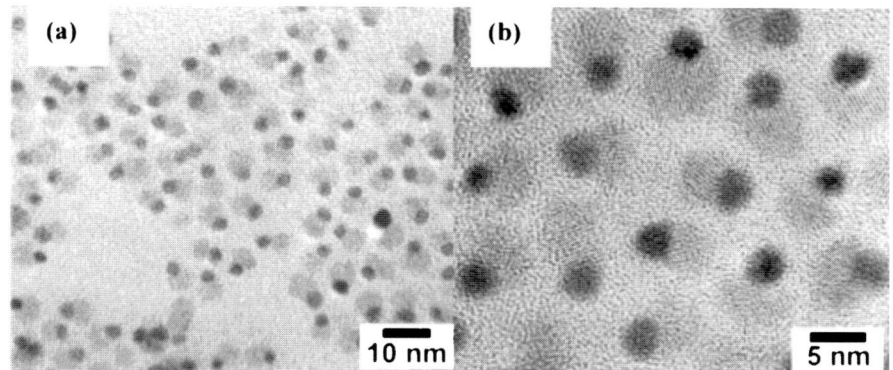

Figure 5.17. (a) A TEM image of FePt–CdS and (b) an HRTEM image. Reproduced with the permission of the American Chemical Society from Gu et al. [66].

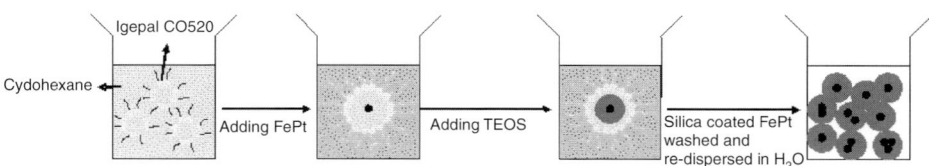

Scheme 5.4. It shows the procedure for encapsulating the as-synthesised FePt from the polyol process in silica [36].

silica-encapsulated magnetic nanoparticles with *fct* FePt structure by combining polyol process, micro-emulsion and sol-gel processes have been developed, which is the subject of the next section.

5.7.1. Synthesis of silica-encapsulated FePt–Fe$_3$O$_4$ (SiO$_2$@FePt–Fe$_3$O$_4$)

In order to encapsulate FePt–Fe$_3$O$_4$ nanoparticle in silica, Yu et al. reported a modified micro-emulsion method using IGEPAL® CO-520, polyoxyethylene(5) nonylphenyl ether, as a surfactant [14,36,79,87]. A more detailed description of the experimental procedure and the preparative scheme can be found in Scheme 5.4.

The TEM image in Figure 5.18(a) clearly shows the as-synthesised FePt–Fe$_3$O$_4$ magnetic nanoparticles from the polyol process, the majority of which are 2.5±0.5 nm in diameter, in good agreement with the XRD result. A close examination of the metal core reveals that the core is composed of at least two different sets of lattice fringes that correspond to *fcc* FePt (1 1 1) and Fe$_3$O$_4$ (3 1 1) with intimate contact. Figure 5.18(b) shows the low-resolution TEM image of FePt–Fe$_3$O$_4$ nanoparticles after subsequent encapsulation with silica, where

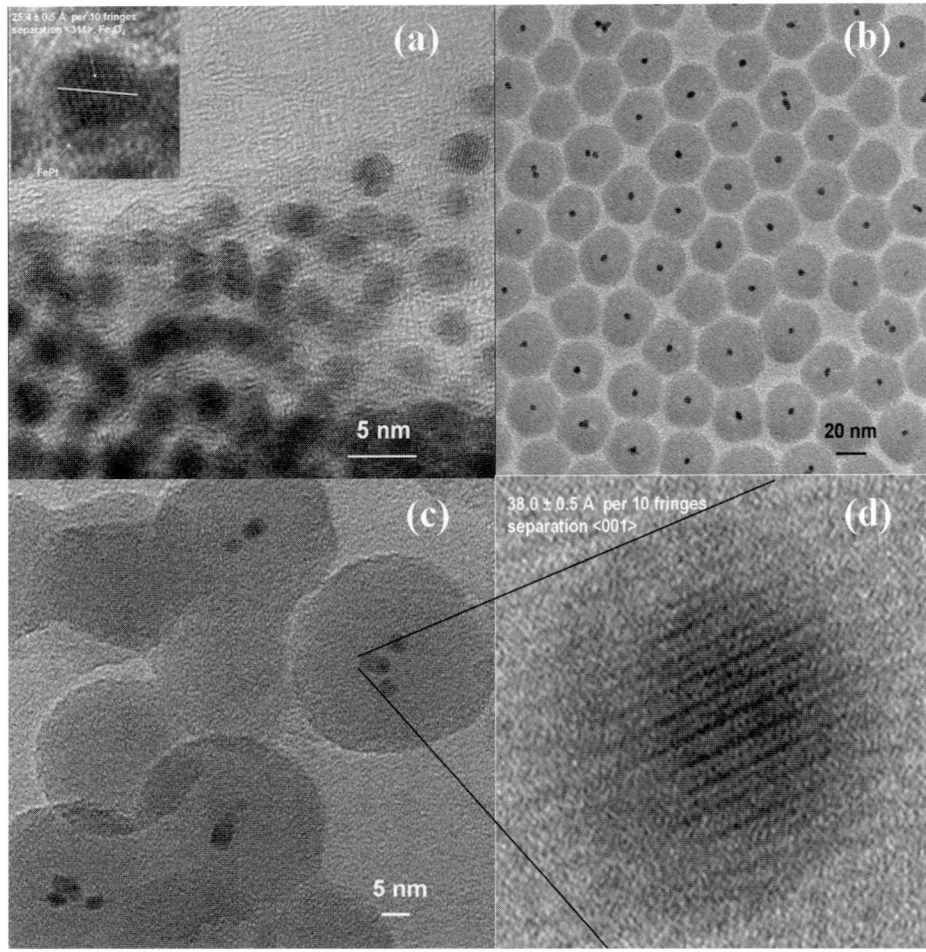

Figure 5.18. TEM images. (a) Monodisperse uncoated mixed-phase FePt–Fe$_3$O$_4$; (b) monodisperse silica-encapsulated FePt–Fe$_3$O$_4$ nanoparticles; (c) core-shell nature of silica-encapsulated *fct* FePt–Fe$_3$Pt after reduction; (d) enlarged image of core region of silica-encapsulated *fct* FePt–Fe$_3$Pt showing segregation of *fct* FePt (inner core) and Fe$_3$Pt (outer core) in the thick silica shell. Reproduced with the permission of the American Chemical Society from Yu et al. [72].

multiple numbers from 0 to 5 particles (a slight aggregation of the metal core is also indicated by the XRD) are enclosed in each silica shell. Extremely uniform silica particle of 35.5±0.5 nm diameter encapsulating the magnetic nanoparticle is therefore clearly evident. The TEM images presented in Figures 5.18(c) and (d) (enlarged core region) show that the sample after H$_2$/Ar treatment clearly displays two sets of lattice fringes with the inner core matching with (1 1 1) FePt (∼2.3 Å) and the outer core corresponding to (0 0 1) Fe$_3$Pt (∼3.8 Å) inside the silica particle. The enlarged TEM image in Figure 5.18(d) undoubtedly displays the darker inner core as an *fct* FePt due to higher electron density and the lighter outer core as an Fe$_3$Pt due to lower electron density in the silica particle. Apparently, the two metal phases are somehow segregated in the core, possibly associated with

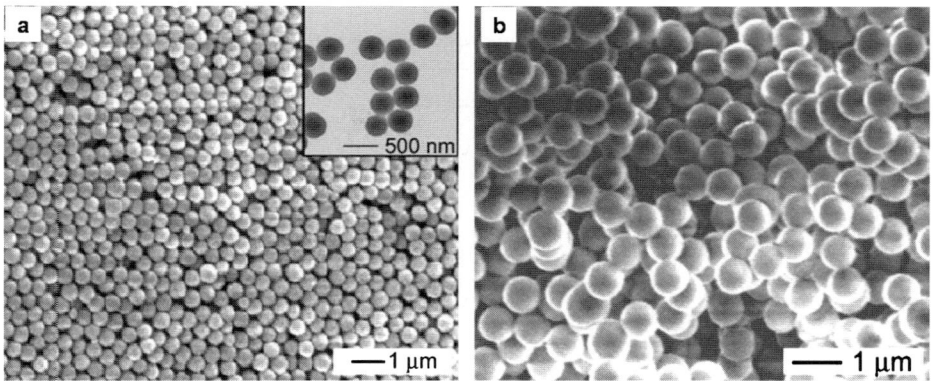

Figure 5.19. (a) SEM image of silica-encapsulated iron oxide with core-shell structure in 300 nm. (b) SEM image of another sample with reducing the concentration of iron oxide nanoparticles in 700 nm. Reproduced with the permission of Elsevier Science from Im et al. [90].

other iron-rich area. Further reduction of the *fcc* FePt–Fe$_3$O$_4$ core leads to segregation of inner *fct* FePt core from Fe$_3$Pt outer core, but with the composite particle placed inside the large silica particle.

According to Sun et al. [88,89], the interaction of soft (Fe$_3$O$_4$ or Fe$_3$Pt) and hard phase (*fct* FePt) in close proximity can provide an exchanged spring coupling, giving enhanced magnetisation. The VSMs indeed show that *fct* FePt–Fe$_3$Pt (after H$_2$/Ar treatment) gives much higher saturated magnetisation (M_s) than *fcc* FePt–Fe$_3$O$_4$ and single phases *fct* FePt and *fcc* FePt with the similar metal core and silica shell dimensions [72].

Im also used the sol-gel process to produce monodisperse superparamagnetic iron oxide nanoparticles (Figure 5.19). Large silica beads can control from 140 nm to 1.5 μm with narrow distribution by varying iron oxide concentration [90].

5.7.2. Synthesis of monodisperse silica-encapsulated iron platinum nanoparticle with external Ti–O–Si coating (TiO$_2$–SiO$_2$@FePt)

The synthesis of magnetic Ti–O–Si (FePt) core-shell nanocatalysts can be implemented stepwise by a bottom-up layer-by-layer approach, see Scheme 5.5.

To prepare the titanium coating on TiO$_2$@FePt, a stock solution of titanium *n*-butoxide (TBOT) in ethanol can be used as chemical precursor for the titania coating (Figure 5.20). Typically, silica-encapsulated iron platinum (SiO$_2$@FePt) nanoparticles in ethanol solution, water, TBOT and ethanol are mixed under stirring. The mixture is repeatedly stirred and then centrifuged. The residue can be dispersed in ethanol. Finally, all samples can be magnetically precipitated by exposure to magnetic field. The ability to engineer a single catalyst particle by this 'bottom-up' approach through layer-by-layer synthesis at a high precision enables a new generation of catalysts with desirable functionalities, including magnetic separable properties, to be achieved [91–93].

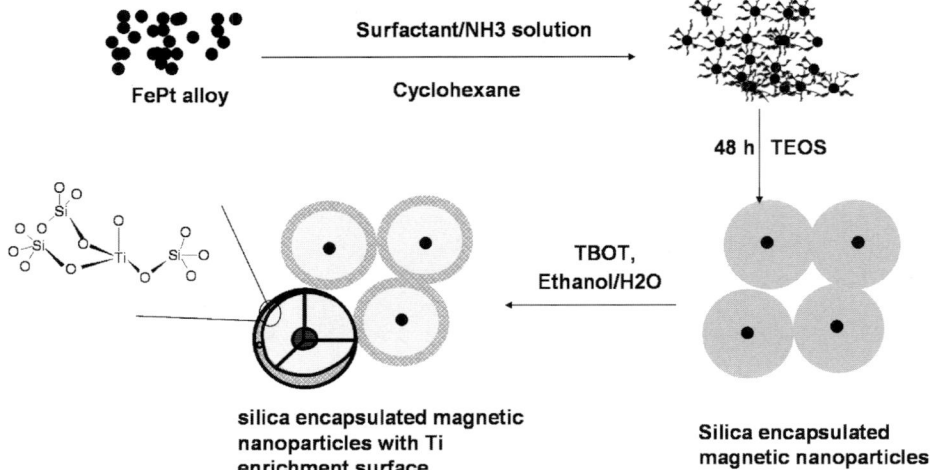

Scheme 5.5. Procedure of stepwise syntheses of monodisperse core-shells magnetic Ti–O–Si nanocatalyst for stilbene epoxidation. Reproduced with the permission of the American Chemical Society from Tang et al. [28].

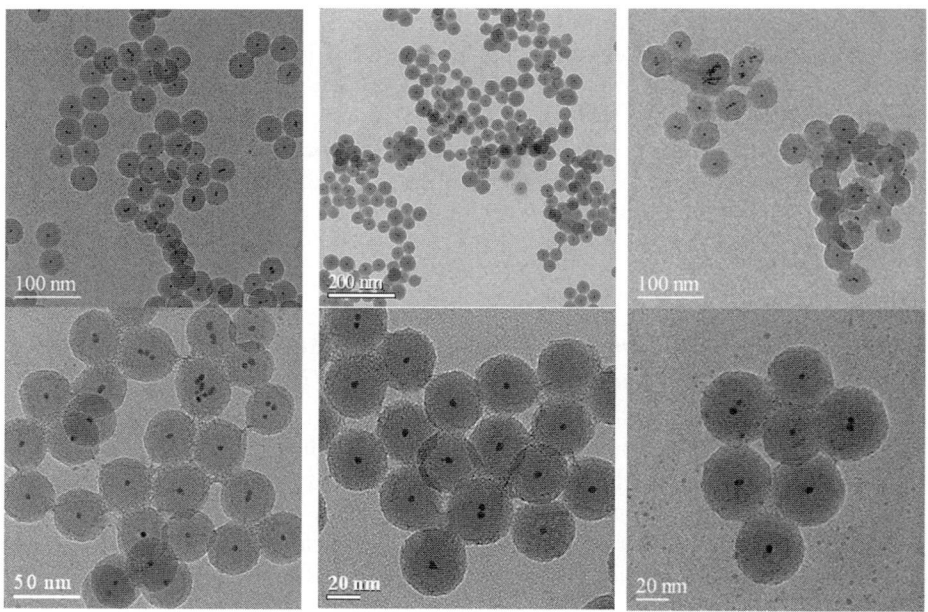

Figure 5.20. HRTEM images of titanium-rich coatings on silica-encapsulated FePt nanoparticles. Reproduced with the permission of the American Chemical Society from Tang et al. [28] and Yu et al. [72].

5.7.3. Synthesis of polymer-encapsulated iron platinum (polymer@FePt)

The creation for interface between metal and polymer is useful for many applications [94–98]. By adopting the synthesis of surfactant-stabilised metal nanoparticle (naked form) described in Sections 5.7.1 and 5.7.2, polymer encapsulation of metal or alloy particles, such as FePt, can also be accomplished, for example, by the addition of poly(vinyl butyral-*co*-vinyl alcohol-*co*-acetate), PVB, at 220°C into the polyol mixture with reflux. After the sample is collected, washed and dried, the resulting polymer@FePt nanoparticles can be calcined to the metal alloy inside carbon coating (Figure 5.21). Scheme 5.6 summarises the details of material synthesis and the general preparative procedure.

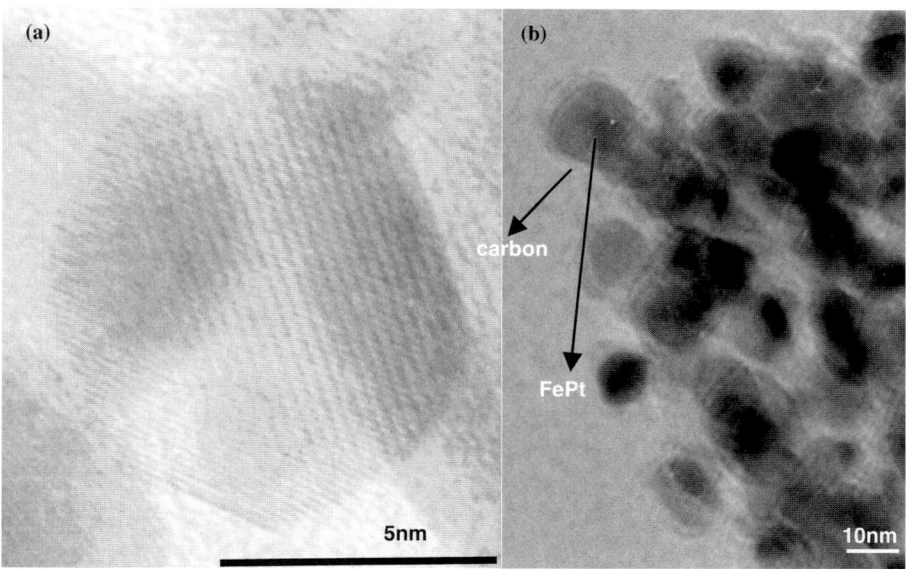

Figure 5.21. HRTEM images of a carbon@FePt multilayer annealed at 700°C for 2 h at different magnifications. (a) Identification of lattice fringes for *fct* structure carbon@FePt on the carbon-coated copper grid. (b) The particles are encapsulated in carbon shell. Reproduced with the permission of Wiley-VCH from Caiulo *et al.* [97].

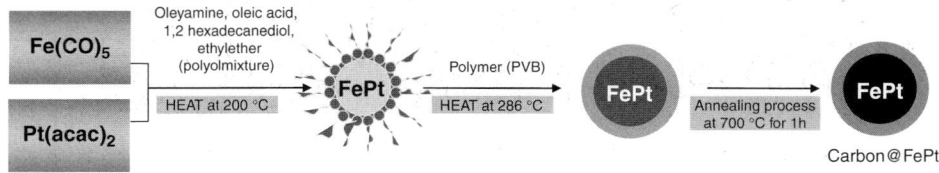

Scheme 5.6. The concept of synthetic procedure for polymer@FePt. Reproduced with the permission of Wiley-VCH from Caiulo *et al.* [97].

5.7.4. Formation of glucose-encapsulated iron platinum spheres by hydrothermal treatment

According to the procedure described by Sun and Li [98], hydrothermal treatment of glucose solution in the presence of the metal precursors can produce encapsulated metal particle in a carbonaceous coating. After calcination of the sample at elevated temperatures, metal particle encapsulated in carbon shells can be made [97]. The schematic outline of the pressure-tight Teflon-coated Parr autoclave is shown in Scheme 5.7; it is important to keep the reaction mixture at 180°C for the hydrothermal synthesis.

Typical TEM images (Figure 5.22) of the glucose-treated sample after heat treatment (calcination) show that there is a high loading of metal nanoparticles with a particle size ranging from 3 to 6 nm in a carbon matrix. HRTEM (Figure 5.22) reveals that most FePt nanoparticles are found decorated with partially graphitic shells (interplanar separation of 3.45 Å that matches with the interlayer separation

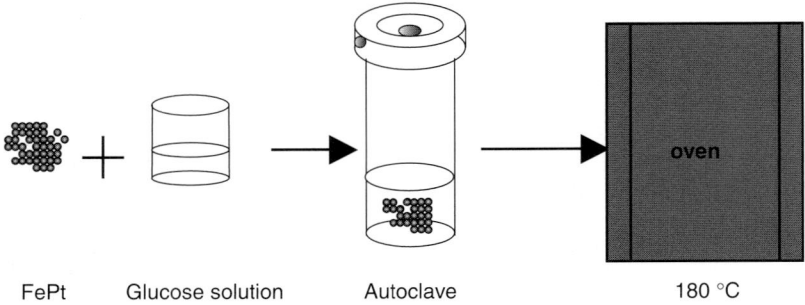

Scheme 5.7. A schematic diagram showing the setup for *in situ* synthesis of glucose-coated metal nanoparticles (as encapsulating spheres) in aqueous phase by hydrothermal treatment [97].

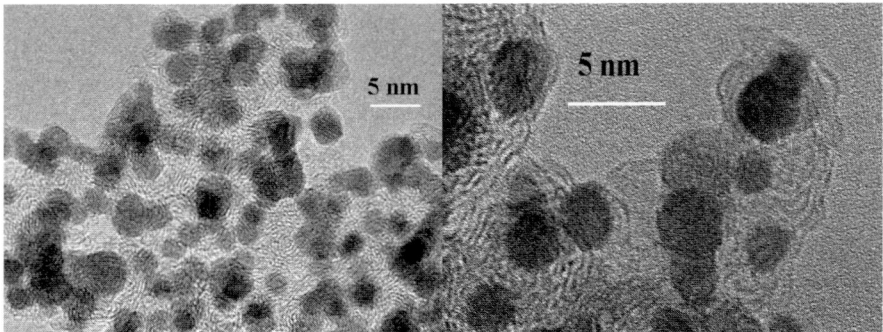

Figure 5.22. Left: Transmission electron micrograph (TEM) showing 3–6 nm FePt particles in carbon matrix; Right: HRTEM showing image of isolated FePt particles encapsulated in partially graphitic coatings of interlayer separation of 3.45 Å. Reproduced with the permission of Wiley-VCH from Caiulo *et al.* [97].

of spheroidal fullerenic carbon shells [6]). These carbon-coated metallic materials could have interesting applications as the entrapped metal nanoparticles in the thermally stable carbon matrix may alter their thermal reactivity and mobility.

References

[1] J.R. Krenn, Nat. Mater. **2** (2003) 210.
[2] C.N.R. Rao, A. Muller and A.K. Cheetham (Eds.), The Chemistry of Nanomaterials: Synthesis, Properties and Applications, Vol. 1, Wiley-VCH, Weinheim, 2004, p. 1.
[3] R.P. Feynman, Science **183** (1974) 601.
[4] T.J. Yoon, J.S. Kim, B.G. Kim, K.N. Yu, M.H. Cho and J.K. Lee, Angew. Chem. Int. Ed. Engl. **44** (2005) 1068.
[5] S.W. Song, K. Hidajat and S. Kawi, Langmuir **21** (2005) 9568.
[6] S.C. Tsang, V. Caps, I. Paraskevas, D. Chadwick and D. Thompsett, Angew. Chem. Int. Ed. Engl. **43** (2004) 5645.
[7] E. Katz and I. Willner, Angew. Chem. Int. Ed. Engl. **43** (2004) 6042.
[8] V.K. La Mer and R.H. Dinegar, J. Am. Chem. Soc. **72** (1950) 4847.
[9] C.B. Murray, C.R. Kagan and M.G. Bawendi, Annu. Rev. Mater. Sci. **30** (2000) 545.
[10] C. Burda, X.B. Chen, R. Narayanan and M.A. El-Sayed, Chem. Rev. **105** (2005) 1025.
[11] B.K. Paul and S.P. Moulik, J. Dispersion Sci. Technol. **18** (1997) 301.
[12] P.A. Winsor, Trans. Faraday Soc. **44** (1948) 376.
[13] S.P. Moulik and B.K. Paul, Adv. Colloid Interface Sci. **78** (1998) 99.
[14] S.C. Tsang, C.H. Yu, X. Gao and K. Tam, J. Phys. Chem. B **110** (2006) 16914.
[15] C.J. Brinker, Y.F. Lu, A. Sellinger and H.Y. Fan, Adv. Mater. **11** (1999) 579.
[16] S.H. Sun, S. Anders, T. Thomson, J.E.E. Baglin, M.F. Toney, H.F. Hamann, C.B. Murray and B.D. Terris, J. Phys. Chem. B **107** (2003) 5419.
[17] S.H. Sun, C.B. Murray, D. Weller, L. Folks and A. Moser, Science **287** (2000) 1989.
[18] S.H. Sun, E.E. Fullerton, D. Weller and C.B. Murray, IEEE Trans. Magn. **37** (2001) 1239.
[19] M. Chen, J.P. Liu and S.H. Sun, J. Am. Chem. Soc. **126** (2004) 8394.
[20] S.S. Kang, Z.Y. Jia, S.F. Shi, D.E. Nikles and J.W. Harrell, J. Appl. Phys. **97** (2005) 10J318.
[21] L.E.M. Howard, H.L. Nguyen, S.R. Giblin, B.K. Tanner, I. Terry, A.K. Hughes and J.S.O. Evans, J. Am. Chem. Soc. **127** (2005) 10140.
[22] J.I. Park and J. Cheon, J. Am. Chem. Soc. **123** (2001) 5743.
[23] M. Chen and D.E. Nikles, J. Appl. Phys. **91** (2002) 8477.
[24] S. Kang, J.W. Harrell and D.E. Nikles, Nano Lett. **2** (2002) 1033.
[25] S.S. Kang, D.E. Nikles and J.W. Harrell, J. Appl. Phys. **93** (2003) 7178.
[26] C. Liu, X.W. Wu, T. Klemmer, N. Shukla and D. Weller, Chem. Mater. **17** (2005) 620.
[27] S.H. Sun, S. Anders, H.F. Hamann, J.U. Thiele, J.E.E. Baglin, T. Thomson, E.E. Fullerton, C.B. Murray and B.D. Terris, J. Am. Chem. Soc. **124** (2002) 2884.
[28] H. Tang, C.H. Yu, W. Oduoro, H. He and S.C. Tsang, Langmuir **24** (2008) 1587.
[29] S.C. Tsang, C.H. Yu, X. Gao and K.Y. Tam, Int. J. Pharm. **327** (2006) 139.
[30] S.I. Stoeva, F.W. Huo, J.S. Lee and C.A. Mirkin, J. Am. Chem. Soc. **127** (2005) 15362.
[31] N.L. Rosi and C.A. Mirkin, Chem. Rev. **105** (2005) 1547.
[32] J. Lyklema and J.F.L. Duval, Adv. Colloid Interface Sci. **114** (2005) 27.
[33] H. Schamel, Phys. Rep. **140** (1986) 161.
[34] R.W. Obrien, D.W. Cannon and W.N. Rowlands, J. Colloid Interface Sci. **173** (1995) 406.
[35] A. Roucoux, J. Schulz and H. Patin, Chem. Rev. **102** (2002) 3757.
[36] C.H. Yu, N. Caiulo, C.C.H. Lo, K. Tam and S.C. Tsang, Adv. Mater. **18** (2006) 2312.
[37] D.A. Doshi, A. Gibaud, V. Goletto, M.C. Lu, H. Gerung, B. Ocko, S.M. Han and C.J. Brinker, J. Am. Chem. Soc. **125** (2003) 11646.

[38] E.V. Shevchenko, D.V. Talapin, C.B. Murray and S. O'Brien, J. Am. Chem. Soc. **128** (2006) 3620.
[39] E.V. Shevchenko, D.V. Talapin, N.A. Kotov, S. O'Brien and C.B. Murray, Nature **439** (2006) 55.
[40] K.M.K. Yu, C.M.Y. Yeung, D. Thompsett and S.C. Tsang, J. Phys. Chem. B **107** (2003) 4515.
[41] S. Santra, R. Tapec, N. Theodoropoulou, J. Dobson, A. Hebard and W.H. Tan, Langmuir **17** (2001) 2900.
[42] Y. Lee, J. Lee, C.J. Bae, J.G. Park, H.J. Noh, J.H. Park and T. Hyeon, Adv. Fun. Mater. **15** (2005) 503.
[43] J. Park, E. Lee, N.M. Hwang, M.S. Kang, S.C. Kim, Y. Hwang, J.G. Park, H.J. Noh, J.Y. Kini, J.H. Park and T. Hyeon, Angew. Chem. Int. Ed. Engl. **44** (2005) 2872.
[44] T. Hyeon, Chem. Commun. (2003) 927.
[45] J. Park, B. Koo, K.Y. Yoon, Y. Hwang, M. Kang, J.G. Park and T. Hyeon, J. Am. Chem. Soc. **127** (2005) 8433.
[46] J. Park, K.J. An, Y.S. Hwang, J.G. Park, H.J. Noh, J.Y. Kim, J.H. Park, N.M. Hwang and T. Hyeon, Nat. Mater. **3** (2004) 891.
[47] K.S. Suslick, M.M. Fang and T. Hyeon, J. Am. Chem. Soc. **118** (1996) 11960.
[48] I. Lisiecki and M.P. Pileni, Langmuir **19** (2003) 9486.
[49] A.H. Lu, W.C. Li, N. Matoussevitch, B. Spliethoff, H. Bonnemann and F. Schuth, Chem. Commun. (2005) 98.
[50] T. Hyeon, Y. Chung, J. Park, S.S. Lee, Y.W. Kim and B.H. Park, J. Phys. Chem. B **106** (2002) 6831.
[51] F.X. Redl, K.S. Cho, C.B. Murray and S. O'Brien, Nature **423** (2003) 968.
[52] F.X. Redl, C.T. Black, G.C. Papaefthymiou, R.L. Sandstrom, M. Yin, H. Zeng, C.B. Murray and S.P. O'Brien, J. Am. Chem. Soc. **126** (2004) 14583.
[53] V. Perez-Dieste, O.M. Castellini, J.N. Crain, M.A. Eriksson, A. Kirakosian, J.L. Lin, J.L. McChesney, F.J. Himpsel, C.T. Black and C.B. Murray, Appl. Phys. Lett. **83** (2003) 5053.
[54] C.B. Murray, S.H. Sun, W. Gaschler, H. Doyle, T.A. Betley and C.R. Kagan, IBM J. Res. Dev. **45** (2001) 47.
[55] C.B. Murray, S.H. Sun, H. Doyle and T. Betley, MRS Bull. **26** (2001) 985.
[56] Y. Hou, H. Kondoh, T. Ohta and S. Gao, Appl. Surf. Sci. **241** (2005) 218.
[57] J. Park, E. Kang, S.U. Son, H.M. Park, M.K. Lee, J. Kim, K.W. Kim, H.J. Noh, J.H. Park, C.J. Bae, J.G. Park and T. Hyeon, Adv. Mater. **17** (2005) 429.
[58] M. Green and P. O'Brien, Chem. Commun. (2001) 1912.
[59] N. Cordente, M. Respaud, F. Senocq, M.J. Casanove, C. Amiens and B. Chaudret, Nano Lett. **1** (2001) 565.
[60] L. Manna, E.C. Scher and A.P. Alivisatos, J. Am. Chem. Soc. **122** (2000) 12700.
[61] Y. Yin and A.P. Alivisatos, Nature **437** (2005) 664.
[62] D.V. Talapin, J.H. Nelson, E.V. Shevchenko, S. Aloni, B. Sadtler and A.P. Alivisatos, Nano Lett. **7** (2007) 2951.
[63] A. Ghezelbash, M.B. Sigman and B.A. Korgel, Nano Lett. **4** (2004) 537.
[64] T.H. Larsen, M. Sigman, A. Ghezelbash, R.C. Doty and B.A. Korgel, J. Am. Chem. Soc. **125** (2003) 5638.
[65] M.B. Sigman, A. Ghezelbash, T. Hanrath, A.E. Saunders, F. Lee and B.A. Korgel, J. Am. Chem. Soc. **125** (2003) 16050.
[66] H.W. Gu, R.K. Zheng, X.X. Zhang and B. Xu, J. Am. Chem. Soc. **126** (2004) 5664.
[67] I.S. Lee, N. Lee, J. Park, B.H. Kim, Y.W. Yi, T. Kim, T.K. Kim, I.H. Lee, S.R. Paik and T. Hyeon, J. Am. Chem. Soc. **128** (2006) 10658.
[68] D.K. Yi, S.S. Lee, G.C. Papaefthymiou and J.Y. Ying, Chem. Mater. **18** (2006) 614.
[69] J.P. Lellouche, G. Senthil, A. Joseph, L. Buzhansky, I. Bruce, E.R. Bauminger and J. Schlesinger, J. Am. Chem. Soc. **127** (2005) 11998.
[70] H.W. Gu, P.L. Ho, K.W.T. Tsang, L. Wang and B. Xu, J. Am. Chem. Soc. **125** (2003) 15702.
[71] P. Tartaj and C.J. Serna, Chem. Mater. **14** (2002) 4396.

[72] C.H. Yu, C.C.H. Lo, K. Tam and S.C. Tsang, J. Phys. Chem. C **111** (2007) 7879.
[73] J. Kim, J.E. Lee, J. Lee, Y. Jang, S.W. Kim, K. An, H.H. Yu and T. Hyeon, Angew. Chem. Int. Ed. Engl. **45** (2006) 4789.
[74] L. Wang, Z.M. Yang, J.H. Gao, K.M. Xu, H.W. Gu, B. Zhang, X.X. Zhang and B. Xu, J. Am. Chem. Soc. **128** (2006) 13358.
[75] J.H. Gao, L. Li, P.L. Ho, G.C. Mak, H.W. Gu and B. Xu, Adv. Mater. **18** (2006) 3145.
[76] H.W. Gu, K.M. Xu, C.J. Xu and B. Xu, Chem. Commun. (2006) 941.
[77] B.F. Pan, F. Gao and H.C. Gu, J. Colloid Interface Sci. **284** (2005) 1.
[78] S. Ko and J. Jang, Angew. Chem. Int. Ed. Engl. **45** (2006) 7564.
[79] D.K. Yi, S.T. Selvan, S.S. Lee, G.C. Papaefthymiou, D. Kundaliya and J.Y. Ying, J. Am. Chem. Soc. **127** (2005) 4990.
[80] J.I. Lai, K. Shafi, A. Ulman, K. Loos, Y.J. Lee, T. Vogt, W.L. Lee and N.P. Ong, J. Phys. Chem. B **109** (2005) 15.
[81] S.J. Park, S. Kim, S. Lee, Z.G. Khim, K. Char and T. Hyeon, J. Am. Chem. Soc. **122** (2000) 8581.
[82] H.G. Boyen, G. Kastle, F. Weigl, B. Koslowski, C. Dietrich, P. Ziemann, J.P. Spatz, S. Riethmuller, C. Hartmann, M. Moller, G. Schmid, M.G. Garnier and P. Oelhafen, Science **297** (2002) 1533.
[83] B.D. Chithrani, A.A. Ghazani and W.C.W. Chan, Nano Lett. **6** (2006) 662.
[84] O.M. Wilson, M.R. Knecht, J.C. Garcia-Martinez and R.M. Crooks, J. Am. Chem. Soc. **128** (2006) 4510.
[85] X. Gao, K.M.K. Yu, K.Y. Tam and S.C. Tsang, Chem. Commun. (2003) 2998.
[86] X. Gao, C.H. Yu, K.Y. Tam and S.C. Tsang, J. Pharm. Biomed. Anal. **38** (2005) 197.
[87] K.M.K. Yu, D. Thompsett and S.C. Tsang, Chem. Commun. (2003) 1522.
[88] H. Zeng, J. Li, J.P. Liu, Z.L. Wang and S.H. Sun, Nature **420** (2002) 395.
[89] H. Zeng, S.H. Sun, J. Li, Z.L. Wang and J.P. Liu, Appl. Phys. Lett. **85** (2004) 792.
[90] S.H. Im, T. Herricks, Y.T. Lee and Y.N. Xia, Chem. Phys. Lett. **401** (2005) 19.
[91] J.M. Fraile, J.I. Garcia, J.A. Mayoral and E. Vispe, J. Catal. **233** (2005) 90.
[92] C. Beck, T. Mallat, T. Burgi and A. Baiker, J. Catal. **204** (2001) 428.
[93] G. Ricchiardi, A. Damin, S. Bordiga, C. Lamberti, G. Spano, F. Rivetti and A. Zecchina, J. Am. Chem. Soc. **123** (2001) 11409.
[94] S. Sivaramakrishnan, P.J. Chia, Y.C. Yeo, L.L. Chua and P.K.H. Ho, Nat. Mater. **6** (2007) 149.
[95] S.A. Haque, S. Koops, N. Tokmoldin, J.R. Durrant, J.S. Huang, D.D.C. Bradley and E. Palomares, Adv. Mater. **19** (2007) 683.
[96] S. Park, J.H. Lim, S.W. Chung and C.A. Mirkin, Science **303** (2004) 348.
[97] N. Caiulo, C.H. Yu, K.M.K. Yu, C.C.H. Lo, W. Oduro, B. Thiebaut, P. Bishop and S.C. Tsang, Adv. Fun. Mater. **17** (2007) 1392.
[98] X.M. Sun and Y.D. Li, Angew. Chem. Int. Ed. Engl. **43** (2004) 597.

Chapter 6

Structure of Isolated Clusters

J.A. Blackman
Department of Physics, University of Reading, Whiteknights, Reading RG6 6AF, UK;
Department of Physics and Astronomy, University of Leicester, University Road, Leicester LE1 7RH, UK

6.1.	Introduction	143
6.2.	Theoretical methodology	144
	6.2.1. Post-Hartree–Fock quantum chemistry methods	144
	6.2.2. Density functional theory	145
	6.2.3. Tight-binding methods	147
	6.2.4. Semi-empirical potentials and global optimisation	149
6.3.	Overview of specific materials	152
	6.3.1. Introduction	152
	6.3.2. Alkali metals	153
	6.3.3. Noble metals	155
	6.3.4. Ferromagnetic transition metals (Fe, Co, Ni)	158
	6.3.5. Transition metals	159
	6.3.6. Divalent and trivalent metals	161
6.4.	Very large clusters	165
	References	168

6.1. Introduction

We have seen, in Chapter 2, that the two shell models (electronic and geometric) give a remarkably good account of the relative stability of clusters of the simple metals. Peaks in the mass abundance spectra can be linked to magic numbers. The electronic shell model applies reasonably well for small clusters while, for large clusters, the geometric shell model generally provides a better description (see Figure 2.10 in Section 2.2.5).

At its basic level, the electronic shell model pictures electrons in a phenomenological spherical well. Ellipsoidal distortions and a self-consistent potential give extra sophistication to the model. Although the model can provide a useful insight into the underlying physics of the clusters and has been quite successful in predicting experimental quantities like the ionisation potential (IP), electron affinity and static polarisability for simple metals, it is obviously desirable to perform full quantum mechanical calculations that take into account the details of the individual atoms of the cluster. This is particularly important for small clusters, which may well adopt a geometry that is far from spherical, and for those metals for which p- and d-electrons play a significant role.

The geometric shell model provides a scheme for identifying clusters of high stability in terms of shell filling. This results in a sequence of magic numbers that one can associate with particular geometric structures: fcc, icosahedral, decahedral, etc. However to predict which structure is preferred requires the injection of some quantum mechanics into the theory, albeit at a much more approximate level than is necessary for small clusters.

The next section (Section 6.2) outlines some of the theoretical approaches to cluster modelling that allow one to address sizes from a few atoms to tens of thousands of atoms. The methods range from quantum chemistry techniques that provide a highly accurate description of small clusters ($N < 20$) to calculations based on semi-classical potentials that allow one to derive structural phase diagrams for cluster sizes of up to tens of thousands of atoms.

This is followed (Section 6.3) with a survey of results of calculations of cluster structures for a range of elements. Very large clusters are discussed in Section 6.4. We confine our attention here to non-magnetic systems. Clusters exhibiting a net magnetic moment are discussed separately in Chapter 8.

There are a number of review articles that are useful for a more detailed description of particular techniques. Bonačić-Koutecký *et al.* [1,2] give an account of the highly accurate quantum chemical techniques that can tackle the properties of small clusters. The review of Baletto and Ferrando [3] considers the more approximate techniques that can explore the structural properties of clusters of hundreds or thousands of atoms.

6.2. Theoretical methodology

6.2.1. *Post-Hartree–Fock quantum chemistry methods*

The Hartree–Fock method is a base-level theory of quantum chemistry, but since it uses a single-determinant wave function it does not describe the correlation of electrons with different spins. Electrons of the same spin are correlated due to the antisymmetric character of the wave function. The neglect of correlations produces energy values that are higher than the true ones and various strategies for including the effect of correlations are employed in post-Hartree–Fock theories.

Exact wave functions cannot be expressed as a single determinant. Configuration interaction (CI) methods proceed by constructing other determinants in which one or more occupied orbitals within the Hartree–Fock determinant are replaced by virtual orbitals [1,2]. Each replacement is equivalent to an electron excitation.

In the full CI method the wave function is formed as a linear combination of the Hartree–Fock determinant and all possible substituted determinants. Except for the smallest systems, it is necessary to truncate the expansion at a fairly low level of substitution (say single and double excitations). This leads to size inconsistency as the deviation from the full CI limit increases with the number of electrons in the system. There are refinements to the method using a subset of orbitals within a full CI calculation. For properly describing clusters where d bonds are important, a multireference-based theory is needed with a large active space and an extended basis set.

Other approaches to adding higher excitations to Hartree–Fock theory include the Møller–Plesset perturbation theory [4] and the coupled cluster (CC) technique [5]. Møller–Plesset perturbation theory is based on Rayleigh–Schrödinger perturbation theory and adds the higher excitations as non-iterative corrections. Taken to order n, it is known as the MPn method. In practice MP4 is the most commonly used.

Coupled with a geometry optimisation technique, these methods generally provide a strategy for a very accurate determination of the ground state properties of clusters, but they are costly computationally and are limited to clusters smaller than about 20 atoms.

6.2.2. Density functional theory

Calculations based on the local density approximation (LDA) in density functional theory (DFT) include the effect of electron correlations much less expensively than the quantum chemistry methods and have extended the accessible size range to clusters of several hundred atoms [6–10].

Usually the exchange-correlation energy functional, that we have already seen in Eq. (2.17) (in Section 2.2.2) in the context of the jellium model, is divided into an exchange and correlation part

$$E_{xc} = E_x + E_c. \tag{6.1}$$

The local exchange functional is generally defined in Dirac-Slater [11] form as

$$E_x^{LDA} = -\frac{3}{2}\left(\frac{3}{4\pi}\right)^{1/3} \int d\vec{r}\left[\rho_\alpha^{4/3} + \rho_\beta^{4/3}\right], \tag{6.2}$$

where we have used the local spin density (LSDA) form in terms of the electron densities for the two spin components, α ($=\uparrow$) and β ($=\downarrow$), and ρ is of course a function of $\vec{r}$. The first term in Eq. (2.19) (in Section 2.2.2) for the jellium model is the Dirac–Slater functional although, of course, the Kohn–Sham wave functions are now atom-centred.

A popular local spin density correlation energy functional is that due to Vosko et al. [12], which has the following form

$$E_c^{LDA}[\rho_\alpha, \rho_\beta] = \int d\vec{r}\rho\varepsilon_c(\rho_\alpha, \rho_\beta), \tag{6.3}$$

where

$$\varepsilon_c = A\left\{\ln\left(\frac{x^2}{X(x)}\right) + \frac{2b}{Q}\tan^{-1}\left(\frac{Q}{2x+b}\right) - \frac{bx_0}{X(x_0)}\right.$$
$$\left. \times \left[\ln\left(\frac{(x-x_0)^2}{X(x)}\right) + \frac{2(b+2x_0)}{Q}\tan^{-1}\left(\frac{Q}{2x+b}\right)\right]\right\}. \quad (6.4)$$

In Eq. (6.4), $X(x) = x^2 + bx + c$, $Q = (4c - b^2)^{1/2}$, $x = (3/4\pi\rho)^{1/6}$, and A, x_0, b and c are constants.

The LSDA replaces the inhomogeneous density of the system with a locally homogeneous electron gas. Typically the exchange energy is underestimated by about 10% and the correlation energy overestimated by $\sim 100\%$. The total energy is too high and the gap between occupied and unoccupied orbitals too low [13]. Perdew and Zunger [14] have proposed self-interaction corrections to the LSDA that have reduced the error.

The LSDA is based on a model of slowly varying density and the error introduced by this approximation will be more serious for highly localised systems than for delocalised free-electron like systems. The generalised-gradient approximation (GGA) attempts to overcome this by including the explicit dependence of the gradient of the charge density in the exchange-correlation functional. For some properties there is a tendency for the GGA to overcorrect the LDA results, and various GGA functionals have been developed to overcome this problem.

One of the most widely used GGA exchange-energy functionals is that introduced by Becke (B88) [15], which can be written as

$$E_x^{GGA} = E_x^{LDA} - \beta \sum_{\sigma=\alpha,\beta} \int d\vec{r}\rho_\sigma^{4/3} \frac{x_\sigma^2}{1 + 6\beta x_\sigma \sinh^{-1}(x_\sigma)}, \quad (6.5)$$

where x_σ is the dimensionless ratio

$$x_\sigma = \rho_\sigma^{-4/3}|\nabla\rho_\sigma|. \quad (6.6)$$

The GGA exchange functional of Perdew and Wang (PW86) [16] is also well known.

The GGA correlation functional of Perdew (P86) [17] is in common use and is often paired with the B88 exchange functional (as a pair: BP86). Other popular exchange and correlation functional pairs were introduced by Perdew and Wang (PW91) [18] and by Perdew, Burke and Ernzerhof (PBE) [19].

Generally GGA exchange-correlation energy functionals take the form

$$E_{xc}^{GGA}[\rho_\uparrow, \rho_\downarrow] = \int d\vec{r} f(\rho_\uparrow, \rho_\downarrow, \nabla\rho_\uparrow, \nabla\rho_\downarrow). \quad (6.7)$$

One way to go beyond this restricted GGA form is to include dependence on the Laplacian $\nabla^2 \rho_\sigma$ or the kinetic energy density [20]. Functionals that include this additional dependence are known as meta-GGAs. The Lee–Yang–Parr (LYP) correlation functional [21] is an early example of this type and is very popular paired with B88 (as a pair: BLYP). The LYP functional is expressed as

$$E_c = -a \int d\vec{r} \frac{\gamma}{1 + d\rho^{-1/3}} \left\{ \rho + 2b\rho^{-5/3} \left[2^{2/3} C_F \left(\rho_\alpha^{8/3} + \rho_\beta^{8/3} \right) - \rho t_W \right.\right.$$
$$\left.\left. + \frac{1}{9} \left(\rho_\alpha t_W^\alpha + \rho_\beta t_W^\beta \right) + \frac{1}{18} \left(\rho_\alpha \nabla^2 \rho_\alpha + \rho_\beta \nabla^2 \rho_\beta \right) \right] \exp\left(-c\rho^{-1/3} \right) \right\} \quad (6.8)$$

where C_F, a, b, c and d are constants,

$$\gamma = 2\left(1 - \frac{\rho_\alpha^2 + \rho_\beta^2}{\rho^2}\right), \quad (6.9)$$

and

$$t_W = \frac{1}{8} \frac{|\nabla \rho|^2}{\rho} - \frac{1}{8} \nabla^2 \rho. \quad (6.10)$$

Hybrid functionals have been proposed by Becke [22] as another way to go beyond the GGA. Basically the functionals are formulated as a mixture of Hartree–Fock and DFT exchange coupled with DFT correlation. B3LYP [23] is the most widely used hybrid scheme.

A number of comparisons have been made of the results of calculations using a wide range of functionals for metallic dimers [24–27] and small clusters [28–30]. There has been work on optimising basis sets for GGA calculations [31], and also a comparison of the predictions of GGA and MP2 and MP4 calculations [32–34]. A number of authors [24,25,27,29] find B3LYP disappointing for transition metals, compared with non-hybrid methods, particularly with regard to bond length prediction, and suggest that hybrid methods are best avoided for metal clusters. However, the non-hybrid methods generally overestimate binding energies. This being said, there is remarkably good agreement between the predictions of geometric configurations of clusters from DFT–GGA and MP4 calculations [32].

Geometry optimisation techniques similar to those of quantum chemistry calculations are used, but there is also the possibility of using DFT within a molecular dynamics procedure such as the Car–Parrinello method [35] to study the ground state; most calculations of this sort have been done on relatively small clusters (<20 atoms). The term *ab initio* molecular dynamics (AIMD) is used to describe techniques of this type (we make no comment on whether or not DFT is really *ab initio*).

6.2.3. *Tight-binding methods*

The tight-binding scheme provides a method that is far less costly computationally than the methods so far discussed, but still describes systems within a quantum mechanical framework. It allows one to study clusters that are inaccessible to *ab initio* calculations up to 1000 atoms or more in size. It was initially introduced to investigate bulk transition metals [36] and the inclusion of spin polarisation is straightforward allowing access to magnetic properties. It has been applied to most elemental bulk metals [37]. The inclusion of a molecular dynamics scheme is sometimes incorporated into tight-binding calculations to obtain relaxed configurations in clusters [38].

The tight-binding (TB) model is used principally, but not exclusively, in studies of the transition metals. It is described by a Hamiltonian expressed in terms of matrix elements in an orthogonal basis set built from the s, $p(x, y, z)$, $d(xy, yz, zx, x^2-y^2, 3z^2-r^2)$ valence atomic orbitals. Intersite matrix elements are determined by the 10 Slater–Koster [36] hopping integrals $ss\sigma$, $sp\sigma$, $sd\sigma$, $pp\sigma$, $pp\pi$, $pd\sigma$, $pd\pi$, $dd\sigma$, $dd\pi$, $dd\delta$, and are assumed to decay exponentially with distance. Three centre integrals are ignored.

Three on-site terms are used which depend just on the orbital angular momentum ($\lambda = s, p, d$). The usual repulsive energy can be incorporated into these elements if we let them vary as a function of the local "density" around each atom. The advantage of doing this is that it allows us to obtain the total energy by summing up the occupied energy levels [39].

$$\varepsilon_{i\lambda} = a_\lambda + b_\lambda \rho_i^{2/3} + c_\lambda \rho_i^{4/3} + d_\lambda \rho_i^2 \tag{6.11}$$

$$\rho_i = \sum_{j \neq i} \exp\left[-\gamma\left(\frac{r_{ij}}{r_0} - 1\right)\right]. \tag{6.12}$$

A cut-off is imposed on the distances included in the sum in Eq. (6.12); r_0 is the interatomic distance in the bulk. The Hamiltonian for the model is written in terms of annihilation and creation operators as

$$H = \sum_{i\lambda\sigma} \varepsilon_{i\lambda} c^\dagger_{i\lambda\sigma} c_{i\lambda\sigma} + \sum_{\substack{i \neq j \\ \lambda\mu\sigma}} \beta^{\lambda\mu}_{ij} c^\dagger_{i\lambda\sigma} c_{j\mu\sigma}, \tag{6.13}$$

where i, j and λ, μ are site and orbital labels respectively, and σ is a spin label; $\beta^{\lambda\mu}_{ij}$ are the hopping integrals.

The various parameters are usually obtained by fitting to bulk properties. The TB method gives unrealistic charge transfers in clusters (or any system with inequivalent atoms). A shift is added to the on-site terms in Eq. (6.13) to artificially produce charge neutrality and a subtraction has to be included in the expression for the total energy to account for this. Generally the results obtained are rather insensitive to the precise value of the shift used.

Minor details vary between calculations [38,40–42] in attempts to improve the fit to data and the transferability between systems. Lathiotakis et al. [38], for example, add an extra orbital (excited s^* state). Mercer and Chou [43] noted that the overlap of wave functions on neighbouring atoms resulted in on-site terms in the Hamiltonian that were off-diagonal in the orbitals. In a cubic bulk solid the terms cancel and they were unable to evaluate them, but Xie and Blackman [41] managed to obtain these elements through DFT calculations on small clusters and include them in TB cluster calculations.

Spin-polarised systems are usually treated using a Hubbard model framework within the Hartree–Fock approximation. This results in on-site terms additional to Eq. (6.13), of which the principal one is

$$H_{\text{int}} = -\bar{J}_\lambda \sum_{i\lambda\sigma} \sigma c^\dagger_{i\lambda\sigma} c_{i\lambda\sigma}. \tag{6.14}$$

where

$$\bar{J}_\lambda = \frac{1}{2}\sum_\mu J_{\lambda\mu}(\bar{n}_{i\mu\uparrow} - \bar{n}_{i\mu\downarrow}). \qquad (6.15)$$

J is an exchange integral and $\bar{n}_{i\mu\sigma}$ is the component of electron density at site i associated with orbital μ and spin σ. Various refinements to this basic picture can be included [44–47].

6.2.4. Semi-empirical potentials and global optimisation

Quantum mechanics in any guise is inadequate for studying very large clusters, and semi-classical potentials along with various geometry optimisation techniques are used to extend the size range to thousands of atoms. It is essential that the methods are economical computationally while maintaining sufficient accuracy over sizes ranging from the very large down to those that overlap with the regime accessible quantum mechanical techniques.

Empirical or semi-empirical potentials have been developed that can be used with Monte Carlo or molecular dynamics simulations. Pair potentials (Lennard-Jones or Morse) have been used extensively for the rare gases and, as we shall see in Section 6.4, the range of the potential can have a marked effect on the structures obtained [48,49]. However, for metals, the interactions are not pairwise and it is necessary to choose potentials containing a many-body contribution. A number of schemes have been developed: embedded atom model (EAM) [50,51], effective medium theory (EMT) [52,53], glue model [54], second moment methods [55–58].

The various methods start with a common functional form for the total energy

$$E_{\text{tot}} = \sum_i \left[\frac{1}{2} \sum_{j \neq i} V(r_{ij}) + F(\bar{\rho}_i) \right], \qquad (6.16)$$

where the first term is a repulsive pair-potential between atoms separated by a distance r_{ij}. In the EAM, the first term is an embedding energy. Each atom of the metal is considered as an impurity embedded in a host provided by the rest of the electrons, and $F(\bar{\rho}_i)$ is the energy required to embed atom i in an electron gas of density $\bar{\rho}_i$ created by its neighbour's electronic charge

$$\bar{\rho}_i = \sum_{j \neq i} \rho(r_{ij}). \qquad (6.17)$$

This many-body energy term, which gives a metallic-bond character to the potential, means that an atom can compensate for a coordination deficit by reducing its interatomic distances. The embedding functions and pair interactions can be obtained only approximately from first principles calculations and so empirical fits to bulk properties have been used to generate them in numerical form [50,51].

The second moment approximations [55,59] for the total energy can be written as

$$E_{\text{tot}} = \varepsilon \sum_i \left[\sum_{j \neq i} V(r_{ij}) - c\sqrt{\rho_i} \right], \tag{6.18}$$

where ε is a parameter with the dimensions of energy and c a dimensionless parameter. Formally Eq. (6.18) is just a special case of Eq. (6.16), with $V(r)$ the pair potential in dimensionless units. The square root form for the many-body term is characteristic of the second moment approximation and ρ_i is now expressed as a sum of pair potentials associated with site i

$$\rho_i = \sum_j \phi(r_{ij}). \tag{6.19}$$

This term, which is the counterpart of the embedding energy in the EAM, can now very explicitly be described as bond energy.

The outline of the argument [55,59] leading to the form for the bond energy is as follows. For simplicity consider just one orbital site and local density of states $d_i(E)$ at site i with centre of gravity ε_i. The bond energy and the second moment of the local density of states are given respectively by

$$E_{\text{bond}} = 2 \sum_i \int_{-\infty}^{E_F} (E - \varepsilon_i) d_i(E) dE. \tag{6.20}$$

and

$$\mu_i^{(2)} = \int_{-\infty}^{\infty} (E - \varepsilon_i)^2 d_i(E) dE. \tag{6.21}$$

The result for the bond energy in an approximation in which no moments higher than the second are used is

$$E_{\text{bond}}^{(i)} = \text{constant} \times \sqrt{\mu_i^{(2)}}. \tag{6.22}$$

This dependence of $E_{\text{bond}}^{(i)}$ on the second moment can be seen easily [59] if we represent the local density of states, $d_i(E)$, by a Gaussian centred at ε_i and with width $\sqrt{\mu_i^{(2)}}$. The condition of charge neutrality is invoked to confirm that other factors arising from the Gaussian integration of Eq. (6.21) are site-independent.

The local density of state can be written as

$$d_i(E) = \langle i | \delta(E - H) | i \rangle, \tag{6.23}$$

where H is the Hamiltonian for the system and $|i\rangle$ the orbital on-site i; the second moment is

$$\mu_i^{(2)} = \int_{-\infty}^{\infty} (E - \varepsilon_i)^2 \langle i | \delta(E - H) | i \rangle dE$$

$$= \langle i | (\varepsilon_i - H)^2 | i \rangle. \tag{6.24}$$

6.2. Theoretical methodology

Writing the Hamiltonian in terms of on-site energies and hopping integrals

$$H = \sum_i \varepsilon_i |i\rangle\langle i| + \sum_{i \neq j} \beta_{ij} |i\rangle\langle j|, \qquad (6.25)$$

it can be seen that the second moment can be expressed in terms of the hopping integrals

$$\mu_i^{(2)} = \sum_{j \neq i} \beta_{ij}^2. \qquad (6.26)$$

No directional dependence is included in the second moment approximation and the hopping integrals β_{ij} depend only on the distance r_{ij}, which is written as $\phi(r_{ij})$ in Eq. (6.19). The important point though is that environmental dependence is included in the bond energy part of the potentials of Eq. (6.18) through Eq. (6.19). The potentials are designed for transition metals where bonding involving the d-orbitals is dominant. The simple single orbital description given above is generalised to an average hopping for the five d-orbitals.

The second moment approximation is very effective for metals that tend to form close-packed structures. However it is not able to account for systems preferring bcc or hcp structures. It is the fourth moment that controls the stability of the bcc structure, and the fifth and sixth moments are required if hcp and fcc packing are to be differentiated [59].

Fairly simple functional forms are generally used to parameterise $V(r)$ and $\phi(r)$ in Eqs. (6.18) and (6.19). Commonly used potentials are those introduced by Sutton and Chen [56]

$$V(r) = \left(\frac{1}{r}\right)^n \qquad (6.27)$$

$$\phi(r) = \left(\frac{1}{r}\right)^m, \qquad (6.28)$$

where m and n are positive integers with $n > m$. Sutton and Chen [56] provide the following set of parameters: $n = 12$, $m = 6$ for Ag and Rh, $n = 10$, $m = 8$ for Au and Pt, $n = 9$, $m = 6$ for Ni and Cu. Results within a (m, n) group are generic and moving from one member of the group to the other simply involves rescaling the ε and a parameters.

The Gupta potential [58] is an alternative to that of Sutton and Chen. Again it is of the second moment type but is based on an exponential parameterisation.

$$E_{\text{tot}} = \sum_i [V_{\text{rep}} + V_{\text{bond}}], \qquad (6.29)$$

$$V_{\text{rep}}(i) = \xi \sum_{j \neq i} \exp\left[-p\left(\frac{r_{ij}}{r_0} - 1\right)\right], \qquad (6.30)$$

$$V_{\text{bond}}(i) = -\zeta \left\{\sum_{j \neq i} \exp\left[-2q\left(\frac{r_{ij}}{r_0} - 1\right)\right]\right\}^{1/2}. \qquad (6.31)$$

Once a potential has been chosen, there are a number of global optimisation techniques that can be used to find the lowest energy of a cluster. Some algorithms can be incorporated into standard Monte Carlo or molecular dynamics procedures. In general, for a given number of atoms, N, and a particular potential energy model, there are a huge number of local energy minima and the task of finding the deepest minimum is formidable. Indeed there are indications that the number of minima increases exponentially with N, reaching at least 10^{12} for a 55-atom cluster [60]. Nonetheless good putative global minima have been obtained for clusters well above 100 atoms.

Techniques include the basin-hopping algorithm [61–64], genetic algorithms [65–71], simulated annealing [72,73] and quantum annealing [73–75]. A good review of the issues surrounding global optimisation is provided by Baletto and Ferrando [3] and a comprehensive account is given in the book by Wales [76].

6.3. Overview of specific materials

6.3.1. Introduction

In this section we summarise the results for structure calculations on a range of metals. Numerous approaches are used and the literature on the subject is vast. The reviews of Bonačić-Koutecký et al. [1,2] provides a survey of highly accurate calculations on small clusters ($N \leq 10$) and Baletto and Ferrando [3] and Alonso [77] discuss the more recent work on a range of metal clusters. The Cambridge Cluster Database [78] lists coordinates of the global minimum energy geometries of clusters using a number of different semi-empirical potentials.

A number of the commonly occurring cluster motifs are illustrated in Figure 6.1. We show symmetric configurations. Generally there is some departure from symmetry. For example, the triangular configuration in 3b is usually isosceles rather than equilateral, and linear and planar clusters generally show some departure from pure one or two dimensionality. Nevertheless the basic motif is usually apparent.

The structural behaviour is strongly element dependent. The size at which three-dimensional (3D) clusters occur varies considerably. Nickel, for example, adopts the tetrahedral structure at $N = 4$, while gold clusters remain planar until N is larger than 10. Alkali metals are intermediate between these two extremes. Cations and anions often display a different structure to the neutral cluster. There is the interesting question as to whether the magic numbers that arise in the jellium model reappear in the detailed structural calculations. There is particular interest at special sizes since the geometric shell model suggests that magic number clusters, such as 13, 55, 147 (icosahedral or cuboctahedral), 38 (truncated octahedral) or 75, 146, 192 (Marks decahedral), should be particularly stable.

We discuss separately the behaviour found in different groups of elements: alkalis, noble metals, transition elements, and divalent and trivalent metals. For each group we provide a basic bibliography (inevitably incomplete) to aid readers who wish to explore particular systems in more detail.

6.3. Overview of specific materials

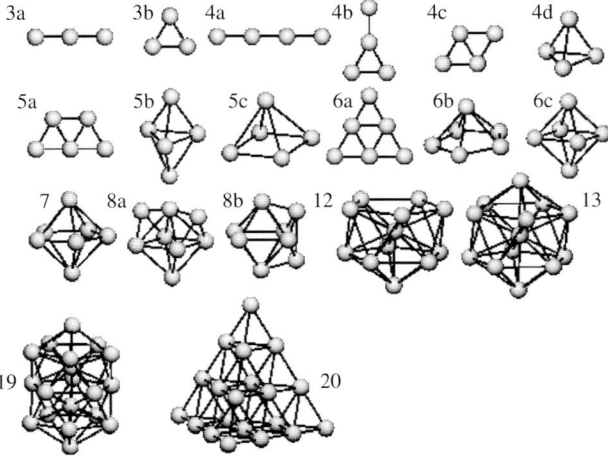

Figure 6.1. Some commonly occurring small clusters (schematic). The 3D figures are: tetrahedron (4d), trigonal bipyramid (5b), square pyramid (5c), pentagonal pyramid (6b), octahedron (6c), pentagonal bipyramid (7), bicapped octahedron (8a), bisdisphenoid (8b), incomplete icosahedron (12), icosahedron (13), double icosahedron (19), large tetrahedron (20).

6.3.2. Alkali metals

Table 6.1 lists representative work on the alkali metals that covers a range of cluster sizes and calculational techniques. The fairly recent work of Solov'yov et al. [32] gives a comprehensive bibliography and a detailed comparison of the results by different methods: MP4 and the BLYP density functional method with gradient-corrected functionals.

For neutral Li, Na and K there is agreement [32,79,80,83,84,93] on planar geometry for $3 \leq N \leq 5$: isosceles triangle, rhombus, trapezoid. The six-atom cluster is an intermediate case between the planar structure of the smaller clusters and a 3D form for the larger ones. A planar triangular isomer and a pentagonal pyramid are very close in energy [32,79,80], while for $N = 7$ the 3D pentagonal bipyramid is clearly the lowest energy structure.

The Li_8, Na_8, Li_{20} and Na_{20} clusters have a higher point group symmetry (tetrahedral: T_d) than their neighbours [32,79,80,86,88], which corresponds to the high symmetry (spherical shape) in the 8 and 20 magic number clusters of the electronic shell model. Peaks are also found in the binding energy per atom at $N = 8, 20$. On the other hand, Röthlisberger and Andreoni [84] and Spiegelmann et al. [93] found structures based on pentagonal symmetry at $N = 20$. At $N = 13$, the most stable isomer is neither an icosahedron nor a cuboctahedron but rather a capped structure based on pentagonal symmetry [32,84]. The double icosahedron is preferred at $N = 19$ however.

Calculations have been performed [32,79,81,83,98] for the ionised clusters. There are a number of marked differences between their structure and those of the neutral clusters for $N < 10$, but the differences are much less pronounced in the larger clusters. The Li_3^- anion is linear, whereas the cation and neutral clusters

Table 6.1. Calculations on structure of alkali clusters.

Authors	Method(s)	Elements	Size range	Ch.	Props.
Boustani et al. [79]	QC	Li	≤ 9	n, c	ip
Bonačić-Koutecký et al. [80]	QC	Na	≤ 9	n	ip
Bonačić-Koutecký et al. [81]	QC	Na	≤ 9	c	ip
Boustani and Koutecký [82]	QC	Li	≤ 9	a	ea
Spiegelmann and Pavolini [83]	QC	Na, K	≤ 6	n, c	ip
Bonačić-Koutecký et al. [1]	QC	Na	≤ 9	n, c	
Röthlisberger and Andreoni [84]	AM	Na	≤ 20	n	
Röthlisberger et al. [85]	AM	Na	13	n	
Bonačić-Koutecký et al. [86]	QC	Na	20	n	
Poteau and Spiegelmann [87]	TB	Na	4–34	n	
Jellinek et al. [88]	AM	Li	8	n	
Sung et al. [89]	AM	Li	≤ 147	n	
Reichardt et al. [90]	AM	Li	8	n	
Jellinek et al. [91]	AM	Li	8	n	
Li et al. [92]	GO	Na, K, Cs, Rb	9–309	n	
Spiegelmann et al. [93]	TB	Na	3–20, 34, 40	n	
Rayanne et al. [94]	DFT	Li, Na	≤ 22	n	sp
Nogueira et al. [95]	DFT	Na	$\leq 9, 21$	n, a, c	ip
Kronik et al. [96]	DFT	Na	≤ 20	n	sp
Kümmel et al. [97]	AM	Na	8–40	n	sp
Kümmel et al. [98]	DFT	Na	≤ 59	c	pa
Ishikawa et al. [99]	QC	Li	6	n	
Lai et al. [100]	GO	Na, K, Cs, Rb	3–56	n	
Solov'yov et al. [32]	QC, DFT	Na	≤ 20	n, c	mm, sp, ip, f
Chandrakumar et al. [33]	QC, DFT	Na	≤ 10	n	sp
Ali et al. [34]	QC, DFT	Cs	≤ 10	n, c	ip, ea

Notes: Elements included in the work and the size range are noted. Abbreviations are: Method: QC (quantum chemistry: CI, MP$_n$, CC); DFT (density functional theory, various GGA); AM (*ab initio* MD/MC such as Car–Parrinello); TB (tight-binding and TBMD); GO (second moment or EAM potential plus some form of global optimisation). Charge (Ch.): neutral (n), cation (c), anion (a). Properties (Props.) calculated in addition to structure and binding energies: ip (ionisation potential), ea (electron affinity), sp (static dipole polarisability), mm (multipole moments), f (frequencies of vibrational modes), pa (photoabsorption spectra), pe (photoemission spectra).

are triangular [79,82]. For $N = 4$, the anion is again linear, the neutral cluster is a rhombus (as noted above) and the cation has a structure like 4b in Figure 6.1 [79,82]. Solov'yov et al. [32] find the 4b structure and the rhombus very close in energy for the Na$_4^+$ cation. For the $N = 5$ cation, the trigonal bipyramid is close in energy to a planar isomer, with the former being preferred for Li$_5^+$ while Na$_5^+$ adopts a planar structure [32,79,81]. The lithium anion is also planar [82].

Most of the work on larger clusters is based on semi-classical potentials and some global optimisation technique. Lai et al. [100] used Gupta potentials and both the basin-hopping method and genetic algorithm technique to study Na, K, Rb and Cs clusters for $3 \leq N \leq 56$. There is some discrepancy for $N \leq 20$ between the structures found by Lai et al. [100] and those from the *ab initio* calculations. In particular Lai et al. find non-planar structures for 4 and 5 atom clusters and a highly stable Mackay icosahedron for 13 atoms. The potential-based methods tend to unduly favour close-packing at small sizes, but we expect them to be reasonably reliable for larger clusters.

A sequence of particularly stable clusters is found [100] for $N = 19, 23, 26, 29, 32, 34$ and 55 for all four alkali metals (Na, K, Rb, Cs). Generally they find that the growth pattern is icosahedral for all four metals and the 19 atom double icosahedron can be regarded as the seeding unit with atoms added to sites so that there is a tendency to form tetrahedral pyramids, pentagonal pyramids or fused pentagonal pyramids. As noted above, the 19 atom double icosahedron is also the favoured structure in the *ab initio* calculations.

There are a few cluster sizes at which there is divergence from a pattern common to all the alkali metals. Cs_{21} is an exception in being formed by adding six-coordinate atoms to a Cs_{15} seed unit, and Na_{44} Na_{45} and Na_{49} have a structure that is different from the other three metals. The only departures from the icosahedral growth pattern occur at $N = 36$ and 38. The former is a distorted incomplete truncated octahedron. For $N = 38$, K and Rb form truncated octahedrons with hexagonal and square faces, while the truncated octahedral of the Na and Cs clusters have hexagonal and rhombus faces.

6.3.3. Noble metals

We list representative work by a range of techniques on Cu, Ag and Au in Table 6.2. The format and abbreviations are as in Table 6.1 for the alkali metals. Copper and silver are rather similar in their behaviour, but gold shows marked differences from the other two metals, not least in the size of cluster at which a transition from 2- to 3D geometry occurs. The minimum size of cluster at which the transition occurs according to various authors is summarised in Table 6.3, which is based on the compilation of Fernández *et al.* [135] with some additions. Fernández *et al.* [135] perform DFT-GGA calculations to study the geometries of Cu, Ag and Au as neutrals, cations and anions for $N \leq 13$ and $N = 20$ and provide a comprehensive comparison of their results with those of other authors. Note that the figures represent the smallest size at which a 3D cluster could occur according to a particular set of calculations and not necessarily the size at which it does occur. For example, Bonačić-Koutecký *et al.* [120] found planar geometry for Au in calculations done over the range $N = 2$–10; in this case therefore Au_{11} represents the smallest size for 3D geometry rather than a definite assignment.

It seems likely that neutral gold clusters are planar to at least $N = 10$, whereas silver and copper are planar only to $N = 5$ or 6. For the Cu and Ag cations and anions, the size at which the transition occurs differs from that for the neutral cluster by 1 at the most, while for the Au cation in particular the transition appears to take place at a significantly lower size than that for the neutral or anionic cluster. The structure of neutral Au clusters to $N = 10$ according to Bonačić-Koutecký *et al.* [120] is shown in Figure 6.2.

The structures are similar up to $N = 5$ for all three metals. A number of authors [117,118,125,126,131,135] find the Cu_6 and Ag_6 take the triangular structure shown for Au_6 in Figure 6.2, with a transition to a pentagonal bipyramid for Cu_7 and Ag_7, while others [108,137] find a pentagonal pyramid at $N = 6$. Generally the structures

Table 6.2. Calculations on structure of noble metal clusters.

Authors	Method(s)	Elements	Size range	Ch.	Props.
Bauschlicher et al. [101]	QC	Cu	≤ 3	n, a	ea, f
Bauschlicher et al. [102]	QC	Cu, Ag, Au	2, 3	n, a	ea
Fujima and Yamaguchi [103]	DFT	Cu	≤ 19	n	
Bauschlicher et al. [104]	QC	Cu, Ag, Au	4, 5	n	ip, ea
Bonačić-Koutecký et al. [105]	QC	Ag	≤ 9	n, c	ip
Bonačić-Koutecký et al. [106]	QC	Ag	≤ 9	a	ip
Santamaria et al. [107]	DFT	Ag	≤ 6	n	ip
Massobrio et al. [108]	AM	Cu	≤ 4, 6, 8, 10	n	
Cleveland et al. [109]	GO	Au	≤ 520		
Jennison et al. [6]	DFT	Ag	55, 135, 140		
Mottet et al. [110]	GO	Cu, Ag, Au	Large	n	
Doye and Wales [63]	GO	Cu, Ag, Au	≤ 80	n	
Garzón et al. [111]	GO	Au	38, 55, 75	n	
Bravo-Pérez et al. [112]	QC	Au	≤ 6	n	f
Bravo-Pérez et al. [113]	QC	Au	3–6	n	
Erkoç and Yilmaz [114]	GO	Ag	3–177	n	
Michaelin et al. [68]	GO	Ag, Au	6–75		
Grönbeck and Andreoni [115]	DFT	Au	≤ 5	n	ea
Häkkinen and Landman [116]	DFT	Au	≤ 10	n, a	
Bonačić-Koutecký et al. [117]	QC	Ag	5–8	n	pa
Fournier [118]	DFT	Ag	≤ 12	n	
Xie and Blackman [41]	DFT, TB	Ag	13, 55, 135		
Baletto et al. [119]	GO	Cu, Ag, Au	Large	n	
Bonačić-Koutecký et al. [120]	DFT	Au	≤ 10	n	ip, f
Furche et al. [121]	AM	Au	3–13	a	
Garzón et al. [122]	GO	Au	28, 55, 75	n	
Gilb et al. [123]	DFT	Au	3–13	c	
Häkkinen et al. [124]	DFT	Cu, Ag, Au	7	a	pa
Jaque and Toro-Labbé [125]	DFT	Cu	≤ 9	n	ip, sp
Jug et al. [126]	DFT	Cu	≤ 10	n, a, c	ip, ea
Oviedo and Palmer [127]	DFT	Cu, Ag, Au	13	n	
Wang et al. [128]	DFT	Au	≤ 20	n	ip
Weis et al. [129]	QC, DFT	Ag	≤ 11	c	
Häkkinen et al. [130]	DFT	Au	4–14	a	pe
Lee et al. [131]	DFT	Ag, Au	≤ 13	n, a	ip, ea
Matulis et al. [132]	DFT	Ag	≤ 10	n, a	ip
Wang et al. [133]	DFT	Cu, Ag, Au	20	n	
Zhao et al. [134]	DFT	Au	≤ 20	n, a, c	mm, sp
Fernández et al. [135]	DFT	Cu, Ag, Au	≤ 13, 20	n, c, a	ip, ea
Häkkinen et al. [136]	GO, DFT	Cu, Ag, Au	53–58	a	pe
Kabir et al. [137]	TB	Cu	≤ 55	n	ip
Shao et al. [138]	GO	Ag	≤ 80	n	
Häkkinen and Moseler [139]	DFT	Ag, Au	55	a	pe
Koskinen et al. [140]	TB, DFT	Au	4–14	a	
Xiao et al. [141]	DFT, GO	Au	≤ 55	n	
Yang et al. [142]	DFT	Cu	8–20	n	ip
Yang et al. [143]	DFT	Ag	9–20	n	ip, sp
Chui et al. [144]	GO	Au	Large	n	

Notes: Elements included in the work and the size range are noted. Abbreviations are: Method: QC (quantum chemistry: CI, MP_n, CC); DFT (density functional theory, various GGA); AM (*ab initio* MD/MC such as Car–Parrinello); TB (tight-binding and TBMD); GO (second moment or EAM potential plus some form of global optimisation). Charge (Ch.): neutral (n), cation (c), anion (a). Properties (Props.) calculated in addition to structure and binding energies: ip (ionisation potential), ea (electron affinity), sp (static dipole polarisability), mm (multipole moments), f (frequencies of vibrational modes), pa (photoabsorption spectra), pe (photoemission spectra).

Table 6.3. Minimum size of cluster at which onset of 3D geometry occurs according to various calculations (as referenced).

Cu⁻	Ag⁻	Au⁻	Cu	Ag	Au	Cu⁺	Ag⁺	Au⁺
6[a,b]	6[a,c], 7[d]	≥14[d,e,f], 13[a], 7[g]	7[a,b,h], 6[i,j]	7[a,d,k,l], 6[m]	11[n], 12[a], ≥14[d,o], 7[p], 8[g]	5[a,b]	5[m,q], 6[a]	8[a], 9[r]

[a]Ref. [135].
[b]Ref. [126].
[c]Ref. [105].
[d]Ref. [131].
[e]Ref. [121].
[f]Ref. [130].
[g]Ref. [116].
[h]Ref. [125].
[i]Ref. [137].
[j]Ref. [108].
[k]Ref. [117].
[l]Ref. [118].
[m]Ref. [106].
[n]Ref. [120].
[o]Ref. [141].
[p]Ref. [128].
[q]Ref. [129].
[r]Ref. [123].

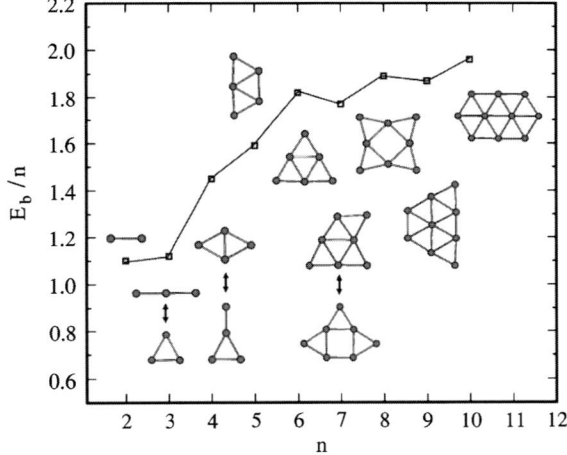

Figure 6.2. Binding energies per atom, $E_b/n = [E(Au_n) - nE(atom)]/n$, for neutral Au_n clusters as a function of cluster size ($n = 2$–10). For $n = 3, 4, 7$ there are two structural isomers with energy differences smaller than 0.1 eV. Both are shown. Reproduced with the permission of the American Institute of Physics from Bonačić-Koutecký et al. [120].

for Cu and Ag are similar but there are occasional differences, for example, at $N = 9$ and 12 [135].

There have been a number of proposals for the origin of the tendency to planar ordering that is peculiar to Au. Perhaps the most compelling indication that gold is special is the reminder by Häkkinen et al. [136] of Pyykkö's [145] comment that gold is the most relativistic element below fermium. The relativistic effects result in a strong contraction of the outer $6s$ shell and a reduced $5d$–$4s$ energy gap. This leads to significant hybridisation of the atomic $5d$ and $6s$ orbitals and direct d–d bonding effects which favour planar structures [124,136].

Another feature of the noble metals that is surprising is exemplified in Figure 2.6 (in Section 2.2.4). Silver and copper show features that are reminiscent of the alkali metals. In particular, there is evidence of magic numbers in the mass spectra that are characteristic of the shell model as can be seen in the Figure 2.6. Fujima and

Yamaguchi [103] have performed DFT calculations on a range of Cu structures. They found that the 3d orbitals are quite localised around the individual atoms, whereas the charge deriving from the 4s and 4p orbitals extends over the whole cluster. The states that have predominantly s–p character have energy levels that show a correspondence with the shell model, while the 3d bands are full and fall between these states. There is a one-to-one correspondence therefore between the filling of shells in the electronic shell model and in the DFT calculations. Massobrio et al. [108] also used DFT calculations to explore the correspondence with the shell model for Cu clusters. They found that the ground state symmetries of Cu_N and Na_N clusters are the same, but the Cu clusters tend to be more compact. They also found significant s–d hybridisation and were able to relate their results to the shell model only qualitatively.

At $N = 13$, the cuboctahedron has lower energy than the icosahedron for Ag [41,127] while, for copper, the icosahedron is the more stable [103]. However, it has been suggested that the lowest energy isomers of Ag_{13} are actually disordered [127]. The icosahedron is the more stable structure for both copper and silver at the next geometric magic number, 55 [6,41,136,137,139]. For Au_{55}, however a low symmetry structure rather than the icosahedron is found to be the ground state [136].

Larger clusters have been studied mainly using semi-empirical potentials. Doye and Wales [63] have used Sutton–Chen potentials and the basin-hopping algorithm to obtain the global minima for clusters of up to 80 atoms. They find the same lowest energy structures for Ag_7, Ag_8 and Ag_9 as Bonačić-Koutecký [105], although the method does not reproduce the planar structures of Au. Doye and Wales [63] were able to make specific structural assignments to a significant number of the Ag clusters. Of the 72 clusters ($N = 9–80$) studied, 30 had icosahedral character, 23 were based on decahedra and 13 were close packed (fcc, hcp or a mixture of stacking sequences); fewer assignments were possible with the other two metals. In all three metals the 38 atom cluster has a high symmetry octahedral structure and is particularly stable. Cu and Ag exhibit magic number icosahedral symmetry at $N = 13$ and 55. The Ag_{13} result is in disagreement with the behaviour coming from DFT calculations. Other approaches have used semi-empirical potentials and optimisation as a starting point followed by local relaxation by the DFT method [111,68,136,139] and have found the low symmetry structures of Au at $N = 38$ and 55.

6.3.4. *Ferromagnetic transition metals (Fe, Co, Ni)*

These materials are ferromagnetic in bulk, and spin polarisation and structure are intimately related in clusters. We will explore these issues in detail in Chapter 8. For the moment it will suffice to note that 3D geometry occurs for each of these elements as early as it can do, namely at $N = 4$. Most calculations propose that the structures ($N = 3–8$) are triangle, tetrahedron, trigonal bipyramid, octahedron, pentagonal bipyramid, bicapped octahedron, with icosahedral symmetry established at $N = 12, 13$. The other elements (Cr and Mn) that show complex magnetic ordering in bulk are also discussed in Chapter 8.

6.3.5. Transition metals

A representative bibliography of some transition metal elements is shown in Table 6.4, with Ti and V from the 3d series, Nb and Pd from 4d and Pt from 5d. Some of studies include a calculation of a magnetic moment and this is noted in the table. There has been considerable interest in the possibility that a material that is non-magnetic in bulk could exhibit a magnetic moment when it is in the form of a small cluster. Despite theoretical predictions of magnetic moments in many 4d and 5d metal clusters, the only proven case experimentally of a moment is in Rh clusters up to about 60 atoms in size. The only other example of 4d magnetism is in Ru monolayers on graphite. The Rh case is discussed in Chapter 8 and for the moment we simply note in Table 6.4 work that includes Rh.

The majority of calculations on neutral clusters predict that 3D geometry occurs with the tetrahedral structure at $N = 4$: Ti [164,169,172,182,184], V [149,160,168], Nb [147,150,159,170,178], Pd [7,148,152,161,162,165,167,175,181], Pt [148,158,174]. There is dissent from this general trend for V and Pt, however. The density-functional calculations of Grönbeck et al. [115,153] yielded planar geometry for V_4, Pt_4 and Pt_5. Similar work by Li et al. [177] also found planar geometry V_4, and Yang et al. [154] even found a planar structure in Pt clusters to $N = 6$.

Tetrahedral or distorted tetrahedral structures have been found for the ionic clusters: Ti_4^- [172], V_4^+ [160,180], Nb_4^- [151,155], Nb_4^+ [159,166], Pd_4^- [162,167]. However Li et al. [177] found planar geometry for the V_4 anion and cation, Majumdar and Balasubramanian [178] obtained planar geometry for Nb_4^-. Efremenko and Sheintuch [162] found that the Pd cation remained planar until $N = 6$ with 3D geometry setting in with a pentagonal bipyramid at Pd_7^+. Fortunelli [158] obtained a tetrahedral structure for Pt_4^- as well as for the neutral cluster, but a planar form for Pt_4^+, while Grönbeck and Andreoni [115] obtained a planar Pt_4^- consistent with their calculations on the neutral cluster.

Icosahedral or distorted icosahedral structures are found for $N = 13, 19, 55$: Ti_{13} [169,172,179,182,184], V_{13} [168], Ti_{19} and Ti_{55} [169,179,184], and a truncated octahedral for Ti_{38} [184] and an fcc structure for Ti_{43} [179]. Taneda et al. [168] find amorphous-like disordered structures for V for $N = 15$–17, 55, 147, 309, and Kumar and Kawazoe [170] find that the icosahedron is not a local minimum for Nb_{13}, but it rather relaxes to a structure with a threefold rotational axis. Wang et al. [171] also identify distortion to threefold symmetry or Ti_{13}.

Icosahedra, generally with some distortion, are found to be the lowest energy state of Pd_{13} [7,41,45,162,167,175], Pd_{55} [6–8,40,41,45,175] and Pd_{135} [6,41]. There is some disagreement above this size: a number of calculations find that Pd_{147} [7,8,40,45] adopts icosahedral geometry; Nava et al. [8] find that Pd_{309} prefers a cuboctahedral structure, while Barreteau et al. [40,45] find an icosahedral Pd_{309} with cuboctahderal geometry appearing at $N = 561$. Jennison et al. [6] find a switch to an fcc structure at Pd_{140}.

The situation with Pt is less clear. A number of calculations find preference for an icosahedral structure at Pt_{13} [63,119,174], while Watari and Ohnishi [157] found that the cuboctahdral structure is preferred (also for Pd_{13}). Yang et al. [154] found several low symmetry structures with energies lower than either the icosahedron or

Table 6.4. Calculations on some transition metal clusters (V, Ti, Nb, Pd, Pt).

Authors	Method(s)	Elements	Size range	Ch.	Props.
Sachdev et al. [146]	GO	Pt	5–60	n	
Goodwin and Salahub [147]	DFT	Nb	3–6	n	ip
Dai and Balasubramanian [148]	QC	Pt, Pd	4	n	
Zhao et al. [149]	TB	V	4–15	n	m
Grönbeck and Rosén [150]	DFT	Nb	≤10	n	ip, ea, pe
Kietzmann et al. [151]	DFT	Nb	3–8	a	pe
Valerio and Toulhoat [152]	DFT	Pd	≤4, 6	n	
Grönbeck and Rosén [153]	DFT	V	≤8	n	ip, ea
Jennison et al. [6]	DFT	Pd	55, 135, 140	n	
Yang et al. [154]	DFT	Pt	≤6, 13	n	f
Doye and Wales [63]	GO	Pt, Rh	≤80	n	
Fournier et al. [155]	DFT	Nb	3–8	a	f
Grönbeck et al. [156]	DFT	Nb	8–10	n, a, c	f
Watari and Ohnishi [157]	DFT	Pt, Pd	13	n	
Fortunelli [158]	DFT	Pt	≤4	n, a, c	
Fowler et al. [159]	DFT	Nb	3, 4	n, c	
Wu et al. [160]	DFT	V	≤9	n, c	ip, m
Zacarias et al. [161]	DFT, QC	Pd	≤4	n, c	ip, ea
Barreteau et al. [45]	TB	Pd, Rh	≤19, 55, 147	n	m
Barreteau et al. [40]	TB	Pd, Rh	13–561	n	
Efremenko and Sheintuch [162]	QC, DFT	Pd	≤13	n, a, c	
Grönbeck and Andreoni [115]	DFT	Pt	≤5	n, a	ea
Guirado-López et al. [163]	TB	Pd, Rh	9–165	n	
Wei et al. [164]	DFT	Ti	≤10	n	ip, m
Krüger et al. [165]	DFT	Pd	4–309	n	
Majumdar and Balasubramanian [166]	QC, DFT	Nb	3–5	c	f
Moseler et al. [167]	DFT	Pd	≤7, 13	n, a	ip
Taneda et al. [168]	TB	V	≤17, 55, 147	n	
Xie and Blackman [41]	DFT, TB	Pd	13, 55, 135	n	
Zhao et al. [169]	DFT	Ti	≤14, 19, 55	n	pe
Baletto et al. [119]	GO	Pd, Pt	Large	n	
Kumar and Kawazoe [7]	DFT	Pd	≤23, 55, 147	n	m
Kumar and Kawazoe [170]	DFT	Nb	≤23	n	
Wang et al. [171]	DFT	Ti	13	n, a, c	m
Castro et al. [172]	DFT	Ti	3–8, 13	n, a	pe, m
Fortunelli and Aprà [173]	DFT	Pt	13, 38, 55	n, a, c	ip, ea
Nava et al. [8]	DFT	Pd	≤309	n	
Sebetci and Güvenc [174]	GO	Pt	≤21	n	
Zhang et al. [175]	DFT	Pd	≤13, 19, 55	n	m
Aprà et al. [176]	DFT	Pt	55	n	
Li et al. [177]	DFT	V	≤8	n, a, c	ip, ea
Majumdar and Balasubramanian [178]	QC, DFT	Nb	4, 5	n, a	ip, ea
Wang et al. [179]	DFT	Ti	13, 19, 43, 55	n	m
Ratsch et al. [180]	DFT	V	3–15	c	f
Rogan et al. [181]	DFT, GO	Pd	≤13	n	
Salazar-Villanueva et al. [182]	DFT	Ti	≤15	n	ea
Walsh [183]	GO, DFT	Nb	10	n, c	f
Joswig and Springborg [184]	GO	Ti	≤100	n	

Notes: Elements included in the work and the size range are noted. Abbreviations are: Method: QC (quantum chemistry: CI, MP_n, CC); DFT (density functional theory, various GGA); AM (*ab initio* MD/MC such as Car–Parrinello); TB (tight-binding and TBMD); GO (second moment or EAM potential plus some form of global optimisation). Charge (Ch.): neutral (n), cation (c), anion (a). Properties (Props.) calculated in addition to structure and binding energies: ip (ionisation potential), ea (electron affinity), sp (static dipole polarisability), mm (multipole moments), f (frequencies of vibrational modes), pa (photoabsorption spectra), pe (photoemission spectra), m (magnetic moment calculation).

the cuboctahedron, which accords with Sachdev et al. [146] who found low symmetry structures for both Pd_{13} and Pt_{55}. Fortunelli and Aprà [173] find a distortion to D_{4h} point group symmetry occurs at $N = 13$ but there is icosahedral symmetry at $N = 55$.

Pt is next to Au in the periodic table and the two elements have some similarities with some tendency towards low symmetry clusters. Doye and Wales [63] used Sutton–Chen potentials and a global optimisation scheme to model a range of metals. In fact, identical Sutton-Chen potentials (with $n = 10$ and $m = 8$ in Eqs. (6.27) and (6.28)) are used for Pt and Au. Doye and Wales [63] found a marked absence of signatures of icosahedral ordering except for Pt_{13} and low symmetry at Pt_{55}. A tendency towards amorphisation has been proposed by Aprà et al. [176]. They observed a rosette-like structural transformation at fivefold vertices in Pt_{55}, which results in low symmetry structures and it is suggested that a similar mechanism operates in Au clusters.

6.3.6. Divalent and trivalent metals

A representative bibliography of some divalent (Mg, Ca, Sr, Zn, Cd, Hg) and trivalent (Al) metal elements is shown in Table 6.5.

There is general agreement that the divalent elements from group 2 (Mg, Ca, Sr) [185,188–190,193–195] and from group 12 (Zn, Cd, Hg) [201,203–206] form 3D clusters with tetrahedral geometry at $N = 4$. The anionic clusters show similar behaviour [192,193], but there are departures from this pattern for the Mg cation [194]: Mg_3^+ and Mg_4^+ are both linear and 3D geometry occurs with a trigonal bipyramid at Mg_5^+.

However the feature that receives most attention in the divalent materials is the size-induced metal-insulator transition [218]. Since the divalent elements have an s^2 closed-shell atomic configuration like the rare gases, their dimers are weakly van der Waals bonded. The reason the elements are metallic in the bulk phase is the overlap of the s and p bands. There is interest then in studying the transition from van der Waals through covalent to metallic behaviour as a function of cluster size, defining a insulator to metal transition with possibly an intermediate semiconducting state.

The concept of an insulator to metal transition is, of course, relevant for all clusters. In the limit of a bulk metal the energy levels are quasi-continuous, but as the size of a cluster is reduced the separation of the discrete energy levels in the neighbourhood of the Fermi energy increases until the average separation becomes of the order of kT, which is a convenient demarcation of the transition between the two regimes. For the transition metals it will occur at somewhat smaller size than in the alkali metals because of the narrowness of the d- in comparison with the s-band. With the divalent elements, the overlap of the s and p bands is also necessary to establish incipient bulk behaviour.

A number of quantities are used to interrogate the transition to the metallic state, the most obvious being the energy gap between the highest occupied molecular orbital (HOMO) and the lowest unoccupied molecular orbital (LUMO). Nearest neighbour bond lengths can be expected to reduce as the character of the bonding

Table 6.5. Calculations on some divalent and trivalent metal clusters (Mg, Ca, Sr, Zn, Cd, Al).

Authors	Method(s)	Elements	Size range	Ch.	Props.
Kumar and Car [185]	AM	Mg	≤13	n	g
Röthlisberger et al. [85]	AM	Mg, Al	13	n	
Hearn and Johnston [186]	GO	Ca, Sr	≤20	n	
Wang et al. [187]	QC	Sr	≤13	n	ip, ea
Köhn et al. [188]	DFT	Mg	≤22	n	
Kumar and Kawazoe [189]	AM	Sr	≤35, 55, 147	n	g
Mirick et al. [190]	DFT	Ca	3–13	n	f
Wang et al. [191]	DFT, GO	Sr	≤63	n	
Acioli and Jellinek [192]	DFT	Mg	≤22	a	g
Jellinek and Acioli [193]	DFT	Mg	≤22	n, a	ip, g
Lyalin et al. [194]	DFT, QC	Mg	≤21	n, c	ip, g
Solov'yov et al. [195]	DFT	Mg	3–8	n	pa
Dong et al. [196]	TB	Ca	32–84	n, c	f, g
Lyalin et al. [197]	DFT	Sr	≤11	n, c	
Singh [198]	DFT	Hg	≤79	n	
Yonezawa and Tanikawa [199]	AM	Cd	≤20	n	ip
Yu and Doig [200]	QC	Zn, Cd, Hg	2	n	sp, ip
Flad et al. [201]	QC	Zn, Cd, Hg	≤6	n	ip, ea, f
Hartke et al. [202]	GO	Hg	7–14	n	
Zhao [203]	GO, DFT	Cd	≤21	n	ip
Michaelian et al. [204]	GO, DFT	Zn, Cd	13, 38, 55, 75, 147	n	
Moyano et al. [205]	DFT, QC	Hg	6, 13, 19, 23, 26, 29	n	sp
Wang et al. [206]	DFT	Zn	≤20	n	ip, f, g
Doye et al. [207]	GO	Zn, Cd	13–125	n	
Akola et al. [208]	DFT	Al	≤23	n	ip
Kumar [209]	AM	Al	14	n	
Rao and Jena [210]	DFT	Al	≤15	n, a, c	ip, ea, m
Akola et al. [211]	AM	Al	12–15	a	pe
Akola et al. [212]	AM	Al	19–102	n, a	ip, pe, g
Lloyd et al. [213]	GO	Al	21–51	n	
Doye et al. [214]	GO	Al	≤190	n	
Manninen et al. [215]	GO	Al	≤110	n	g
Yao et al. [216]	DFT	Al	19	n	g
Chuang et al. [217]	GO	Al	≤23	n	

Notes: Elements included in the work and the size range are noted. Abbreviations are: Method: QC (quantum chemistry: CI, MP$_n$, CC); DFT (density functional theory, various GGA); AM (*ab initio* MD/MC such as Car–Parrinello); TB (tight-binding and TBMD); GO (second moment or EAM potential plus some form of global optimisation). Charge (Ch.): neutral (n), cation (c), anion (a). Properties (Props.) calculated in addition to structure and binding energies: ip (ionisation potential), ea (electron affinity), sp (static dipole polarisability), mm (multipole moments), f (frequencies of vibrational modes), pa (photoabsorption spectra), pe (photoemission spectra), g (a HOMO–LUMO gap calculation).

changes from van der Waals to metallic. In Hg, for example, the interatomic distance reduces from 3.63 Å in the dimer to 3.60 Å in the bulk. The degree of *p*-character of the valence electronic charge will increase with the *s–p* hybridisation. Behaviour consistent with that predicted by the jellium model could be taken as indication of free electron (i.e. metallic-like behaviour), and so departures from it at small sizes are often used to locate the size at which the transition occurs. Properties studied from this point of view include the abundance spectra and the IP.

Experimentally, energy gaps can be extracted from photoelectron spectroscopy (PES) measurements and attempts are often made to relate these to calculations of the HOMO–LUMO gap. The IP and PES techniques are discussed in Chapter 7.

There is a reasonable consensus that the transition to metallic behaviour occurs around $N = 20$ for the majority of the divalent elements. Hg is an exception though and the estimates for this material vary wildly: $N = 80$ and 135 in the calculations of Singh [198] and Pastor et al. [219,220] respectively, and $N = 70$ and ~ 400 respectively in the experiments of Rademann [221] and Busani et al. [222]. Part of the difficulty in assigning a definite size for the transition is absence of a clear transition in a physical quantity. The HOMO–LUMO gap does fall off with cluster size but it does not do so monotonically. The fractional p-character of the valence charge electronic charge increases with cluster size, but there is no definite signature to indicate at what point the metallic state has been reached.

The PES experiments of Thomas et al. [223] on Mg possibly give a clearer picture however (see Section 7.2.3(iv)). As in the work of Busani et al. [222] on Hg clusters, they performed experiments on anionic clusters. The electronic structure of the Mg_N^- clusters involves the $3s^2$ molecular orbitals and a single electron in the otherwise empty $3p$ orbitals. The energy difference between the highest of the $3s^2$ orbitals (HOMO) and the singly occupied $3p$ orbital (LUMO for Mg_N) is a gap that can be monitored by the separation of peaks in the photoelectron spectra. The gap becomes smaller with increasing cluster size and the point at which it vanishes marks the overlap of the s- and p-bands. Thomas et al. [223] observe that gap closure occurs at $N = 18$, but with a number of reopenings at certain larger values of N, in particular 20 and 35. The experiments are also correlated with magic numbers in the mass spectra, with peaks at many of the positions predicted by the jellium model. This makes sense if we regard the $3s^2$ electrons as ones that can be treated by the electronic shell model. It also offers an explanation of the gap reopenings observed at $N = 20$ and 35. Adding one electron to a closed electronic shell involves a significantly bigger energy gap than the addition to a partially filled shell. Since Mg is divalent material, 20 and 35 correspond to the shell closing magic numbers of 40 and 70 of the electronic shell model (see Section 2.2.1). The metallic transition at a similar cluster size has also been deduced by Diederich et al. [224] by interpreting features in their Mg cluster mass spectra in terms of electronic shell model magic numbers.

Theoretical calculations by Jellinek and Acioli [192,193] provide support for the model. They propose that identifying the observed gap with a calculated HOMO–LUMO gap is not correct and that certain corrected terms should be included. They also point out that it is important to perform the calculation in the structure of the anion. Their results are compared with the experiments [223] in Figure 6.3(a) and it can be seen that the theory reproduces the vanishing gap. The HOMO–LUMO gap, as normally calculated, is shown in panel (b).

We now consider briefly trivalent Al clusters from group 13. The transition to 3D geometry at small sizes appears to occur at $N = 6$. It is reported than Al_N adopts planar rhomboidal and trapezoidal structures at $N = 4$ and $N = 5$ respectively and Al_6 is octahedral [208,210,217].

Band overlap is important also in Al clusters. The electronic structure of an Al atom is $[Ne]3s^2 3p^1$, and the energy separation between the $3s$ and $3p$ orbitals is

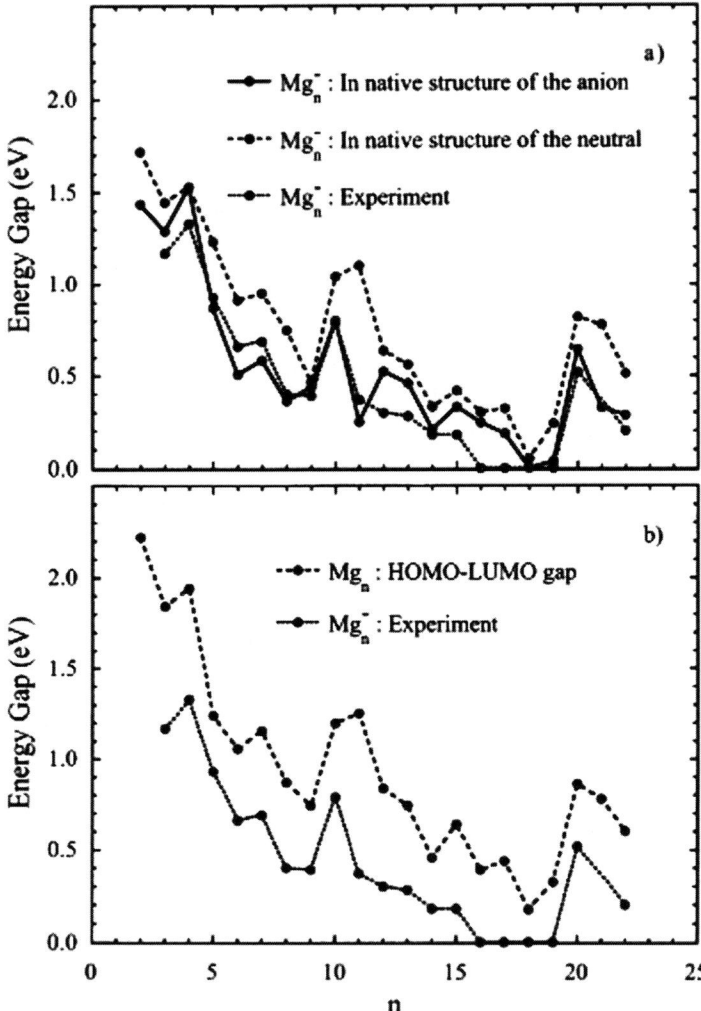

Figure 6.3. (a) Calculated difference in binding energies of two most external electrons in Mg_N^- structure and gap from PES experiments; (b) calculated HOMO–LUMO gap for neutral Mg_N^- clusters and experimental gap. Reproduced with the permission of the American Chemical Society from Jellinek and Acioli [193].

3.6 eV. In bulk the s- and p-bands overlap and there is interest in knowing the critical cluster size for which the overlap first occurs. Li et al. [225] have observed peaks in photoelectron spectra that can be identified as originating from the $3s$ and $3p$ orbitals. The gap between the peaks becomes smaller as the cluster size increases and by $N = 9$ the peaks have merged indicating band overlap. The fact that the electronic structure of bulk Al has a reasonably free electron character suggested that once s–p overlap has occurred in clusters there is a reasonable chance that the electronic shell model could apply. The PES results [225] gave strong evidence of shell closure for Al_{13}^- which corresponds to the magic electron number of 40. There was also some indication of shell closure for larger clusters but it is weaker, and there is none for N above 75.

Calculations indicate that at $N = 13$, structures with icosahedral motifs are favoured [211,217], but at around $N = 19$ isosahedral, decahedral and cuboctahedral structures become very close in energy [212,216,217]. No clear structural pattern appears to emerge for larger sizes although fcc and decahedral motifs do appear [212,215] and there is some tendency towards fcc with an octahedral growth pattern [212]. Doye [214] has suggested that a model of polytetrahedral clusters may be relevant. Attempts to relate model calculations to PES experiments have been reasonably successful for clusters smaller than about 30 atoms, but not at larger sizes [210,211].

6.4. Very large clusters

The icosahedral structure has been a frequently occurring motif in the discussion of Section 6.3, which has focused mainly on clusters smaller than about 100 atoms. This is not a structure that can appear in bulk materials and so there must be a critical size above which the icosahedron is no longer favoured energetically. The semi-empirical many-body potentials outlined in Section 6.2.4 provide the basis for investigations into the size regime where such a transition is likely to occur. However it is useful first to discuss the behaviour obtained with a simpler potential to get a qualitative understanding of the factors influencing the structure.

Doye et al. [48] studied the minimum energy configurations within various cluster geometries to determine which are the preferred structures. They used the classical Morse potential [226] to model the interatomic interactions and molecular dynamics to achieve the configuration of lowest energy. The Morse potential is written

$$V_M = \varepsilon \sum_{i<j} e^{\rho_0(1-r_{ij}/r_0)}[e^{\rho_0(1-r_{ij}/r_0)} - 2], \qquad (6.32)$$

where r_{ij} is the separation of atoms i and j, ε and r_0 define the energy and length scales respectively, and ρ_0 determines the range of the attractive part of the potential as illustrated in Figure 6.4.

Doye et al. [48] compare the energies of fcc, icosahedral and decahedral structures for a range of values of the cluster size, N, and the parameter ρ_0 and obtain a "phase diagram" for the lowest energy states as shown in Figure 6.5. The energies of Mackay and anti-Mackay overlayers for sizes between icosahedral shell filling magic numbers are also considered. It can be seen that, even for clusters smaller than 100 atoms, a narrow potential well (large ρ_0) favours decahedral or fcc ordering over icosahedral.

Following Doye et al. [48], we can understand the trends in the following way. The Morse potential energy can be partitioned into three parts

$$V_M = \varepsilon(-n_{nn} + E_{strain} + E_{nnn}), \qquad (6.33)$$

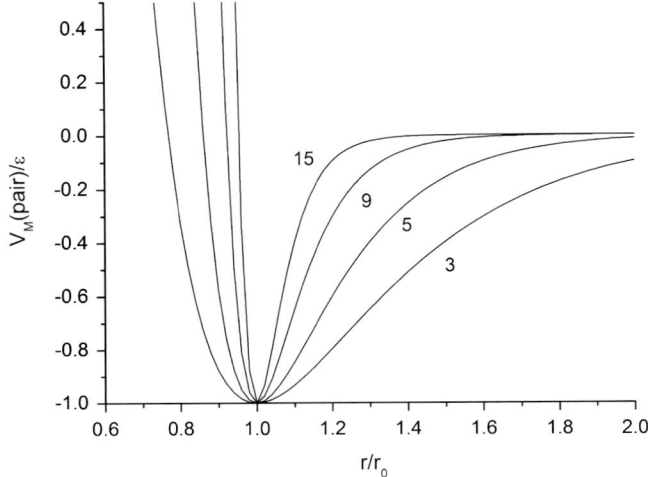

Figure 6.4. Morse potential for a pair of atoms as a function of their separation. Labels on lines are values of ρ_0.

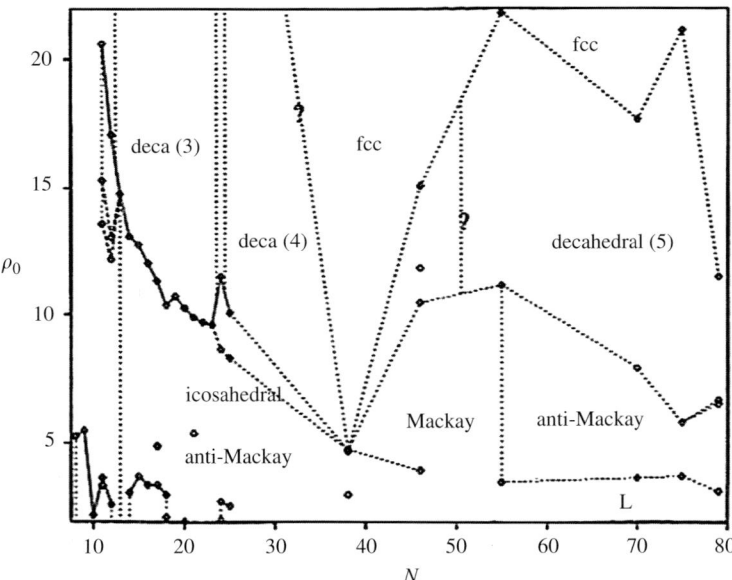

Figure 6.5. "Phase diagram" for the lowest energy structural types as a function of N and ρ_0. The decahedral structures are labelled by the number of atoms along the fivefold axis. Reproduced with the permission of American Institute of Physics from Doye et al. [48].

with n_{nn}, the number of nearest neighbours, given by

$$n_{nn} = \sum_{i<j; x_{ij}<x_0} 1, \tag{6.34}$$

where $x_{ij} = r_{ij}/r_0 - 1$, and x_0 defines a criterion for nearest neighbours. x_{ij} is a measure of the strain in contact between atoms i and j and the strain energy is given by

$$E_{\text{strain}} = \sum_{i<j; x_{ij}<x_0} (e^{-\rho_0 x_{ij}} - 1)^2, \tag{6.35}$$

This leaves the contribution from non-nearest neighbours

$$E_{\text{nnn}} = \sum_{i<j; x_{ij}>x_0} e^{-\rho_0 x_{ij}} (e^{-\rho_0 x_{ij}} - 2). \tag{6.36}$$

The dominant term is n_{nn}, and for atoms internal to the cluster it is just the coordination number, which is 12 for each of the three structures. The contribution from the surface is different for each of the structures. The $\{1\,1\,1\}$ faces are more closely packed that the $\{1\,0\,0\}$ faces, which means that overall icosahedral structures have the largest n_{nn}, and fcc generally have the smallest.

The non-nearest neighbour interactions cause a contraction of the nearest neighbour distances and this increases the strain energy, E_{strain}. This effect increases with the range of the potential, that is it is largest at small values of ρ_0. The icosahedral structure is the most highly strained while the fcc structure is essentially unstrained. We see that n_{nn} and E_{strain} are in competition and the balance between the two will be determined by the range of the interaction (value of ρ_0). A small value of ρ_0 favours the icosahedral structure, while fcc is preferred for large ρ_0. The decahedron is an intermediate case and, of course, the precise position of the phase boundaries depends strongly on N.

The analysis with the Morse potential nicely illustrates the sensitivity of structure to the shape of potential, but the Morse potential is inadequate for describing accurately the behaviour of real systems. Baletto et al. [3,119] have extended the study to real metals by use of quenched molecular dynamics simulations and semi-empirical potentials (Gupta and EAM) to investigate the sizes at which the crossovers from one energetically favoured structure to another occur for a range of materials. Their results are summarised in Table 6.6. For Cu there is a wide icosahedral region (to $N \approx 1000$) followed by a very wide decahedral window (to $N \approx 53000$). This contrasts markedly with Au for which there is a tendency to establish the fcc structure at small sizes; it exhibits very narrow icosahedral and

Table 6.6. Crossover sizes for six metals from Baletto et al. [3,119].

Metal	$N(\text{icos} \to \text{deca})$	$N(\text{deca} \to \text{fcc})$
Cu	1000	53000
Ag	<300	20000
Au	<100	500
Ni	1200	60000
Pd	<100	6500
Pt	<100	6500

The columns refer to the icosahedral to decahedral and the decahedral to fcc (truncated octahedral) transitions.

decahedral windows. Ag and Ni are similar to Cu, while Pd and Pt are more similar to Au. This comparison of the three basic structures was performed by allowing the atoms in the cluster to relax to an energy minimum starting from an initial structural motif.

The crossover for divalent metals has been investigated by Hearn and Johnston [186] using Murrell–Mottram potentials [227]. They find transitions from icosahedral to fcc structures for N around 32000 and 128000 for Ca and Sr respectively.

References

[1] V. Bonačić-Koutecký, P. Fantucci and J. Koutecký, Chem. Rev. **91** (1991) 1035.
[2] V. Bonačić-Koutecký, P. Fantucci and J. Koutecký, Quantum Chemistry of Clusters, in: H. Haberland (Ed.), Clusters of Atoms and Molecules, Springer Series in Chemical Physics, Vol. 52, Springer, Berlin, 1994, pp. 15–49.
[3] F. Baletto and R. Ferrando, Rev. Mod. Phys. **77** (2005) 371.
[4] C. Møller and M.S. Plesset, Phys. Rev. **46** (1934) 618.
[5] J. Čížek, Adv. Chem. Phys. **14** (1969) 35.
[6] D.R. Jennison, P.A. Schultz and M.P. Sears, J. Chem. Phys. **106** (1997) 1856.
[7] V. Kumar and Y. Kawazoe, Phys. Rev. B **66** (2002) 144413.
[8] P. Nava, M. Sierka and R. Ahlrichs, Phys. Chem. Chem. Phys. **5** (2003) 3372.
[9] M.L. Tiago, Y. Zhou, M.M.G. Alemany, Y. Saad and J.R. Chelikowsky, Phys. Rev. Lett. **97** (2006) 147201.
[10] G. Rollmann, M.E. Gruner, A. Hucht, R. Meyer, P. Entel, M.L. Tiago and J.R. Chelikowsky, Phys. Rev. Lett. **99** (2007) 083402.
[11] J.C. Slater, Quantum Theory of Molecules and Solids, Vol. 4: The Self-Consistent Field for Molecules and Solids, McGraw-Hill, New York, 1974.
[12] S.J. Vosko, L. Wilk and M. Nusair, Can. J. Phys. **58** (1980) 1200.
[13] S. Chrétien and D.R. Salahub, Density functional theory, methods, techniques, and applications (course 4), in: C. Guet, P. Hobza, F. Spiegelman and F. David (Eds.), Atomic Clusters and Nanoparticle, Proceedings of the Les Houches Summer School Session LXXII. Springer, Berlin, 2001, pp. 105–160.
[14] J.P. Perdew and A. Zunger, Phys. Rev. B **23** (1981) 5048.
[15] A.D. Becke, Phys. Rev. A **38** (1988) 3098.
[16] J.P. Perdew and Y. Wang, Phys. Rev. B **33** (1986) 8800; **40** (1989) 3399.
[17] J.P. Perdew, Phys. Rev. B **33** (1986) 8822; **34** (1986) 7406.
[18] J.P. Perdew and Y. Wang, Phys. Rev. B **45** (1992) 13244.
[19] J.P. Perdew, K. Burke and M. Ernzerhof, Phys. Rev. Lett. **77** (1996) 3865.
[20] J.P. Perdew, S. Kurth, A. Zupan and P. Blaha, Phys. Rev. Lett. **82** (1999) 2544.
[21] C. Lee, W. Yang and R.G. Parr, Phys. Rev. **37** (1988) 785.
[22] A.D. Becke, J. Chem. Phys. **98** (1993) 5648.
[23] P.J. Stevens, F.J. Devlin, C.F. Chabalowski and M.J. Frisch, J. Phys. Chem. **98** (1994) 11623.
[24] S. Yanagisawa, T. Tsuneda and K. Hirao, J. Chem. Phys. **112** (2000) 545.
[25] C.J. Barden, J.C. Rienstra-Kiracofe and H.F. Schaefer III, J. Chem. Phys. **113** (2000) 690.
[26] Z.J. Wu, Chem. Phys. Lett. **383** (2004) 251.
[27] F. Furche and J.P. Perdew, J. Chem. Phys. **124** (2006) 044103.
[28] S. Chrétien and D.R. Salahub, Phys. Rev. B **66** (2002) 155425.
[29] G.L. Gutsev and C.W. Bauschlicher Jr., J. Phys. Chem. A **107** (2003) 7013.
[30] S. Zhao, Z.-H. Li, W.-N. Wang, Z.-P. Liu, K.-N. Fan, Y. Xie and H.F. Schaefer III, J. Chem. Phys. **124** (2006) 184102.

[31] P. Calaminici, F. Janetzko, A.M. Köster, R. Mejia-Olvera and B. Zuniga-Guttierrez, J. Chem. Phys. **126** (2007) 044108.
[32] I.A. Solov'yov, A.V. Solov'yov and W. Greiner, Phys. Rev. A **65** (2002) 053203.
[33] K.R.S. Chandrakumar, T.K. Ghanty and S.K. Ghosh, J. Chem. Phys. **120** (2004) 6487.
[34] M. Ali, D.K. Maity, D. Das and T. Mukherjee, J. Chem. Phys. **124** (2006) 024325.
[35] R. Car and M. Parrinello, Phys. Rev. Lett. **55** (1985) 2471.
[36] J.C. Slater and K.F. Koster, Phys. Rev. **94** (1954) 1498.
[37] D.A. Papaconstantopoulos, Handbook of the Band Structure of Elemental Solids, Plenum, New York, 1986.
[38] N.N. Lathiotakis, A.N. Andriotis, M. Menon and J. Connolly, J. Chem. Phys. **104** (1996) 992.
[39] M.J. Mehl and D.A. Papaconstantopoulos, Phys. Rev. B **54** (1996) 4519.
[40] C. Barreteau, M.C. Desjonquères and D. Spanjaard, Eur. Phys. J. D **11** (2000) 395.
[41] Y. Xie and J.A. Blackman, Phys. Rev. B **64** (2001) 195115.
[42] Y. Xie and J.A. Blackman, J. Phys. Condens. Matter **16** (2004) 8589.
[43] J.L. Mercer and M.Y. Chou, Phys. Rev. B **49** (1994) 8506.
[44] A.N. Andriotis and M. Menon, Phys. Rev. B **57** (1998) 10069.
[45] C. Barreteau, R. Guirado-López, D. Spanjaard, M.C. Desjonquères and A.M. Oleś, Phys. Rev. B **61** (2000) 7781.
[46] Y. Xie and J.A. Blackman, Phys. Rev. B **66** (2002) 085410.
[47] G.M. Pastor, Theory of cluster magnetism (course 8), in: C. Guet, P. Hobza, F. Spiegelman and F. David (Eds.), Atomic Clusters and Nanoparticle, Proceedings of the Les Houches Summer School Session LXXII. Springer, Berlin, 2001, pp. 335–400.
[48] J.P.K. Doye, D.J. Wales and R.S. Berry, J. Chem. Phys. **103** (1995) 4234.
[49] J.P.K. Doye and D.J. Wales, J. Chem. Soc. Faraday Trans. **93** (1997) 4233.
[50] M.S. Daw and M.J. Baskes, Phys. Rev. B **29** (1984) 6443.
[51] S.M. Foiles, M.S. Daw and M.J. Baskes, Phys. Rev. B **33** (1986) 7983.
[52] J.K. Nørskov, Phys. Rev. B **26** (1982) 2875.
[53] K.W. Jacobsen, J.K. Nørskov and M.J. Puska, Phys. Rev. B **35** (1987) 7423.
[54] F. Ercolessi, W. Andreoni and E. Tosatti, Phys. Rev. Lett. **66** (1991) 911.
[55] M.W. Finnis and J.E. Sinclair, Philos. Mag. A **50** (1984) 45.
[56] A.P. Sutton and J. Chen, Philos. Mag. Lett. **61** (1990) 139.
[57] V. Rosato, M. Guillopé and B. Legrand, Philos. Mag. A **59** (1989) 321.
[58] R.P. Gupta, Phys. Rev. B **23** (1981) 6265.
[59] A.P. Sutton, Electronic Structure of Materials, Oxford University Press, Oxford, 1993.
[60] J.P.K. Doye and D.J. Wales, J. Chem. Phys. **102** (1995) 9659.
[61] Z. Li and H.A. Scheraga, Proc. Natl. Acad. Sci. U.S.A. **84** (1987) 6611.
[62] J.P.K. Doye and D.J. Wales, Phys. Rev. Lett. **80** (1998) 1357.
[63] J.P.K. Doye and D.J. Wales, New J. Chem. **22** (1998) 733.
[64] L. Zhan, J.Z.Y. Chen, W.-K. Liu and S.K. Lai, J. Chem. Phys. **122** (2005) 244707.
[65] B. Hartke, J. Phys. Chem. **97** (1993) 9973.
[66] R.S. Judson, M.E. Colvin, J.C. Meza, A. Huffer and D. Gutierrez, Int. J. Quantum Chem. **44** (1992) 277.
[67] K. Michaelian, Chem. Phys. Lett. **293** (1998) 202.
[68] K. Michaelian, N. Rendón and I.L. Garzón, Phys. Rev. B **60** (1999) 2000.
[69] S. Darby, T.V. Mortimer-Jones, R.L. Johnston and C. Roberts, J. Chem. Phys. **116** (2002) 1536.
[70] G. Rossi, A. Rapallo, C. Mottet, A. Fortunelli, F. Baletto and R. Ferrando, Phys. Rev. Lett. **93** (2004) 105503.
[71] R.L. Johnston, J. Chem. Soc. Dalton Trans. **2003** (2003) 4193.
[72] S. Kirkpatrick, C.D. Gelatt and M.P. Vecchi, Science **220** (1983) 671.
[73] D.L. Freeman and J.D. Doll, Annu. Rev. Phys. Chem. **47** (1996) 43.
[74] P. Amara, D. Hsu and J.E. Straub, J. Phys. Chem. **97** (1993) 6715.
[75] A.B. Finnila, M.A. Gomez, C. Sebenik, C. Stenson and J.D. Doll, Chem. Phys. Lett. **219** (1994) 343.

[76] D.J. Wales, Energy Landscapes with Applications to Clusters, Biomolecules and Glasses, Cambridge University Press, Cambridge, 2003.
[77] J.A. Alonso, Chem. Rev. **100** (2000) 637.
[78] The Cambridge Cluster Database: http://www-wales.ch.cam.ac.uk/CCD.html
[79] I. Boustani, W. Pewestorf, P. Fantucci, V. Bonačić-Koutecký and J. Koutecký, Phys. Rev. B **35** (1987) 9437.
[80] V. Bonačić-Koutecký, P. Fantucci and J. Koutecký, Phys. Rev. **37** (1988) 4369.
[81] V. Bonačić-Koutecký, I. Boustani, M. Guest and J. Koutecký, J. Chem. Phys. **89** (1988) 4861.
[82] I. Boustani and J. Koutecký, J. Chem. Phys. **88** (1988) 5657.
[83] F. Spiegelmann and D. Pavolini, J. Chem. Phys. **89** (1988) 4954.
[84] U. Röthlisberger and W. Andreoni, J. Chem. Phys. **94** (1991) 8129.
[85] U. Röthlisberger, W. Andreoni and P. Gianozzi, J. Chem. Phys. **96** (1992) 1248.
[86] V. Bonačić-Koutecký, P. Fantucci, C. Fuchs, C. Gatti, J. Pittner and S. Polezzo, Chem. Phys. Lett. **213** (1993) 522.
[87] R. Poteau and F. Spiegelmann, J. Chem. Phys. **98** (1993) 6540; **99** (1993) 10089.
[88] J. Jellinek, V. Bonačić-Koutecký, P. Fantucci and M. Wiechert, J. Chem. Phys. **101** (1994) 10092.
[89] M.-W. Sung, R. Kawai and J.H. Weare, Phys. Rev. Lett. **73** (1994) 3552.
[90] D. Reichardt, V. Bonačić-Koutecký, P. Fantucci and J. Jellinek, Chem. Phys. Lett. **279** (1997) 129.
[91] J. Jellinek, Chem. Phys. Lett. **288** (1998) 705.
[92] Y. Li, E. Blaisten-Barojas and D.A. Papaconstantopoulos, Phys. Rev. B **57** (1998) 15519.
[93] F. Spiegelmann, R. Poteau, B. Montag and P.-G. Reinhard, Phys. Lett. A **242** (1998) 163.
[94] D. Rayane, A.R. Allouche, E. Benichou, R. Antoine, M. Aubert-Frécon, P. Dugourd, M. Broyer, C. Ristori, F. Chandezon, B.A. Huber and C. Guet, Eur. Phys. J. D **9** (1999) 243.
[95] F. Nogueira, J.L. Martins and C. Fiolhais, Eur. Phys. J. D **9** (1999) 229.
[96] L. Kronik, I. Vasiliev and J.R. Chelikosky, Phys. Rev. B **62** (2000) 9992.
[97] S. Kümmel, J. Akola and M. Manninen, Phys. Rev. Lett. **84** (2000) 3827.
[98] S. Kümmel, M. Brack and P.G. Reinhard, Phys. Rev. B **62** (2000) 7602; **63** (2001) 129902.
[99] Y. Ishikawa, Y. Sugita, T. Nihikawa and Y. Okamoto, Phys. Lett. A **333** (2001) 199.
[100] S.K. Lai, P.J. Hsu, K.L. Wu, W.K. Liu and M. Iwamatsu, J. Chem. Phys. **117** (2002) 10715.
[101] C.W. Bauschlicher, S.R. Langhoff and P.R. Taylor, J. Chem. Phys. **88** (1988) 1041.
[102] C.W. Bauschlicher, S.R. Langhoff and H. Partridge, J. Chem. Phys. **91** (1989) 2412.
[103] N. Fujima and T. Yamaguchi, J. Phys. Soc. Jpn. **58** (1989) 1334.
[104] C.W. Bauschlicher, S.R. Langhoff and H. Partridge, J. Chem. Phys. **93** (1990) 8133.
[105] V. Bonačić-Koutecký, L. Češpiva, P. Fantucci and J. Koutecký, J. Chem. Phys. **98** (1993) 7981.
[106] V. Bonačić-Koutecký, L. Češpiva, P. Fantucci, J. Pittner and J. Koutecký, J. Chem. Phys. **100** (1994) 490.
[107] R. Santamaria, I.G. Kaplan and O. Novaro, Chem. Phys. Lett. **218** (1994) 395.
[108] C. Massobrio, A. Pasquarello and R. Car, Chem. Phys. Lett. **238** (1995) 215.
[109] C.L. Cleveland, U. Landman, T.G. Schaaff, M.N. Shafigullin, P.W. Stephens and R.L. Whetten, Phys. Rev. Lett. **79** (1997) 1873.
[110] C. Mottet, G. Tréglia and B. Legrand, Surf. Sci. **383** (1997) L719.
[111] I.L. Garzón, K. Michaelian, M.R. Beltrán, A. Posada-Amarillas, P. Ordejón, E. Artacho, D. Sánchez-Portal and J.M. Soler, Phys. Rev. Lett. **81** (1998) 1600.
[112] G. Bravo-Pérez, I.L. Garzón and O. Novaro, J. Mol. Struct. Theochem. **493** (1999) 225.
[113] G. Bravo-Pérez, I.L. Garzón and O. Novaro, Chem. Phys. Lett. **313** (1999) 655.
[114] Ş. Erkoç and T. Yilmaz, Physica E **5** (1999) 1.

[115] H. Grönbeck and W. Andreoni, Chem. Phys. **262** (2000) 1.
[116] H. Häkkinen and U. Landman, Phys. Rev. B **62** (2000) R2287.
[117] V. Bonačić-Koutecký, V. Veyret and R. Mitrić, J. Chem. Phys. **115** (2001) 10450.
[118] R. Fournier, J. Chem. Phys. **115** (2001) 2165.
[119] F. Baletto, R. Ferrando, A. Fortunelli, F. Montalenti and C. Mottet, J. Chem. Phys. **116** (2002) 3856.
[120] V. Bonačić-Koutecký, J. Burda, R. Mitrić, M. Ge, G. Zampella and P. Fantucci, J. Chem. Phys. **117** (2002) 3120.
[121] F. Furche, R. Ahlrich, P. Weis, C. Jacob, S. Gilb, T. Bienweiler and M. Kappes, J. Chem. Phys. **117** (2002) 6982.
[122] I.L. Garzón, J.A. Reyes-Nava, J.I. Rodríguez-Hernández, I. Sigal, M.R. Beltrán and K. Michaelian, Phys. Rev. B **66** (2002) 073403.
[123] S. Gilb, P. Weis, F. Furche, R. Ahlrichs and M.M. Kappes, J. Chem. Phys. **116** (2002) 4094.
[124] H. Häkkinen, M. Moseler and U. Landman, Phys. Rev. Lett. **89** (2002) 033401.
[125] P. Jaque and A. Toro-Labbé, J. Chem. Phys. **117** (2002) 3208.
[126] K. Jug, B. Zimmerman, P. Calaminici and A.M. Köster, J. Chem. Phys. **116** (2002) 4497.
[127] J. Oviedo and R.E. Palmer, J. Chem. Phys. **117** (2002) 9548.
[128] J. Wang, G. Wang and J. Zhao, Phys. Rev. B **66** (2002) 035418.
[129] P. Weis, T. Bierweiler, S. Gilb and M.M. Kappes, Chem. Phys. Lett. **355** (2002) 355.
[130] H. Häkkinen, B. Yoon, U. Landman, X. Li, H.-J. Zhai and L.-S. Wang, J. Phys. Chem. A **107** (2003) 6168.
[131] H.M. Lee, M. Ge, B.R. Sahu, P. Tarakeshwar and K.S. Kim, J. Phys. Chem. B **107** (2003) 9994.
[132] V.E. Matulis, O.A. Ivashkevich and V.S. Gurin, J. Mol. Struct. Theochem. **664–665** (2003) 291.
[133] J. Wang, G. Wang and J. Zhao, Chem. Phys. Lett. **380** (2003) 716.
[134] J. Zhao, J. Yang and J.G. Hou, Phys. Rev. B **67** (2003) 085404.
[135] E.M. Fernández, J.M. Soler, I.L. Garzón and L.C. Balbás, Phys. Rev. B **70** (2004) 165403.
[136] H. Häkkinen, M. Moseler, O. Kostko, N. Morgner, M.A. Hoffmann and B.V. Issendorff, Phys. Rev. Lett. **93** (2004) 093401.
[137] M. Kabir, A. Mookerjee and A.K. Bhattacharya, Phys. Rev. A **69** (2004) 043203.
[138] X. Shao, X. Liu and W. Cai, J. Chem. Theory Comput. **1** (2005) 762.
[139] H. Häkkinen and M. Moseler, Comput. Mater. Sci. **35** (2006) 332.
[140] P. Koskinen, H. Häkkinen, G. Seifert, S. Sanna, T. Frauenheim and M. Moseler, New J. Phys. **8** (2006) 9.
[141] L. Xiao, B. Tolberg, X. Hu and L. Wang, J. Chem. Phys. **124** (2006) 114309.
[142] M. Yang, K.A. Jackson, C. Koehler, T. Frauenheim and J. Jellinek, J. Chem. Phys. **124** (2006) 024308.
[143] M. Yang, K.A. Jackson and J. Jellinek, J. Chem. Phys. **125** (2006) 144308.
[144] Y.H. Chui, G. Grochola, I.K. Snook and S.P. Russo, Phys. Rev. B **75** (2007) 033404.
[145] P. Pyykkö, Chem. Rev. **88** (1988) 563.
[146] A. Sachdev, R.I. Masel and J.B. Adams, Catal. Lett. **15** (1992) 57.
[147] L. Goodwin and D.R. Salahub, Phys. Rev. A **47** (1993) R774.
[148] D. Dai and K. Balasubramanian, J. Chem. Phys. **103** (1995) 648.
[149] J. Zhao, X. Chen, Q. Sun, F. Liu, G. Wang and K.D. Lain, Physica B **215** (1995) 377.
[150] H. Grönbeck and A. Rosén, Phys. Rev. B **54** (1996) 1549.
[151] H. Kietzmann, J. Morenzin, P.S. Bechthold, G. Ganteför, W. Eberhardt, D.-S. Yang, P.A. Hackett, R. Fournier, T. Pang and C. Chen, Phys. Rev. Lett. **77** (1996) 4528.
[152] G. Valerio and H. Toulhoat, J. Phys. Chem. A **100** (1996) 10827.
[153] H. Grönbeck and A. Rosén, J. Chem. Phys. **107** (1997) 10620.
[154] S.H. Yang, D.A. Drabold, J.B. Adams, P. Ordejón and K. Glassford, J. Phys. Condens. Matter **9** (1997) L39.

[155] R. Fournier, T. Pang and C. Chen, Phys. Rev. A **57** (1998) 3683.
[156] H. Grönbeck, A. Rosén and W. Andreoni, Phys. Rev. A **58**.
[157] N. Watari and S. Ohnishi, Phys. Rev. B **58** (1998) 1665.
[158] A. Fortunelli, J. Mol. Struct. Theochem. **493** (1999) 233.
[159] J.E. Fowler, A. García and M. Ugalde, Phys. Rev. A **60** (1999) 3058.
[160] X. Wu and A.K. Ray, J. Chem. Phys. **110** (1999) 2437.
[161] A.G. Zacarias, M. Castro, J.M. Tour and J.M. Seminario, J. Phys. Chem. A **103** (1999) 7692.
[162] I. Efremenko and M. Sheintuch, J. Mol. Catal. A Chem. **160** (2000) 445.
[163] R. Guirado-López, M.C. Desjonquères and D. Spanjaard, Phys. Rev. B **62** (2000) 13188.
[164] S.H. Wei, Z. Zeng, J.Q. You, X.H. Yau and X.G. Gong, J. Chem. Phys. **113** (2000) 11127.
[165] J. Krüger, S.S. Vent, F. Nörtemann, M. Staufer and N. Rösch, J. Chem. Phys. **115** (2001) 2082.
[166] D. Majumdar and K. Balasubramanian, J. Chem. Phys. **115** (2001) 885.
[167] M. Moseler, H. Häkkinen, R.N. Barnett and U. Landman, Phys. Rev. Lett. **86** (2001) 2545.
[168] A. Taneda, T. Shimizu and Y. Kawazoe, J. Phys. Condens. Matter **13** (2001) L305.
[169] J. Zhao, Q. Qiu, B. Wang, J. Wang and G. Wang, Solid State Commun. **118** (2001) 157.
[170] V. Kumar and Y. Kawazoe, Phys. Rev. B **65** (2002) 125403.
[171] S.-Y. Wang, W. Duan, D.-L. Zhao and C.-Y. Wang, Phys. Rev. B **65** (2002) 165425.
[172] M. Castro, S.-R. Liu, H.-J. Zhai and L.-S. Wang, J. Chem. Phys. **118** (2003) 2116.
[173] A. Fortunelli and E. Aprà, J. Phys. Chem. A **107** (2003) 2934.
[174] A. Sebetci and Z.B. Güvenc, Surf. Sci. **525** (2003) 66.
[175] W. Zhang, Q. Ge and L. Wang, J. Chem. Phys. **118** (2003) 5793.
[176] E. Aprà, F. Baletto, R. Ferrando and A. Fortunelli, Phys. Rev. Lett. **93** (2004) 065502.
[177] S. Li, M.M.G. Alemany and J.R. Chelikowsky, J. Chem. Phys. **121** (2004) 5893.
[178] D. Majumdar and K. Balasubramanian, J. Chem. Phys. **121** (2004) 4014.
[179] S.-Y. Wang, J.-Z. Yu, H. Mizuseki, J.-A. Yan, Y. Kawazoe and C.-Y. Wang, J. Chem. Phys. **120** (2004) 8463.
[180] C. Ratsch, A. Fielicke, A. Kirilyuk, J. Behler, G. von Helden, G. Meijer and M. Scheffler, J. Chem. Phys. **122** (2005) 124302.
[181] J. Rogan, G. García, J.A. Valdivia, W. Orellana, A.H. Romero, R. Ramírez and M. Kiwi, Phys. Rev. B **72** (2005) 115421.
[182] M. Salazar-Villanueva, P.H. Hernández Tejeda, U. Pal, J.F. Rivas-Silva, J.I. Rodrigues Mora and J.A. Ascencio, J. Phys. Chem. A **110** (2006) 10274.
[183] T.R. Walsh, J. Chem. Phys. **124** (2006) 204317.
[184] J.-O. Joswig and M. Springborg, J. Phys. Condens. Matter **19** (2007) 106207.
[185] V. Kumar and R. Car, Phys. Rev. B **44** (1991) 8243.
[186] J.E. Hearn and R.L. Johnston, J. Chem. Phys. **107** (1997) 4674.
[187] Y. Wang, H.-J. Flad and M. Doig, J. Phys. Chem. A **104** (2000) 5558.
[188] A. Köhn, F. Weigend and R. Ahlrichs, Phys. Chem. Chem. Phys. **3** (2001) 711.
[189] V. Kumar and Y. Kawazoe, Phys. Rev. B **63** (2001) 075410.
[190] J.W. Mirick, C.-H. Chien and E. Blaisten-Barojas, Phys. Rev. A **63** (2001) 023202.
[191] G.M. Wang, E. Blaisten-Barojas, A.E. Roitberg and T.P. Martin, J. Chem. Phys. **115** (2001) 3640.
[192] P.H. Acioli and J. Jellinek, Phys. Rev. Lett. **89** (2002) 213402.
[193] J. Jellinek and P.H. Acioli, J. Phys. Chem. A **106** (2002) 10919; **107** (2003) 1670.
[194] A. Lyalin, I.A. Solov'yov, A.V. Solov'yov and W. Greiner, Phys. Rev. A **67** (2003) 063203.
[195] I.A. Solov'yov, A.V. Solov'yov and W. Greiner, J. Phys. B **37** (2004) L137.
[196] X. Dong, G.M. Wang and E. Blaisten-Barojas, Phys. Rev. B **70** (2004) 205409.
[197] A. Lyalin, A.V. Solov'yov, C. Bréchignac and W. Greiner, J. Phys. B **38** (2005) L129.
[198] P.P. Singh, Phys. Rev. B **49** (1994) 4954.
[199] F. Yonezawa and H. Tanikawa, J. Non-Cryst. Solids **205–207** (1996) 793.

[200] M. Yu and M. Dolg, Chem. Phys. Lett. **273** (1997) 329.
[201] H.J. Flad, F. Schutz, Y. Wang, M. Dolg and A. Savin, Eur. Phys. J. D **6** (1999) 243.
[202] B. Hartke, H.-J. Flad and M. Dolg, Phys. Chem. Chem. Phys. **3** (2001) 5121.
[203] J. Zhao, Phys. Rev. A **64** (2001) 043204.
[204] K. Michaelian, M. Beltrán and I.L. Garzón, Phys. Rev. B **65** (2002) 041403.
[205] G.E. Moyano, R. Wesendrup, T. Söhnel and P. Schwerdtfeger, Phys. Rev. Lett. **89** (2002) 103401.
[206] J. Wang, G. Wang and J. Zhao, Phys. Rev. A **68** (2003) 013201.
[207] J.P.K. Doye, Phys. Rev. B **68** (2003) 195418.
[208] J. Akola, H. Häkkinen and M. Manninen, Phys. Rev. B **58** (1998) 3601.
[209] V. Kumar, Phys. Rev. B **57** (1998) 8827.
[210] B.K. Rao and P. Jena, J. Chem. Phys. **111** (1999) 1890.
[211] J. Akola, M. Manninen, H. Häkkinen, U. Landman, X. Li and L.-S. Wang, Phys. Rev. B **60** (1999) 13216.
[212] J. Akola, M. Manninen, H. Häkkinen, U. Landman, X. Li and L.-S. Wang, Phys. Rev. B **62** (2000) R11297.
[213] L.D. Lloyd, R.L. Johnston, C. Roberts and T.V. Mortimer-Jones, Chemphyschem **3** (2002) 408.
[214] J.P.K. Doye, J. Chem. Phys. **119** (2003) 1136.
[215] K. Manninen, J. Akola and M. Manninen, Phys. Rev. B **68** (2003) 235412.
[216] C.-H. Yao, B. Song and P.-L. Cao, Phys. Rev. B **70** (2004) 195431.
[217] F.-C. Chuang, C.Z. Wang and K.H. Ho, Phys. Rev. B **73** (2006) 12541.
[218] R.L. Johnston, Phil. Trans. R. Soc. London A **356** (1998) 211.
[219] G.M. Pastor, P. Stampfli and K.H. Bennemann, Phys. Scr. **38** (1988) 623.
[220] G.M. Pastor and K.H. Bennemann, Transition from van der Waals to metallic bonding in mercury clusters, in: H. Haberland (Ed.), Clusters of Atoms and Molecules, Springer Series in Chemical Physics, Vol. 52, Springer, Berlin, 1994, pp. 86–113.
[221] K. Rademann, Z. Phys. D **19** (1991) 161.
[222] R. Busani, M. Folkers and O. Cheshnovsky, Phys. Rev. Lett. **81** (1998) 3836.
[223] O.C. Thomas, W. Zheng, S. Xu and K.H. Bowen Jr., Phys. Rev. Lett. **89** (2002) 213403.
[224] T. Diederich, T. Döppner, J. Braune, J. Tiggesbäumker and K.-H. Meiwes-Broer, Phys. Rev. Lett. **86** (2001) 4807.
[225] X. Li, H. Wu, X.-B. Wang and L.-S. Wang, Phys. Rev. Lett. **81** (1998) 1909.
[226] P.M. Morse, Phys. Rev. **34** (1929) 57.
[227] J.N. Murrell and P.E. Mottram, Mol. Phys. **69** (1990) 571.

Chapter 7

Photoexcitation and Optical Absorption

J.A. Blackman
Department of Physics, University of Reading, Whiteknights, Reading RG6 6AF, UK;
Department of Physics and Astronomy, University of Leicester, University Road,
Leicester LE1 7RH, UK

7.1.	Introduction	176
7.2.	Electron excitation techniques	177
	7.2.1. Ionisation potential and electron affinity	177
	7.2.2. Photoionisation and the ionisation potential	178
	7.2.2.1. Alkali metals	180
	7.2.2.2. Noble metals	181
	7.2.2.3. Trivalent metals	183
	7.2.3. Photoemission and electron affinity	184
	7.2.3.1. Alkali metals	188
	7.2.3.2. Noble metals	189
	7.2.3.3. Transition metals	194
	7.2.3.4. Divalent and trivalent metals	195
7.3.	Optical spectroscopy	196
	7.3.1. Introduction	196
	7.3.2. Bulk dielectric function	199
	7.3.3. The Mie theory	201
	7.3.4. Experimental methods	205
	7.3.4.1. Beam depletion spectroscopies	205
	7.3.4.2. Techniques for single clusters	206
	7.3.5. Specific materials	207
	7.3.5.1. Small clusters	208
	7.3.5.2. Large clusters	216
	7.3.6. Effective medium theories	218
	References	223

HANDBOOK OF METAL PHYSICS
ISSN 1570-002X/DOI 10.1016/S1570-002X(08)00207-3

© 2009 ELSEVIER B.V.
ALL RIGHTS RESERVED

7.1. Introduction

Laser light is a powerful and widely used probe of the properties of metallic clusters. Photoionisation, photoemission and optical absorption are laser-based techniques, each of which yields different information. The simplest thing one can do with the light is to excite an electron out of a cluster and measure the lowest photon energy required (threshold) for the ionisation to occur. The threshold energy for a bulk material is the work function, and for clusters, there are size-dependent deviations from the bulk work function.

Two ionisation thresholds are defined: ionisation potential (IP) and electron affinity (EA). The former is the threshold energy required to ionise a neutral cluster (leaving it positively charged after the ionisation). The latter is the energy to remove an electron from a negatively charged (anionic) cluster, leaving it neutral after the process. A cluster may relax to a different geometry after an electron has been removed. Strictly speaking, the IP and the EA are defined in terms of energy differences of the relaxed nanoparticle before and after excitation. However, the relaxation is a slow process, and it takes some time after ionisation for the cluster to relax to its new geometry. Negligible relaxation takes place over the timescale of threshold energy measurements, and often it is the vertical detachment energy (VDE) that is reported rather than the true IP or EA. The difference between the VDE and the IP or EA is important only for small clusters for which significant geometry relaxation occurs.

The variation in the ionisation energies with cluster size is of principal interest. Particularly, stable clusters require higher energies for excitation by photons, and this is reflected by peaks in the threshold energies, giving us a way of monitoring magic numbers analogous to the abundance spectra discussed in Chapter 2.

Further information about the electronic structure of clusters can be obtained if the kinetic energy of the emitted electrons is measured (photoelectron spectroscopy). The energy spectrum of the emitted electrons relates closely to the energy dependence of the density of states of the electronic energies of the clusters. This allows a comparison with the results from theoretical calculations.

Insight into certain properties can be obtained directly from the appearance of the photoemission spectra. Features in the spectra of single atoms or very small clusters can be assigned to excitations from particular energy levels (with s, p or d character). As cluster size increases, these energy levels hybridise and form bands, with corresponding changes that can be observed in the spectra. This is of particular interest for $3d$ magnetism (s–d hybridisation) and the metal–insulator transition (s–p hybridisation).

Photoionisation and photoemission spectroscopies are reviewed in Section 7.2. Following this, in Section 7.3, we discuss optical absorption. The main feature here is the surface plasmon, a collective resonance, which dominates the absorption of energy from an incident light beam. As long as the size of the particle is small compared with the wavelength of light, a simple theory predicts a resonant frequency that is independent of particle size. However, there are departures from the simple theory, and there has been a lot of effort at understanding the observed behaviour. For larger particles, additional resonances can be excited. The existence

7.2. Electron excitation techniques

7.2.1. *Ionisation potential and electron affinity*

The IP corresponds to the difference in ground state energies between neutral and ionised clusters (one electron stripped off for the first IP), while the EA is the energy difference between a neutral cluster and the corresponding anion. For an N-atom cluster,

$$\text{IP} = E_N^+ - E_N^0, \quad \text{EA} = E_N^0 - E_N^-, \tag{7.1}$$

where E_N is the ground state energy of the cluster and the superscript indicates its charge state.

According to the liquid drop model (in which the cluster is approximated by a uniform conducting sphere), the IP and EA of a metal cluster of radius R asymptotically approach the work function W of a planar metal surface as

$$\text{IP}(R) = W + \alpha \frac{e^2}{R} + O(R^{-2}), \tag{7.2}$$

$$\text{EA}(R) = W - \beta \frac{e^2}{R} + O(R^{-2}), \tag{7.3}$$

where α and β are constants that depend rather weakly on the material. The relation between R and N is given by $R = r_s N^{1/3}$, where r_s is the Wigner–Seitz radius.

For alkali metal clusters that are not too small, the equations for IP and EA usually give a reasonable fit to experimental data. An example for potassium IP data is shown in Figure 7.1. The solid line is a best fit to

$$\text{IP}(N) = W + AN^{-1/3}, \tag{7.4}$$

which, apart from the neglect of the $O(R^{-2})$ term, is equivalent to Eq. (7.2).

There has been extensive theoretical work on the constants α and β in Eqs. (7.2) and (7.3). They are related [2,3] and can be expressed as

$$\alpha = \frac{1}{2} + c, \quad \beta = \frac{1}{2} - c. \tag{7.5}$$

Classical theories based on image-potential models claim a value $c = -1/8$, and experimental work (such as the data in Figure 7.1) seems to give reasonable agreement with that result. However, the basis of the classical derivation has been challenged [4,5] with the assertion that any classical theory gives $\alpha = \beta = 1/2$, and c should be regarded as a quantum correction that requires a more realistic approach for its evaluation.

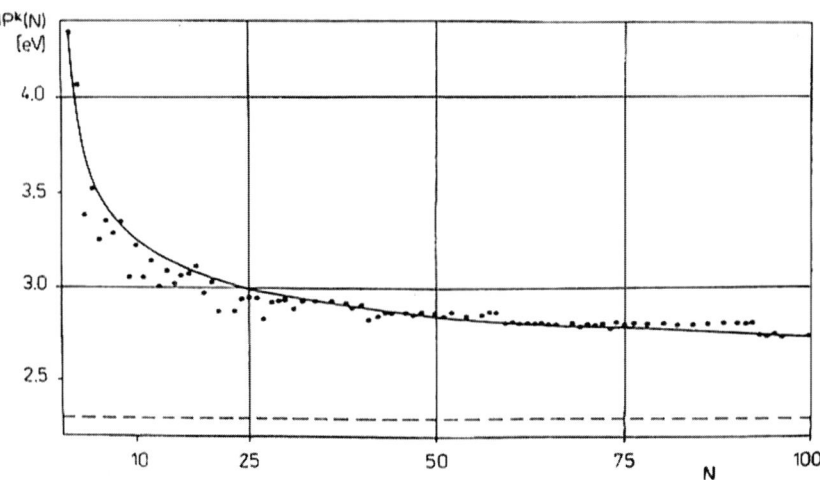

Figure 7.1. The ionisation potential for potassium clusters as a function of cluster size N. The dots are experimental data, and the solid line is a fit to Eq. (7.4). The broken line indicates the bulk limit, W. Reproduced with the permission of Springer Science+Business Media from Müller et al. [1].

There have been a number of calculations for the work functions W and the size-effect coefficients c (treating it as a quantum correction), based on the jellium or the stabilised jellium model. Both give a value of c of about -0.08. Plots of these quantities as a function of r_s from a stabilised jellium calculation [3] are shown in Figure 7.2. The value of c is essentially constant at about -0.08 except at high densities ($r_s < 3$). In any case, there is considerable structure in the IP for small clusters, and care has to be taken in assigning an accurate value to c from the experimental plots.

7.2.2. *Photoionisation and the ionisation potential*

The IP is usually measured in photoionisation experiments in which an electron in a neutral cluster is removed by a photon from a tunable laser. The resulting cluster ions are mass selected, and the ion signal is recorded as a function of photon energy. The photoionisation intensity shows a gradual onset at the threshold, followed by a relatively steep rise to a maximum [6]. At higher photon energies, there is a gradual decline in intensity. If a cooled cluster source is used, a much sharper threshold occurs with a considerably reduced low-energy tail [7]. Electron impact ionisation provides an alternative method for measuring IPs and generally produces similar results to the photoionisation [8,9].

The IP is defined as the difference between the energies of ionised and neutral clusters. Particularly for small clusters, the ground state geometries of the two are generally different. However, on the timescale of an electronic transition, there is essentially no change in the positions of the atomic nuclei, so the photoionisation process leaves an ionised cluster with the same geometry as the neutral cluster. It is in a vibrationally excited state, and a naive analysis of the threshold would indicate

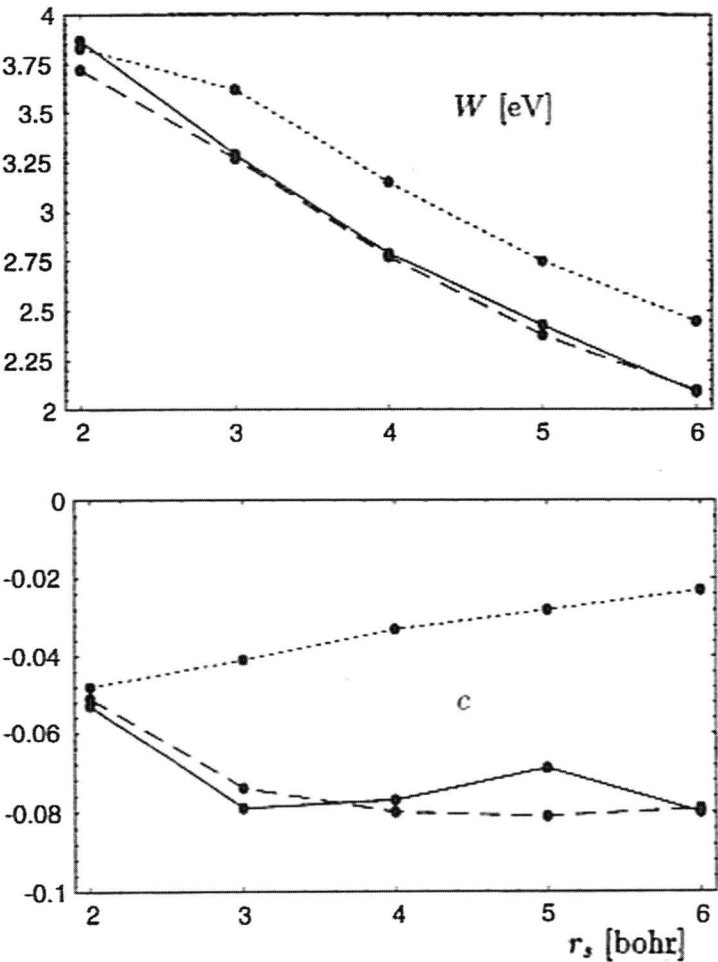

Figure 7.2. Work function W (upper panel) and size-effect coefficient c (lower panel) for stabilised jellium as a function of r_s. Two schemes are used that give very similar results (full and broken lines). The dotted lines are results using Koopmans expression, which, as noted in Section 2.2.2, is unreliable. Reproduced with the permission of the American Institute of Physics from Seidl et al. [3].

a very high value for the IP. The energy extracted directly from the photoionisation threshold is known as the vertical ionisation potential (VIP) or vertical detachment energy (VDE) in distinction to the true (or adiabatic) IP defined by Eq. (7.1). The shape of the photoionisation efficiency plot depends on temperature through the vibrationally excited states and is also influenced by electron final-state effects. The distinction between adiabatic and VIP is really significant only for small clusters as the difference in geometry between the neutral and ionic states is small for large clusters.

Strategies for extracting IPs from experimental data are discussed by de Heer [4]. A commonly used approach is based on the scheme used by Guyon and Berkowitz

[10] for polyatomic molecules, which takes the threshold at the intercept on the photon energy axis of a straight line fitted to the first rise of the photoionisation profile, generally with some correction for the effect of finite instrumental bandwidth. Usually, experiments are performed on a series of clusters, and one is interested in features in the IP that reflect a trend over the series rather than highly accurate individual values, and the procedure described is quite adequate without further refinement.

7.2.2.1. Alkali metals

IPs for sodium and potassium clusters extracted from photoionisation efficiency spectra are shown in Figure 7.3. The figure assembles data from a number of authors [4]. The general trend of an IP decreasing with cluster size as represented by Eq. (7.2) is observed. Major steps can be seen at 8, 18, 20, 40, 58 and 92, corresponding to the electronic spherical shell closing of the jellium model with fine structure associated with sub-shell closings of the ellipsoidally distorted structures that was discussed in Section 2.2. The fine structure is similar in both spectra but is more pronounced for sodium. The odd–even alternations can be seen in both spectra.

There have been a number of attempts to theoretically reproduce the structure in the IP plots. Yannouleas and Landman [11] use a refinement of the ellipsoidally distorted jellium model to calculate the IP of sodium for $3 \leq N \leq 105$ and obtained a rather good agreement with experiment. They noted that the lifting of the

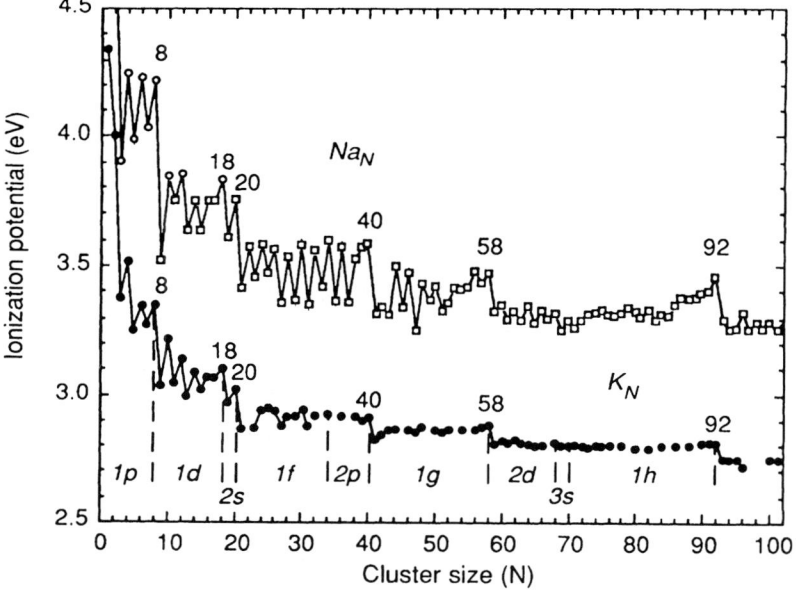

Figure 7.3. Ionisation potentials for sodium and potassium clusters as a function of cluster size. The labelling below the potassium plot corresponds to the filled shell levels (see Figure 2.2 in Section 2.2). Reproduced with the permission of the American Physical Society from de Heer [4].

degeneracies via the triaxial-shape distortions underlies the appearance of the odd–even structure in the IP spectra. Yannouleas and Landman [12] performed a similar calculation for potassium. The agreement between theory and experiment is not as good as that for sodium, but the introduction of finite temperatures into the calculation brings the calculated spectra for potassium more in line with the experimental data. At low temperatures, the calculations tend to produce more pronounced structure than is observed experimentally, but at room temperature, the calculations give reasonable agreement with experiment.

Most of the more recent work uses *ab initio* methods within an atomic basis to study the detailed atomic structure of small clusters, where there is a pronounced structure in the size dependence of the IP. The calculations search for the arrangements of atoms that give the lowest energy. Solov'yov *et al.* [13], for example, perform calculations on sodium clusters with $N \leq 20$ and compare the results obtained with various *ab initio* techniques. Other theoretical work on alkali metals is noted in Table 6.1 in Section 6.3.2.

7.2.2.2. Noble metals

The IPs for clusters of the noble metals Cu [14], Ag [15] and Au [16] have been measured by photoionisation and, for Ag, by electron impact ionisation [7,8]. The results of Alameddin *et al.* [15] for Ag are shown in Figure 7.4. The structure in the IP only manifests itself when a cold source is used. Significant features in the plot are labelled with the cluster sizes. A number of the features can be related to the jellium model: peaks at 18, 34, 40, 58 and 92 and steps down at 9 and 21 after shell fillings at 8 and 20, with odd–even steps up to $N = 49$ similar to sodium.

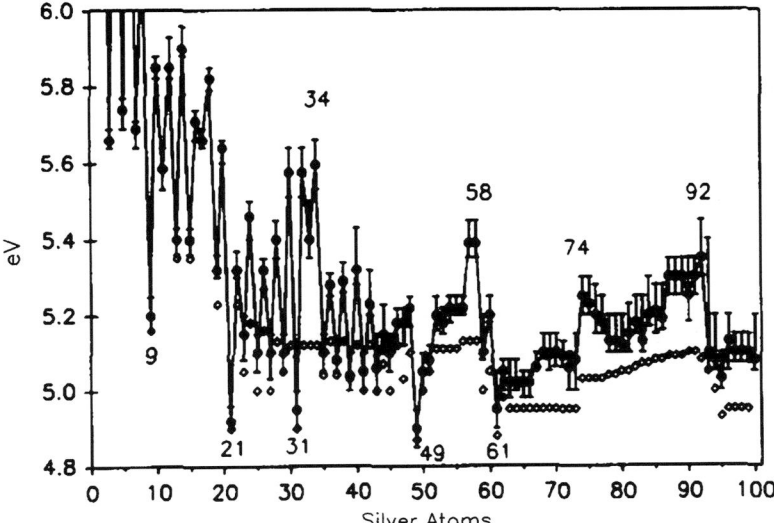

Figure 7.4. Vertical ionisation potentials for Ag clusters as a function of cluster size for liquid nitrogen source (filled circles) and room temperature source (open diamonds). Reproduced with the permission of Elsevier Science from Alameddin *et al.* [15].

Calculations of the IP of noble metals that use an atomic basis are noted in Table 6.2 in Section 6.3.3. The results of generalised gradient approximation density functional calculations for both the IP and the EA of all three noble metals are shown in Figure 7.5 for clusters of up to 13 atoms. The structure in the plots agrees well with the experimental data. Kabir et al. [18] have used a tight-binding molecular dynamics approach to optimise the geometry of copper clusters up to $N = 55$. They calculate the IP and the gap between the highest occupied molecular orbital (HOMO) and the lowest unoccupied molecular orbital (LUMO) (which is related to IP−EA). The geometry is therefore treated explicitly, but the features in

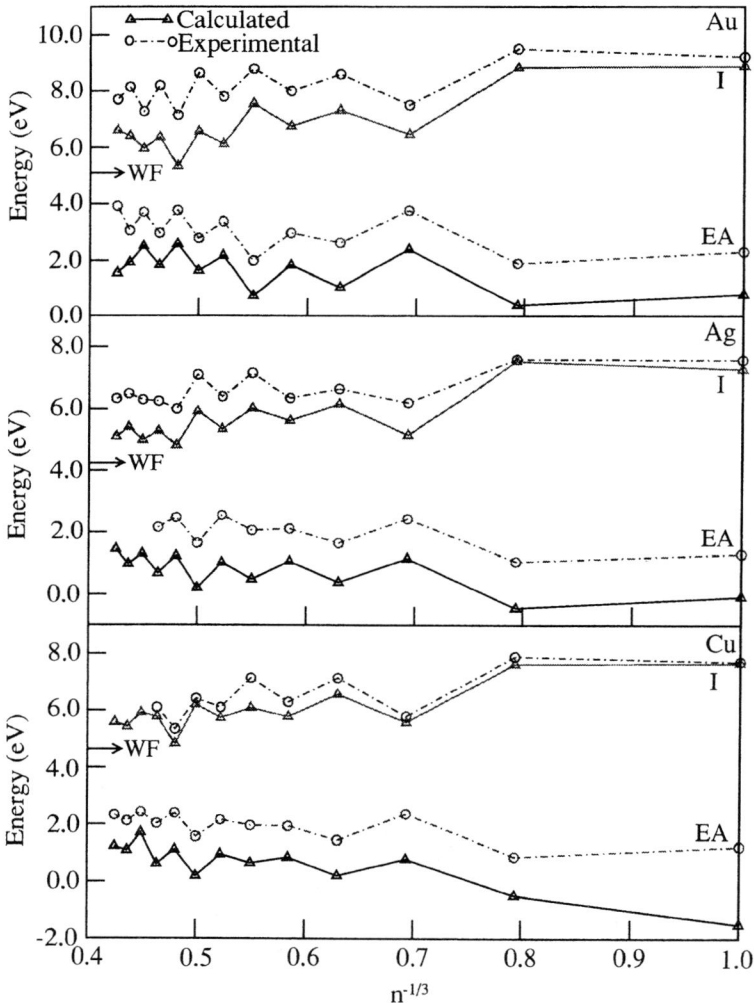

Figure 7.5. Calculated ionisation potential and electron affinity for Au_N, Ag_N and Cu_N clusters versus $N^{-1/3}$ (for cluster sizes $N \leq 13$) compared with experimental values. The experimental work function is indicated by an arrow. Reproduced with the permission of the American Physical Society from Fernández et al. [17].

both the IP and the HOMO–LUMO gap are characteristic of the electronic shell model. The even–odd alternation due to electron pairing is observed, and there are peaks at $N = 2, 8, 18, 20, 34$ and 40, corresponding to the shell closings of the electronic shell model. They find that the structure of the copper clusters is icosahedral except in the range 40–44, where a decahedral configuration is preferred. They relate the loss of odd–even alternation above $N = 40$ to this structural transition, suggesting that the decahedral structure is less sensitive to electron pairing effects.

7.2.2.3. Trivalent metals

Measurements of the IPs of Al clusters have been made by Persson et al. [19] and Schriver et al. [20], and of In clusters by Persson et al. [19] and Pellarin et al. [21]. The photoionisation mass spectra of In and Al in the $N = 40$–120 range from Persson et al. [19] are shown in Figure 7.6. The IP can be measured directly by varying the energy of the incident photons to locate the ionisation threshold. An alternative approach is to use a fixed photon energy and record the mass spectrum. This reflects the cluster-size dependence of the ionisation cross section, and since clusters with filled electronic shells are particularly stable, one expects to see an increase in the cross section *after* the shell closing.

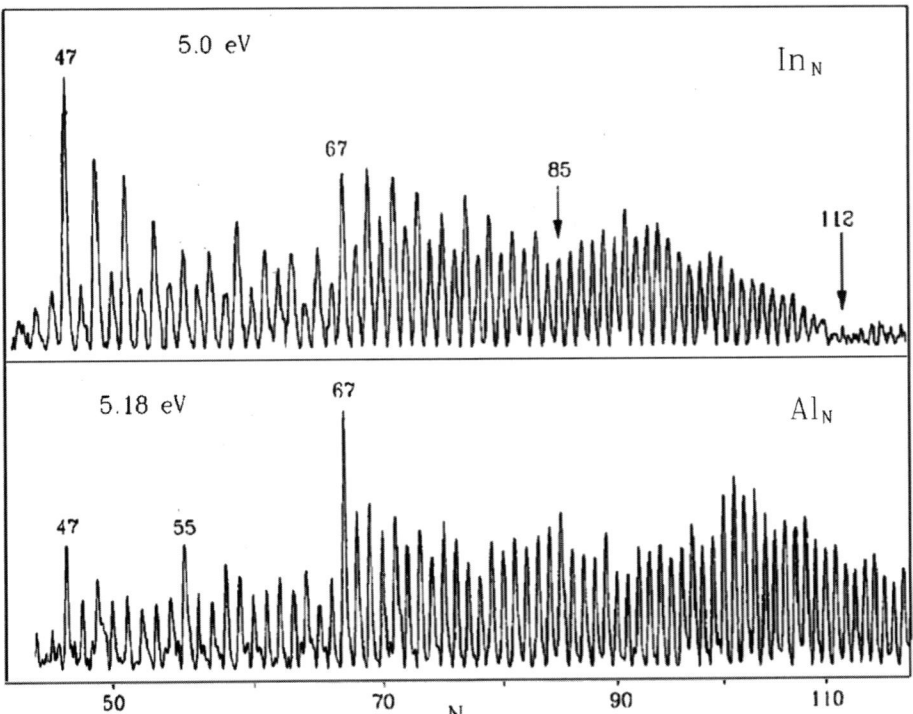

Figure 7.6. Photoionisation mass spectra of In and Al as a function of cluster size (on the lower horizontal axis). The photon energies are indicated on the plots. Reproduced with the permission of Elsevier Science from Persson et al. [19].

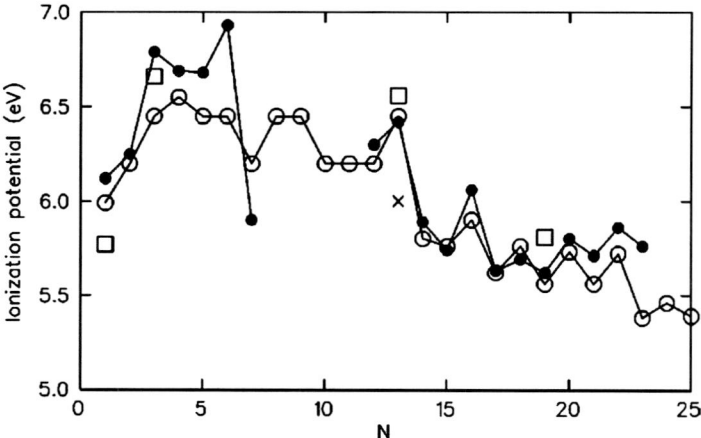

Figure 7.7. Calculated ionisation potential (solid dots) of Al clusters as a function of cluster size [22]. Experimental results [20] are indicated by open circles. Open squares and crosses are from other calculations. Reproduced with the permission of the American Physical Society from Akola et al. [22].

Thus the increase at $N = 47$ in Figure 7.6 can be associated with electronic shell filling at $N = 46$, corresponding to an electron number of 138 (see Figure 2.2 of Section 2.2.1), and a similar increase at 66–67 can be ascribed to a higher magic number at 198. Features at several other sizes [19–21] can be associated with electronic shell filling, but there are too many anomalies for that model to be really a useful one for trivalent materials. It is better to go directly to calculations based on the atoms in the cluster. Calculations of the IP of both divalent and trivalent metals are noted in Table 6.5 of Section 6.3.6, and Figure 7.7 shows the results of a density functional calculation on Al by Akola et al. [22] that reproduces the experimental data for clusters in the 10–20 atom range rather well.

7.2.3. Photoemission and electron affinity

In contrast to photoionisation experiments, photoemission spectroscopy measures the kinetic energy of the emitted electron. Usually, mass-selected ionic clusters are used in the experiments, but photoelectron spectroscopy can also be performed on neutral cluster beams. We focus here on anionic clusters in which the photoemission process is a transition from the cluster in an initial anionic state to a neutral final state. The photoemission intensity is generally plotted against the difference between the photon energy and the kinetic energy of the emitted electron, known as the binding energy

$$BE = h\nu - E_{\text{kin}}. \tag{7.6}$$

We illustrate the process and the obtainable information with the simple schematic example shown in Figure 7.8, which could represent photoemission from

7.2. *Electron excitation techniques* 185

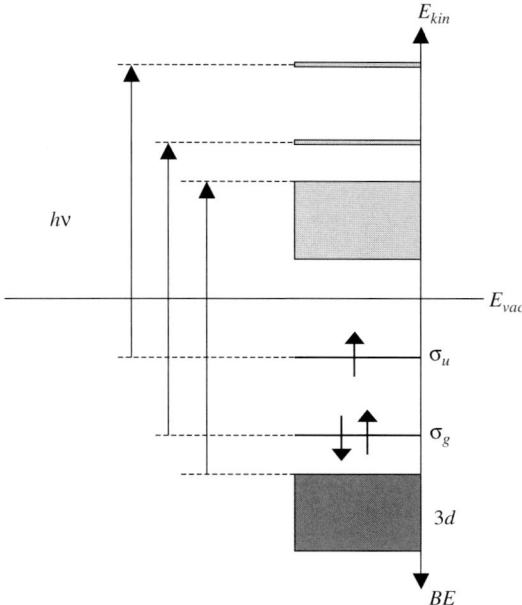

Figure 7.8. Schematic representation of the photoemission process from a Cu_2^- anion within a single-particle picture. The photon energy ($h\nu$), electron kinetic energy (E_{kin}) and binding energy (BE) are related by Eq. (7.6). Binding energies are measured with respect to vacuum energy (E_{vac}). The resulting photoemission spectrum is indicated in light grey.

an anionic copper dimer. Copper atoms have a filled $3d$ shell and a singly occupied $4s$ orbital. In the neutral Cu_2 molecule, the $4s$ orbitals form a bonding σ_g and an antibonding σ_u molecular orbital. The σ_g orbital is filled with two electrons and the σ_u orbital is empty. The extra electron in the Cu_2^- anion goes into the σ_u orbital. The $3d$ energy levels are represented by a band lying below the energy of the bonding orbital. Two sharp features in the kinetic energy spectrum are expected, originating from excitations from the σ_g and σ_u energy levels, and a broad one is expected from the $3d$ orbitals (the transition from the edge is shown in the diagram).

In this example, the neutral cluster has a closed electronic shell, the HOMO, so the extra electron of the anion must occupy the LUMO of the neutral cluster. The difference in kinetic energy of the two sharp features in the photoemission spectrum provides a direct measurement of the HOMO–LUMO gap of the neutral cluster, which is an important parameter in assessing cluster stability; a large gap indicates high stability.

The EA is the energy difference between the ground states of the anion and the neutral cluster, so for transitions between excited states, we can write in principle

$$h\nu - E_{kin} = EA + E^0 - E^- \qquad (7.7)$$

where E^0 and E^- are the excitation energies of the neutral and anionic clusters, respectively. Usually one can assume that the anion is in its ground state [$E^- = 0$ in Eq. (7.7)], so the feature at the lowest binding energy corresponds to the transition to the ground state of the neutral cluster, i.e. the EA. However, this statement is valid only for large clusters. The timescale of the transition is fast compared with the relaxation of the nuclei, so the features relate to the electronic states of the neutral cluster in the ground state geometry of the anion, and the transitions are vibronic (involving a change in both the electronic and vibrational states). This is particularly important for small clusters as we have seen in an analogous situation with the IP.

The quantity that is generally measured is the VDE, which corresponds to the transition between the ground state of the anion and a vibrationally excited state of the neutral. We can also call it the VIP of the anion. We should also distinguish the VDE from the vertical EA, which is the energy difference between the neutral and the anion, both in the ground state geometry of the neutral. The adiabatic EA of the neutral and the IP of the anion are of course equivalent. This issue is discussed in detail by Cha et al. [23].

A typical experimental arrangement [24,25] for photoelectron studies on mass-selected anionic clusters is illustrated in Figure 7.9, which is from a review by Eberhardt [26]. It consists essentially of three stages: cluster production, separation of anions from neutrals and mass selection and measurement of kinetic energy of detached photoelectrons. In this illustration, the first stage is an intensely focused laser pulse that is used to vaporise material from the metal target. This is a common method for noble metals, as is electric arc vaporisation. The various types of cluster sources are discussed in Chapter 3. A pulse of helium gas is introduced synchronously, and the metal atoms and ions lose energy through multiple collisions with the He atoms and condense into aggregates. The helium gas also acts as a transport medium. The cluster–helium mixture then expands adiabatically into the vacuum vessel, forming a supersonic molecular beam.

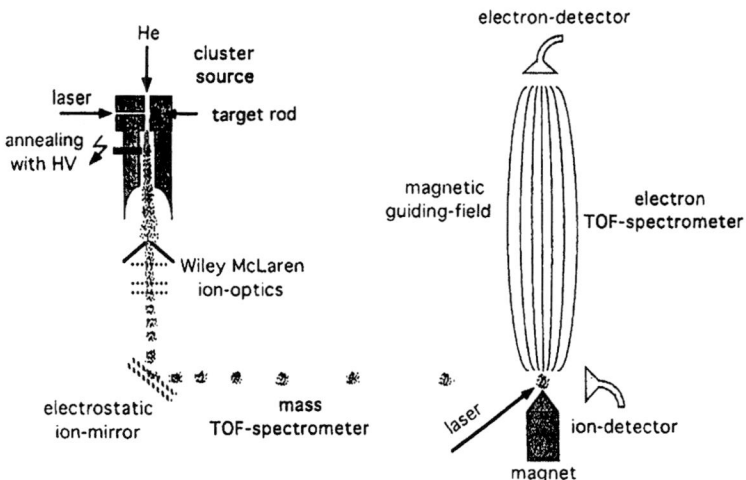

Figure 7.9. The experimental setup for photoelectron spectroscopy of mass-selected cluster anions comprises three parts: cluster source, mass time-of-flight spectrometer and electron time-of-flight spectrometer. Reproduced with the permission of Elsevier Science from Eberhardt [26].

The cluster beam contains neutrals and charged clusters of various sizes. The negatively charged clusters are accelerated in a pulsed electric field of a Wiley–McLaren-type time-of-flight setup and deflected away from the neutral beam by an electrostatic mirror. The clusters acquire velocities that depend on their size, so they arrive at the next stage in the sequence of their masses, with the smallest clusters (fastest) arriving first, thus allowing the selection of particular cluster sizes. The cluster anions are decelerated prior to the photodetachment process in order to improve the overall resolution by reducing the Doppler broadening of the emitted photoelectrons.

The selected cluster anions are then irradiated by a laser pulse, and the kinetic energy of the detached electrons is measured using a "magnetic bottle" time-of-flight electron spectrometer [24,25,27]. The electrons are ejected in random directions, but a strongly inhomogeneous magnetic field in the region of electron emission acts as a magnetic mirror and deflects all the electrons in the direction of the electron detector, resulting in a very large detection efficiency.

Some caution has to be exercised in interpreting photoemission spectra in terms of a single-particle electronic shell structure such as that in Figure 7.8. This is particularly relevant for higher binding energies. Shake-up processes can occur in which electron detachment can cause the excitation of another bound electron into a higher bound single-particle orbit. The detachment of a high-binding electron can also cause a major disruption to the electronic structure so that the final state can no longer be regarded as the initial configuration, with one electron missing. These effects generally occur at higher energies. The lower energy features usually correspond to single-electron transitions, and major configurational changes are absent so that the single-particle picture provides a useful qualitative interpretation. For a more quantitative analysis, however, one has to take into account an increase in the binding energies of single-particle states due to the change in the charge state of the cluster. Differences in the multiplet structures of the anion and the neutral can also affect the energies [28].

Photoelectron spectroscopy provides a wealth of information about metallic clusters. The EA (or VDE) plotted against cluster size shows oscillations similar to those observed in abundance spectra and VIPs, and provides an indication of the relative stability of the clusters; for simple metals the oscillations can be related to the electronic shell model. The full photoemission spectrum reflects the electronic density of states and is a monitor of the increasing electron delocalisation and the merging of different orbitals (e.g. s and p orbitals in divalent metals) as the cluster size is increased. The power of the technique is restricted to isolated clusters (those in the gas phase). Photoemission work on deposited clusters provides very limited information because of the dominance of the substrate. In addition, there is significant fragmentation of the clusters because of the necessity of using a high photon flux to get a decent signal. The status of and prospects for experimental work on the electronic structure of deposited clusters is surveyed by Meiwes-Broer [29].

We now discuss representative results from photoemission experiments on a range of cluster materials in the gas phase. VDEs are determined, in principle, for small clusters from the position of the lowest energy peak in the spectra and, for larger clusters, from the position of the signal onset. Finite resolution has to be compensated for [30,31], and sometimes, theoretical calculations are used to deduce the EA from the VDE; the difference can be 0.1 eV or more in the small clusters [30].

7.2.3.1. Alkali metals

Photoemission spectra have been published for Na_N^- ($N \leq 7$) [30,32], K_N^- ($N \leq 19$) [33], and Rb_N^- and Cs_N^- ($N \leq 3$) [34]. The photoemission data for potassium shown in the upper panels of Figure 7.10 are typical of a simple metal and illustrate the evolution of the spectra with cluster size. The spectra are from Eaton et al. [33] and are very similar to the results of McHugh et al. [30]. The lower panels display an analysis [4] based on the electronic ellipsoidal shell model (see Section 2.2.3). It is noticed that there is a correspondence between the low-energy peaks of the experimental spectra and the single-particle energy levels of the ellipsoidal shell model. These are represented by the vertical lines in Figure 7.10, which are then Gaussian broadened to simulate the behaviour at 250 K.

The higher energy peaks in the experimental data (those below which there is no corresponding feature in the model spectra) are ascribed to more complex processes such as photodetachment accompanied by the excitation to a higher level of an electron remaining in the cluster. The simple single-particle picture gives a

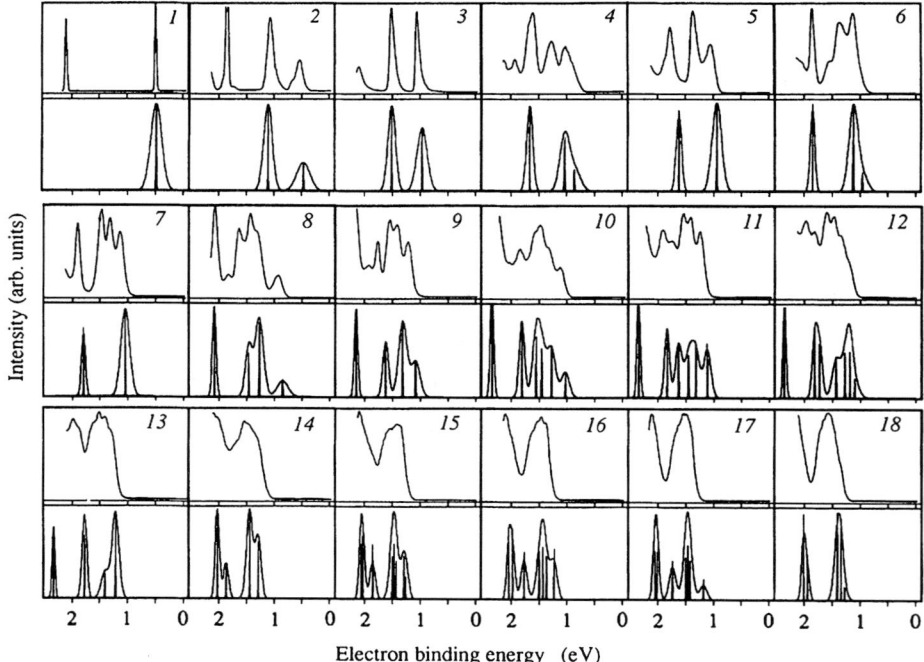

Figure 7.10. Measured photoelectron spectra of K_N^- clusters. Upper panels: experiment. Lower panels: three-dimensional oscillator model (vertical lines), with the calculated spectra obtained by broadening to $T = 250$ K. Cluster sizes are indicated. Reproduced with the permission of American Physical Society from de Heer et al. [4].

remarkably good account of the observed spectra. Generally, more structures are observed in the experimental spectra than occur in those calculated. A good example of this is the K_7 cluster. The anion contains eight valence electrons, has a filled shell configuration (1s and 1p in Figure 2.2; note we are using the electronic shell model energy level notation here, not atomic orbitals) and is therefore spherical with a triply degenerate 1p level. It has been demonstrated [34] that the inclusion of the ionic cores splits this degeneracy due to crystal field effects, which brings the calculated spectra more in line with an experiment in which the transitions associated with this is resolved.

The jellium model has also been used to calculate the EA for potassium clusters [11], and if spheroidal or ellipsoidal distortion is included, the results compare reasonably with experimental values. However, the smallest clusters ($N \leq 5$) adopt a planar geometry (Section 6.3.2), and one obviously needs to go beyond the jellium model for a realistic description. A detailed analysis of the photoelectron spectra of small alkali clusters ($N \leq 5$), using *ab initio* quantum chemical methods, is discussed by Bonačić-Koutecký *et al.* [35].

The preceding discussion has centred on anionic clusters. More recently, Wrigge *et al.* [36] have reported photoemission work on Na_N^+ clusters with $31 \leq N \leq 500$. They find that the electronic shell model predictions show excellent overall agreement with their results, apart from the case of Na_{55}^+ for which the jellium model predicts a prolate shape, while the experimental evidence points to an icosahedral structure. Shell model characteristics for both positive and negative Na ions have also been found in photoemission measurements on cluster sizes around the magic numbers of 20 and 40 [37].

7.2.3.2. Noble metals

The noble metals, copper, silver and gold, are monovalent and are expected to have some features in common with the alkali materials, but with additional features associated with photoemission from the filled *d* orbitals. The former, but not the latter, may be susceptible to interpretation in terms of the electronic shell model. There has been extensive work using photoelectron spectroscopy on clusters of all three metals: Cu [31,38,40,42,43,46], Ag [28,31,39–41,46] and Au [31,40,44–46].

The photoemission spectra for copper anions ($N \leq 410$) were recorded by Cheshnovsky *et al.* [42]. The work was followed by a comprehensive study of the spectra from Cu, Ag and Au anions [31], together with a comparison with ellipsoidal shell model predictions. The photoemission plots of Cheshnovsky *et al.* [31,42] are shown in Figure 7.11(a) and nicely display the evolution from single atom to what is essentially bulk behaviour. Two bands can be clearly identified, one growing out of the atomic 4s levels and eventually forming the conduction band, while the other narrower one arises from the atomic 3d levels and gradually converges to the bulk 3d band with the sharp onset.

The EA extracted from the Cu anion spectra [31,42] is shown in Figure 7.11(b). Two plots are seen: one for transitions involving *s* electrons (the EA) and the other marking the onset of transitions involving the *d* electrons. Linear behaviour as a function of R^{-1} is seen in both plots in accordance with Eq. (7.3), but with distinct

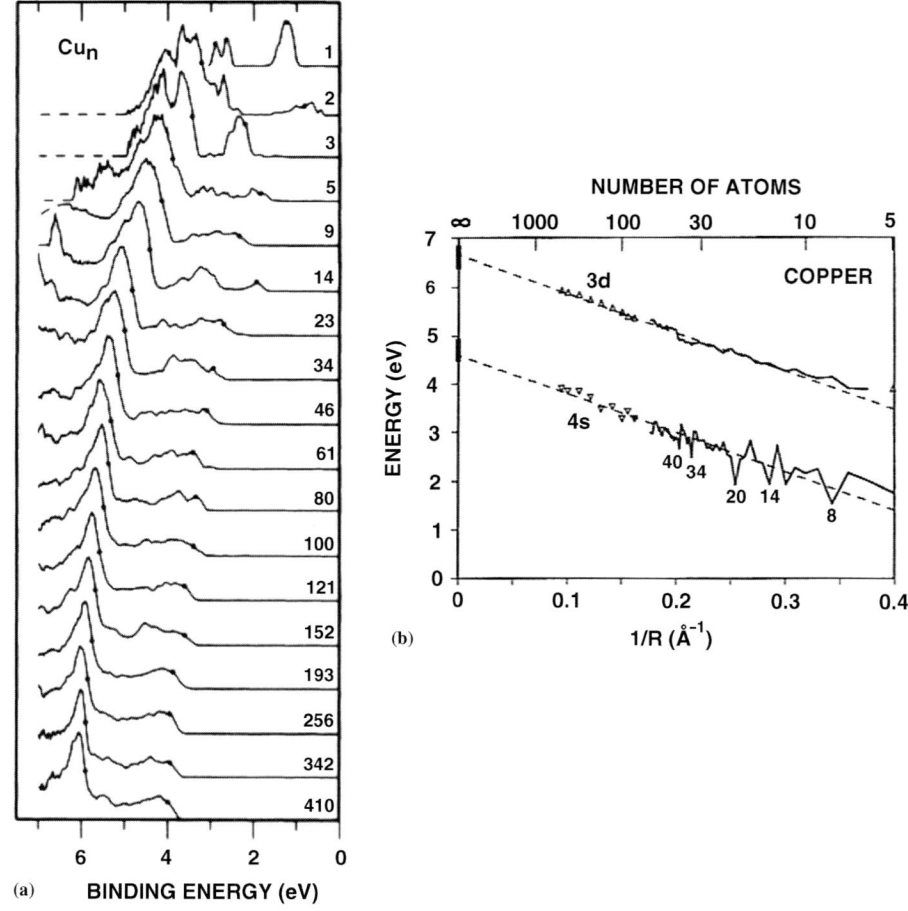

Figure 7.11. (a) Photoelectron spectra of copper anions in the size range of 1–410 atoms, taken with an excimer laser at 7.9 eV. Reproduced with the permission of American Physical Society from Cheshnovsky et al. (1990) [42]. (b) Electron affinities and 3d-band-energy onsets of copper clusters as a function of $1/R$. The black solid bars on the left-hand axis indicate the range of work functions for the various surface planes of bulk copper. Reproduced with the permission of American Institute of Physics from Taylor et al. [31].

structure in the EA for the smaller clusters with features that could be attributed to the electronic shell model. More detailed studies of the structure and a comparison with the shell model predictions have been made for Cu [31,38], Ag [31,39] and Au [31].

In bulk, the top of the 3d band density of states of copper is located about 2 eV below the Fermi energy, while in silver, the top of the corresponding 4d band is about 4 eV below the Fermi level. The photoelectron spectra for clusters of the two metals are similar, but with a greater separation of the features associated with the s and d electrons in Ag. It also means that a higher photon energy is needed to see all the features in Ag. Gold, as we shall see, is rather different from the other two metals because of relativistic effects.

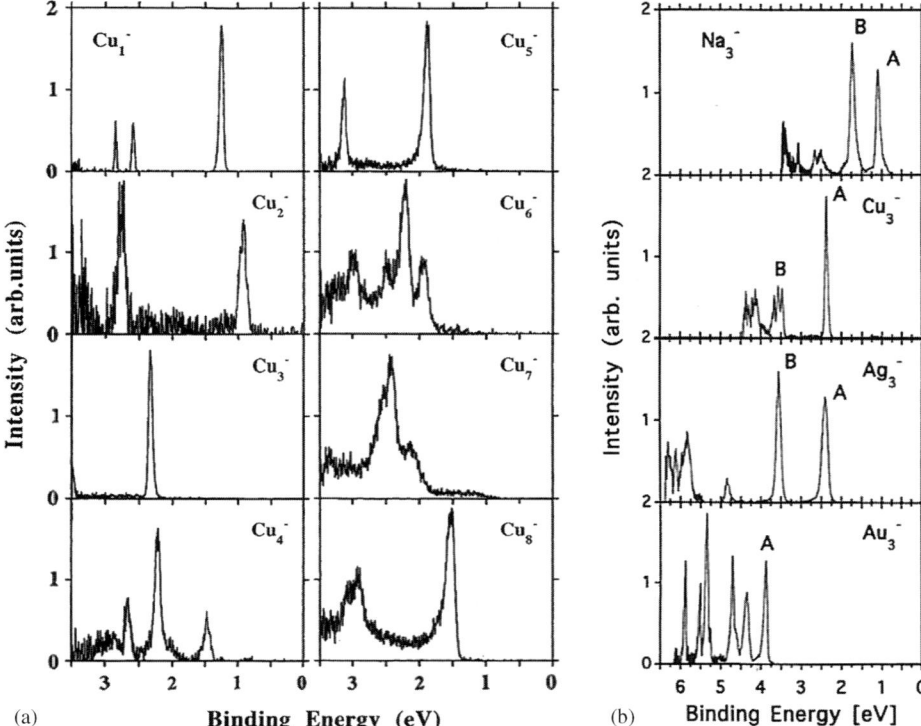

Figure 7.12. (a) Photoelectron spectra of Cu_N^- ($N = 1$–8) anions excited with a photon energy of $h\nu = 3.5$ eV. Reproduced with the permission of Elsevier Science from Eberhardt [26]. (b) Comparison of photoelectron spectra of Na_3^-, Cu_3^-, Ag_3^- and Au_3^-. The respective photon energies are $h\nu = 3.49$ eV (Na), $h\nu = 5.0$ eV (Cu) and $h\nu = 6.424$ eV (Ag and Au). Reproduced with the permission of the American Institute of Physics from Handschuh et al. [28].

To understand the features in the photoemission spectra, it is instructive to consider small clusters and the evolution of the spectra with size. A detailed analysis of the structure of the photoemission spectra for Cu and Ag for clusters of up to about 20 atoms has been carried out by Cha et al. [43] and Handschuh et al. [28]. Some typical spectra are reproduced in Figure 7.12. The diagram on the left is from Eberhardt [26] and summarises collaborative work [43,47] on Cu anions. The diagram on the right compares three-atom anionic clusters of Na and the noble metals [28].

The feature at the lowest binding energy gives the VDE. A striking aspect of Figure 7.12(a) is the dependence of the VDE on cluster size, with even-numbered clusters (with an odd number of electrons in the anion) having a low VDE, while the odd-numbered ones have a high VDE. The even–odd alternation of the VDE matches that of the abundance spectra and the IP.

We now consider the individual spectra. The lowest energy peak (at 1.24 eV) for the single-atom anion, Cu_1^-, corresponds to emission from the doubly occupied 4s orbital. The feature at 2.7 eV is assigned to photoemission from the $3d^{10}4s^2$ configuration of the anion into the excited state, $3d^9 4s^2$, of neutral Cu_1. The nine d electrons combine to states with total angular momenta $J = 3/2$ or $J = 5/2$, which are split by the spin–orbit interaction yielding the two peaks in the spectrum. We can begin the correspondence to the electronic shell model (Figures 2.2 and 2.4 of Chapter 2) by assigning the low-energy peak to excitations from the 1s shell. The peak arising from the 3d orbitals has no counterpart in the shell model.

Cu_2^- has already been discussed as a schematic introduction in Figure 7.8, and the two main peaks can be seen in the experimental spectrum. The broad feature appearing beyond 3.3 eV is associated with transitions from the 3d states. In shell model language, we have a filled 1s and a singly occupied 1p shell, and the two peaks are assigned to emission from these shells.

It is now convenient to compare the spectra of Na_3^-, Cu_3^-, Ag_3^- and Au_3^-. There are more details in the Cu_3^- spectrum in Figure 7.12(b) because a higher photon energy was used in the experiments. The three s orbitals of the trimer form three molecular orbitals: bonding σ_g, nonbonding σ_u and antibonding $2\sigma_g$. The configuration of the anion ground state can be written as $\sigma_g^2 \sigma_u^2$ and that of the neutral ground state as $\sigma_g^2 \sigma_u$. Feature A then is assigned to photoemission from the nonbonding orbital σ_u.

The first excited state has the configuration $\sigma_g \sigma_u^2$, and feature B can be assigned to photoemission from the bonding orbital σ_g. The two peaks are very clear in Na_3^- and Ag_3^-, but peak B shows splitting in the case of Cu_3^-. The reason proposed for this splitting is partial hybridisation with the 3d-derived orbitals. This is consistent with the absence of splitting of peak B in Ag_3^-. As already noted, the separation of the s and d bands is considerably greater in bulk Ag than in Cu, so hybridisation effects would be much less significant.

Both peaks are at similar energies for copper and silver because of the similarity of their s-derived states. Relativistic effects contract the 6s orbitals of gold and push peak A to a higher energy than that for the other transition metals. Strong hybridisation with 5d orbitals prevents the assignment of a B-type peak. The remaining features in the spectra of the transition metals are due to shake-up processes.

The correspondence of the peaks with energy levels in the shell model is, as for Cu_2^-, with 1s and 1p, the latter now being doubly occupied.

As we go to larger clusters, the shell model is useful in interpreting gross features. The shell model 1p level is filled at Cu_7^-, and occupation of the 1d level begins at Cu_8^-, which is marked by a clearly split off sharp peak at about 1.6 eV. Cha et al. [43] continue the study to Cu_{18}^- and find a similar split-off peak that is not present in the Cu_{17}^- spectrum. The 1d shell filling is completed at the Cu_{17}^- anion, and occupancy of the next level (2s) starts at Cu_{18}^-. Mechanisms that are not integral to the shell model, such as multiplet splitting, have to be invoked to explain other features of the spectra, such as the split peaks in Cu_4^- and Cu_6^-.

We have already been alerted by Figure 7.12(b) to the fact the Au appears to behave somewhat differently from the other two metals. The noble metals tend to form linear or planar structures for small clusters as we have discussed in Section 6.3.3. The Cu_N^- and Ag_N^- anions adopt three-dimensional geometries at $N=6$, while Au_N^- remains planar till around $N=14$ (Table 6.3 in Section 6.3.3), a situation somewhat removed from the assumptions made in the spherical jellium model. However, the fact that Cu_N^- appears to be a three-dimensional cluster for $N \geq 6$

lends plausibility to the shell model discussion of the spectra of Cu_7^- and Cu_8^-. Even up to about 20 atoms, and as three-dimensional clusters, the structures (particularly Au) generally show rather low symmetry. An interesting special case is Au_{20}, which exhibits a high symmetry tetrahedral structure [48].

The size range $N = 50$–60 is interesting because it encompasses magic numbers of both the electronic and geometric shell models at 58 and 55, respectively. Häkkinen et al. [46] have studied the noble metals in this size region in what is something of a *tour de force* in the complementary use of photoelectron spectroscopy and theory to gain insight into the structure of clusters.

The photoemission spectra for the noble metal anions ($N = 53, 55, 57, 58$) are shown in Figure 7.13. The electrons from the atomic s orbitals form a half-filled free-electron-like band that will overlap with the d band but only at some electron volts below the Fermi level. Low-binding-energy features in the photoelectron spectra can be assigned to emission from the s band as we have seen already. According to the electronic shell model, clusters with 58 valence s electrons should have a closed-shell

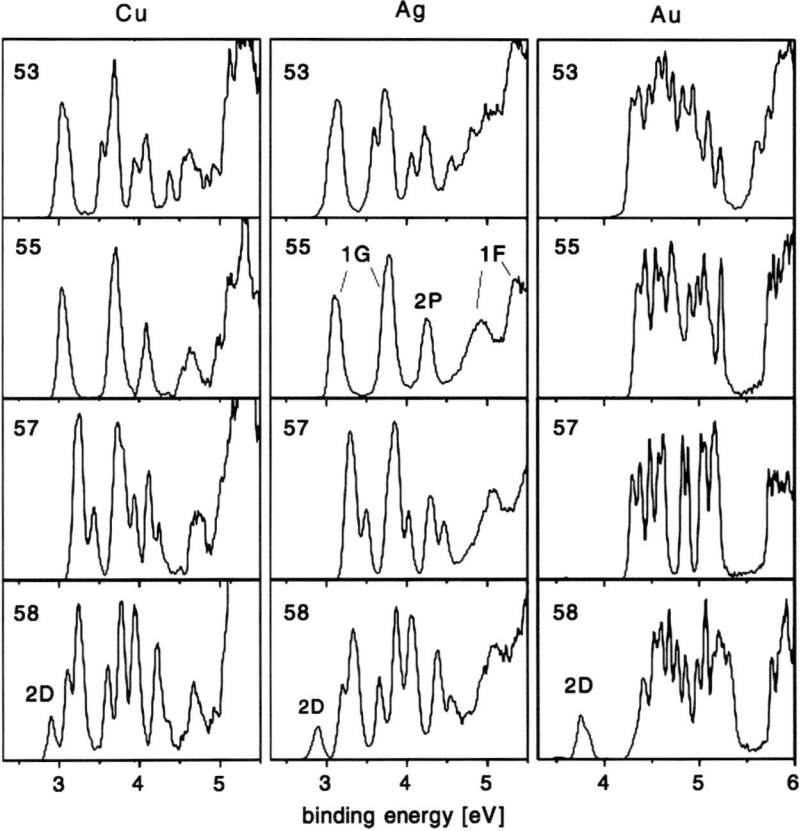

Figure 7.13. Photoemission spectra of Cu_N^-, Ag_N^- and Au_N^- for $N = 53, 55, 57, 58$ obtained at a photon energy of 6.42 eV. Some of the peaks are associated with emission from particular electronic shell levels and labelled accordingly. Reproduced with the permission of the American Physical Society from Häkkinen et al. [46].

configuration. This occurs at $N = 57$ for the anion and, in all three plots, it can be seen that going from $N = 57$ to $N = 58$ introduces a new peak corresponding to single occupation of the $2d$ level (see Figure 2.2 of Section 2.2.1).

There are two other striking features of the plots. The most obvious is the difference in appearance of the Au spectrum from those of Cu and Ag, which are rather similar to each other. The second is the absence of any splitting of the peaks in the Cu_{55}^- and Ag_{55}^- spectra, which indicates a highly symmetric structure for these clusters, something characteristic of a geometric shell model magic number.

It seems then that there are aspects of both models at play here: the new peak at $N = 58$ for all three metals receives a ready explanation from the electronic shell model, while the evidence of high symmetry at $N = 55$ for copper and silver is best understood through the geometric shell model. Gold, on the other hand, shows no evidence at all of any symmetries.

Häkkinen et al. [46] also performed density functional theory calculations of the electronic density of states of silver and gold clusters for several different cluster structures and compared their results with the experimental photoemission spectra. For Ag_{55}^-, only the cuboctahedral and icosahedral structures yielded densities of states that were reminiscent of the experimental data, and energy considerations identified the icosahedron as the preferred geometry.

Scalar relativistic corrections are necessary for a realistic calculation of the Au_{55}^- density of states. A non-relativistic calculation resulted in behaviour similar to the other two metals. In the relativistic calculation, Au_{55}^- did not adopt either cuboctahedral or icosahedral geometry but rather a low-symmetry ground state whose density of states has a qualitative resemblance to the measured spectra.

7.2.3.3. Transition metals

We now move to the next level of complexity, the transition metals with unfilled $3d$, $4d$ or $5d$ bands. Photoelectron spectroscopy has been used to study vanadium [49], titanium [50,51], niobium [52], chromium [53], palladium [54] and platinum [54] and, of those that show ferromagnetism in bulk, iron [55,56], cobalt [57–60] and nickel [51,54,58,60].

It is instructive to compare the photoemission spectrum of a transition metal cluster with the spectrum of a noble metal cluster, as has been done by Ganteför and Eberhardt [54] for Ni_N^- and Cu_N^- ($N = 2$–9). Their results are reproduced in Figure 7.14. Although there are minor differences due to changed experimental conditions (such as different photon energies), the essential features in the Cu_N^- spectra are the same as in Figure 7.12(a). We compare, first, the spectra for Ni_2^- and Cu_2^-. The peaks marked A and B, which are due to emission from the bonding and antibonding states derived from the atomic $4s$ orbitals, occur in both. The structure associated with the $3d$-derived orbitals is clearly at a higher energy than both peaks for Cu_2^- but lies in between the two in the case of Ni_2^-. Even so, there, both peaks are still quite sharp in Ni_2^-, so there is very little hybridisation between the s and d orbitals.

The low-energy peak (A) persists for larger clusters of copper and for nickel also up to $N = 6$, but at Ni_7^-, it has disappeared as a distinct entity, and the shape of the spectra then changes rather little as we move to larger clusters. The behaviour is interpreted as an indication of the onset of s–d hybridisation and an explanation of

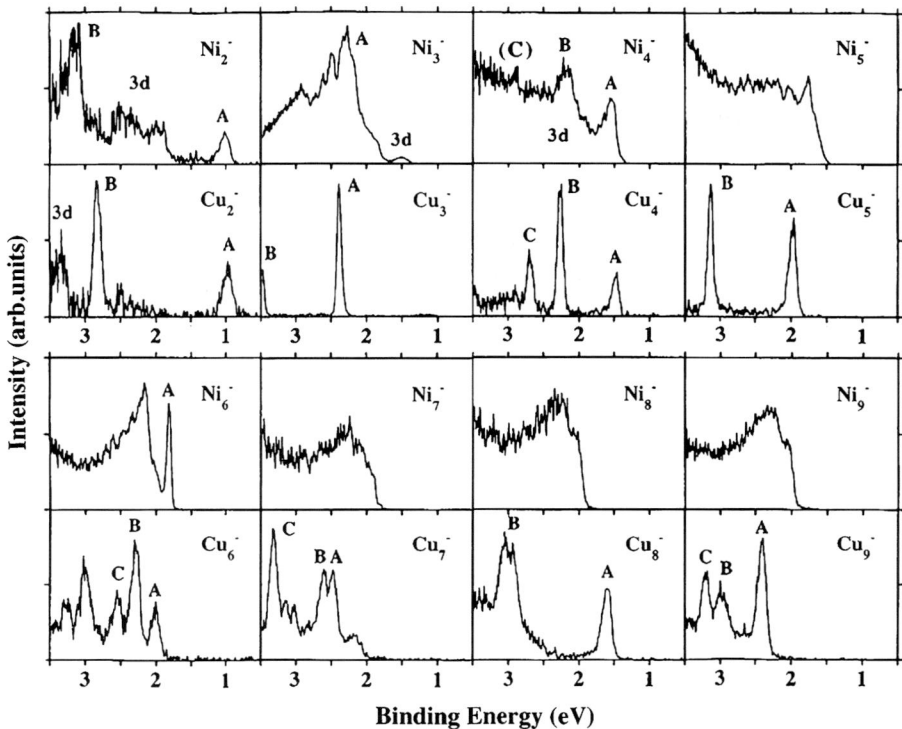

Figure 7.14. Comparison of the photoelectron spectra of Ni_N^- and Cu_N^- for $N = 2$–9. The photon energy is 4.0 eV. For marked features, see text. Reproduced with the permission of the American Physical Society from Ganteför and Eberhardt [54].

the rapid initial fall in the magnetic moment from its atomic value as the size of the cluster increases. This is discussed in Section 8.4.1.

Similar behaviour has been observed in iron by Wang et al. [56] and in cobalt and nickel by Liu et al. [60]. They use low photon energies to enhance the resolution of the threshold features and higher photon energies to access a wider range of binding energies. Photoemission spectra for Fe_N^- [56] at two photon energies are shown in Figure 7.15. The convergence of the spectra to the form measured in the bulk can be seen from the high-photon-energy data in Figure 7.15(b). A distinct low-energy peak is apparent in the low-photon-energy data till $N = 15$, and from the high-photon-energy data, convergence to the bulk spectrum is complete at about $N = 25$.

7.2.3.4. Divalent and trivalent metals

The principal focus of interest in the divalent material is the size-induced metal–insulator transition [37]. These elements, as atoms, have an s^2 closed-shell configuration; in the bulk, the bands derived from the s and p atomic orbitals overlap to form a conduction band. In aluminium, as an example of a trivalent element, the atomic configuration is $[Ne]3s^23p^1$, with an energy separation between the $3s$ and $3p$ orbitals of 3.6 eV. Again in bulk, there is a merging of the s and p

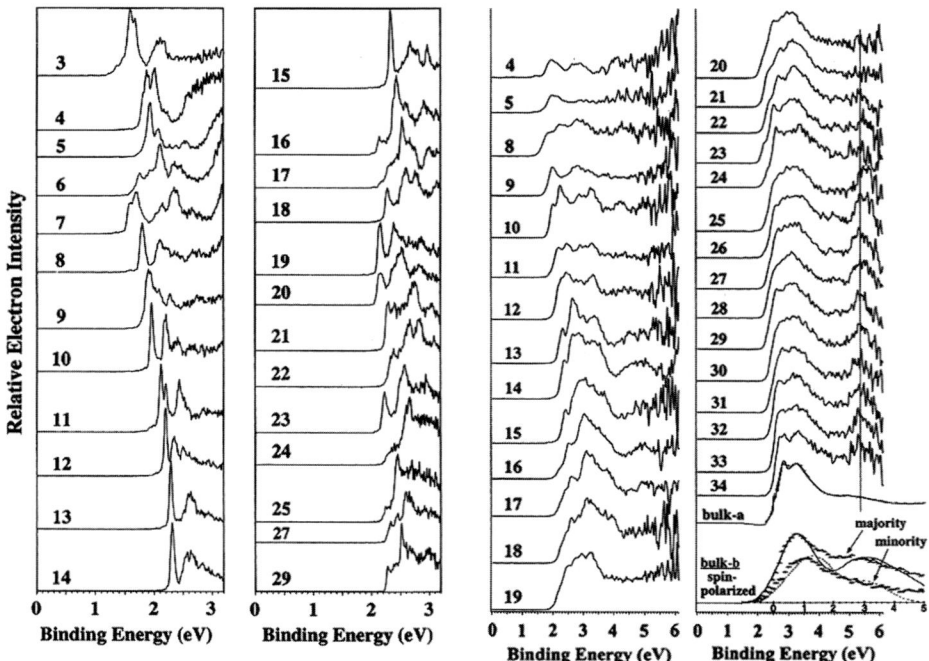

Figure 7.15. Photoemission spectra of Fe_N^- clusters: (a) $N = 2$–29 using photon energy of 3.496 eV and (b) $N = 4$–34 with photon energy of 6.424 eV; the spectra from two bulk measurements are included for comparison. Reproduced with the permission of Elsevier Science from Wang et al. [56].

bands. Peaks in the photoemission spectra of divalent and trivalent materials can be assigned to an s or p origin and the evolution of the gap between the peaks as a function of cluster size allows one to monitor the overlap process. Photoemission data are available for Hg [37,61,62], Mg [63], Zn [37], Al [23,39,64] and Ga [23].

Photoelectron spectra for Mg_N^- [63] and Hg_N^- [37] are displayed in Figure 7.16. In both plots, one can clearly see two well-resolved peaks for small clusters and their gradual approach as the cluster sizes increase. For Hg_N^-, the gap between the peaks is discernible throughout the spectra to $N = 250$, putting the insulator–metal transition at a significantly larger cluster size than that deduced by other techniques. In contrast, for Mg_N^-, all trace of two distinct peaks has vanished by $N = 16$. Relating photoemission data to the gaps predicted by theoretical calculations is a matter of some subtlety [65,66]. A further discussion of the divalent and trivalent materials is given in Section 6.3.6.

7.3. Optical spectroscopy

7.3.1. Introduction

The optical properties of small metal particles are very different from those of the bulk, a fact that has been employed with spectacular effect by the artists of

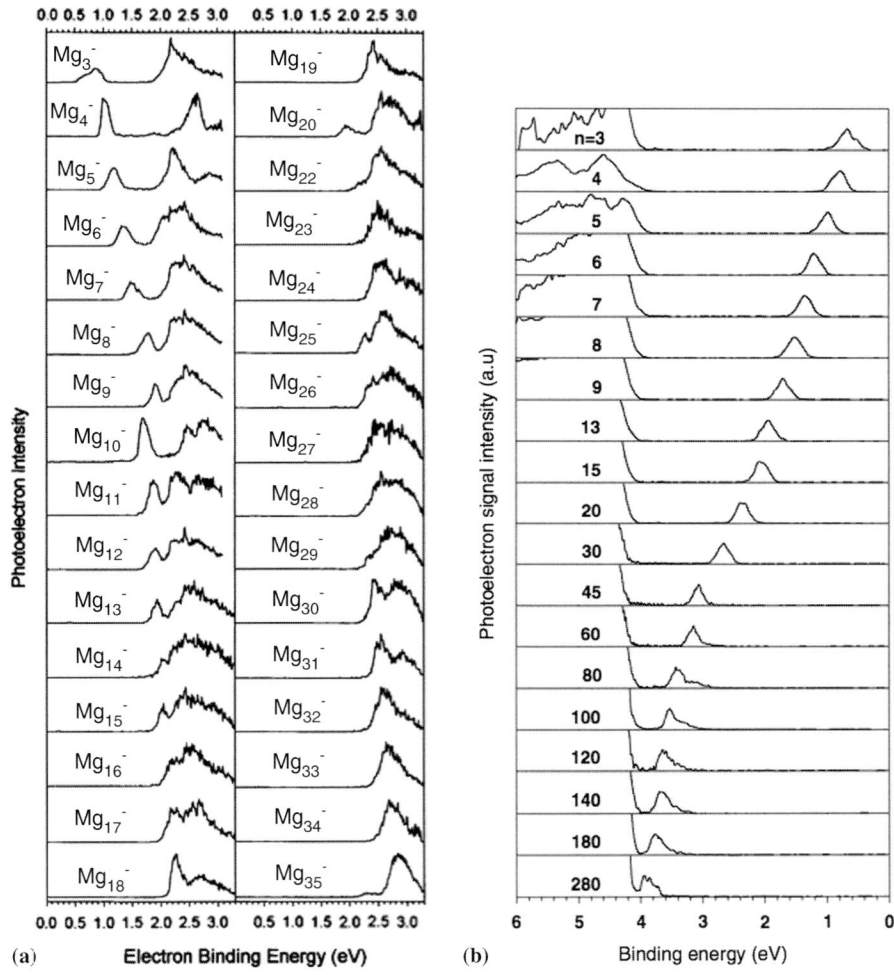

Figure 7.16. Photoelectron spectra of: (a) Mg_N^- with photon energy of 3.493 eV, reproduced with the permission of the American Physical Society from Thomas et al. [63], and (b) Hg_N^- with photon energy of 7.9 eV, reproduced with the permission of Annual Reviews from von Issendorff and Cheshnovsky [37].

mediaeval church windows. Colloidal particles of gold, for instance, have a deep-red colour, as discussed by Michael Faraday [67] in a lecture in 1857 entitled "Experimental Relations of Gold (and Other Metals) to Light". Gold, by contrast, is yellow in reflection and green in transmission. A nice example of the red colouration, which is probably due to embedded gold nanoparticles, is seen in the stained glass roundels of the Labours of the Months dated 1480-1500 from Norwich, England [see the Victoria and Albert Museum Web site http://www.vam.ac.uk/vastatic/microsites/1220_gothic/pdf/stained_glass_roundels.pdf]. Perhaps the most famous of the ancient examples of the use of metallic nanoparticles to produce a dramatic effect is the Lycurgus cup, a Roman artefact dating from the fourth century [see [26] or the British Museum Web site http://www.britishmuseum.org/explore/highlights/highlight_objects/pe_mla/t/the_lycurgus_cup.aspx]. The vessel appears red in transmission and pale green in reflected light, an effect that arises from the 70-nm particles of a gold–silver alloy embedded in the glass [68,69].

Nanoparticles were the essential element in the decoration of mediaeval and renaissance ceramics with lustre [70,71]. The technique was first used in the ninth century in Mesopotamia and gradually spread along the Mediterranean basin, reaching its highest level of development in Gubbio and other towns in Umbria, Italy, in the 15th and 16th centuries. Lustre consists of a thin film of silver and copper clusters with diameters up to a few tens of nanometres deposited onto a tin-opacified lead glaze. Lustre produces iridescent reflections of different colours.

The principal difference between the nanoparticle and the bulk is the so-called surface plasmon polariton, which can be excited in a small particle with diameter less than the wavelength of light, but not in the bulk material. The electrons in the metal are set into oscillation with respect to the positive-ion background, with a restoring force due to the surface polarisation of the particle. The resonant behaviour typically occurs in the visible part of the spectrum. The absorption associated with the resonance is further increased due to surface scattering of the electrons if the size of the particle is less than the bulk mean free path of the electrons (typically some tens of nanometres). The theory for the behaviour was first developed by Gustav Mie in 1908 [72].

There are a number of aspects of the resonant behaviour that are now of considerable interest because of potential applications. The large optical absorption and, for the larger particles, significant scattering of light means that their presence is highly visible. Strong absorption also implies a significant rise in temperature of the particle. Near resonance, there is strong enhancement of the electric field both in the particle and in its immediate environment.

Raman spectroscopy involves the inelastic scattering of light by molecules, with a shift in frequency due to the excitation of vibrational modes. Detailed structural information about a molecule can be gleaned by the technique. The main difficulty with the method is the extremely small scattering cross sections, so that a large number of molecules are required to yield useful spectra. If the molecule is adsorbed on a metallic nanocluster, the enhanced electric field near the surface can be exploited to produce a very large magnification in the field exciting the Raman effect. A similar enhancement occurs in the emitted field, so that the overall magnification is huge and single-molecule spectroscopy is possible.

Surface-enhanced Raman scattering (SERS) was first observed using colloidal particles by Creighton *et al.* [73] a few years after its detection from roughened metal electrodes [74], and the theory of SERS was developed by Messinger *et al.* [75] and Wang and Kerker [76]. Enhancements of about eight orders of magnitude can be achieved [77–84].

For applications, it is essential to be able to tune the surface plasmon resonance to a desired frequency. Single particles are rather inflexible in this regard, and greater versatility has been introduced with the development of core–shell particles comprising a dielectric core within a metallic shell [85–89]. Alternative approaches use nanorods [90–93] rather than nanospheres or link two or more particles together to hybridise the resonances of individual clusters [94–96]. The resulting plasmon band is very sensitive to interparticle distance. Linear chains of nanoparticles have also been proposed as possible conduits for electromagnetic energy transport (optical interconnects) [97–101] and as an efficient nanolens [102]. One way of linking metallic nanoparticles together is to functionalise them with single-stranded DNA oligonucleotides. The addition of a complementary single-stranded DNA template results in the self-assembly of the nanoparticles into dimers or more complex entities via the Watson–Crick base-pairing interactions. There are a number of schemes for the self-assembly of functionalised nanoparticles using DNA [103–105].

The colour change that results from bringing two particles into close proximity is visible to the naked eye. This is the basis of one of a number of strategies for using functionalised nanoparticles as sensors of biological or chemical molecules [79,94,106–112]. Another exciting prospect is in tumour therapy [113,114]. Optical penetration through tissue is optimal in the near-infrared part of the spectrum. Core–shell particles can be tuned to absorb in this frequency region, and if they can be functionalised for delivery at the site of a tumour, the resultant temperature rise in the particle should cause tumour ablation without unwanted tissue damage.

Some applications and potential applications of metallic nanoclusters will be discussed in some detail in Chapter 10. Our objective here is to outline the optical properties of single-element clusters, the theory of which forms the basis for subsequent developments. A comprehensive account of most aspects of the subject through to the mid-1990s has been given by Kreibig and Vollmer [115]. There are a number of reviews [105,116–120] that survey subsequent developments, including some of the science behind the potential applications alluded to above.

The dielectric function of the bulk material plays an essential role in the optical behaviour of metallic nanoparticles, so we begin by recalling the essential features.

7.3.2. *Bulk dielectric function*

The Drude–Lorentz–Sommerfeld model [121,122] of a free electron gas provides the simplest basis for a discussion of the optical properties of a metal. The response of a free electron of mass m and charge $-e$ to an external field $E = E_0 \exp(-i\omega t)$ is given by

$$m\frac{d^2\mathbf{r}}{dt^2} + m\Gamma\frac{d\mathbf{r}}{dt} = -eE, \qquad (7.8)$$

where Γ is a phenomenological damping constant. The polarisation is given by $P = -ne\mathbf{r}$, where n is the electron density, and assuming the system is isotropic, the dielectric function is related to the polarisation by $\varepsilon = 1 + 4\pi P/E$ [$\varepsilon = 1 + P/(\varepsilon_0 E)$ in SI units]. The frequency-dependent dielectric function is straightforwardly obtained from these equations

$$\varepsilon = 1 - \frac{\omega_p^2}{\omega^2 + i\Gamma\omega}, \qquad (7.9)$$

where ω_p is the plasma frequency

$$\omega_p^2 = \frac{4\pi ne^2}{m} \text{ or } \omega_p^2 = \frac{ne^2}{\varepsilon_0 m} \text{ in SI units.} \qquad (7.10)$$

We give the expressions in both Gaussian and SI units, but note that the former is more commonly used in the literature.

For small damping ($\Gamma \ll \omega$) the real and imaginary parts of the dielectric function ($\varepsilon = \varepsilon_1 + i\varepsilon_2$) can be written as

$$\varepsilon_1 \approx 1 - \frac{\omega_p^2}{\omega^2}, \quad \varepsilon_2 \approx \frac{\omega_p^2 \Gamma}{\omega^3}. \tag{7.11}$$

The damping constant can be related to the electron mean path ℓ by $\Gamma = v_F/\ell$, where v_F is the Fermi velocity.

The dielectric function is related to the complex index of refraction by $n + ik = \sqrt{\varepsilon}$. The reflectivity is given by

$$\mathcal{R} = \frac{(n-1)^2 + k^2}{(n+1)^2 + k^2}. \tag{7.12}$$

The absorption coefficient α, which describes the fractional decrease in intensity I with distance, $\alpha = -I^{-1} dI/dr$, is given by

$$\alpha = \frac{2\omega k}{c}. \tag{7.13}$$

The free electron plasma frequency manifests itself in two ways in bulk materials. The reflectivity is close to 100% for frequencies below ω_p and drops rapidly for higher frequencies. The cross section for electron energy loss experiments is proportional to $-\text{Im}(1/\varepsilon)$, and consequently, a sharp peak is observed at the plasma frequency.

The behaviour in alkali metals like sodium is close to the free electron model. The volume plasmon energy from Eq. (7.10) is $\hbar\omega_p = 5.95$ eV. Corrections accounting for ion core polarisation reduce [121] this to 5.58 eV. The corresponding figures for potassium are 4.29 eV (free electron) and 3.86 eV (corrected or core polarisation). However, the departure from free electron behaviour is much more dramatic for those elements for which interband transitions are significant [123].

Silver, for example, has a free electron plasmon energy of about 9.2 eV, but the effect of interband transitions reduce that to around 3.8 eV. This is illustrated in Figure 7.17. The top panel of Figure 7.17(a) displays the real and imaginary parts of the dielectric function deduced from experimental reflectance measurements. It shows reasonable free electron like for energies less than about 3 eV but departs strongly from Drude behaviour due to excitations of the 4d electrons to the unfilled 5s band. It is convenient to decompose the dielectric function into two contributions:

$$\varepsilon = \varepsilon^{(f)} + \delta\varepsilon^{(b)}, \tag{7.14}$$

where $\varepsilon^{(f)}$ represents the free electron part (from the s electrons) and $\delta\varepsilon^{(b)}$ the interband part. Eq. (7.14) is often written in terms of free electron and interband susceptibilities as $\varepsilon = 1 + \chi^s + \chi^d$. The decomposition of the real part, ε_1, according to Eq. (7.14), is shown in Figure 7.17(b). The threshold energy for interband transitions is also shown in the figure. If interband transitions were absent, the bulk

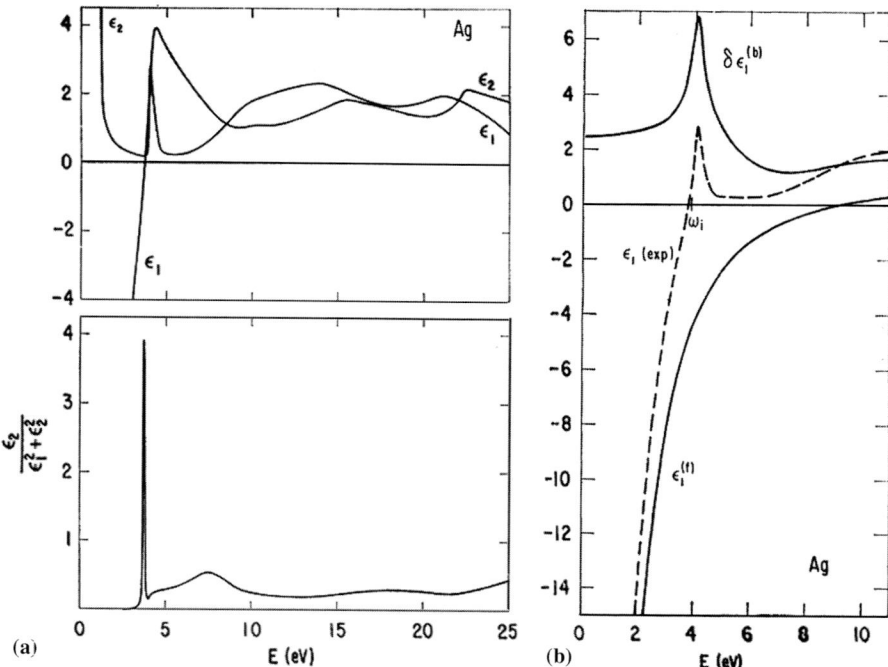

Figure 7.17. (a) Real and imaginary parts of the dielectric function for bulk silver obtained by a Kramers–Kronig analysis of experimental reflectance data (top panel); energy loss function, $-\mathrm{Im}(1/\varepsilon)$ (lower panel). (b) The real part of the dielectric function (broken line) from (a) has been decomposed into free and bound contributions according to Eq. (7.14). The threshold energy for interband transitions is indicated by ω_i. Reproduced with the permission of the American Physical Society from Ehrenreich and Philipp [123].

plasma resonance would be given by Eq. (7.11) (i.e. where $\varepsilon_1^{(f)} = 0$). Figure 7.17(b) indicates that this is at 9.2 eV, but the condition $\varepsilon_1 = 0$ pushes it to much lower energy, and this occurs where ε_2 is small, resulting in a very sharp energy loss peak, as is seen in the lower panel of Figure 7.17(a). The plasma resonance can be regarded as a hybrid cooperative effect involving both the d electrons and the conduction electrons or alternatively as due to the conduction electrons screened by a frequency-dependent dielectric constant.

7.3.3. The Mie theory

We now consider the response of a metal cluster to an external electric field. The positive charges are assumed to be static and the negative charges free to move under the influence of the field, resulting in a polarisation of the cluster. If the diameter of the cluster is much less than the wavelength of light, the electric field at any instant of time is uniform across the cluster, and a quasi-static approach can be used. The treatment is appropriate as long as the energy of the incident photons is less than the ionisation threshold. It is also assumed that the particle is big enough that macroscopic optics (Mie theory) will dominate over quantum effects. The size down to which the Mie theory is conceptually valid has been a topic of much debate, and we will return to this point later.

A simple quasi-static treatment is as follows. Consider the cluster as a sphere with dielectric constant ε in a medium with dielectric constant ε_m subject to a homogeneous static electric field E_0. The field, E, inside the sphere is given by [124]

$$\frac{E}{E_0} = \frac{3\varepsilon_m}{\varepsilon + 2\varepsilon_m} \tag{7.15}$$

For future reference, we note that the polarisability, α, of the particle in vacuum is given by

$$\alpha = R^3 \frac{(\varepsilon - 1)}{(\varepsilon + 2)}, \tag{7.16}$$

which comes from Eq. (7.15) by the standard relations of electrostatics: $p = VP = \alpha E_0$ and $P = (\varepsilon - 1)E/4\pi$, where V = volume of the particle. The static polarisability of a metal is $\alpha = R^3$ since ε diverges in the $\omega \to 0$ limit.

For a simple metal, we use the Drude formula, Eq. (7.11), for the frequency-dependent dielectric function in Eq. (7.15). The resonant (or Mie) frequency, ω_M, is given by setting the real part of the denominator in Eq. (7.15) to zero, yielding

$$\omega_M^2 = \frac{\omega_p^2}{1 + 2\varepsilon_m} \tag{7.17}$$

or, if the sphere is in a vacuum, $\omega_M = \omega_p/\sqrt{3}$. The collective excitation is generally referred to as a surface plasmon.

The above is a back-of-envelope derivation of the resonant frequency. Formally, the effect of spheres of arbitrary size on an incident electric field is done by a multipole expansion of scattering electric and magnetic fields [72,125–127]. The extinction cross section, σ_{ext}, can be written in terms of the forward-scattering amplitude $S(0)$ as

$$\sigma_{ext} = \frac{4\pi c^2}{\varepsilon_m \omega^2} \text{Re} S(0). \tag{7.18}$$

The scattering amplitude is expressed as a sum over the Mie coefficients, a_j and b_j, which respectively represent the electric and magnetic multipole terms:

$$S(0) = \frac{1}{2} \sum_{j=1}^{\infty} (2j + 1)(a_j + b_j). \tag{7.19}$$

Each term can be written in terms of Riccati–Bessel cylindrical functions with argument x, where $x = \sqrt{\varepsilon_m} R\omega/c$ and R is the radius of the sphere. For $x \ll 1$, we can write the terms as a power series in x. The leading terms for the electric dipole, magnetic dipole and electric quadrupole components are respectively

$$a_1 \approx -\frac{i2x^3}{3}\frac{\varepsilon - \varepsilon_m}{\varepsilon + 2\varepsilon_m}, \qquad (7.20)$$

$$b_1 \approx \frac{ix^5}{45}\frac{(\varepsilon - \varepsilon_m)}{\varepsilon_m}, \qquad (7.21)$$

$$a_2 \approx -\frac{ix^5}{15}\frac{\varepsilon - \varepsilon_m}{2\varepsilon + 3\varepsilon_m}. \qquad (7.22)$$

All other terms are $O(x^7)$ or higher.

Since x is essentially the ratio of the size of the particle to the wavelength of the incident light, we retain just the leading term in Eq. (7.20) for small particles, which (assuming a loss-free embedding medium, i.e. ε_m is real) leads to the result for the extinction cross section for a particle of

$$\begin{aligned}\sigma_{ext} &= 3\sqrt{3\varepsilon_m}\left(\frac{\omega}{c}\right) V \operatorname{Im}\left(\frac{\varepsilon - \varepsilon_m}{\varepsilon + 2\varepsilon_m}\right) \\ &= 9\varepsilon_m^{3/2}\left(\frac{\omega}{c}\right) V \frac{\varepsilon_2}{(\varepsilon_1 + 2\varepsilon_m)^2 + \varepsilon_2^2}\end{aligned} \qquad (7.23)$$

where V is the particle volume. The denominator, which determines the resonant frequency, is the same as in the simple derivation of Eq. (7.15). Not surprisingly, Eq. (7.23) is proportional to the polarisability [see Eq. (7.16) for the $\varepsilon_m = 1$ expression].

The extinction cross section comprises contributions from both the absorption and scattering of light, i.e. $\sigma_{ext} = \sigma_{abs} + \sigma_{scatt}$. The Mie theory also gives an expression for the scattering cross section. The electric dipolar contribution is proportional to x^6 (or V^2) unlike the extinction cross section, which is proportional to x^3 (or V). For small particles, therefore, the contribution from the scattering is very small, and the absorption is usually the dominant cause of attenuation. In noble metals, for example, σ_{scatt} is typically two orders of magnitude smaller than σ_{abs} for clusters with diameter ~ 20 nm. The reduction in the intensity of light in travelling a distance z through a medium containing clusters is given by the Lambert–Beer law

$$I = I_0 \exp(-\gamma z). \qquad (7.24)$$

The extinction constant is related to the cross section by $\gamma = n_{clust}\sigma_{ext}$, where n_{clust} is the number density of clusters.

Close to ω_M, we can use Eq. (7.11) to write Eq. (7.23) in terms of a simple Lorentzian

$$\sigma_{ext} = \sigma_{ext}^{max} \frac{\omega^2 \Gamma^2}{(\omega^2 - \omega_M^2)^2 + \omega^2 \Gamma^2}, \qquad (7.25)$$

where $\sigma_{ext}^{max} = 9\varepsilon_m^{3/2}\omega V/(c\varepsilon_2)$, the maximum value of the cross section. It can be seen that the cross section scales with the volume of the particle, but the position and width of the absorption peak are independent of particle size.

For large particles, additional resonances can be excited through higher order terms in the Mie expansion. Possible resonant frequencies are

$$\omega_j^2 = \frac{\omega_p^2}{1 + \varepsilon_m(j+1)/j}. \tag{7.26}$$

The $j=1$ term is the leading electric dipole resonance just discussed. For large R, terms with higher values of j become relevant, which in the limit of large j and for $\varepsilon_m = 1$ yield the familiar frequency $\omega_M = \omega_p/\sqrt{2}$ for the plasmon associated with a planar surface.

The treatment for $x \ll 1$ can be extended to ellipsoidal particles [127–129]. The depolarisation factor for a sphere is $1/3$. For an arbitrary depolarisation factor, L, Eq. (7.15) is modified to

$$\frac{E}{E_0} = \frac{\varepsilon_m}{\varepsilon_m + L(\varepsilon - \varepsilon_m)}, \tag{7.27}$$

which yields a resonant frequency

$$\omega_M^2 = \frac{\omega_p^2}{1 + \varepsilon_m(L^{-1} - 1)}. \tag{7.28}$$

For an ellipsoid with principal axes R_1, R_2 and R_3, the associated depolarisation factors are

$$L_i = \frac{R_1 R_2 R_3}{2} \int_0^\infty \frac{ds}{(s + R_i^2)\sqrt{(s + R_1^2)(s + R_2^2)(s + R_3^2)}}. \tag{7.29}$$

It is straightforward to show that

$$L_1 + L_2 + L_3 = 1. \tag{7.30}$$

Assuming a random orientation of the particles with respect to the incident beam, we have the possibility of three resonant frequencies for ellipsoidal particles. If $\varepsilon_m = 1$, there is a particularly simple relation between these frequencies,

$$\omega_1^2 + \omega_2^2 + \omega_3^2 = \omega_p^2. \tag{7.31}$$

More generally, as long as the principal axes do not differ in length too much from the radius R_0 of the related spherical particle, then approximately

$$\omega_i = \omega_M^0 \left[1 - \frac{3}{5}\frac{R_i - R_0}{R_0}\right], \tag{7.32}$$

where ω_M^0 is the resonant frequency of the spherical particle.

It should be noted that, for noble metals like silver, the simple relation, Eq. (7.17), between the volume and surface plasmon frequencies no longer applies. The volume and surface plasmon resonances are given by $\varepsilon_1 = 0$ and $\varepsilon_1 = -2\varepsilon_m$, respectively.

7.3. Optical spectroscopy

Using Eq. (7.14), the frequency, ω_V, of the volume plasmon can be expressed as

$$\omega_V^2 = \frac{\omega_p^2}{1 + \delta\varepsilon_1^{(b)}(\omega_V)}, \tag{7.33}$$

and the Mie frequency is now

$$\omega_M^2 = \frac{\omega_p^2}{1 + \delta\varepsilon_1^{(b)}(\omega_M) + 2\varepsilon_m}, \tag{7.34}$$

where ω_p is, of course, the bulk free electron plasmon frequency.

The frequency of the surface plasmon resonance for a spherical particle, predicted by the Mie theory, is independent of the particle size. For many applications, one would like to be able to fabricate a nanoparticle tuned to a particular resonant frequency. One can, of course, vary the shape of the particle, the metal from which it is made or the dielectric embedding material, but the use of core–shell particles provides additional versatility.

We consider a spherical particle comprising a core ($0 < r < qR$) and a shell ($qR < r < R$) with dielectric functions ε_1 and ε_2, respectively. For a particle in a vacuum, the polarisability is given by [127]

$$\alpha = R^3 \frac{(\varepsilon_2 - 1)(\varepsilon_1 + 2\varepsilon_2) + q^3(2\varepsilon_2 + 1)(\varepsilon_1 - \varepsilon_2)}{(\varepsilon_2 + 2)(\varepsilon_1 + 2\varepsilon_2) + q^3(2\varepsilon_2 - 2)(\varepsilon_1 - \varepsilon_2)}. \tag{7.35}$$

The expression for the homogeneous sphere, Eq. (7.16), is recovered by setting $\varepsilon = 0$ and $\varepsilon_2 = \varepsilon$. It can be seen from Eq. (7.35) that the ratio, q, of the two radii provides an additional tuning parameter in core–shell particles.

Numerical methods have been developed for arbitrary particle shapes and environments such as the finite-element discrete dipole approximation (DDA) theory [130–132].

7.3.4. Experimental methods

Kreibig and Vollmer [115] have reviewed various techniques for measuring the optical properties of clusters and summarised the advantages and drawbacks of each. We briefly discuss just two approaches here. The first is appropriate for clusters in beams and can probe small clusters (up to about 1000 atoms). The second is relevant for larger clusters ($R > 5$ nm) and can probe individual particles.

7.3.4.1. Beam depletion spectroscopies

The method is based on the fact that the absorption of a photon by a cluster creates electronic excitations that relax very rapidly among the vibrational modes providing heating of the cluster. For small clusters, the temperature rise is sufficient to cause the evaporation of an atom from the cluster, and the number of clusters fragmenting is a measure of the probability of photon absorption (i.e. the cross section). It is estimated [133], for example, that the absorption of a 2-eV photon will cause a temperature rise of about 430 K in a 20-atom sodium cluster, and that the average time for the evaporation of an atom is $\sim 30\,\mu s$.

In depletion spectroscopy [4], the clusters are produced in a supersonic expansion of a mixture of the metal vapour and an inert gas, through a nozzle, into a vacuum. The molecular beam is highly collimated and enters a detector downstream for mass analysis. The cluster beam is illuminated along its whole length by pulsed laser light that enters through a window in the detector. In the resulting fragmentation, the evaporating atom and the daughter cluster recoil out of the collimated beam. The beam depletion (reduction in cluster intensity at the detector due to the photon absorption) is a measure of the absorption cross section. The ratio, r, of the count rates at the detector with the laser on and off is measured. The depletion results from single-photon absorption, and r is related to ϕ, the number of photons per unit area per unit time in the laser pulse, by $r = \exp(-\sigma_{abs}\phi t)$, where σ_{abs} is the absorption cross section and t the exposure time. Only single-photon absorption takes place in this method.

An alternative approach is to allow multi-photon absorption and a variety of daughter fragments (not just those resulting from the loss of a single atom). In this case, size selection of the clusters is done initially and then mass analysis performed after the photon absorption to determine the size distribution of the fragments. For large clusters (greater than about 40 atoms), the probability of evaporation by single-photon excitation becomes vanishingly small, so the single-photon technique is unavailable and one has to use multistep photon excitation. For large clusters, the fragment size distribution has the form of a Poisson distribution. The absorption cross section, σ_{abs}, can be deduced from the dependence of the position of the peak of the distribution on the laser power [134]. Both methods have been successfully employed for small clusters (up to about 40 atoms), and the multistep photon method has been used for clusters of up to 1500 atoms. Variable temperature versions of the technique have also been developed [135].

7.3.4.2. *Techniques for single clusters*

Recent interest for large clusters ($2R = 5$–150 nm; $N = 4 \times 10^3$–1×10^8 atoms) has focused on the optical properties of single particles, especially those of the noble metals. The reason for this is as follows. The surface plasmons are excited by light, and this results in a resonant enhancement of the local field in and near the particle. It can be easily seen from Eq. (7.15) that the enhancement is inversely proportional to the damping constant ($\propto \Gamma^{-1}$). As we noted in Section 7.3.1, this enhancement can be exploited in a number of applications such as SERS. The enhancement is usually discussed in terms of the dephasing time (i.e. decay time for collective oscillations) T_2, which is related to Γ by $T_2 = 2\hbar/\Gamma$, so the enhancement is proportional to T_2. The SERS cross section, for example, is proportional to the fourth power of the enhancement, i.e. T_2^4, so there is considerable motivation for accurate measurements of the dephasing time.

We need to distinguish the homogeneous linewidth Γ_{hom} of a surface plasmon resonance from the inhomogeneous linewidth Γ_{inhom}. The former characterises an individual particle. There will be variations in Γ_{hom} and the resonant frequency among the individual particles in an ensemble due to differences in size, shape, surface structure and dielectric environment, and measurements on an ensemble will yield a linewidth Γ_{inhom} reflecting these fluctuations that is generally larger than Γ_{hom}. It is the local field enhancement in individual particles, however, that determine the application potential and hence the focus on single particles and the efforts to measure Γ_{hom}.

Most experiments have been on Ag or Au particles embedded in a thin film of a transparent material of known refractive index. Various methods have been employed. Particles in the size range of 20–150 nm have been studied by scanning near-field microscopy [136] and by monitoring the light scattered by the particle, using dark-field spectroscopic techniques [92,137–139]. A strong dependence of the optical properties on the particle shape has been found [138,139].

As noted in Section 7.3.3, the scattering cross section scales with V^2, whereas absorption is proportional to V, the volume of the particle. As a consequence, scattering techniques are generally restricted to particles larger than about 20 nm. Methods based on photothermal far-field absorption have been developed [140–142] to extend the range to smaller particles. Berciaud et al. [141,142] use two laser beams; one is an intensity-modulated heating beam and the other a probe beam. Absorption of the heating beam causes a time-modulated temperature increase and thus a time-modulated index of refraction in the vicinity of the particle. The propagation of the probe beam through this region produces a frequency-shifted scattered field, which is detected by its beat note with the probe field, the signal being directly proportional to the absorption cross section. Berciaud et al. [141] claim to be able to probe metallic clusters as small as 67 atoms by this method. An alternative approach [143] using a far-field optical technique is based on the modulation of the position of the particle within the intensity spatial profile of the laser beam. A technique that combines confocal microscopy and interferometry has been developed [144] to overcome the V^2 problem of scattering methods. The key to the idea is the cross-term between scattered and reflected fields, which is the quantity measured in the interferometry that scales with V.

The motivation for measuring the homogeneous linewidth can be illustrated by the results of Klar et al. [136]. They measured the linewidths of gold clusters with diameters 32–48 nm and found that $\Gamma_{hom} \approx 160$ meV, reduced from $\Gamma_{inhom} \approx 300$ meV. Clearly, the difference is significant if one is interested in an application that depends on the fourth power of the linewidth. Finally, in this review of techniques, we note that persistent spectral hole burning methods have been used [145,146] to deduce homogeneous linewidths. The technique is rather different from the ones we just discussed, not least in the fact that measurements are made on an ensemble of clusters rather than on a single particle.

7.3.5. Specific materials

The designation of clusters as large or small depends on whether they are liable to exhibit *extrinsic* or *intrinsic* size effects [115]. Most of the preceding discussion has been in terms of the quasi-static approximation in which only the electric dipole mode in the Mie theory is excited but, as the particle size increases, higher order modes [Eq. (7.19)] become more dominant, causing the plasmon absorption band to redshift and broaden. Retardation effects cause the excitation of the higher order modes because the light cannot homogeneously polarise the cluster. This is known as an extrinsic size effect and is relevant to particles with diameters larger than about 20 nm. The excitation of multipolar resonances can be induced by a reduced symmetry environment, either through the shape of the cluster itself or by the presence of a substrate [147,148].

For smaller clusters, intrinsic size effects become important. The "spill-out" effect (see below) is quantum mechanical in origin and results in a shift of the position of the surface plasmon. The size of the particle is generally smaller than the electron mean free path in the bulk material, and this will cause an increase in the damping constant Γ and a broadening of the resonance. If the particle is very small (less than about 20 atoms), the question arises as to whether it is legitimate to use the concept of plasmon. Certainly at the level of trimers or tetramers, a rich single-particle excitation spectrum is seen for which a quantum chemical description is more appropriate, but there is a grey area in which both approaches claim success despite their apparent conceptual differences. The issue has been the subject of controversy and intense discussion [5,35,115,149,150–156].

We now discuss some results from experiments on alkali and noble metals – first on small clusters with $N<100$, which are clearly in the intrinsic size-effect regime, and then on larger particles ($R \geq 2.5$ nm) – illustrating some of the work on individual clusters referred to in Section 7.3.4.

There are only a few metals, such as the alkalis, Ag, Al, Hg and Mg that show a well-developed surface plasmon resonance in vacuum ($\varepsilon_m = 1$). The requirement is a small ε_2 at a frequency where $\varepsilon_1 = -2$, a condition that is satisfied by Ag but not Cu, as can be seen clearly from plots of the dielectric function [123]. The immersion of particles in a medium of judiciously chosen refractive index can shift the resonance to a position of small ε_2. A range of colours of the light transmitted through such a solution can be obtained by suitably tuning the refractive index [157].

7.3.5.1. Small clusters

7.3.5.1.1. Sodium. Most experimental studies of optical absorption in alkali metals have concentrated on sodium as the canonical example of a simple metal. Predominantly, the work [133,135,158–170] is on small clusters with $N<100$, although there have been some reports [171,172] on large clusters ($5<R<150$ nm). Potassium [134,173], lithium [174–176] and caesium [177] have also been studied, mostly for small clusters, but with data up to $N=900$ for K [134] and $N=1500$ for Li [175].

The early work [4,133,158] showed absorption spectra that could be decomposed into one, two or three peaks, and this gave some indication that the surface plasmon picture coupled with the electronic shell model, with possible spheroidal or ellipsoidal distortions, could capture the essential physics. However, a number of difficulties with the simple model soon became apparent.

The bulk plasmon energies [from Eq. (7.10)] for Na and K are respectively 5.95 eV and 4.29 eV [121]. These values predict [Eq. (7.17)] Mie resonances for spherical clusters in vacuum at 3.44 eV for Na and 2.48 eV for K. The peaks in the experimental absorption spectra for the small clusters are significantly lower in energy (red shifted) than the classical values: 2.1–2.8 eV for Na and 1.9–2.1 eV for K. The experimental bulk plasmon energies are actually slightly smaller than those predicted, due to the neglect of ion core polarisation effects (see Section 7.3.2). However, even if the experimental values of the bulk plasmon frequency are used in Eq. (7.17), the calculated Mie resonance energies still lie above the observed values.

The red shift is usually attributed to the quantum mechanical spill-out of the electron density beyond the geometrical cluster radius given by $R = r_s N^{1/3}$. If we assume that the electrons are spread uniformly over a sphere of radius $R+\delta$ instead of one of radius R, then the density in Eq. (7.10) is reduced by a factor $[R/(R+\delta)]^3$ and the Mie frequency $\omega_M = \omega_p/\sqrt{3}$ is reduced to $\bar{\omega}_M$, where

$$\bar{\omega}_M = \omega_M \left(\frac{R}{R+\delta}\right)^{3/2} \approx \omega_M \left(1 - \frac{3}{2}\frac{\delta}{R}\right). \tag{7.36}$$

The spill-out parameter δ depends only weakly on r_s and R, so that the frequency of the surface plasmon is predicted to undergo a red shift from its classical value that scales inversely with the particle radius. Evidence of this is displayed in Figure 7.18. The plots in the figure on the left show the absorption cross section σ_{abs} for five cationic clusters that correspond to electronic closed-shell magic numbers (number of 3s electrons: 8, 20, 40, 58, 93) and should be spherical according to the model and thus simple to analyse.

All the peaks show a shoulder on the high-energy side, so the root mean square (rms) frequencies, $\omega_{rms} = \langle \omega^2 \rangle^{1/2}$, were also evaluated, where

$$\langle \omega^2 \rangle = \frac{\int_0^\infty \omega^2 \sigma(\omega) d\omega}{\int_0^\infty \sigma(\omega) d\omega}. \tag{7.37}$$

The peak positions and the rms energies normalised to ω_M are plotted in Figure 7.18(b). The trend defined by Eq. (7.36) appears to operate for $N \geq 41$, but interestingly, the shift is in the opposite direction for the smallest two clusters. The absorption in very small clusters is sensitive to their detailed atomic structure, and it is likely that their behaviour will not follow a trend that is a simple function of their radii. The normalised peak positions of the potassium clusters, K_{500}^+ and K_{900}^+, (from [134]) are also included in the plot. Both the peak positions and the rms energies extrapolate rather convincingly to the $R \to \infty$ limit. The three additional lines on the plot are theoretical predictions. "Spill-out 1" is a plot using the theoretical value of $\delta = 0.54$ Å [178] and "spill-out 2" a best fit to experimental data obtained with $\delta = 0.60$ Å (although it should be noted that this is an upper limit because the experimental spectrum was integrated in Eq. (7.37) over the energy range observed, rather than 0 to ∞). "Dyn screen" is from an alternative theory [179] based on dynamic screening that also shows an R^{-1} dependence of the peak position.

An important check on the collective nature of the resonance is the Thomas–Reiche–Kuhn or f-sum rule

$$\int \sigma(\omega) d\omega = 2\pi^2 \frac{Ne^2}{mc}, \tag{7.38}$$

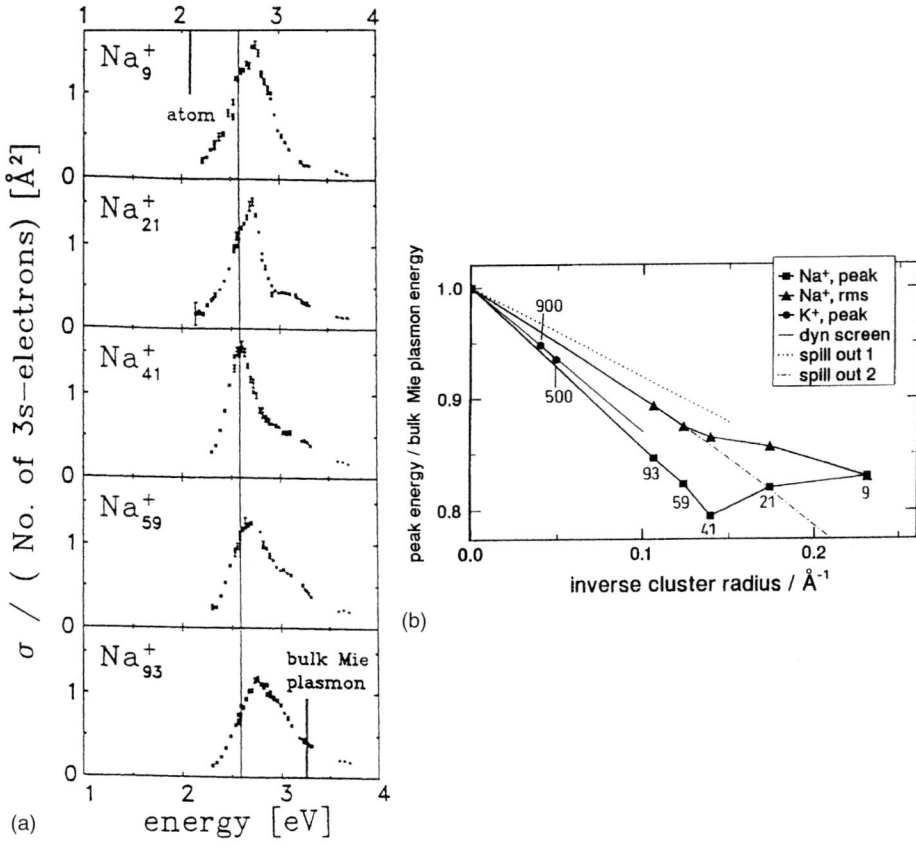

Figure 7.18. (a) Optical response of five sodium cluster ions that are spherical according to the electronic shell model. The vertical line gives the peak position for Na_{41}^+. The positions of the atomic and Mie resonances are also indicated. (b) Normalised positions of peak maxima and rms energies (square and triangular data points) versus inverse cluster radius. Normalised peak positions for K_{500}^+ and K_{900}^+ (from [134]) are also included. The significance of the three lines is discussed in the text. Reproduced with the permission of the American Physical Society from Reiners et al. [165].

which relates the integrated absorption cross section to the total number of valence electrons, N. Typically, the experimentally determined cross sections exhaust 70–90% of the sum rule [158,163], which is contrary to the predictions of the electronic shell model for which all the weight is in the surface plasmon resonance. Various refinements of the jellium model have been developed (see Brack [5]). They are mostly based on the time-dependent local density approximation (TDLDA) [152,180] or the random phase approximation (RPA) [153,154,181–183]. Note that the RPA based on the jellium model should be distinguished from the *ab initio* RPA [35].

One of the discrepancies between experiments and predictions of the simple jellium model is the splitting of the peak at electronic shell magic numbers of 20 and 40. The shoulder appearing in the absorption spectrum of Na_{21}^+ in Figure 7.18(a) could be interpreted as evidence of oscillator strength to the right of the main peak, and Bréchignac et al. [134] report a just-resolved two-peak structure for Na_{21}^+. A double-peak structure is more clearly resolved in the corresponding neutral clusters Na_{20} [160] and Na_{40} [158]. As noted above, the jellium model predicts a spherical cluster and a single peak.

An explanation of this behaviour was proposed within the frameworks of the TDLDA and RPA refinements to the jellium model in terms of the so-called "fragmentation" of the surface plasmon resonance. The fragmentation (i.e. splitting of the plasmon peak into several components) is attributed to the interference of particle–hole excitations (single-particle transitions) with the collective excitation. The degree of fragmentation depends on the proximity of the single-particle excitations, and this can vary with the charge state of the cluster. The effect is also much smaller in potassium than in sodium, so that the calculated absorption spectrum of K_{20} still shows a single peak [154] while that of Na_{20} splits into two through the fragmentation [182], both in agreement with experiment [160,173].

As noted earlier, the spill-out effect offers a potential explanation of the discrepancy between the positions of the main absorption peak observed in the experiments and that predicted by the Mie theory. The electron density is calculated self-consistently in the TDLDA and RPA theories, allowing an estimate of the shift in position due to spill-out. Unfortunately, most calculations yield a shift that is too small to bring the theory in line with experiments. Exceptions to this are the RPA calculations of Kresin [184,185] that use a Thomas–Fermi description of the valence electrons and the TDLDA approach of Saito et al. [155] that includes self-interaction corrections. For a comparison of different estimates of the resonance position, see Bréchignac et al. [134].

The damping constant, Γ, for a bulk free electron metal in Eq. (7.9) is typically about 0.02 eV, whereas the full width at half maximum of the surface plasmon resonance that appears in Eq. (7.25) is generally in the range of 0.3–0.5 eV. It is not surprising that a naive assignment of the bulk free electron value to Γ is inappropriate since the electron mean free path in the bulk is larger than the diameter of a metallic nanoparticle. Two other principal mechanisms have been proposed to explain the observed width of the resonance. One is based on the coupling of the surface plasmon resonance to ionic vibrations (electron–phonon coupling), and the other considers interference of the resonance with particle–hole states that lie close in energy (the analogue of Landau damping in solids).

The Landau damping type of broadening is inversely proportional to the cluster radius [186–189], and this is often added to the bulk damping constant, Γ_0, as

$$\Gamma = \Gamma_0 + \frac{A v_F}{R}, \tag{7.39}$$

where v_F is the Fermi velocity and A a parameter. Surface scattering, which shortens the electron mean free path, can also be included in this functional form. Evidence of this characteristic radius dependence has been observed in larger supported or embedded clusters (see Section 7.3.5.2). For the small alkali metal clusters for which measurements on free particles have been performed, line broadening through electron–phonon coupling appears likely as the dominant mechanism.

A number of theoretical approaches have treated this type of line broadening. Bertsch and Tománek [156] consider small ellipsoidal distortions of the jellium sphere. The distortion results in a shift of the plasmon energy and a change in the surface energy of the cluster, from which the electron–phonon coupling can be determined. The thermal fluctuations of the cluster surface are treated adiabatically and lead to a broadening of the surface plasmon resonance, with temperature dependence $\Gamma \propto \sqrt{T}$. Akulin et al. [190] use the local density method for the spherical jellium model in the RPA, so that the effect of the ionic cores is included in an average sense as a positively charged background. Parameterised distortions to this average charged background are introduced in the "random matrix model" [191], and again an rms coupling to the electronic terms that is proportional to $\sqrt{T}$ appears. The results compare with experimental observations on Na_{21}^+ [190].

The next level of sophistication is the work of Pacheco and Schöne [192], who include the ionic structure explicitly into the calculations and perform a Monte Carlo sampling of phase space. Their model is based on the self-consistent jellium model, to which the effect of the ionic structure is introduced by second-order pseudopotential perturbation theory [193]. This enables the energy of a cluster to be determined for arbitrary geometry and hence also the positions of the atoms in the ground state configuration. The optical absorption is calculated via the TDLDA equations [194]. The canonical Monte Carlo simulation explores the available phase space (the set of coordinates of the ions in the cluster). The line shape for the optical absorption is determined by the aggregated response over the Monte Carlo trajectory (typically $\sim 10^4$ steps). Pacheco and Schöne [192] performed a detailed study of Na_8 and showed that two peaks in absorption plot at low temperatures merged for temperatures of 100 K to produce a single broad peak that is consistent with room temperature measurements.

There have been series of measurements to determine the temperature dependence of the absorption spectra of sodium in particular [135,164,166,167]. Schmidt et al. [135] reported results for Na_N^+ ($N = 4$–16) over a wide temperature range. A representative selection ($N = 8$–12) is reproduced in Figure 7.19. Measurements on two of the cluster sizes, Na_9^+ and Na_{11}^+, have been made at $T = 35$ K, and a significant structure in the cross sections can be seen, which is not resolved at higher temperatures due to the line broadening. The two-peak structure in the Na_9^+ spectrum at low temperatures merges into a single peak at room temperature (c.f. the theoretical predictions of Pacheco and Schöne [192] for Na_8).

Schmidt et al. [135] note that the resonance peaks in the smaller clusters are observable over the full temperature range and one can follow the broadening of the individual peaks, whereas in the larger clusters, the peaks tend to merge and fewer can be resolved at high temperatures. They examine the temperature dependence of linewidths Γ for Na_4^+ and Na_5^+ and find that these are compatible with a $\sqrt{T}$ behaviour [156], although not a conclusive verification of this dependence.

Although the simple jellium model gives a reasonable qualitative account of absorption cross sections, it is limited because of the neglect of the ionic structural geometry. By contrast, calculations based on the methods of quantum chemistry [35,195] show that the structure in the absorption spectra is sensitive to the positions of the atoms in a cluster, and one should perform a geometry optimisation or at least determine the structural isomers that are close in energy if one wishes to

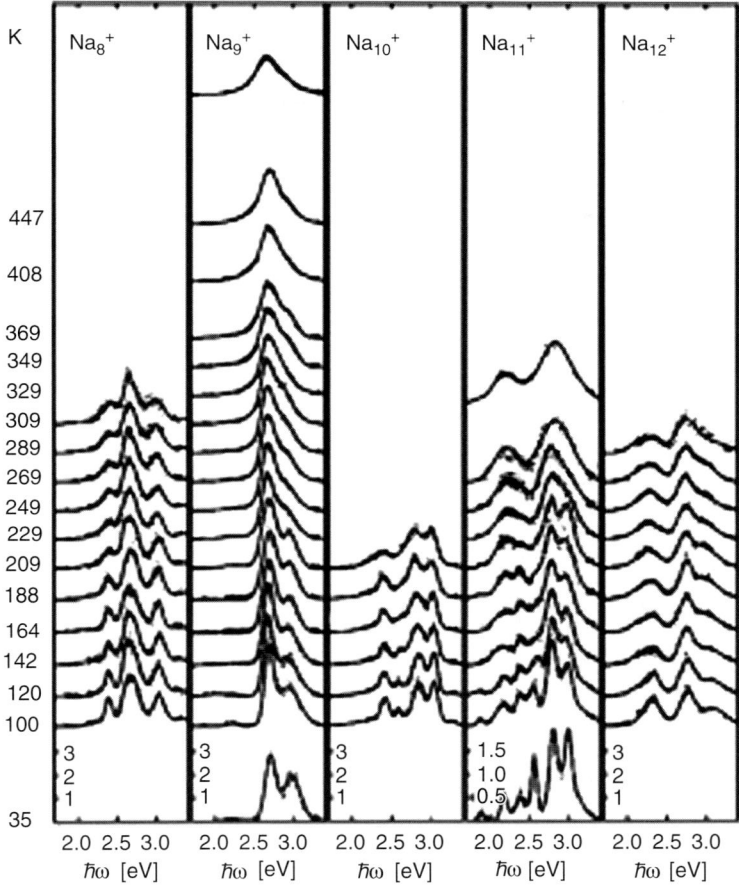

Figure 7.19. Photoabsorption cross sections in Å^2 for Na$_N^+$ ($N=8$–12) as a function of photon energy. The cross section scale is shown at the bottom left of each column. The temperature in kelvin is on the left side. Reproduced with the permission of the American Physical Society from Schmidt et al. [135].

reproduce the observed spectra. The quantum chemistry methods are computationally expensive, but studies have been done up to Na$_{21}^+$ [195].

There have been a number of studies that have attempted to include an account of the atomic structure within the jellium model [192–194,196–199]. Generally, the starting point is the structureless self-consistent jellium model treated in the local density approximation of density functional theory to which the ionic structure is introduced via the second-order pseudopotential perturbation theory. In most cases, a local pseudopotential is used, parameterised in a convenient model form.

This is the procedure adopted by Kümmel et al. [199] in calculations of the absorption spectra for sodium clusters up to $N=59$. Their results for Na$_9^+$ are reproduced in Figure 7.20. According to the basic jellium model, this is a magic number cluster (with eight valence electrons) and should therefore be spherically symmetric with a single absorption peak. However, the experimental plot in Figure 7.20 shows that the main peak is split into two, and there is an additional

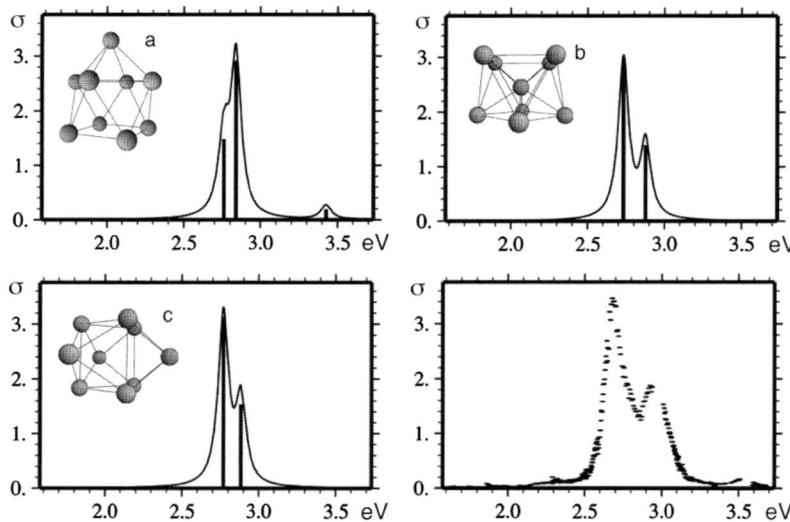

Figure 7.20. Low-energy configurations for Na_9^+ and corresponding optical absorption spectra from the collective model. A phenomenological line broadening of width 0.08 eV has been applied. The dotted curve (bottom right) is the experimental cross section for $T = 105$ K [164,170]. Reproduced with the permission of the American Physical Society from Kümmel et al. [199].

small feature at about 3.5 eV. The calculations find three isomers close in energy, as shown in the figure. Those indicated by (a) and (c) are degenerate ground states, and (b) lies just 0.12 eV higher in energy. Importantly, the positions of the principal resonances are close in energy to the experimental observations. The electron densities are very nearly spherical, supporting the electronic shell model as a description of the basic physics, but the ionic configurations are nonspherical, which accounts for the structure in the absorption spectrum. Kümmel et al. [199] propose that the experimental observations can be interpreted in terms of a mixture of isomers (a) and (c), with (c) dominating as far as the relative heights of the main peaks are concerned and (a) contributing the small feature near 3.5 eV.

Figure 7.20 demonstrates the sensitivity of the absorption to ionic structure. Similar behaviour has been found by a number of other authors, both by the jellium+pseudopotential approach [192–194,196–198] and by the *ab initio* methods of quantum chemistry [35,195].

7.3.5.1.2. Silver. As noted earlier, interband transitions in the noble metals shift the bulk plasma frequency to a value much lower than the Drude prediction. Since silver is the only noble metal to exhibit a well-defined surface plasmon resonance in vacuum, it is not surprising that most of the experimental work on noble metal clusters has focused on silver. Many of the optical absorption measurements have been on small neutral clusters embedded in rare gas matrices [200–208], but a few have addressed free ionic silver clusters [209–211]. The use of superfluid helium nanodroplets has been found to be an alternative route to forming metal clusters of well-defined composition, and the optical properties of clusters formed in this way have also been studied [212–214].

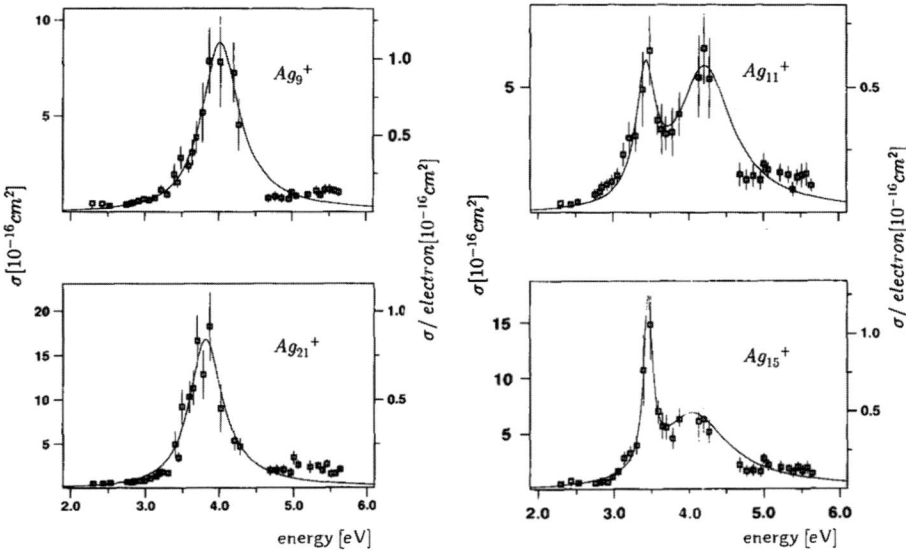

Figure 7.21. Absorption spectra for silver clusters. The data have been fitted to a single Lorentzian function (Ag_9^+, Ag_{21}^+) or two Lorentzians (Ag_{11}^+ and Ag_{15}^+). Reproduced with the permission of Elsevier Science from Tiggesbäumker et al. [209].

It is evidently inappropriate to use Eq. (7.10) and $\omega_M = \omega_p/\sqrt{3}$ for the energy of the surface plasmon in silver. Rather, one should base the analysis on Eq. (7.23) and set $\varepsilon_1 + 2\varepsilon_m = 0$, using the full bulk dielectric function. Using this condition for free clusters ($\varepsilon_m = 1$) and the plot of ε_1 in Figure 7.17 yields a resonance energy of $\sim 3.5\,\text{eV}$. The collinear ion-beam depletion technique has been used by Tiggesbäumker et al. [209,210] to measure the optical absorption of Ag_N^+ clusters up to $N = 70$. Their results for four clusters are reproduced in Figure 7.21.

There are well-defined single main peaks for Ag_9^+ and Ag_{21}^+, and double peaks for Ag_{11}^+ and Ag_{15}^+, which is consistent with the spherical and spheroidal configurations, respectively, that would be predicted by the basic electronic shell model description. Furthermore, the peaks are all in the region of the above estimate of 3.5 eV. We have seen that, in the case of alkali clusters, there is a red shift in the resonance frequency with particle size [Eq. (7.36) and Figure 7.18(b) in the $N > 40$ regime]. By contrast, the silver anions were found to exhibit a blue shift [210]; the experimental data [210,215] are displayed in Figure 7.22.

The spill-out effect discussed above provided an explanation for the red shift observed in alkali clusters. Spill-out is also present in silver but affects only the $5s$ electrons. The $4d$ electrons are much more confined to the core of the cluster [210,215,217], so there is a central region of the particle with a finite density of both s and d electrons and an outer region typically about 1 Å thick where only the conduction electrons are present and the d-electron screening is ineffective. As we have seen, the resonance frequency of free electrons is higher than that when screening is present, so the effect of this unscreened outer region is to push the collective resonance of the particle to higher frequencies. Since the influence of the

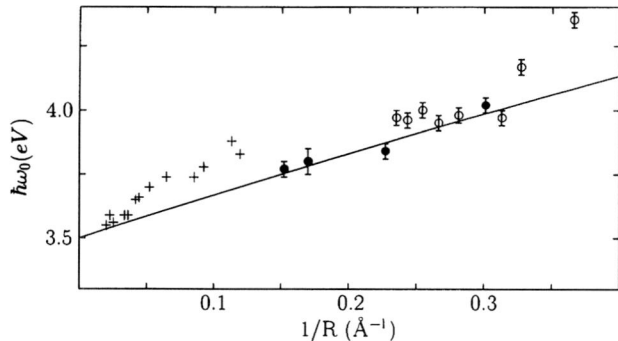

Figure 7.22. Average resonance frequencies of silver clusters as a function of the inverse of particle radius. The open and filled circles are data from experiments on free charged particles in the gas phase [209,210], and the crosses are from studies [216] of large Ag particles ($10 \leq R \leq 50$ Å) in an Ar matrix (corrected for the dielectric constant of the matrix). The linear fit is through the data points for spherical clusters (filled circles) and the Mie frequency of 3.5 eV. Reproduced with the permission of the American Physical Society from Liebsch [215].

outer region increases as the size of the particle decreases, a blue shift is predicted in agreement with observation.

We have two competing effects: a blue shift due to the unscreened outer shell of the cluster and a red shift from the spill-out effect. The dominant behaviour depends on the detailed density distributions of the s and d electrons and in turn on the charge state of the cluster. The positively charged silver ions display a blue shift [209,210], as illustrated in Figure 7.22. However, a red shift occurs in the negatively charged ions [211]. We are confining our discussion to the alkali and noble metals, but note as an aside that the resonance frequency has also been studied in mercury clusters [218]. It was found to be virtually independent of the size and the charge state of the cluster and accorded well with the calculated Mie frequency.

The effect of the cluster geometry was found to be essential for a detailed description of the absorption spectra of small alkali clusters. It is particularly important for the noble metals since there is a transition from two- to three-dimensional geometry that has to be considered (see Section 6.3.3). A number of calculations have been performed for $2 \leq N \leq 8$ using either *ab initio* methods of quantum chemistry [219,220] or density functional procedures with pseudopotentials [221]; the results have been related to experimental data on clusters embedded in a rare gas matrix.

7.3.5.2. Large clusters

Kreibig and Vollmer [115] provide a comprehensive review of research through to the mid-1990s. We conclude here with a description of some of the results of more recent work that employs the techniques for probing individual particles that were mentioned in Section 7.3.4.

Figure 7.23 displays the results of optical absorption measurements on gold clusters by Berciaud *et al.* [142] using photothermal far-field absorption. The particles, which had average diameters of 5, 10, 20 and 33 nm, were in aqueous

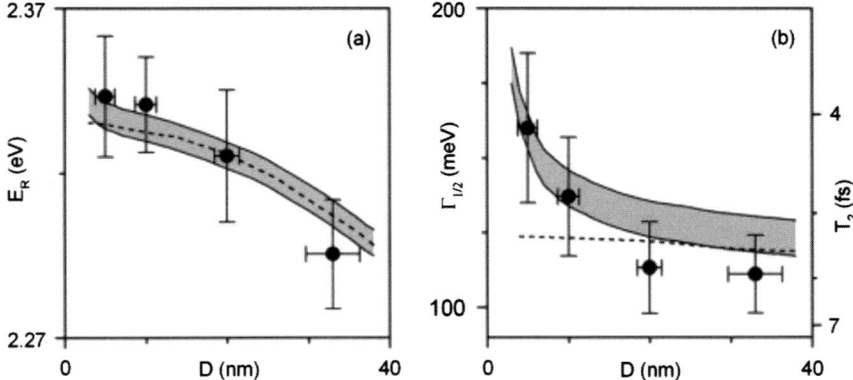

Figure 7.23. Size dependence of the peak position (E_R) (a) and half width ($\Gamma_{1/2}$) (b) of the absorption spectra of gold particles. The data points are the experimental results. See text for the definition of the dotted line and the grey area. Reproduced with the permission of the American Chemical Society from Berciaud et al. [142].

polyvinyl solutions. The positions and half-widths of the peaks in the absorption spectra are plotted against the particle diameters. The dephasing time is also shown.

The experimental data were compared with the predictions of the Mie theory [Eq. (7.23)], using the experimentally determined bulk dielectric function of gold including interband transitions [Eq. (7.14)] and a phenomenological damping constant of the form of Eq. (7.39) ($\Gamma_{1/2} = \Gamma/2$). The dotted lines in Figure 7.23 are the results with A in the expression for the damping constant set equal to zero. A best fit to the experimental data was obtained with a non-zero value of A, as shown by the grey area, which represents the uncertainties in the bulk dielectric function of gold. The agreement between theory and experiment using a fixed value of A is rather good.

The effect of the shape of clusters on the optical properties has been studied by light-scattering techniques using dark-field spectroscopy. Some results of Mock et al. [139] for the surface plasmon resonance position are shown in Figure 7.24 for spherical particles and particles that are triangular or pentagonal in a projection of their morphology onto a two-dimensional surface. There is a considerable variation in the position of the resonance with size and shape, and indeed, even for particles of the same nominal shape and size, there is a significant spread in results. This is explained in an analysis of the morphology of the triangular clusters; some clusters are essentially two-dimensional platelets while others are believed to be more like truncated tetrahedral.

Nanospheres and nanorods represent the two extremes of particle shape. Sönnichsen et al. [92] have used light-scattering spectroscopy to study such particles with the results reproduced in Figure 7.25. The linewidth Γ is plotted against the resonant energy for gold spheres and rods. Nominal sphere sizes ranged from 20 nm to 100 nm. The results of calculations using the Mie theory are also shown in Figure 7.25 and agree well with the experimental data. The Mie theory can be adapted for ellipsoidal particles, and the rods were simulated with spheroids with various aspect ratios. The plots illustrate well the sensitivity of the linewidth and dephasing time to the shape of the nanoparticle.

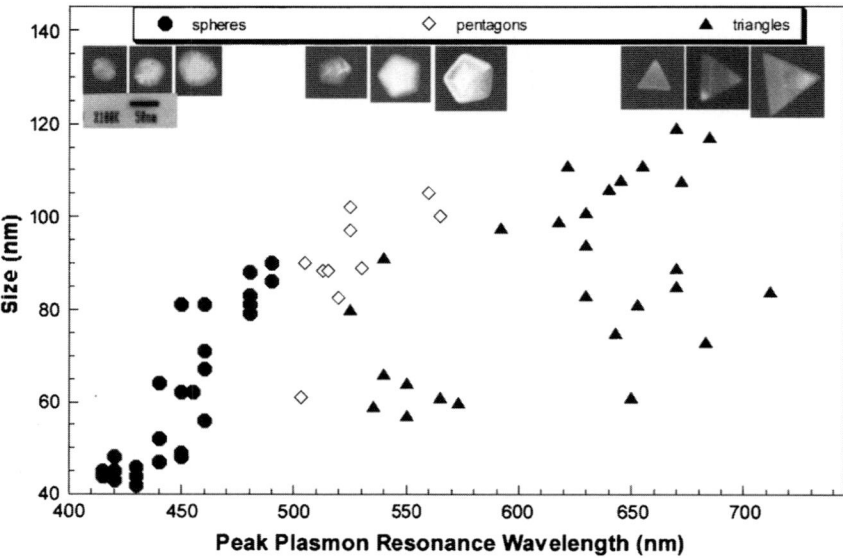

Figure 7.24. Plot of size of silver particles from TEM images against the spectral peak wavelength for spherical, pentagonal and triangular particles. Representative TEM images and key to symbols are shown at the top of the figure. Measurements are made on individual particles. Reproduced with the permission of the American Institute of Physics from Mock *et al.* [139].

7.3.6. *Effective medium theories*

The Mie theory is applicable at very low particle density. Approximate "effective medium" theories have been developed for higher densities. The basic idea is as follows. The dielectric function ε relates the polarisation of a single component medium to the field as

$$P = \frac{1}{4\pi}(\varepsilon - 1)E. \tag{7.40}$$

Following the comment about units in Section 7.3.2, we note that $1/4\pi$ should be replaced by ε_0 to convert to SI units. In a composite medium, we should replace the quantities in Eq. (7.40) by averages denoted by angular brackets

$$\langle P \rangle = \frac{1}{4\pi}(\langle \varepsilon E \rangle - \langle E \rangle). \tag{7.41}$$

Clearly, it would be convenient to have a quantity that relates the mean polarisation to the mean field, and this defines an effective dielectric function

$$\langle P \rangle = \frac{1}{4\pi}(\varepsilon_{\text{eff}} - 1)\langle E \rangle. \tag{7.42}$$

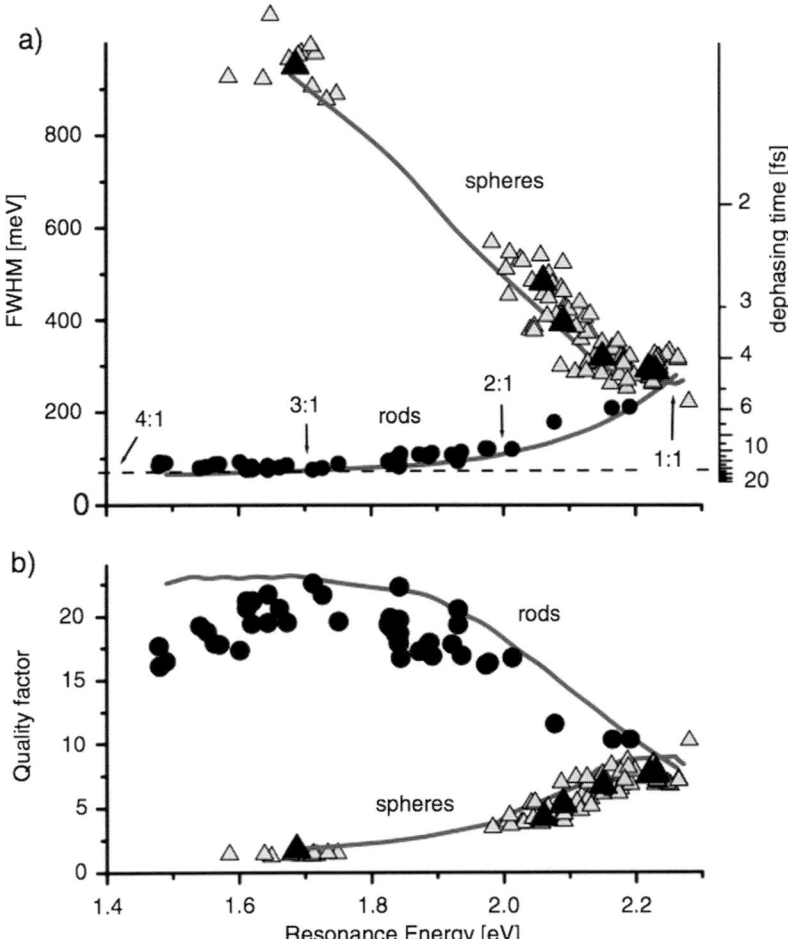

Figure 7.25. (a) Linewidth Γ of plasmon resonances for individual nanospheres (grey triangles) and nanorods (dots) versus resonance energy. The nominal sphere sizes are 150, 100, 80, 60, 40 and 20 nm from left to right. The black triangles are averages for spherical particles of the same nominal size. The lines are from theoretical calculations using the Mie theory. The rods are approximated as spheroids. Some aspect ratios of the rods are indicated. (b) Same data plotted as a quality factor $Q = E_{\text{res}}/\Gamma$. Reproduced with the permission of the American Physical Society from Sönnichsen et al. [92].

Comparing Eqs. (7.41) and (7.42) provides us with an expression from which ε_{eff} can be calculated, at least in principle,

$$\varepsilon_{\text{eff}} = \frac{\langle \varepsilon E \rangle}{\langle E \rangle} \tag{7.43}$$

One of the most widely used effective medium theories dates back to the work of Maxwell Garnett [222,223] over a century ago. Consider a material of dielectric constant ε_m in which there is an electric field E_0. If a single metallic sphere of radius R and dielectric constant ε is introduced into the material, the field E_{sph} inside the sphere is given by

$$\frac{E_{sph}}{E_0} = 1 - A, \quad (7.44)$$

where

$$A = \frac{\varepsilon - \varepsilon_m}{\varepsilon + 2\varepsilon_m}. \quad (7.45)$$

We have just rewritten Eq. (7.15). In the medium, at a distance $r\ (>R)$ from the centre of the sphere, the field is modified to

$$\frac{E_{med}}{E_0} = 1 - A\left(\frac{R}{r}\right)^3 (1 - 3\cos^2\theta), \quad (7.46)$$

where θ is the angle between $\vec{r}$ and $\vec{E}_0$. The sphere behaves as a dipole as far as the region exterior to it is concerned. Averaging over all space, the second term in Eq. (7.46) is identically zero, so $\langle E_{med}\rangle = E_0$.

Now if there is an ensemble of spheres in the medium of volume fraction f, the relevant average quantities can be written as

$$\frac{\langle E\rangle}{E_0} = 1 - fA, \quad (7.47)$$

$$\frac{\langle \varepsilon E\rangle}{E_0} = (1-f)\varepsilon_m + f\varepsilon(1-A). \quad (7.48)$$

Using Eq. (7.43), the Maxwell Garnett result trivially follows

$$\varepsilon_{eff} = \varepsilon_m \left(\frac{1 + 2fA}{1 - fA}\right). \quad (7.49)$$

Interactions between particles have been neglected, so rigorously speaking, it is an approximation valid only at low f. However, interestingly, it yields the correct limit at $f \to 1$ as well as $f \to 0$ and has often applied outside its strict range of validity.

Linearising Eq. (7.49) for small f, the expression reduces to

$$\varepsilon_{eff} \approx \varepsilon_m \left[1 + 3f\frac{\varepsilon - \varepsilon_m}{\varepsilon + 2\varepsilon_m}\right]. \quad (7.50)$$

The complex refractive index (see Section 7.3.2) for the effective medium is given by $n + ik = \sqrt{\varepsilon_{eff}}$ and $k = \text{Im}(\varepsilon_{eff})/(2n)$. In this low-density limit, $n \approx \sqrt{\varepsilon_m}$, and Eq. (7.13) yields the following expression for the absorption coefficient:

$$\alpha = 9f\varepsilon_m^{3/2}\left(\frac{\omega}{c}\right)\frac{\varepsilon_2}{(\varepsilon_1 + 2\varepsilon_m)^2 + \varepsilon_2^2}. \quad (7.51)$$

Since f is the product of the volume V and the number density of the particles, this verifies that the Maxwell Garnett expression correctly reduces to the Mie result [Eq. (7.23)] in the $f \to 0$ limit.

It is useful to put Eq. (7.49) in the context of what is known about the packing of spheres. The maximum density of equal spheres close packed as an ordered array is $f = 0.74$ and, as a random assembly [224], it is $f \approx 0.645$. Of course, one can increase the density with a carefully chosen distribution of sizes [225]. However, the percolation threshold [226] is at a much lower density, $f \approx 0.247$, so one needs to be at a significantly smaller filling fraction than that to be able to ignore the particle–particle interactions, which are not included in the Maxwell Garnett theory. A reasonable assumption for the applicability of Eq. (7.49) is $f < 0.1$.

Bruggeman [227] introduced an alternative approach that attempts to include, in a self-consistent way, the effect on a particle of its composite environment. The theory is symmetric in the two components of the composite, so it is clearer to express it in terms of components a and b with volume densities and dielectric functions f_a, ε_a and f_b, ε_b, respectively ($f_a + f_b = 1$), rather than as a matrix medium with inclusions. One starts with the effective medium itself, with dielectric function ε_{eff}, and E_0 is now the field in the effective medium. Consider a sphere of material i ($=a$ or b) introduced into the medium. Adapting Eqs. (7.44) and (7.45), the field E_i inside the sphere is

$$\frac{E_i}{E_0} = 1 - \frac{\varepsilon_i - \varepsilon_{\text{eff}}}{\varepsilon_i + 2\varepsilon_{\text{eff}}}. \qquad (7.52)$$

The condition for self-consistency is taken as $\langle E_i \rangle = E_0$, which yields the Bruggeman result for the effective dielectric function

$$f_a \frac{\varepsilon_a - \varepsilon_{\text{eff}}}{\varepsilon_a + 2\varepsilon_{\text{eff}}} + f_b \frac{\varepsilon_b - \varepsilon_{\text{eff}}}{\varepsilon_b + 2\varepsilon_{\text{eff}}} = 0. \qquad (7.53)$$

Eq. (7.53) reduces to the same low-density limit as the Maxwell Garnett theory, Eq. (7.50), if we set $\varepsilon_a = \varepsilon_m$, $\varepsilon_b = \varepsilon$, $f_a = 1 - f$ and $f_b = f \to 0$.

Effective medium theories appear in many areas of physics and, although appearing superficially different, often turn out to be equivalent mathematically. The Maxwell Garnett and Bruggeman theories are respectively equivalent to the average t-matrix approximation and the coherent potential approximation widely used in the theory of disordered alloys [228,229]. This equivalence becomes most transparent when the formalism is expressed in terms of a Green's function scattering theory [230,231].

Except at $f \ll 1$, the two effective medium theories predict strongly different absorption spectra, with the Bruggeman theory displaying much broader structures. The Maxwell Garnett theory is likely to yield better results for metallic inclusions embedded in a dielectric matrix when there is a sharp interface between the two materials (provided $f < 0.1$). The Bruggeman theory blurs the sharp interface, and this will broaden the surface plasmon resonance – probably too much at small f. However, the theory has the advantage of applicability for arbitrary f and, in fact, correctly predicts a percolation threshold to metallic conduction (the effective medium value for the threshold is $f = 1/3$). It may well yield better results for nanocomposites of less regularly shaped particles or aggregates.

The main defect of any effective medium theory is the neglect of fluctuations in the net dipolar field felt by a particle due to neighbours in its immediate environment. We have seen an analogous problem in Chapter 4 in the theory of the formation of islands on a surface through diffusion and aggregation. Nevertheless, the theories have been widely applied to the interpretation of experimental data, and there have been a number attempts to develop the treatment of topological effects [232–239] (see [115,240] for reviews).

We conclude with a couple of examples that provide useful illustrations. Palpant et al. [241] studied gold clusters in the size range of 2–4 nm embedded in an alumina matrix. The filling factors were in the range of 1.5–6.5% and so, in principle, within the range of applicability of the Maxwell Garnett theory. However, they found that the positions of the absorption peaks were blueshifted with respect to the theoretical predictions and considerably broadened. They concluded that the discrepancy lay with the use of the bulk dielectric function in the Maxwell Garnett theory, that the behaviour observed was predominantly an intrinsic size effect that manifests itself in the absorption spectra of individual particles and that ensemble effects are small by comparison.

The second example concerns core–shell particles. Ung et al. [86] prepared 13.2 nm gold particles coated in silica of various shell thicknesses (0.2–2 nm) prior to forming them into thin films. Intrinsic size effects should be small for particles of this size. The shell acts as an inorganic spacer and allows a much greater range of packing fractions than would be allowed with bare particles. One can approach closer to the density for random sphere packing [224] without encountering the percolation threshold on the way. Optical absorption spectra were measured and the peak positions compared with an analysis using the Maxwell Garnett theory.

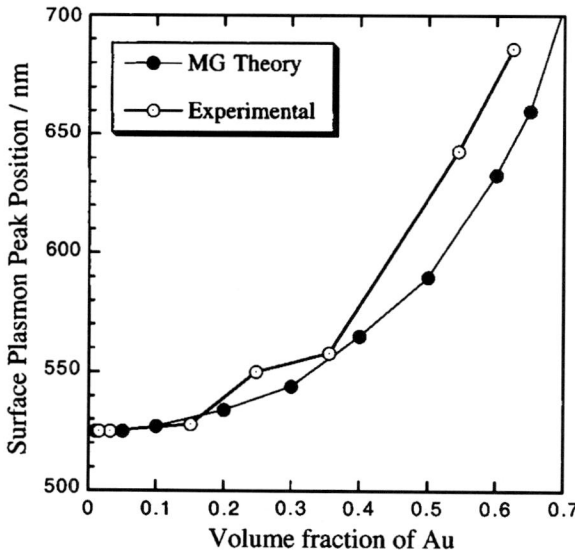

Figure 7.26. Positions of peaks in optical absorption spectra of silica-coated gold particles and predictions of the Maxwell Garnett theory. Reproduced with the permission of the American Chemical Society from Ung et al. [86].

The comparison is reproduced in Figure 7.26 and the agreement is excellent up to a volume fraction of gold of about 40%.

References

[1] H. Müller, H.-G. Fritsche and L. Skala, Analytic cluster models and interpolation formulae for cluster properties, in: H. Haberland (Ed.), Clusters of Atoms and Molecules, Vol. 52, Springer, Berlin, 1994, pp. 114–140. Springer Series in Chemical Physics.
[2] E. Engel and J.P. Perdew, Phys. Rev. B **43** (1991) 1331.
[3] M. Seidl, J.P. Perdew, M. Brajczewska and C. Fiolhais, J. Chem. Phys. **108** (1998) 8182.
[4] W.A. de Heer, Rev. Mod. Phys. **65** (1993) 611.
[5] M. Brack, Rev. Mod. Phys. **65** (1993) 677.
[6] W.A. Saunders, K. Clemenger, W.A. de Heer and W.D. Knight, Phys. Rev. B **32** (1985) 1366.
[7] M.L. Homer, J.L. Persson, E.C. Honea and R.L. Whetten, Z. Phys. D **22** (1991) 441.
[8] C. Jackschath, I. Rabin and W. Schulze, Z. Phys. D **22** (1992) 517.
[9] C. Jackschath, I. Rabin and W. Schulze, Ber. Bunsenges. Phys. Chem. **96** (1992) 1200.
[10] P.M. Guyon and J. Berkowitz, J. Chem. Phys. **54** (1971) 1814.
[11] C. Yannouleas and U. Landman, Phys. Rev. B **51** (1995) 1902.
[12] C. Yannouleas and U. Landman, Phys. Rev. Lett. **78** (1997) 1424.
[13] I.A. Solov'yov, A.V. Solov'yov and W. Greiner, Phys. Rev. A **65** (2002) 053203.
[14] M.B. Knickelbein, Chem. Phys. Lett. **19** (1992) 2129.
[15] G. Alameddin, J. Hunter, D. Cameron and M.M. Kappes, Chem. Phys. Lett. **192** (1992) 122.
[16] G.A. Bishea and M.D. Morse, J. Chem. Phys. **95** (1991) 5646.
[17] E.M. Fernández, J.M. Soler, I.L. Garzón and L.C. Balbás, Phys. Rev. B **70** (2004) 165403.
[18] M. Kabir, A. Mookerjee and A.K. Bhattacharya, Phys. Rev. A **69** (2004) 043203.
[19] J.L. Persson, R.L. Whetten, H.-P. Cheng and R.S. Berry, Chem. Phys. Lett. **186** (1991) 215.
[20] K.E. Schriver, J.L. Persson, E.C. Honea and R.L. Whetten, Phys. Rev. Lett. **64** (1990) 2539.
[21] M. Pellarin, B. Baguenard, C. Bordas, M. Broyer, J. Lermé and J.L. Vialle, Z. Phys. D **26** (1993) S137.
[22] J. Akola, H. Häkkinen and M. Manninen, Phys. Rev. B **58** (1998) 58.
[23] C.-Y. Cha, G. Ganteför and W. Eberhardt, J. Chem. Phys. **100** (1994) 995.
[24] C.-Y. Cha, G. Ganteför and W. Eberhardt, Rev. Sci. Instrum. **63** (1992) 5661.
[25] H. Handschuh, G. Ganteför and W. Eberhardt, Rev. Sci. Instrum. **66** (1995) 3838.
[26] W. Eberhardt, Surf. Sci. **500** (2002) 242.
[27] P. Kruit and F.H. Read, J. Phys. [E] **16** (1983) 313.
[28] H. Handschuh, C.-Y. Cha, P.S. Bechthold, G. Ganteför and W. Eberhardt, J. Chem. Phys. **102** (1995) 6406.
[29] K.-H. Meiwes-Broer, Electronic level structure of metal clusters at surfaces, in: K.-H. Meiwes-Broer (Ed.), Metal Clusters at Surfaces Springer, Berlin, 2000, pp. 151–173. Clusters of Atoms and Molecules.
[30] K.M. McHugh, J.G. Eaton, G.H. Lee, H.W. Sarkas, L.H. Kidder, J.T. Snodgrass, M.R. Manaa and K.W. Bowen, J. Chem. Phys. **91** (1989) 3792.
[31] K.J. Taylor, C.L. Pettiette-Hall, O. Chesnovsky and R.E. Smalley, J. Chem. Phys. **96** (1992) 3319.
[32] G. Ganteför, H. Handschuh, H. Möller, C.-Y. Cha, P.S. Bechthold and W. Eberhardt, Surf. Rev. Lett. **3** (1996) 399.

[33] J.G. Eaton, L.H. Kidder, H.W. Sarkas, K.M. McHugh and K.H. Bowen, in: P. Jena, S.N. Khanna and B.K.N. Rao (Eds.), Physics and Chemistry of Finite Systems: From Clusters to Crystals, Vol. I, Kluwer Academic, Dordrecht, 1992, pp. 493–507. NATO Advanced Summer Institute, Series C: Mathematical and Physical Sciences.
[34] U. Röthslisberger and W. Andreoni, J. Chem. Phys. **94** (1991) 8129.
[35] V. Bonačić-Koutecký, P. Fantucci and J. Koutecký, Chem. Rev. **91** (1991) 1035.
[36] G. Wrigge, M. Astruc Hoffmann and B.V. Issendorff, Phys. Rev. A **65** (2002) 063201.
[37] B. von Issendorff and O. Cheshnovsky, Annu. Rev. Phys. Chem. **56** (2005) 549.
[38] C.L. Pettiette, S.H. Lang, M.J. Craycraft, J. Conceicao, R.T. Laaksonen, O. Cheshnovsky and R.E. Smalley, J. Chem. Phys. **88** (1988) 5377.
[39] G. Ganteför, K.-H. Meiwes-Broer and H.O. Lutz, Phys. Rev. A **37** (1988) 2716.
[40] J. Ho, K.M. Ervin and W.C. Lineberger, J. Chem. Phys. **93** (1990) 6987.
[41] G. Ganteför, M. Gausa, K.-H. Meiwes-Broer and H.O. Lutz, Faraday Discuss. Chem. Soc. **86** (1988) 197.
[42] O. Cheshnovsky, K.J. Taylor, J. Conceicao and R.E. Smalley, Phys. Rev. Lett. **64** (1990) 1785.
[43] C.-Y. Cha, G. Ganteför and W. Eberhardt, J. Chem. Phys. **99** (1993) 6308.
[44] H. Handschuh, G. Ganteför, P.S. Bechthold and W. Eberhardt, J. Chem. Phys. **100** (1994) 7093.
[45] H. Häkkinen, B. Yoon, U. Landman, X. Li, H.-J. Zhai and L.-S. Wang, J. Phys. Chem. A **107** (2003) 6168.
[46] H. Häkkinen, M. Moseler, O. Kostko, N. Morgner, M.A. Hoffmann and B.V. Issendorff, Phys. Rev. Lett. **93** (2004) 093401.
[47] G. Ganteför, C.-Y. Cha, H. Handschuh, G. Schulze Icking-Konert, B. Kessler, O. Gunnarsson and W. Eberhardt, J. Electron Spectrosc. Relat. Phenom. **76** (1995) 37.
[48] J. Li, X. Li, H.-J. Zhai and L.-S. Wang, Science **299** (2003) 864.
[49] H. Wu, S.R. Desai and L.S. Wang, Phys. Rev. Lett. **77** (1996) 2436.
[50] H. Wu, S.R. Desai and L.S. Wang, Phys. Rev. Lett. **76** (1996) 212.
[51] M. Castro, S.-R. Liu, H.-J. Zhai and L.-S. Wang, J. Chem. Phys. **118** (2003) 2116.
[52] H. Kietzmann, J. Morenzin, P.S. Bechthold, G. Ganteför, W. Eberhardt, D.-S. Yang, P.A. Hackett, R. Fournier, T. Pang and C. Chen, Phys. Rev. Lett. **77** (1996) 4528.
[53] L.S. Wang, H. Wu and H. Cheng, Phys. Rev. B **55** (1997) 12884.
[54] G. Ganteför and W. Eberhardt, Phys. Rev. Lett. **76** (1996) 4975.
[55] L.-S. Wang, H.-S. Cheng and J. Fan, J. Chem. Phys. **102** (1995) 9480.
[56] L.-S. Wang, X. Li and H.-F. Zhang, Chem. Phys. Lett. **262** (2000) 53.
[57] H. Yoshida, A. Terasaki, K. Kobayashi, M. Tsukada and T. Kondow, J. Chem. Phys. **102** (1995) 5960.
[58] J. Morenzin, H. Kietzmann, P.S. Bechthold, G. Ganteför and W. Eberhardt, Pure Appl. Chem. **72** (2000) 2149.
[59] S.-R. Liu, H.-J. Zhai and L.-S. Wang, Phys. Rev. B **63** (2001) 153402.
[60] S.-R. Liu, H.-J. Zhai and L.-S. Wang, Phys. Rev. B **65** (2002) 113401.
[61] B. Kaiser and K. Rademann, Phys. Rev. Lett. **69** (1992) 3204.
[62] R. Busani, M. Folkers and O. Cheshnovsky, Phys. Rev. Lett. **81** (1998) 3836.
[63] O.C. Thomas, W. Zheng, S. Xu and K.H. Bowen, Phys. Rev. Lett. **89** (2002) 213403.
[64] X. Li, H. Wu, X.-B. Wang and L.-S. Wang, Phys. Rev. Lett. **81** (1998) 1909.
[65] P.H. Acioli and J. Jellinek, Phys. Rev. Lett. **89** (2002) 213402.
[66] J. Jellinek and P.H. Acioli, J. Phys. Chem. A **106** (2002) 10919; **107** (2003) 1670.
[67] M. Faraday, Philos. Trans. R. Soc. Lond. **147** (1857) 145.
[68] D.J. Barber and I.C. Freestone, Archaeometry **32** (1990) 33.
[69] G.L. Hornyak, C.J. Patrissi, E.B. Oberhauser, C.R. Martin, J.C. Valmalette, L. Lamaire, J. Dutta and H. Hoffmann, Nanostruct. Mater. **9** (1997) 571.
[70] I. Borgia, B. Brunetti, I. Mariani, A. Sgamellotti, F. Cariati, P. Fermo, M. Mellini, C. Viti and G. Padeletti, Appl. Surf. Sci. **185** (2002) 206.
[71] E. Bontempi, P. Colombi, L.E. Depero, L. Cartechini, F. Presciutti, B.G. Brunetti and A. Sgamellotti, Appl. Phys. A: Mater. Sci. Process. **83** (2006) 543.
[72] G. Mie, Ann. Phys. (Leipzig) **25** (1908) 377.

[73] J.A. Creighton, C.G. Blatchford and M.G. Albrecht, J. Chem. Soc. Faraday Trans. 2: Mol. Chem. Phys. **75** (1979) 790.
[74] M. Fleishmann, P.J. Hendra and A.J. McQuillan, Chem. Phys. Lett. **23** (1974) 123.
[75] B.J. Messinger, K.U. von Raben, R.K. Chang and P.W. Barber, Phys. Rev. B **24** (1981) 649.
[76] D.-S. Wang and M. Kerker, Phys. Rev. B **24** (1981) 1777.
[77] A. Otto, J. Raman Spectrosc. **27** (1991) 743.
[78] A. Otto, I. Mrozek, H. Grabhorn and W. Akeman, J. Phys. Condens. Matter **4** (1992) 1143.
[79] K. Kneipp, Y. Wang, H. Kneipp, L.T. Perelman, I. Itzkan, R.R. Dasari and M.S. Feld, Phys. Rev. Lett. **78** (1997) 1667.
[80] S. Nie and S.R. Emory, Science **275** (1997) 1102.
[81] A.M. Michaels, M. Nirmal and L.F. Brus, J. Am. Chem. Soc. **121** (1999) 9932.
[82] H. Xu, E.J. Bjerneld, M. Käll and L. Börjesson, Phys. Rev. Lett. **83** (1999) 4357.
[83] M. Moskovits and D.H. Jeong, Chem. Phys. Lett. **397** (2004) 91.
[84] J.B. Jackson and N.J. Halas, Proc. Natl. Acad. Sci. U.S.A. **101** (2004) 17930.
[85] S.J. Oldenburg, R.D. Averitt, S.L. Westcott and N.J. Halas, Chem. Phys. Lett. **288** (1998) 243.
[86] T. Ung, L.M. Liz-Marzán and P. Mulvaney, J. Phys. Chem. B **105** (2001) 3441.
[87] C. Oubre and P. Nordlander, J. Phys. Chem. B **108** (2004) 17740.
[88] E. Prodan and P. Nordlander, J. Chem. Phys. **120** (2004) 5445.
[89] L.M. Liz-Marzin, M. Giersig and P. Mulvaney, Langmuir **12** (1996) 4329.
[90] S. Link, M.B. Mohamed and M.A. El-Sayed, J. Phys. Chem. B **103** (1999) 3073.
[91] M.B. Mohamed, V. Volkov, S. Link and M.A. El-Sayed, Chem. Phys. Lett. **317** (2000) 517.
[92] C. Sönnichsen, T. Franzl, T. Wilk, G. von Plessen, J. Feldmann, O. Wilson and P. Mulvaney, Phys. Rev. Lett. **88** (2002) 077402.
[93] K.G. Thomas, S. Barazzouk, B.I. Ipe, S.T.S. Joseph and P.V. Kamat, J. Phys. Chem. B **108** (2004) 13066.
[94] R. Elghanian, J.J. Storhoff, R.C. Mucic, R.L. Letsinger and C.A. Mirkin, Science **277** (1997) 1078.
[95] K.-H. Su, Q.-H. Wei, X. Zhang, J.J. Mock, D.R. Smith and S. Schultz, Nano Lett. **3** (2003) 1087.
[96] P. Nordlander, C. Oubre, E. Prodan, K. Li and M.I. Stockman, Nano Lett. **4** (2004) 899.
[97] M. Quinten, A. Leitner, J.R. Krenn and F.R. Aussenegg, Opt. Lett. **23** (1998) 1331.
[98] J.R. Krenn, A. Dereux, J.C. Weeber, E. Bourillot, Y. Lacroute, J.P. Goudonnet, G. Schider, W. Gotschy, A. Leitner, F.R. Aussenegg and C. Girard, Phys. Rev. Lett. **82** (1999) 2590.
[99] M. Brongersma, J.W. Hartman and H.A. Atwater, Phys. Rev. B **62** (2000) 16356.
[100] S.A. Maier, P.G. Kik, H.A. Atwater, S. Meltzer, E. Harel, B.E. Koel and A.A.G. Requicha, Nat. Mater. **2** (2003) 229.
[101] S.Y. Park and D. Stroud, Phys. Rev. B **69** (2004) 125418.
[102] K. Li, M.I. Stockman and D.J. Bergmann, Phys. Rev. Lett. **91** (2003) 227402.
[103] C.A. Mirkin, R.L. Letsinger, R.C. Mucic and J.J. Storhoff, Nature **382** (1996) 607.
[104] A.P. Alivasatos, K.P. Johnsson, X. Peng, T.E. Wilson, C.J. Loweth, M.P. Bruchez Jr. and P.G. Schultz, Nature **382** (1996) 609.
[105] J.J. Storhoff and C.A. Mirkin, Chem. Rev. **99** (1999) 1849.
[106] J.J. Storhoff, R. Elghanian, R.C. Mucic, C.A. Mirkin and R.L. Letsinger, J. Am. Chem. Soc. **120** (1998) 1959.
[107] T.A. Taton, C.A. Mirkin and R.L. Letsinger, Science **289** (2000) 1757.
[108] Y. Kim, R.C. Johnson and J.T. Hupp, Nano Lett. **1** (2001) 116.
[109] S. Schultz, D.R. Smith, J.J. Mock and D.A. Schultz, Proc. Natl. Acad. Sci. U.S.A. **97** (2000) 996.
[110] L. Cognet, C. Tardin, D. Boyer, D. Choquet, P. Tamarat and B. Lounis, Proc. Natl. Acad. Sci. U.S.A. **100** (2003) 11350.

[111] L.R. Hirsch, J.B. Jackson, A. Lee, N.J. Halas and J.L. West, Anal. Chem. **75** (2003) 2377.
[112] S.-J. Park, T.A. Taton and C.A. Mirkin, Science **295** (2002) 1503.
[113] L.R. Hirsch, R.J. Stafford, J.A. Bankson, S.R. Sershen, B. Rivera, R.E. Price, J.D. Hazle, N.J. Halas and J.L. West, Proc. Natl. Acad. Sci. U.S.A. **22** (2003) 47.
[114] D.P. O'Neal, L.R. Hirsch, N.J. Halas, J.D. Payne and J.L. West, Cancer Lett. **209** (2004) 171.
[115] U. Kreibig and M. Vollmer, Optical Properties of Metal Clusters, Springer-Verlag, Berlin, 1995.
[116] P. Mulvaney, Langmuir **12** (1996) 788.
[117] C.P. Collier, T. Vossmeyer and J.R. Heath, Annu. Rev. Phys. Chem. **49** (1998) 371.
[118] S. Link and M.A. El-Sayed, Annu. Rev. Phys. Chem. **54** (2003) 331.
[119] K.L. Kelly, E. Coronado, L.L. Zhao and G.C. Schatz, J. Phys. Chem. B **107** (2003) 668.
[120] A. Moores and F. Goettmann, New J. Chem. **30** (2006) 1121.
[121] C. Kittel, Introduction to Solid State Physics, Wiley, New York, 1971.
[122] F. Wooten, Optical Properties of Solids, Academic Press, New York, 1972.
[123] H. Ehrenreich and H.R. Philipp, Phys. Rev. **128** (1962) 1622.
[124] J.D. Jackson, Classical Electrodynamics, Wiley, New York, 1975.
[125] J.A. Stratton, Electromagnetic Theory, McGraw-Hill, New York, 1941.
[126] C.F. Bohren and D.R. Huffman, Absorption and Scattering of Light by Small Particles, Wiley, New York, 1983.
[127] H.C. van de Hulst, Light Scattering by Small Particles, Wiley, New York, 1957.
[128] R. Gans, Ann. Phys. **342** (1912) 881.
[129] R. Gans, Ann. Phys. **352** (1915) 270.
[130] E.M. Purcell and C.R. Pennypacker, Astrophys. J. **186** (1973) 705.
[131] B.T. Draine and P.J. Flatau, J. Opt. Soc. Am. A **11** (1994) 1491.
[132] G.C. Schatz, J. Mol. Struct. (Theochem) **573** (2001) 73.
[133] W.A. de Heer, K. Selby, V. Kresin, J. Masui, M. Vollmer, A. Châtelain and W.D. Knight, Phys. Rev. Lett. **59** (1987) 1805.
[134] C. Bréchignac, P. Cahuzac, N. Kebaïli, J. Leygnier and A. Sarfati, Phys. Rev. Lett. **68** (1992) 3916.
[135] M. Schmidt, C. Ellert, W. Kronmüller and H. Haberland, Phys. Rev. B **59** (1999) 10970.
[136] T. Klar, M. Perner, S. Grosse, G. von Plessen, W. Spirkl and J. Feldmann, Phys. Rev. Lett. **80** (1998) 4249.
[137] C. Sönnichsen, S. Geier, N.E. Hecker, G. von Plessen, J. Feldmann, H. Ditlbacher, B. Lamprecht, J.R. Krenn, F.R. Aussenegg, V.-H. Chan, J.P. Spatz and M. Möller, Appl. Phys. Lett. **77** (2000) 2949.
[138] C. Sönnichsen, T. Franzl, T. Wilk, G. von Plessen and J. Feldmann, New J. Phys. **4** (2002) 93.
[139] J.J. Mock, M. Barbic, D.R. Smith, D.A. Schultz and S. Schultz, J. Chem. Phys. **116** (2002) 6755.
[140] D. Boyer, P. Tamarat, A. Maali, B. Lounis and M. Orrit, Science **297** (2002) 1160.
[141] S. Berciaud, L. Cognet, G.A. Blab and B. Lounis, Phys. Rev. Lett. **93** (2004) 257402.
[142] S. Berciaud, L. Cognet, P. Tamarat and B. Lounis, Nano Lett. **5** (2005) 515.
[143] A. Arbouet, D. Christofilos, N. Del Fatti, F. Vallée, J.R. Huntzinger, L. Arnaud, P. Billaud and M. Broyer, Phys. Rev. Lett. **93** (2004) 127401.
[144] K. Lindfors, T. Kalkbrenner, P. Stoller and V. Sandoghdar, Phys. Rev. Lett. **93** (2004) 037401.
[145] F. Stietz, J. Bosbach, T. Wenzel, T. Vartanyan, A. Goldmann and F. Träger, Phys. Rev. Lett. **84** (2000) 5644.
[146] J. Bosbach, C. Hendrich, F. Stietz, T. Vartanyan and F. Träger, Phys. Rev. Lett. **89** (2002) 257404.
[147] R. Lazzari, S. Roux, I. Simonsen, J. Jupille, D. Bedeaux and J. Vlieger, Phys. Rev. B **65** (2002) 235424.

[148] J.E. Millstone, S. Park, K.L. Shuford, L. Qin, G.C. Schatz and C.A. Mirkin, J. Am. Chem. Soc. **127** (2005) 5312.
[149] V. Kresin, Phys. Rep. **220** (1992) 1.
[150] M. Brack, Phys. Rev. B **39** (1989) 3533.
[151] Z. Penzar, W. Ekardt and A. Rubio, Phys. Rev. B **42** (1990) 5040.
[152] W. Ekardt, Phys. Rev. Lett. **52** (1984) 1925.
[153] C. Yannouleas, R.A. Broglia, M. Brack and P.F. Bortignon, Phys. Rev. Lett. **63** (1989) 255.
[154] C. Yannouleas, J.M. Pacheco and R.A. Broglia, Phys. Rev. B **41** (1990) 6088.
[155] S. Saito, G.F. Bertsch and D. Tománek, Phys. Rev. B **43** (1991) 6804.
[156] G.F. Bertsch and D. Tománek, Phys. Rev. B **40** (1989) 2749.
[157] S. Underwood and P. Mulvaney, Langmuir **10** (1994) 3427.
[158] K. Selby, V. Kresin, J. Masui, M. Vollmer, W.A. de Heer, A. Scheidemann and W.D. Knight, Phys. Rev. B **43** (1991) 4565.
[159] C.R.C. Wang, S. Pollack and M.M. Kappes, Chem. Phys. Lett. **166** (1990) 26.
[160] S. Pollack, C.R.C. Wang and M.M. Kappes, J. Chem. Phys. **94** (1991) 2496.
[161] C. Bréchignac, P. Cahuzac, F. Carlier, M. de Frutos and J. Leygnier, Z. Phys. D **19** (1991) 1.
[162] C. Bréchignac, P. Cahuzac, F. Carlier, M. de Frutos and J. Leygnier, Chem. Phys. Lett. **189** (1992) 28.
[163] J. Borggreen, P. Chowdhury, N. Kebaïli, L. Lundsberg-Nielsen, K. Lützenkirchen, M.B. Nielsen, J. Pedersen and H.D. Rasmussen, Phys. Rev. B **48** (1993) 17507.
[164] T. Reiners, W. Orlik, C. Ellert, M. Schmidt and H. Haberland, Chem. Phys. Lett. **215** (1993) 357.
[165] T. Reiners, C. Ellert, M. Schmidt and H. Haberland, Phys. Rev. Lett. **74** (1995) 1558.
[166] C. Ellert, M. Schmidt, C. Schmitt, T. Reiners and H. Haberland, Phys. Rev. Lett. **75** (1995) 1731.
[167] C. Ellert, M. Schmidt, T. Reiners and H. Haberland, Z. Phys. D **39** (1997) 317.
[168] C. Schmitt, C. Ellert, M. Schmidt and H. Haberland, Z. Phys. D **42** (1997) 145.
[169] R. Schlipper, R. Kusche, B.v. Issendorff and H. Haberland, Phys. Rev. Lett. **80** (1998) 1194.
[170] M. Schmidt and H. Haberland, Eur. Phys. J. D **6** (1999) 109.
[171] P. Markowicz, K. Kolwas and M. Kolwas, Phys. Lett. A **236** (1997) 543.
[172] J.-H. Klein-Wiele, P. Simon and H.-G. Rubahn, Phys. Rev. Lett. **80** (1998) 45.
[173] C. Bréchignac, P. Cahuzac, F. Carlier and J. Leygnier, Chem. Phys. Lett. **164** (1989) 433.
[174] J. Blanc, M. Broyer, J. Chevaleyre, P. Dugourd, H. Kuhling, P. Labastie, M. Ulbricht, J.P. Wolf and L. Wöste, Z. Phys. D **19** (1991) 7.
[175] C. Bréchignac, P. Cahuzac, J. Leygnier and A. Sarfati, Phys. Rev. Lett. **70** (1993) 2036.
[176] C. Ellert, M. Schmidt, C. Schmitt, H. Haberland and C. Guet, Phys. Rev. B **59** (1999) 7841.
[177] H. Fallgren and T.P. Martin, Chem. Phys. Lett. **168** (1990) 233.
[178] W. Ekardt and G. Penzar, Phys. Rev. B **43** (1991) 1322.
[179] P. Apell and Å. Ljungbert, Solid State Commun. **44** (1982) 1367.
[180] W. Ekardt, Phys. Rev. B **31** (1985) 6360.
[181] P.-G. Reinhard, M. Brack and O. Genzen, Phys. Rev. A **41** (1990) 5568.
[182] C. Yannouleas and R.A. Broglia, Phys. Rev. A **44** (1991) 5793.
[183] C. Yannouleas, E. Vigezzi and R.A. Broglia, Phys. Rev. B **47** (1993) 9849.
[184] V. Kresin, Phys. Rev. B **39** (1989) 3042.
[185] V. Kresin, Phys. Rev. B **40** (1989) 12507.
[186] A. Kawabata and R. Kubo, J. Phys. Soc. Jpn. **21** (1966) 1765.
[187] B. Barma and V. Subrahmanyam, J. Phys. Condens. Matter **1** (1989) 7681.
[188] C. Yannouleas and R.A. Broglia, Ann. Phys. (N.Y.) **217** (1992) 105.
[189] G. Weick, G.-L. Ingold, R.A. Jalabert and D. Weinmann, Phys. Rev. B **74** (2006) 165421.
[190] V.M. Akulin, C. Bréchignac and A. Sarfati, Phys. Rev. B **55** (1997) 1372.

[191] V.M. Akulin, C. Bréchignac and A. Sarfati, Phys. Rev. Lett. **75** (1995) 220.
[192] J.M. Pacheco and W.D. Schöne, Phys. Rev. Lett. **79** (1997) 4986.
[193] W.D. Schöne, E. Ekardt and J.M. Pacheco, Phys. Rev. B **50** (1994) 11079.
[194] W.D. Schöne, E. Ekardt and J.M. Pacheco, Z. Phys. D **36** (1996) 65.
[195] V. Bonačić-Koutecký, J. Pittner, C. Fuchs, P. Fantucci, M.F. Guest and J. Koutecký, J. Chem. Phys. **104** (1996) 1427.
[196] K. Yabana and G.F. Bertsch, Phys. Rev. B **54** (1996) 4484.
[197] A. Rubio, J.A. Alonso, X. Blasé, L.C. Balbás and S.G. Louie, Phys. Rev. Lett. **77** (1996) 247.
[198] F. Calvayrac, P.-G. Reinhard and E. Suraud, J. Phys. B At. Mol. Opt. Phys. **31** (1998) 1367.
[199] S. Kümmel, M. Brack and P.-G. Reinhard, Phys. Rev. B **62** (2000) 7602; **63** (2001) 129902.
[200] W. Harbich, S. Fedrigo and J. Buttet, Chem. Phys. Lett. **195** (1992) 613.
[201] S. Fedrigo, W. Harbich and J. Buttet, Phys. Rev. B **47** (1993) 10706.
[202] S. Federigo, W. Harbich, J. Belyaev and J. Buttet, Chem. Phys. Lett. **211** (1993) 166.
[203] C. Félix, C. Sieber, W. Harbich, J. Buttet, I. Rabin, W. Schulze and G. Ertl, Chem. Phys. Lett. **313** (1999) 105.
[204] I. Rabin, W. Schulze and G. Ertl, Chem. Phys. Lett. **312** (1999) 394.
[205] I. Rabin, W. Schulze, G. Ertl, C. Félix, C. Sieber, W. Harbich and J. Buttet, Chem. Phys. Lett. **320** (2000) 59.
[206] C. Félix, C. Sieber, W. Harbich, J. Buttet, I. Rabin, W. Schulze and G. Ertl, Phys. Rev. Lett. **86** (2001) 2992.
[207] C. Sieber, J. Buttet, W. Harbich, C. Félix, R. Mitrić and V. Bonačić-Koutecký, Phys. Rev. A **70** (2004) 041201.
[208] F. Conus, V. Rodrigues, S. Lecoultre, A. Rydlo and C. Félix, J. Chem. Phys. **125** (2006) 024511.
[209] J. Tiggesbäumker, L. Köller, H.O. Lutz and K.-H. Meiwes-Broer, Chem. Phys. Lett. **190** (1992) 42.
[210] J. Tiggesbäumker, L. Köller, K.-H. Meiwes-Broer and A. Liebsch, Phys. Rev. A **48** (1993) 1749.
[211] J. Tiggesbäumker, L. Köller and K.-H. Meiwes-Broer, Chem. Phys. Lett. **260** (1996) 428.
[212] J. Tiggesbäumker and F. Stienkemeier, Phys. Chem. Chem. Phys. **9** (2007) 4748.
[213] F. Federmann, K. Hoffmann, N. Quaas and J.-P. Toennies, Eur. Phys. J. D **9** (1999) 11.
[214] T. Diederich, J. Tiggesbäumker and K.-H. Meiwes-Broer, J. Chem. Phys. **116** (2002) 3263.
[215] A. Liebsch, Phys. Rev. B **48** (1993) 11317.
[216] K.P. Charlé, W. Schulze and B. Winter, Z. Phys. D **12** (1989) 471.
[217] V.V. Kresin, Phys. Rev. B **51** (1995) 1844.
[218] H. Haberland, B. von Issendorff, J. Yufeng and T. Kolar, Phys. Rev. Lett. **69** (1992) 3212.
[219] V. Bonačić-Koutecký, J. Pittner, M. Boiron and P. Fantucci, J. Chem. Phys. **110** (1999) 3876.
[220] V. Bonačić-Koutecký, V. Vevret and R. Mitrić, J. Chem. Phys. **115** (2001) 10450.
[221] J.C. Idrobo, S. Öğüt and J. Jellinek, Phys. Rev. B **72** (2005) 085445.
[222] J.C. Maxwell Garnett, Philos. Trans. R. Soc. Lond. **203** (1904) 385.
[223] J.C. Maxwell Garnett, Philos. Trans. R. Soc. Lond. **205** (1906) 237.
[224] T. Aste and D. Weaire, The Pursuit of Perfect Packing, Institute of Physics, Bristol, 2000.
[225] P.S. Dodds and J.S. Weitz, Phys. Rev. E **65** (2002) 056108.
[226] D. Stauffer and A. Aharony, Introduction to Percolation Theory, Taylor & Francis, London, 1994.
[227] D.A.G. Bruggeman, Ann. Phys. (Leipzig) **24** (1935) 636.

[228] J.A. Krumhansl, It's a random world in: H.O. Hooper and A.M. de Graaf (Eds.), Amorphous Magnetism, Plenum Press, New York, 1973, pp. 15–25.
[229] R.J. Elliott, J.A. Krumhansl and P.L. Leath, Rev. Mod. Phys. **46** (1974) 465.
[230] M. Hori and F. Yonezawa, J. Math. Phys. **16** (1975) 352.
[231] G. Ahmed and J.A. Blackman, J. Phys. C: Solid State Phys. **12** (1979) 837.
[232] C.G. Granqvist and O. Hunderi, Phys. Rev. B **16** (1977) 3513.
[233] C.G. Granqvist and O. Hunderi, Phys. Rev. B **18** (1978) 1554.
[234] D.J. Bergmann, Phys. Rep. **43** (1978) 377.
[235] A. Liebsch and B.N.J. Persson, J. Phys. C: Solid State Phys. **16** (1983) 5375.
[236] A. Liebsch and P. Villaseñor-González, Phys. Rev. B **29** (1984) 6907.
[237] V.A. Davis and L. Schwartz, Phys. Rev. B **31** (1985) 5155.
[238] V.A. Davis and L. Schwartz, Phys. Rev. B **33** (1986) 6627.
[239] B.U. Felderhof and R.B. Jones, Phys. Rev. B **39** (1989) 5669.
[240] D.J. Bergman and D. Stroud, Physical properties of macroscopically inhomogeneous media, in: H. Ehrenreich and D. Turnbull (Eds.), Solid State Physics, Vol. 46, Academic Press, San Diego, 1992, pp. 147–269.
[241] B. Palpant, B. Prével, J. Lermé, E. Cottancin, M. Pellarin, M. Treilleux, A. Perez, J.L. Vialle and M. Broyer, Phys. Rev. **57** (1998) 1963.

Chapter 8

Magnetism in Isolated Clusters

C. Binns[1] and J.A. Blackman[1,2]
[1]*Department of Physics and Astronomy, University of Leicester, University Road, Leicester LE1 7RH, UK*
[2]*Department of Physics, University of Reading, Whiteknights, Reading RG6 6AF, UK*

8.1.	Introduction	231
8.2.	Background	233
	8.2.1. Magnetism in atoms and solids	233
	8.2.2. Magnetic anisotropy	238
	8.2.3. Implications for cluster magnetism	243
8.3.	Two experimental techniques	244
	8.3.1. Chemical probes and reactivity	244
	8.3.2. Gradient-field deflection (Stern–Gerlach)	251
8.4.	Magnetism in free clusters	254
	8.4.1. Ferromagnetic 3d transition metals	255
	8.4.2. Orbital moment and magnetic anisotropy in ferromagnetic clusters	261
	8.4.3. Antiferromagnetic 3d transition metals	263
	8.4.3.1. Chromium	263
	8.4.3.2. Manganese	265
	8.4.4. Non-magnetic 3d and 4d metals	267
	8.4.5. Rare-earths	269
	8.4.6. Magnetic ordering temperature	270
	References	271

8.1. Introduction

In recent years a great deal of attention has focused on the magnetic behaviour of small metal clusters. Magnetism in clusters is not as well understood as in either the atomic or bulk states, and considerable effort has been directed towards an understanding of the fundamental science in this mesoscopic regime. There has also been a growing realisation of the enormous potential of nanostructures formed from magnetic clusters in the development of high performance magnetic materials and devices. Consequently there has also been a strong motivation behind the development of technologies that can manufacture tightly controlled magnetic nanostructures including quantum dots, monolayers, self-organised islands, quantum wires and deposited nanoclusters.

The key features determining the magnetic behaviour are the interatomic exchange interaction (the energy responsible for the ferromagnetic alignment of spins on neighbouring atoms) and the magnetic anisotropy (responsible for the alignment of spins along a particular crystal axis). Nanoclusters (particles smaller than 10 nm in diameter) are single-domain particles; their size is well below the critical radius R_{SD} above which it is energetically favourable to form domain walls. The critical size is strongly geometry-dependent. For a sphere, for example, it is given by [1]

$$R_{SD} = \frac{36\sqrt{AK}}{\mu_0 M_s^2}, \tag{8.1}$$

where M_s is the spontaneous magnetisation, K an anisotropy constant and A is related to the exchange stiffness.

Single-domain particles have held enormous fascination since the classic work of Stoner and Wohlfarth [2] and Néel [3] over half a century ago. At temperatures well below those required to disorder the internal atomic magnetic alignment (i.e. the equivalent of the Curie temperature), the particles can be considered as giant moments of ferromagnetically coupled atomic spins. We have the phenomenon of 'superparamagnetism'. Any reorientation of this giant moment, such as occurs in an external field, will take place coherently with the individual atomic moments remaining aligned with each other. There are complications such as canted spins at the surface [4–7] but these can be absorbed into the general picture.

At $T=0$ K, reversing the direction of the cluster magnetisation requires an external field to drive the magnetisation vector across the anisotropy boundary KV separating different magnetic alignments, where K is the anisotropy constant and V is the particle volume. At elevated temperatures when $kT \gg KV$ the anisotropy barrier becomes unimportant but the external field, which tends to orient the particles, must compete with thermal fluctuations of the magnetic moments, which tend to magnetically disorder the array. In general, when a saturating field is removed from a particle (or an assembly) at temperature, T, the magnetisation decays with a relaxation rate, τ, that can be approximated by the Arrhenius relationship:

$$\frac{1}{\tau} = f_0 \exp\left(-\frac{KV}{kT}\right), \tag{8.2}$$

where f_0 is the natural gyromagnetic frequency of the particle.

For volumes typical in deposited transition metal clusters ($<10^{-25}$ m^3), at room temperature, the thermal energy kT is much greater than the anisotropy energy of each particle so that all magnetisation directions are energetically almost equal.

The average component of magnetisation M of a cluster along the direction of an applied field H is then described by the classical Langevin function $\mathcal{L}$:

$$M = \mu_{\text{clus}} \mathcal{L}\left(\frac{\mu_{\text{clus}} H}{kT}\right), \qquad (8.3)$$

where $\mathcal{L}(x) = \coth(x) - 1/x$ and μ_{clus} is the (giant) magnetic moment of each cluster.

A nanocluster typically comprises several hundred atomic spins and so, unlike isolated atoms where very low temperatures or very high fields are required to achieve saturation, assemblies of clusters can be saturated easily. At low temperatures when $kT \sim KV$ (typically less than 50 K) the magnetisation in a given field deviates from superparamagnetism and for sufficiently low temperatures the moment in each particle becomes static. The temperature at which half the cluster moments have relaxed during the time of a measurement is known as the blocking temperature, T_B, and only a narrow temperature region around T_B separates essentially permanently frozen moments from superparamagnetic behaviour. The quest to find deposited clusters smaller than 3 nm in which T_B is above room temperature is an important challenge for future generations of magnetic recording technology.

At low temperatures and for very small particles, there is also interest in the possibility of the tunnelling of the magnetisation vector through the barrier – one of the few examples of macroscopic quantum tunnelling [8].

The deposition of clusters on a surface or their embedding in a matrix is, of course, essential for practical applications. The substrate or embedding material will generally also influence the magnetic properties of the clusters and, at high densities of deposited clusters, cluster–cluster interactions can be important. We will address these issues in the Chapter 9, concentrating here just on isolated clusters, and in Chapter 10 will survey some applications of magnetic clusters.

In Section 8.2 we provide a background survey of the magnetic behaviour of atoms and bulk, with a particular focus on the $3d$ transition metals. The main method of characterisation of free nanoclusters is the gradient-field deflection method based on the classic Stern–Gerlach experiment. This is described in Section 8.3, together with a method for probing the geometry of $3d$ clusters. A detailed account of the magnetisation behaviour from the experiments and from calculations is given in Section 8.4.

8.2. Background

8.2.1. Magnetism in atoms and solids

In isolated atoms, most elements (those without all electronic shells closed) show a non-vanishing magnetic moment, while in the solid state it is only some of the $3d$ transition metals, the rare-earths and the actinides that exhibit non-vanishing magnetisation. For the atoms, Hund's rules give the ground state angular momenta. For example, the electronic configurations of the ground states of those $3d$ transition metals atoms that exhibit magnetic ordering in bulk are Cr ($3d^5 4s$, $^7S_3$),

Mn ($3d^54s^2$, $^6S_{5/2}$), Fe ($3d^64s^2$, $^5D_4$), Co ($3d^74s^2$, $^4F_{9/2}$), Ni ($3d^84s^2$, $^3F_4$) [9]. Only Fe, Co and Ni order ferromagnetically in bulk; Cr and Mn display complex non-ferromagnetic configurations.

In bulk, the rare-earths (lanthanides) generally retain their atomic moments because the 4f-shell radii are so small that there is insignificant overlap of the wave functions of neighbouring atoms. There is strong spin–orbit coupling within the 4f shell, and J (the total angular momentum) rather than S is a constant of the motion as in the isolated atom. The direct exchange between spins on neighbouring atoms is weak, and the dominant coupling is via the indirect Ruderman–Kittel–Kasuya–Yosida (RKKY) interaction [10–12], which is mediated by the conduction electrons. The indirect exchange is determined by the projection of S onto J, i.e. $(g-1)J$, and the energy is proportional to $(g-1)^2 J(J+1)$. Competition between the RKKY interaction, which is an oscillatory function of distance, and the large magnetic anisotropy determines the rich variety of magnetic ordering in these systems [13].

By contrast, the 3d electrons in the bulk transition metals are itinerant because there is significant overlap between the wave functions of neighbouring atoms. The d-electrons form bands and contribute to the conductivity and low temperature linear specific heat. The d-electrons are still comparatively well localised, however. The 4s electrons form broad free-electron-like bands of width 20–30 eV, while the 3d electron bandwidths are much smaller, typically 5–10 eV. Hybridisation and the crystal environment lead to a transfer of one electron from s to d states so that the number of electrons in the s band is close to 1. This statement applies to quite small clusters as well as to the bulk. Consequently there are respectively 7, 8 and 9 electrons per atom in the d-bands of Fe, Co and Ni and, from Hund's rule the predicted spin moments are $\mu_{spin}(Fe) = 3\,\mu_B$, $\mu_{spin}(Co) = 2\,\mu_B$ and $\mu_{spin}(Ni) = 1\,\mu_B$. The bulk values are smaller and non-integer at $\mu_{spin}(Fe) = 2.12\,\mu_B$, $\mu_{spin}(Co) = 1.58\,\mu_B$ and $\mu_{spin}(Ni) = 0.56\,\mu_B$ because of the partial delocalisation. The increase in delocalisation with cluster size has been observed in photoemission experiments (as discussed in Chapter 7).

In the transition metals there is competition between two effects. The exchange interaction, which in isolated atoms leads to Hund's first rule, tends to maximise the total spin through the differential population of the two spin states. However depopulation of one spin state in favour of the other incurs a cost in kinetic and potential energy. The size of the penalty is related to the inverse of the density of states at the Fermi energy, $N(\varepsilon_F)$; with a high density of states there are many unoccupied states just above the Fermi energy allowing the promotion of electrons from minority to majority states at a relatively modest energy cost.

This can be expressed in simple form in a model due to Stoner [14]. Suppose the spin-up and spin-down states are occupied up to energies $\varepsilon_F + \Delta$ and $\varepsilon_F - \Delta$ respectively. The resulting magnetisation, M, can be written

$$\int_{\varepsilon_F}^{\varepsilon_F + \Delta} N(\varepsilon)d\varepsilon = \int_{\varepsilon_F - \Delta}^{\varepsilon_F} N(\varepsilon)d\varepsilon = \frac{1}{2}M, \tag{8.4}$$

where $N(\varepsilon)$ is the density of states per spin. Assuming the density of states is roughly constant over the relevant energy range then

$$\Delta = \frac{M}{2N(\varepsilon_F)}. \tag{8.5}$$

8.2. Background

The exchange energy, which is proportional to M^2, is parameterised by a constant I, and the energy difference between the magnetised and unmagnetised states is

$$E(M) - E(0) = \int_{\varepsilon_F}^{\varepsilon_F+\Delta} \varepsilon N(\varepsilon)d\varepsilon - \int_{\varepsilon_F-\Delta}^{\varepsilon_F} \varepsilon N(\varepsilon)d\varepsilon - \frac{1}{4}IM^2 \quad (8.6)$$
$$= N(\varepsilon_F)\Delta^2 - \frac{1}{4}IM^2$$

again assuming constancy of the density of states. Substituting from Eq. (8.5) yields

$$E(M) - E(0) = \frac{M^2}{4N(\varepsilon_F)}[1 - N(\varepsilon_F)I], \quad (8.7)$$

which is, as expected, even in M since it is arbitrary whether we take spin-up or spin-down as the majority band. Thus we have the Stoner criterion for ferromagnetic instability

$$1 - N(\varepsilon_F)I < 0. \quad (8.8)$$

The susceptibility (χ^{-1} is the second derivative of energy with respect to M) is enhanced by the exchange lowering of the energy and, in the non-magnetic phase (i.e. at $M=0$), is given by

$$\chi = \frac{2N(\varepsilon_F)}{1 - N(\varepsilon_F)I}. \quad (8.9)$$

The Stoner parameter, I, can be evaluated by perturbation theory based on non-spin polarised solutions of the Kohn–Sham equations of density functional theory [15–17]. For infinitesimal Stoner splitting we can write [15,17]

$$I = \int d\vec{r} \gamma^2(\vec{r})|K(\vec{r})|, \quad (8.10)$$

where $\gamma(\vec{r})$ is essentially a normalised local density of states at the Fermi energy

$$\gamma(\vec{r}) = \sum_i \frac{\delta(\varepsilon_F - \varepsilon_i)|\psi_i(\vec{r})|^2}{N(\varepsilon_F)}, \quad (8.11)$$

and ε_i and $\psi_i(\vec{r})$ are the self-consistent energies and wave functions from the Kohn–Sham equations. The quantity $K(\vec{r})$, which expresses the exchange-correlation enhancement to an external field due to the magnetisation, is defined by

$$\left\{\frac{\delta^2 E_{xc}[n;m]}{\delta m(\vec{r})\delta m(\vec{r}')}\right\}_{m=0} = 2K(\vec{r})\delta(\vec{r} - \vec{r}'), \quad (8.12)$$

where $E_{xc}[n;m]$ is the exchange-correlation functional and is proportional to a δ-function only in the local approximation, and n and m are defined in the usual way

$$n_\sigma(\vec{r}) = \sum_i \theta(\varepsilon_F - \varepsilon_{i\sigma})|\psi_{i\sigma}(\vec{r})|^2 \quad (8.13)$$

$$n(\vec{r}) = n_\uparrow(\vec{r}) + n_\downarrow(\vec{r}) \quad (8.14)$$

$$m(\vec{r}) = n_\uparrow(\vec{r}) - n_\downarrow(\vec{r}). \quad (8.15)$$

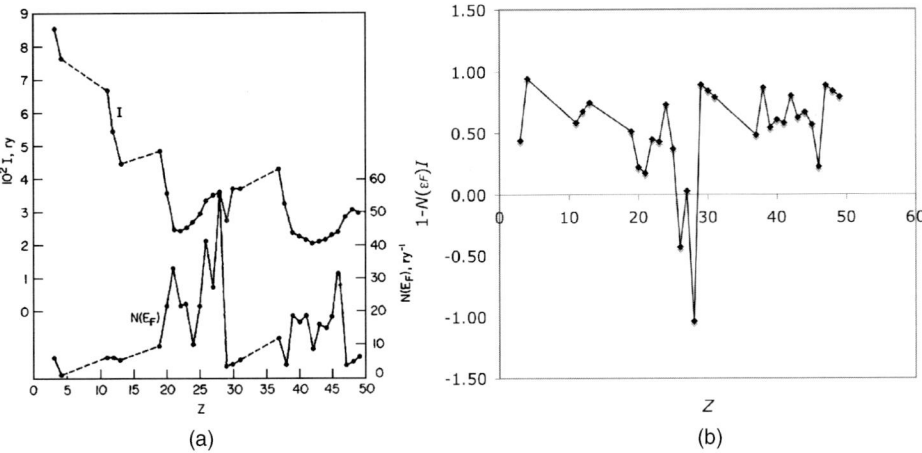

Figure 8.1. (a) Exchange-correlation integral I and density of states $N(\varepsilon_F)$ as functions of atomic number Z. Straight lines are drawn between data point which represent the calculated values. Reproduced with the permission of the American Physical Society from Janak [17]; (b) $[1 - N(\varepsilon_F)I]$ plotted against Z using data from (a).

Janak [17] presents results for I and $N(\varepsilon_F)$ for a wide range of elements (with fcc or bcc structures) across the periodic table as shown in Figure 8.1(a). We have also plotted in Figure 8.1(b) the quantity $1 - N(\varepsilon_F)I$ in the Stoner criterion of Eq. (8.8) from that data. It can be seen clearly that in this calculation only two elements, Fe ($Z = 26$) and Ni ($Z = 28$) satisfy the Stoner criterion for ferromagnetism. In Janak's [17] calculation, cobalt narrowly misses the criterion with $N(\varepsilon_F)I = 0.97$ and, of the remaining elements, the largest susceptibility enhancements are in Ca ($Z = 20$), bcc Sc ($Z = 21$) and Pd ($Z = 46$) with values of $N(\varepsilon_F)I$ of 0.78, 0.84 and 0.76 respectively. Gunnarsson's [16] results are in good agreement for Fe and Ni but he finds that Co clearly satisfies the Stoner criterion also with $N(\varepsilon_F)I$ in the range 1.6–1.8. It is perhaps worth noting that some authors define $N(\varepsilon_F)$ as the total density of states rather than the density of states per spin so that there can be a factor of 2 between the numerical results for $N(\varepsilon_F)$ and I appearing in different papers.

A generalised Stoner criterion was developed by Marcus and Moruzzi [18] by starting with the total energy and lifting the constraint of constancy on I, allowing it to be a function of M and the volume of the unit cell. These authors [19] also obtain excellent agreement between calculated and experimental spin moments: Fe 2.15 (2.12), Co 1.56 (1.58), Ni 0.60 (0.56). The moments are in Bohr magnetons (μ_B), with the experimental moments in parentheses; the experimental moments are corrected to spin only values using the measured g factors (see below). Cr and Mn also have magnetic ground states but these are complex and involve many atoms in the unit cell and non-collinear spin configurations (see Fawcett [20,21] for Cr and Hobbs et al. [22] and Hafner and Hobbs [23] for Mn).

By contrast to the situation in the rare-earths, the orbital moment in the transition metals is almost quenched and the gyromagnetic ratio, g, is close to 2. The difference is due to the overlap of the d-orbitals on different atoms in the transition metals. For the rare-earths we can regard the spin–orbit coupling as strong compared with the effect of the overlap of the f orbitals. In the transition metals, the overlap of the d orbitals dominates. One forms Bloch states using local basis functions determined by the point group symmetry of the lattice. For the d orbitals in a cubic environment these are labelled as xy, yz, zx (t_{2g}) and x^2-y^2, $3z^2-r^2$ (e_g), and each of these is a mixture of opposite orbital moments (m_ℓ and $-m_\ell$) so that the net orbital moment of each state is zero. The spin–orbit coupling then provides a small perturbation and contributes just a small additional moment typically of the order of 0.1 μ_B. The relation between the g factor and the spin, orbital and total moments is

$$\frac{M_{\text{spin}}}{M_{\text{tot}}} = \frac{2}{g} \tag{8.16}$$

$$\frac{M_{\text{orb}}}{M_{\text{tot}}} = \frac{(g-2)}{g}. \tag{8.17}$$

Experimental values for g and the spin and orbital contributions to the total moment for the bulk ferromagnetic transition metals are given in Table 8.1. The values of g and M_{tot} for Fe, hcp Co and Ni are from Stearns [24]; the moment comes from the spontaneous magnetisation extrapolated to 0 K and g is deduced from ferromagnetic resonance experiments. M_{spin} and M_{orb} are obtained from Eqs. (8.16) and (8.17). There is some variation between different authors in quoted values but it is quite small. For example, Gubanov et al. [27] list an Fe total moment of 2.12 μ_B and a g factor for hcp Co of 2.25, Skomski [1] has an hcp Co total moment of 1.76 μ_B, and Stearns [24] notes that a total Ni moment obtained from neutron scattering form factor measurements is 0.58 μ_B. Co undergoes a structural phase transition from hcp to fcc at 722 K. The experimental moment for fcc Co was deduced [25] by extrapolating the data obtained between the transition and the Curie temperature (1390 K) back to zero temperature. The calculated values are from Trygg et al. [26], both excluding and including orbital polarisation (see Section 8.2.2). A comparison of results from several different calculations is given by Beiden et al. [28].

Table 8.1. Values of g and components of the total moment (in μ_B) for bcc Fe, hcp and fcc Co, and fcc Ni [24,25,26].

	g	M_{spin}	M_{orb}	M_{tot}
Fe (bcc)	2.09	2.12	0.10	2.22
Co (hcp)	2.18	1.58	0.14	1.72
				1.75[a]
Co (fcc)	2.09	1.62	0.07	1.69[b]
	2.15	1.62	0.12	1.74[c]
Ni (fcc)	2.2	0.56	0.06	0.62

The key to the rows of fcc Co data is:
[a] Experiment [25].
[b] Calculation including spin–orbit interaction.
[c] Calculation including spin–orbit interaction and orbital polarisation [26].

8.2.2. *Magnetic anisotropy*

In non-relativistic calculations the choice of spin quantization axis is arbitrary; the Hamiltonian describing the system is rotationally invariant. However we know from experience that the magnetisation does generally lie along a preferred direction with respect to the crystalline axes, the details of which depend on a number of factors including the material, the shape of the system and any induced stress on it. The magnetic anisotropy responsible for this preferred alignment arises from the spin–orbit coupling and from the dipole–dipole interaction, both relativistic corrections. The topic is the subject of a number of reviews [29,30].

The anisotropy energy is characterised phenomenologically in terms of constants (K) and, for cubic systems, the direction cosines of the magnetisation, $\alpha_1, \alpha_2, \alpha_3$. The inclusion of anisotropy leaves the Hamiltonian invariant under time reversal; the energy is unchanged if the direction of magnetisation is reversed and so the anisotropy energy is an even function of the direction cosines. The leading terms for cubic systems can be written as

$$E_{\text{anis}} = K_0 + K_1(\alpha_1^2\alpha_2^2 + \alpha_2^2\alpha_3^2 + \alpha_3^2\alpha_1^2) + K_2\alpha_1^2\alpha_2^2\alpha_3^2 + \cdots. \tag{8.18}$$

The second order term can be subsumed into the constant since $\alpha_1^2 + \alpha_2^2 + \alpha_3^2 = 1$, and so the lowest angular dependence occurs at fourth order.

For a uniaxial system one generally uses the polar angle θ between the magnetisation direction and the symmetry axis, and an azimuthal angle ϕ. The leading term is now second order: $K_1\sin^2\theta$. For an hcp system such as Co, the dependence of the anisotropy on the azimuthal angle does not appear until the sixth order

$$E_{\text{anis}} = K_0 + K_1\sin^2\theta + K_2\sin^4\theta + [K_3 + K_3'\cos(6\phi)]\sin^6\theta + \cdots. \tag{8.19}$$

The presence of a surface generally lowers the symmetry. For example, the anisotropy energy for a cubic system with a (0 0 1) surface (such as a slab or a monolayer) can be written as

$$E_{\text{anis}} = K_0 + K_1\sin^2\theta + [K_2 + K_2'\cos(4\phi)]\sin^4\theta + \cdots, \tag{8.20}$$

where the symmetry axis is perpendicular to the surface. Experimental values of the anisotropy constants of the three transition metals that display bulk ferromagnetism are given in Table 8.2.

Assuming that higher order terms in Eq. (8.18) can be neglected, the $\langle 1 0 0 \rangle$ axes are the easy axes for cubic systems if

$$K_1 > 0 \text{ and } K_1 > -\tfrac{1}{9}K_2, \tag{8.21}$$

while the $\langle 1 1 1 \rangle$ axes are the easy axes if

$$K_1 < -\tfrac{4}{9}K_2 \text{ if } K_1 < 0 \text{ or } K_2 < -9K_1 \text{ if } K_1 > 0. \tag{8.22}$$

Usually it is sufficient to take a positive or negative K_1 respectively as the signal for $\langle 1 0 0 \rangle$ or $\langle 1 1 1 \rangle$ easy axes, as indeed is the case for iron and nickel

Table 8.2. Lowest order anisotropy constants for bulk Fe, hcp and fcc Co and Ni [24,26,31].

	Fe (bcc)	Co (hcp)	Co (fcc)	Ni (fcc)
K_1 MJm^{-3} (μeV atom^{-1})	0.052 (3.8)	0.76 (53.1)	-0.057 (-4.0)[a] -0.021 (-1.5)[b] -0.094 (-6.6)[c]	-0.126 (-8.63)

The key to the rows of fcc Co data is:
[a]Experiment [31].
[b]Calculation including spin–orbit interaction.
[c]Calculation including spin–orbit interaction and orbital polarisation [26].
K_1 is defined in Eq. (8.18) for the cubic materials and in Eq. (8.19) for hcp Co.

(Table 8.2). Measurements of K_1 are fairly consistent, but there is significant variation in the higher order constants as determined by different methods [24]. The anisotropy constants can vary greatly with temperature. Those appearing Table 8.2 are zero temperature values. As with the magnetic moments earlier, we give several estimates of K_1 for fcc Co in Table 8.2. The experimental anisotropy constant was deduced [31] from polar Kerr effect measurements on thick films (~ 1000 Å) of Co deposited on Cu substrates. The calculated values are from Trygg et al. [26], both excluding and including orbital polarisation (see later in this section).

We consider now the microscopic origins of the magnetic anisotropy. The first calculations within the itinerant-electron model were performed by Brooks [32]. He considered the spin–orbit coupling terms in the kinetic energy operator of the Dirac equation, which concerns single particle states.

However we will discuss first the other contribution to the anisotropy, which arises from terms in the relativistic two-electron Hamiltonian connected with the Breit interaction; photons, which mediate the electromagnetic interactions, are included in the Hamiltonian. This many-body effect gives rise to an effective dipole–dipole interaction, which has been discussed by Jansen [33] in the language of density functional theory. Treating the many-body Hamiltonian in a Hartree approximation yields the dipolar energy

$$E_{\text{dip}} = \frac{\mu_B^2}{2} \iint d\vec{r} d\vec{r}' \left(\frac{\vec{m}(\vec{r}) \cdot \vec{m}(\vec{r}')}{|\vec{r} - \vec{r}'|^3} - 3 \frac{[\vec{m}(\vec{r}) \cdot (\vec{r} - \vec{r}')][\vec{m}(\vec{r}') \cdot (\vec{r} - \vec{r}')]}{|\vec{r} - \vec{r}'|^5} \right), \quad (8.23)$$

where $\vec{m}(\vec{r})$ is the expectation value of the magnetisation density at position $\vec{r}$. For transition metals, whose magnetisation distribution is roughly spherical, we can write the dipolar energy as the sum over pair interactions

$$E_{\text{dip}} = \frac{\mu_B^2}{2} \sum_{i \neq j} \left[\frac{\vec{m}_i \cdot \vec{m}_j}{r_{ij}^3} - 3 \frac{(\vec{m}_i \cdot \vec{r}_{ij})(\vec{m}_j \cdot \vec{r}_{ij})}{r_{ij}^5} \right], \quad (8.24)$$

where $\vec{m}_i$ and $\vec{m}_j$ are the local moments on sites i and j. For a given pair the dipolar energy is a minimum when the moments are parallel to the vector $\vec{r}_{ij}$ joining them.

The dipole–dipole interaction gives rise to the well-known shape anisotropy [30,34] whose dominant behaviour is expressed in terms of a demagnetising field, which results from the pseudo-charges on the surface of the sample

$$H_d = -4\pi \vec{D} \cdot \vec{M}. \quad (8.25)$$

$\vec{M}$ is the bulk magnetisation and $\vec{D}$ is the demagnetising tensor. For systems such as ultrathin films, the component of $\vec{D}$ perpendicular to the surface is 1 (the other components are zero) and the shape anisotropy energy is given by

$$E_{\text{shape}} = -2\pi M^2 \sin^2\theta, \tag{8.26}$$

where θ is the angle between the magnetisation axis and the normal to the plane. In terms of an anisotropy constant, $K_1 = -2\pi M^2$. Generally the shape anisotropy dominates over other contributions, but for films just a few layers thick the spin–orbit contribution to the anisotropy can sometimes win over the dipolar contribution to produce an easy axis normal to the plane [35].

The spin–orbit interaction takes a convenient form in the Pauli Hamiltonian, which is obtained from the Dirac equation by retaining only terms of order $(v/c)^2$. There are terms in the Pauli Hamiltonian that are independent of spin, and these are often combined with the non-relativistic terms to form the 'scalar-relativistic' Hamiltonian. The spin-dependent term in the Pauli equation (the spin–orbit interaction) is written as

$$H_{\text{SO}} = \frac{e\hbar}{4m^2c^2}\vec{\sigma}\cdot(\vec{E}\times\vec{p}), \tag{8.27}$$

where $\vec{p}$ and $\vec{\sigma}$ are the electron momentum and Pauli spin matrices respectively. The electric field is strongest close to the nucleus where the potential is approximately spherically symmetric so we can write

$$\vec{E} = -\frac{\vec{r}}{r}\frac{dV}{dr}. \tag{8.28}$$

In terms of orbital and spin ($\vec{s} = \hbar\vec{\sigma}/2$) angular momentum operators

$$H_{\text{SO}} = -\frac{e\hbar^2}{2m^2c^2 r}\frac{dV}{dr}\vec{\ell}\cdot\vec{s}$$
$$= \xi(r)\vec{\ell}\cdot\vec{s}. \tag{8.29}$$

For d-orbitals, it is a good approximation to consider only the intra-atomic terms and H_{SO} is written as

$$H_{\text{SO}} = \xi\vec{\ell}\cdot\vec{s}, \tag{8.30}$$

where ξ is the average of $\xi(r)$ over d-orbitals. Calculated values of ξ are shown in Figure 8.2 and it can be seen that, within a particular series, they scale linearly with the square of the atomic number.

A simple picture for the anisotropy has been given by Bruno [30,36]. The correction to the ground state energy due to the spin–orbit interaction for systems having uniaxial anisotropy such as ultrathin films can be written in second order perturbation theory as

$$\Delta E_{\text{SO}} = \sum_{\text{exc}} \frac{|<\text{exc}|H_{\text{SO}}|\text{gr}>|^2}{E_{\text{gr}} - E_{\text{exc}}}, \tag{8.31}$$

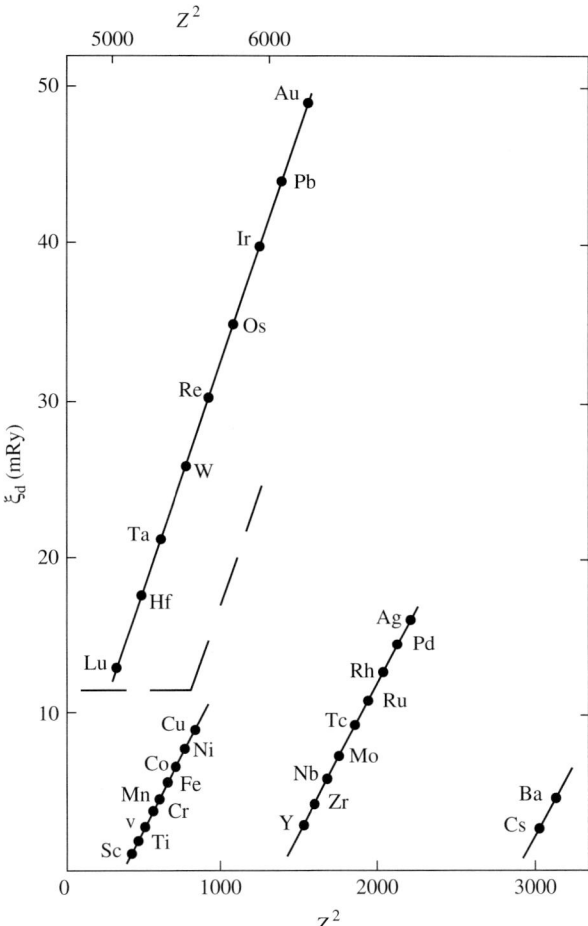

Figure 8.2. Spin–orbit coupling constant, ξ_d, for d-electrons (in mRyd) as a function of the square of the atomic number Z. The values are calculated at the centre of the d-band. Reproduced with the permission of Cambridge University Press from Mackintosh and Andersen [37].

where the labels 'gr' and 'exc' refer to the unperturbed ground and excited states respectively. From this one can make a rough estimate [30] of the order of magnitude of K_1 for uniaxial systems: $K_1 \sim \xi^2/W$; W is the d-electron bandwidth. For cubic crystals the anisotropy occurs in fourth order perturbation theory leading to a rough estimate of K_1, in this case $K_1 \sim \xi^4/W^3$. Taking $\xi \approx 75$ meV and $W \approx 5$ eV (values appropriate for the $3d$ ferromagnets), we obtain $K_1 \approx 1$ meV/atom for a uniaxial system and $K_1 \approx 0.3$ μeV/atom for a cubic system. These values are in order of magnitude agreement with observations on ultrathin films and cubic ferromagnets.

We can also make a phenomenological argument [30] for the direction of the orbital moment. If we have exchange splitting which is larger than the bandwidth, we can neglect the matrix elements of the spin–orbit interaction that couple spin-up and spin-down states. Let $<L_\zeta>$ represent the average component of angular momentum per atom along the direction of spin quantisation. The contributions to the angular momentum from the majority and minority spin are denoted respectively by $<L_{\zeta\uparrow}>$ and $<L_{\zeta\downarrow}>$. The number of d electrons per atom is denoted by n_d. If $n_d<5$ the minority (spin-down) band is empty and $<L_\zeta> = <L_{\zeta\uparrow}>$, whereas if $n_d>5$ the contribution to $<L_\zeta>$ from the majority band is zero and $<L_\zeta> = <L_{\zeta\downarrow}>$. In terms of the component of the orbital moment $<M_\zeta>(=\mu_B<L_\zeta>)$ along ζ we can write

$$<H_{SO}> = \mp \frac{\xi<M_\zeta>}{2\mu_B}, \tag{8.32}$$

where the minus (plus) sign refers to $n_d>5$ ($n_d<5$) respectively. From Eq. (8.32) we can see that the preferred spin direction is that yielding the largest orbital moment and, for $n_d>5$, the orbital moment is parallel to the spin moment, giving a value of g greater than 2.

There has been extensive work on the anisotropy in multilayers. Particular interest has focused on interfacing a ferromagnetic material such as Co with a high anisotropy system like Pd or Pt, which favours magnetisation perpendicular to the layers [35]. Also adding Pd or Pt atoms to a $3d$ transition metal system can strongly enhance the orbital moment over its bulk value [38].

Ab initio calculations are generally in fair agreement with experiment for multilayer systems, but computation is much more problematic for bulk cubic materials where total energy differences for spin alignment along inequivalent directions are very small ($\sim 10^{-6}$ eV/atom). There have been a number of calculations [26,28,39–43] based on the local spin-density approximation (LSDA) to density functional theory. However the orbital moments are generally too small compared with experiment. There was doubt whether it was possible for the LSDA to describe the anisotropy correctly, since the wrong easy axis was obtained for hcp Co and fcc Ni [39]. Improved numerical techniques [26–41,41] did produce the correct easy axis for hcp Co but not for Ni. It is well known, however, that the LSDA is less successful in describing Ni than Fe or Co [44]. To obtain reasonable numerical accuracy a huge number of k-points is required. The anisotropy calculation can be approached using either the force theorem [39] or the total energy [26]; the methods have been compared by Halilov *et al.* [42]. An alternative real space approach has been introduced by Beiden *et al.* [28], which has the potential for also tackling disordered systems.

The magnitudes of the calculated orbital moments can be brought closer to experiment [26,45] by the inclusion of the orbital polarisation term [46], which mimics Hund's second rule. In the LSDA the total energy is a functional of the charge and spin magnetisation densities. The concept of orbital polarisation is driven by the Racah parameter B, with an extra term in the energy functional related to the orbital moment $<\hat{L}>$, given by $\Delta E_{OP} = -\frac{1}{2}<\hat{L}>$ [26,43]. However, although orbital moments are improved, the easy axis of Ni is still incorrect [26].

It has been argued that the key parameter responsible for the exchange-correlation enhancement of the orbital moments is the Hubbard U rather than the intra-atomic Hund's second rule coupling and this is consistent with a more general concept of orbital polarisation [47]. This leads to the LDA+U functional which provides a rotationally invariant prescription for the orbital magnetism [47–49]. The LDA+U method [50] and a related approach [51] have been used in calculations on Fe, Co and Ni. Physically reasonable values of U yield values of anisotropy in agreement with experiment and also the correct easy axis for Ni [50,51]. Good agreement with experiment for the orbital moments is also obtained [51]. Calculation of the interaction parameter U in a metallic environment, of course, remains an open problem.

8.2.3. Implications for cluster magnetism

Atoms at the surface of a cluster have a reduced coordination and are in a lower symmetry environment compared to atoms in the bulk. There is the potential in small clusters for a significant modification of the bulk magnetic properties: e.g. in a 309 atom cuboctahedral cluster, 162 of the atoms (52%) lie at the surface (see Section 1.3).

As far as the spin moment is concerned, a qualitative indication of the effect can be obtained from the second moment approximation (see Section 6.2.4). The bandwidth W is proportional to $z^{1/2}$ where z is the coordination number. If we consider Friedel's model [52,53] of a rectangular band for the d electrons, the density of states scales as $1/W$. The reduced coordination of the surface atoms will result in an enhanced local density of states with, we may expect, the possibility of a ferromagnetic instability through the Stoner criterion, Eq. (8.8), which was absent in bulk. Similarly, from Eq. (8.5), if the exchange splitting Δ is not very sensitive to the environment, a reduced z will cause an increase in the local moments μ_i on atom i on the surface compared with the bulk value μ_{bulk}

$$\mu_i = \left(\frac{z_{bulk}}{z_i}\right)^{1/2} \mu_{bulk}, \tag{8.33}$$

as long as z_i is not too small. One has to be careful in these crude arguments however. The second moment approximation also tells us than a reduced coordination causes a contraction in the interatomic distances [52], and this will lead to a reduction in the magnetic moment. Clearly this points to the desirability of including geometry optimisation in theoretical calculations. An enhanced magnetic moment is indeed observed in clusters and is of considerable interest. Jensen and Bennemann [54] have proposed a simple expression for the average moment $\bar{\mu}_N$ of a cluster of N atoms in terms of bulk and surface contributions

$$\bar{\mu}_N = \mu_{bulk} + (\mu_{surf} - \mu_{bulk})N^{-1/3}. \tag{8.34}$$

Although experiments show more complex size dependence than is implied by this expression, it does display the trend of a decrease in moment towards the bulk value with increasing size.

Symmetry is a determining factor in the magnitude of magnetic anisotropy and the orbital moment. Generally the presence of a surface will lower the symmetry resulting in anisotropy and orbital moments that are enhanced compared with the bulk. Of course the magic number clusters (discussed in Section 2.3.3) have high symmetry and so will exhibit a fourth order anisotropy like the bulk.

8.3. Two experimental techniques

8.3.1. *Chemical probes and reactivity*

The chemical probe method has been employed extensively in the investigation of the geometric structure of clusters of transition metals (principally Fe, Co, Ni) up to sizes of about 200 atoms. There has also been work on other metals such as Cu, Pt and Rh. Atoms and small molecules react with transition metal clusters in ways that are analogous to the physisorption and chemisorption processes that occur on metal surfaces. However, for clusters, the reactivity is dependent on the environment of the atoms on which adsorption takes place and this can be employed in inferring the geometric structure of the cluster. The reactions are, of course, relevant to many heterogeneous catalytic processes, but we will be concerned here with what can be deduced about cluster structure rather than with catalysis *per se* (a discussion of heterogeneous catalysis is given in Chapter 10). The main probe molecules that have been used are NH_3, N_2, H_2O and H_2 (or D_2), and the determination of the number of binding sites on a cluster for a particular adsorbate molecule can give important clues as to possible structures – or eliminate other candidates. We are mostly considering clusters of elements with partially filled *d*-shells. The *d*-orbitals have significant spatial extent and are often involved in the transition metal bonding. The chemical properties are sensitive to both the number and the configuration of the *d* electrons and as a result there is considerable variation across the transition metal series.

A flow-tube reactor (FTR) using helium carrier gas is generally used to study chemical reactions [55,56]. The clusters are produced by pulsed laser vaporisation of a metal target in a flow tube upstream of the FTR. A narrow flow tube and low helium pressure ensures a rapid decrease in metal atom density along the tube such that the cluster growth has finished and the clusters have cooled to ambient temperature before the clusters enter the FTR and, at this point, the reagent gas is introduced. The reaction between the cluster, M_N, and the reagent molecule, A, produces an internally excited complex that requires a collision with a third body (a carrier gas atom, X) to stabilise it otherwise dissociation occurs.

$$M_N + A \rightleftharpoons M_N A^*, \quad M_N A^* + X \to M_N A + X \tag{8.35}$$

Gas flows are continuous and reaction conditions such as pressure, temperature and interaction time can be controlled and measured. The clusters and reaction products expand out of a nozzle and are formed into a molecular beam. The clusters are then pulsed laser ionised and mass analysed in a time-of-flight mass spectrometer.

8.3. *Two experimental techniques* 245

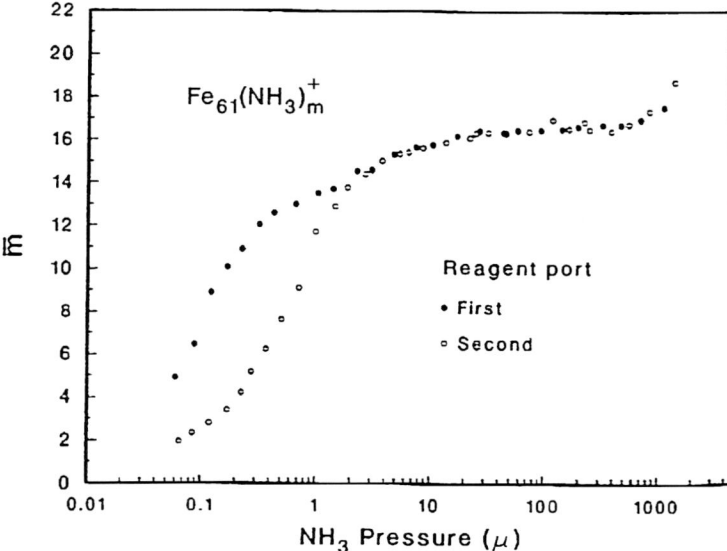

Figure 8.3. The average number of NH_3 molecules bound to a Fe_{61} cluster plotted against partial pressure in flow-tube reactor. Filled/open symbols are for long/short interaction times (first/second port). Reproduced with the permission of the American Institute of Physics from Parks *et al.* [57].

The main quantity of interest is the mean number of reagent molecules bound to a cluster. This is illustrated in Figure 8.3 for Fe_{61} clusters with NH_3 as the adsorbed molecules [57]. The average number of bound molecules, $\bar{m}$, is plotted as a function of the partial pressure of the reagent gas. The interaction time of the NH_3 molecules can be varied by introducing the gas at different ports of the FTR. At low pressures (below about 0.2 Pa in this example), the coverage for a given partial pressure depends on the interaction time; it is a kinetically controlled reaction. At higher pressures the coverage loses its interaction time dependence; it is an equilibrium reaction. The uptake pattern is fairly steep in the kinetically controlled regime but much slower under equilibrium conditions. The transition from kinetically controlled to equilibrium conditions is due to the fact that the binding energy decreases with surface coverage. At low coverage the desorption time is longer than the time that the clusters spend in the FTR, so if a molecule reacts with a cluster it stays on that cluster and the number of molecules bound to the cluster at a given partial pressure depends on the interaction time. At higher coverages the binding energy becomes so low that the desorption time becomes less than the time the clusters spent in the FTR. Equilibrium exists between the adsorption and desorption rates and the coverage no longer depends on the interaction time.

At higher pressures one often sees a levelling off and a region where $\bar{m}$ is independent of pressure. The plateau region generally signals that a certain number of binding sites have been filled (16 in the case of Fe_{61}) and the cluster has been saturated with chemisorbed molecules. A final upturn in the plot indicates that further molecules are being adsorbed but the process now is one of physisorption and the molecules are forming a second layer. It is the value of $\bar{m}$ in the plateau region that gives us information about the number of binding sites of a particular type and a clue to details of the structure of the cluster.

There are different binding rules for the various reagent molecules. Surface studies indicate that the NH_3 molecule binds to the surface through the donation of the N-localised lone pair to a metal atom. The same mechanism is assumed to hold for the cluster but it is found that binding preferentially occurs on the low coordination metal atoms, which means that saturation with ammonia counts the number of vertex atoms on a cluster. Water has similar binding characteristics. As we shall see, the number of vertex atoms does not define a cluster uniquely and complementary experiments using a reagent with different binding rules are necessary to narrow the choice of possible cluster morphologies. Molecular nitrogen is useful in this regard. The N_2 molecule, like NH_3 binds to a metal atom via the lone pair on one of the nitrogen atoms. Unlike H_2, the nitrogen molecule does not dissociate because of the strong nitrogen–nitrogen triple bond. Molecular nitrogen has a more complex set of binding rules than NH_3. Generally all metal atoms on the surface of a cluster can bind a N_2 molecule and some can bind two. The precise details depend on the coordination number of the metal atom [58].

Reactions of molecular hydrogen (H_2 or D_2) are dissociative: $M_N + H_2 \rightarrow M_N HH$. The rate constant, k, for the process is defined by the rate equation

$$\frac{d[M_N]}{dt} = -k[M_N][H_2], \tag{8.36}$$

where the quantities in square brackets are the concentrations of the species. Under steady flow conditions, $[H_2]$ is constant and the equation integrates to

$$\ln[M_N] - \ln[M_N]_0 = -k\Delta t[H_2], \tag{8.37}$$

where Δt is the interaction time. Measurements of $[M_N]$ for different values of $[H_2]$ allow one to determine the rate constant, which shows a strong variation with cluster size. There also appears [59] to be some correlation between reactivity and ionisation potential (IP). Presumably any barrier to the reaction would involve the breaking of the H–H bond. Whetten et al. [59] proposed a model in which there is electron donation from the HOMO of the metal cluster to the antibonding σ^* orbital of H_2. This, it was argued, is the rate-determining step, and the lower the cluster's IP then the lower the activation barrier for the reaction. However, this correlation is only approximate in iron clusters and is not followed in the other transition metals. A much stronger correlation appears to hold between the reactivity and the difference between the IP and the electron affinity (EA) rather than with the IP alone [60]. This is shown in Figure 8.4. The relative reactivity is defined as the ratio of the rate constant of a cluster with N atoms to that of the least reactive cluster (on a logarithmic scale). Plot (a) compares the relative reactivity to the IP. The promotion energy

$$E_p = \frac{IP - EA - e^2}{R}, \tag{8.38}$$

is a measure of the energy required to promote an electron from the HOMO to the LUMO of the cluster. It is assumed that the clusters are spherical with effective radius R, and the Coulomb energy between the positively charged cluster and the

8.3. *Two experimental techniques* 247

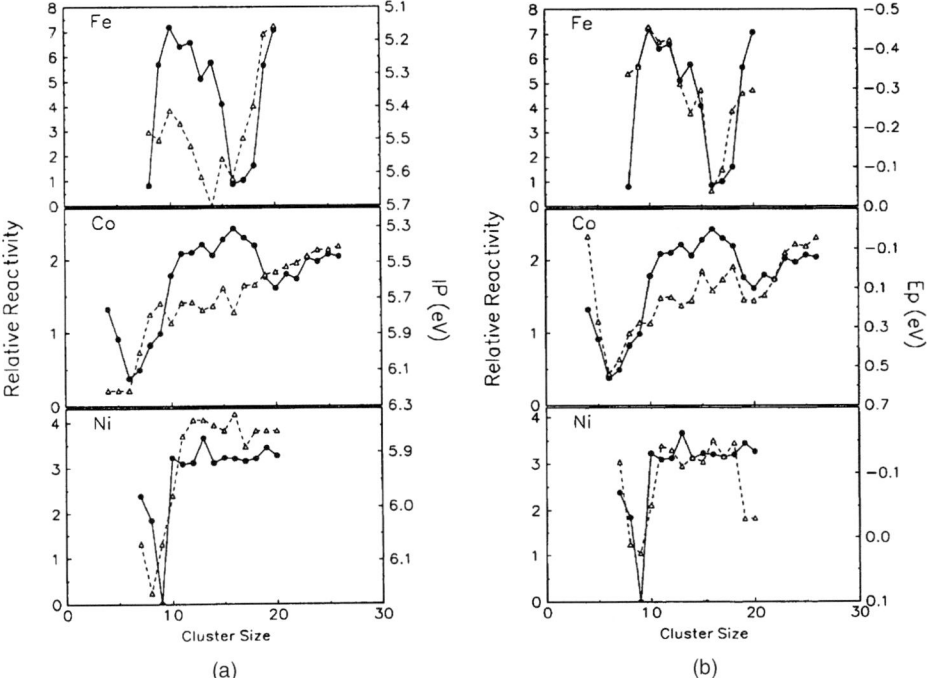

Figure 8.4. Relative reactivity (full line) of Fe, Co and Ni clusters with hydrogen as a function of cluster size [both plots], and ionisation potential (IP) [broken line on plot (a)] and promotion energy (E_p) [broken line on plot (b)]. Reproduced with the permission of the American Physical Society from Conceicao *et al.* [60].

electron is subtracted in Eq. (8.38). The promotion energy is shown in plot (b) and it can be seen that there is significantly closer correlation with the reactivity particularly for Fe and Ni. The model [60] assumes that the activation barrier is due to Pauli repulsion between the valence electrons of the H_2 molecule and the HOMO of the cluster. The barrier is overcome by the HOMO→LUMO promotion.

The dissociated hydrogen atoms are adsorbed in a bridging mode with a single H atom binding to two, three or four metal atoms in contrast to the binding to a single metal atom, which is the reaction mode of NH_3 for example. H and NH_3 bind non-competitively as demonstrated by the fact that hydrogenated metal clusters can adsorb the same number of ammonia molecules as unhydrogenated clusters.

Clusters of cobalt and nickel, which adopt close-packed structures in bulk, are more straightforward than iron in yielding clues about their structure. Iron and other elements towards the middle of the transition series have less filled *d* orbitals and the binding of the atoms in the cluster is much more directional. Although cobalt and nickel are in many ways similar structurally, there are significant differences particularly for the smaller clusters. This is illustrated particularly in the NH_3 uptake plots for Co_{19} and Ni_{19} [58] in Figure 8.5.

There is a clear plateau at $\bar{m} = 6$ for Co_{19} and a less extensive one at $\bar{m} = 12$ at higher ammonia pressure for Ni_{19}. The obvious conclusion is that Co_{19} has 6 apex

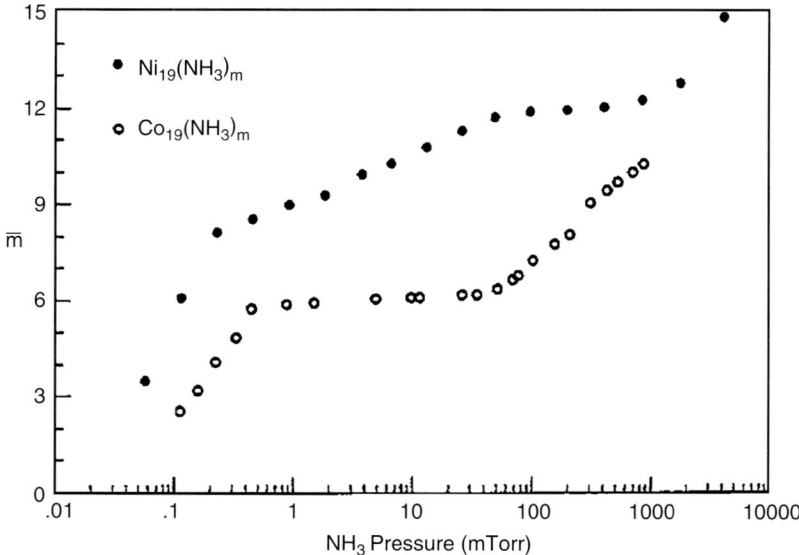

Figure 8.5. Uptake plots for NH_3 reacting with Ni_{19} and Co_{19}. Reproduced with the permission of Elsevier Science from Riley [58].

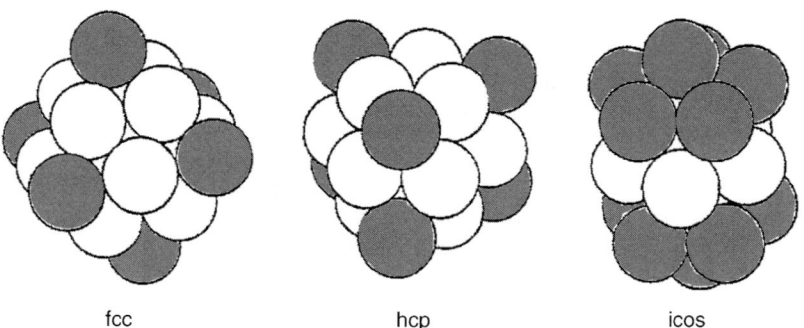

Figure 8.6. Models of three 19-atom clusters. Primary ammonia binding sites are shown in grey. Reproduced with the permission of Elsevier Science from Riley [58].

atoms and Ni_{19} has 12. Of the many possible structures for 19-atom clusters, the 3 illustrated in Figure 8.6 are clearly candidates in this case. The fcc octahedron and the hcp structure have six apex atoms each with four nearest-neighbours. The remaining 12 atoms on the cluster surfaces (there is 1 internal atom) have 7 neighbours, which provide a further distinction between them and the apex atoms. Other evidence suggests that the hcp structure is the most likely choice for Co_{19}.

The double icosahedron in Figure 8.6 has 12 cap atoms each with 6 nearest-neighbours. The remaining five surface atoms have eight (there are two internal atoms). This appears to be a good candidate for the structure of Ni_{19}. The cap atoms are quite close together and repulsion between the bound NH_3 molecules will require a higher ammonia pressure for saturation as is seen in Figure 8.5. Further support

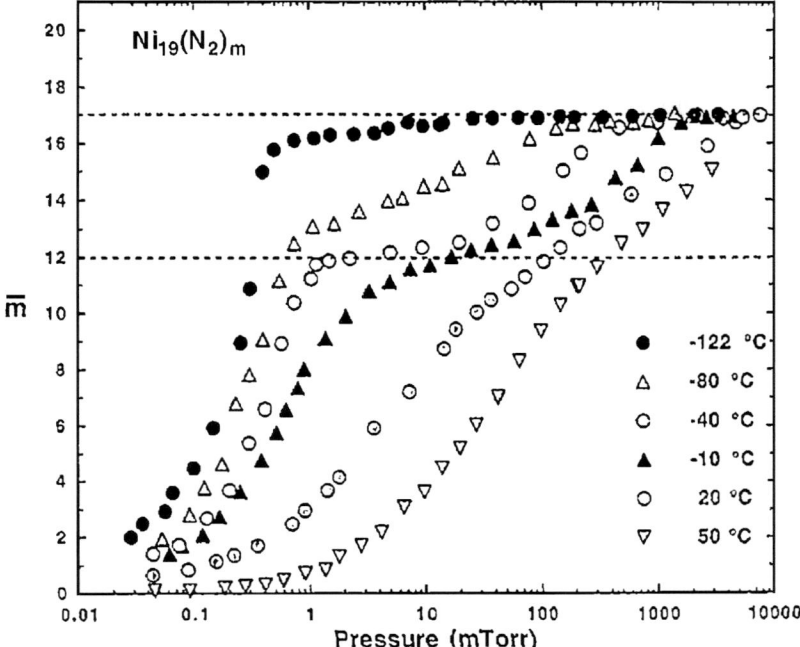

Figure 8.7. Nitrogen uptake plots for Ni$_{19}$ at several different temperatures. Dashed lines are shown at $\bar{m} = 12$ and 17 (see text). Reproduced with the permission of Elsevier Science from Riley [58].

for assigning this structure to Ni$_{19}$ is the nitrogen uptake behaviour displayed in Figure 8.7. A plateau can be seen at $\bar{m} = 17$ and, at low temperatures, a shorter one at $\bar{m} = 12$. The plateau at $\bar{m} = 17$ indicates the total number of surface atoms and points to the double icosahedron as opposed to either of the close-packed structures which each have 18 surface atoms. The secondary plateau at $\bar{m} = 12$ implies that 12 of the sites bind N$_2$ more strongly than the remaining 5.

The structures inferred from similar measurements [61,62] on cobalt and nickel clusters of up to 19 atoms are summarised by Riley [58] and are shown in Figure 8.8. Both elements show similar structures up to $N = 7$, but for larger sizes Ni tends to adopt a structure based on pentagonal packing while Co is more likely to show packing characteristic of the bulk.

Monitoring the number of adsorbed ammonia molecules at saturation seems to indicate that larger clusters (50–150 atoms) of nickel and cobalt adopt structures based on icosahedral packing. The plateau values of NH$_3$ coverage are shown in Figure 8.9 for a range of sizes of cobalt clusters. The plots show a minimum near to $\bar{m} = 12$ for clusters sizes that include $n = 55$, 71, 80, 92, 101 and 116. Results have also been presented [63] for the uptake of NH$_3$ on hydrogenated cobalt clusters. These show a minimum at $\bar{m} = 12$ for $n = 147$ that is not apparent in the data for unhydrogenated clusters. Hydrogenation lowers the cluster–ammonia binding energy resulting in the onset of the equilibrium regime at ammonia coverages at which the reaction is still kinetically controlled. This results in a more extended plateau at saturation [63] and a clearer identification of $\bar{m}$.

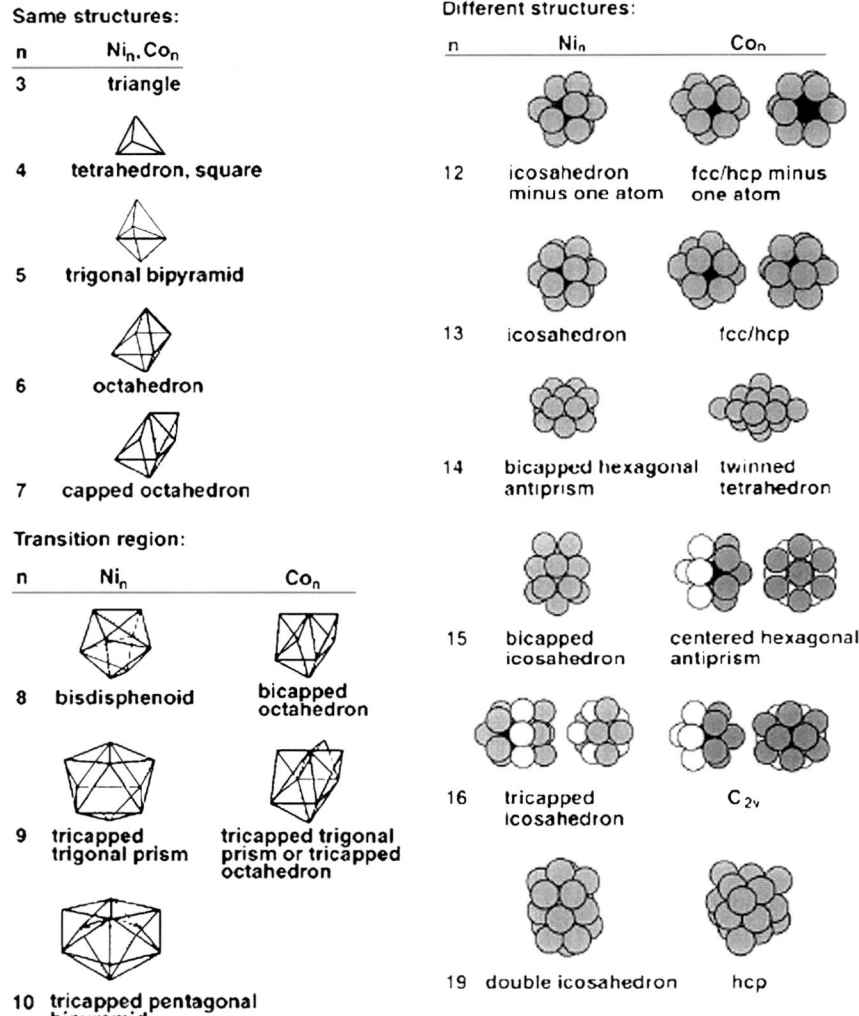

Figure 8.8. Comparison of structures deduced for 3–19 atom nickel and cobalt clusters. Two views are shown for some of the clusters. Reproduced with the permission of Elsevier Science from Riley [58].

Cluster sizes of 55 and 147 are, of course, magic numbers for icosahedral packing and the number of vertices for icosahedra are 12 independent of cluster size. This fact alone is insufficient for the identification. The same number of vertices occurs with the fcc cuboctahedral structure. However it is argued [63] that the cluster sizes between 55 and 147 that display a minimum at $\bar{m} = 12$ correspond to the filling of sub-shells between the main shell fillings of the magic number clusters. Sub-shell filling occurs at different sizes for the cuboctahedral structure and this allows us to identify the icosahedron-based configurations as the preferred structures for the Co clusters for sizes between 70 and 200 atoms.

Generally similar results have been obtained in chemical probe experiments on nickel clusters but there are a number of puzzling gaps in the general trend.

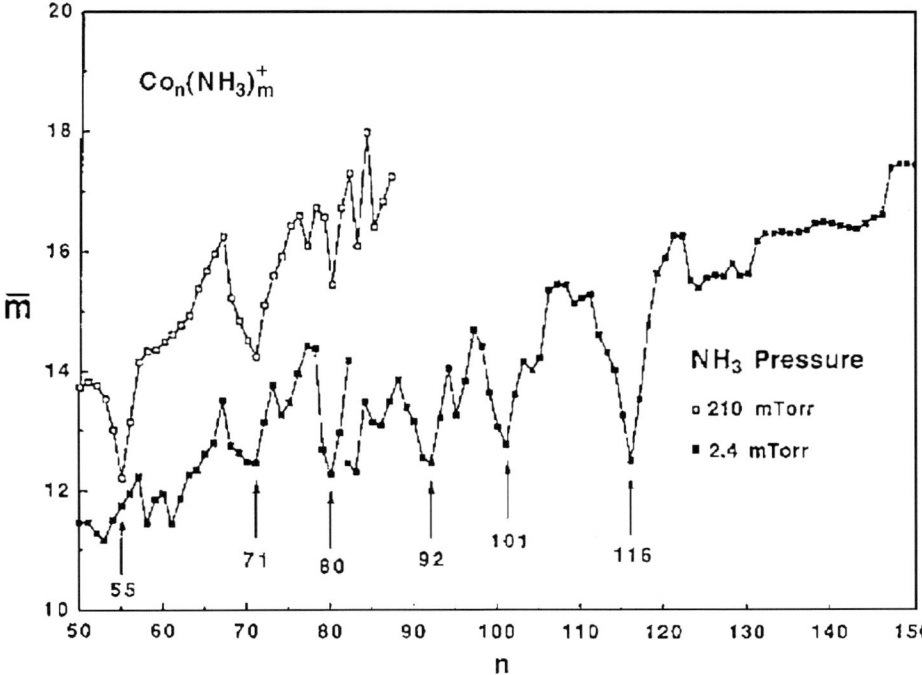

Figure 8.9. Plot of $\bar{m}$ against cluster size for the reaction of ammonia with cobalt clusters in the 5–150 size regime. Two ammonia partial pressures are used. Reproduced with the permission of Elsevier Science from Riley [58].

In particular Ni_{147} does not appear to display an icosahedral structure. Concern has been expressed that saturating a cluster with adsorbate molecules could change the crystal structure (and also incidentally the magnetic moment [61]). Water molecules bond in a similar way to NH_3 but with weaker binding, which if it is an issue should presumably be less likely to affect the cluster structure. Even with H_2O adsorption [63] there was no signature of icosahedral structure for Ni_{147}. There have been a number of detailed studies on intermediate size nickel clusters and a preference for fcc packing identified for Ni_{38}, Ni_{46}, Ni_{47} and Ni_{48} [62,64]. The assignment of structures to iron clusters is much more tentative.

8.3.2. Gradient-field deflection (Stern–Gerlach)

The founding experiments on cluster magnetism were carried out on free particles and determined the magnitude of the magnetic moment as a function of cluster size [65–69]. The first set-up capable of this measurement was reported by Cox et al. [67] and is based on the classic Stern–Gerlach experiment that first detected the electron spin [70]. Figure 8.10 shows, schematically, a typical arrangement. A collimated cluster beam generated by a pulsed laser evaporation source with a variable temperature nozzle is guided into a magnetic field gradient dB/dz that will deflect

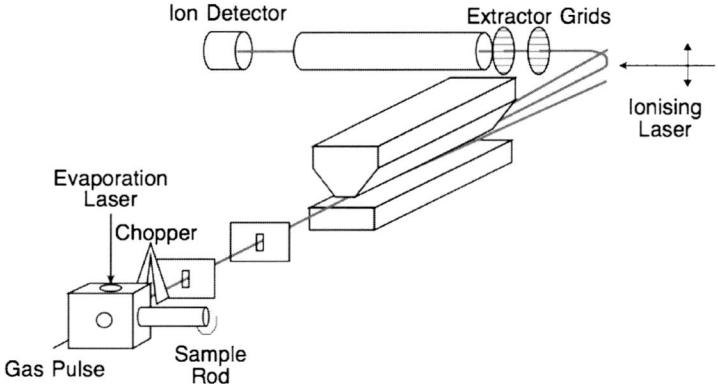

Figure 8.10. Schematic of experiment for measuring the magnetic moment of free clusters.

vertically any particle with a magnetic moment M by an amount:

$$d = M \frac{dB}{dz} L^2 \frac{(1 + 2D/L)}{2mv_x^2} \tag{8.39}$$

where L is the length of the magnet, D the distance from the end of the magnet to the detector, m the cluster mass and v_x its velocity when it enters the magnet. A pulsed ionising laser probes the beam profile after the magnet and is the source for a time-of-flight mass filter. Thus it is possible to measure the deflection of a given mass and determine its magnetic moment as long as v_x is known. This can be defined by a controlled delay between the evaporation laser and the ionising laser. Since the duration of the pulse of clusters can be several milliseconds wide, a mechanical chopper in front of the source (also shown in Figure 8.10) is used to define better the pulse start time. The chopper can be phased with the vaporising laser to define clusters emerging from the nozzle at a specific time relative to the vaporising pulse, a feature that was found to be important in order to define the cluster temperature.

Interpreting the data from clusters provided by the instrument is however not straightforward and requires a careful consideration of the detailed operation of the source. The problem is in accurately defining the cluster vibrational temperature, T_{vib}. It is found for transition metals that they always deflect towards the direction of highest field, which is indicative of superparamagnetism. That is, clusters enter the magnet with a rapidly fluctuating moment and within the field become magnetised to a degree depending on the temperature and the field, as described by Eq. (8.3). Since the residence time in the magnet ($\sim$ms) is orders of magnitude longer that the timescale of the magnetic fluctuation ($\sim$ns), clusters of a given size will attain the same mean magnetisation and deflect by the same amount. Typical deflection profiles are shown in Figure 8.11.

During the initial experiments on free cluster magnetism, significant deviations from the predictions of the superparamagnetic model were found for small residence times of the clusters in the source. The original proposal to explain

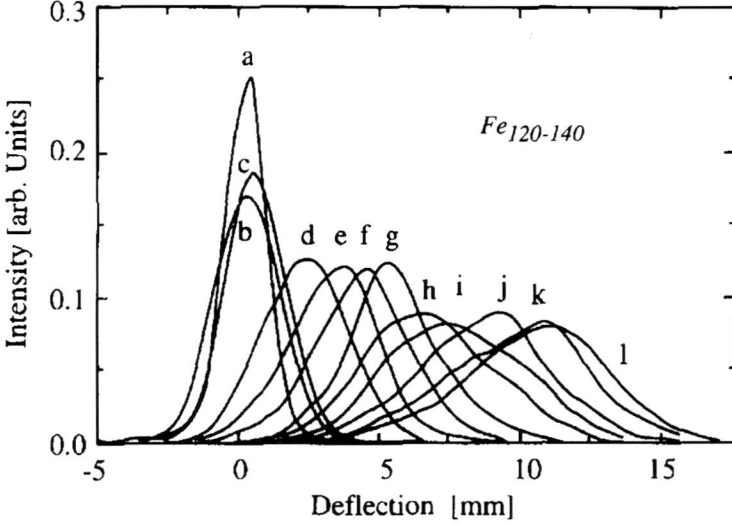

Figure 8.11. Typical Stern–Gerlach deflection profiles for $N=120$–140 iron clusters for a range of magnetic field strengths from 0 kG (a) to 6.7 kG (l). Note the deflections are only in one direction corresponding to the increasing magnetic field. Reproduced with the permission of Elsevier Science from Billas et al. [71].

this [72] was that for short times (<1 ms) after the laser pulse, the gas pressure within the source is still high and the clusters are cooled well below the nozzle temperature by the adiabatic expansion. This was based on the observation that Ar added to the He bath gas was adsorbed onto the surface of the clusters indicating that they were much colder than the temperature within the source. The non-equilibrium conditions can create a situation where the rotational temperature (T_{rot}) is less than the vibrational temperature (T_{vib}) and may be of the same order of magnitude as the magnetic precessional frequency in the applied field, which can lead to a resonant spin rotation coupling that reduces the magnetisation [65,73]. In order for the clusters to be thermalised with the nozzle temperature, the chopper must select clusters produced after delays that are sufficient for the gas pressure to drop below the values where significant cooling occurs in the free jet expansion. The cluster temperature is then the nozzle temperature and the cluster magnetisation is superparamagnetic in the case of Fe, Co and Ni clusters. An alternative mechanism proposed by Douglass et al. [66] is that at short residence times the metal vapour is still hot and requires time to thermalise with the carrier gas. They found no significant supercooling in their free jet expansion and were unable to produce the rare gas adsorption results obtained by Milani and de Heer [72]. In either case, for sufficiently long residence times, the clusters attain the nozzle temperature, which can be controlled and then all results for the transition metals are consistent with the superparamagnetic model. A more subtle consideration is whether a nanocluster containing a few tens of atoms has a sufficient density of states to constitute a heat bath so that thermodynamically based laws are valid. This is assumed to be the case based on the success of the superparamagnetic model in describing the results after the above technical considerations are addressed [74].

In rare-earth clusters, rotational effects can again become important since the spin–orbit interaction can be sufficiently strong to lock the cluster magnetic moment to its atomic lattice. Douglass et al. [66] showed that whether Gd clusters display superparamagnetic or locked-moment behaviour is a sensitive function of their size.

8.4. Magnetism in free clusters

Studying mass-selected magnetic clusters as free particles enables one to address fundamental questions regarding the development of magnetism in materials as they are built atom by atom from the monomer. Various novel behaviours have been observed in free clusters including enhanced magnetic moments in ferromagnetic metals [65], magnetism in metals that are non-magnetic in the bulk [69], ferrimagnetism in antiferromagnetic metals [75], canted spin arrangements in ferromagnetic rare-earths [66,76] and lowered [65,77] or increased [78] Curie temperatures. Some of these properties are expected from the qualitative arguments presented above but to fully understand the range of novel behaviour requires the application of detailed calculations. The density functional methods outlined in Section 6.2.2 are used widely to probe magnetic properties as well as the geometric structure of clusters. There are a few calculations on clusters of up to a few hundred atoms using spin-polarised DFT–GGA, while more approximate calculations on larger clusters are possible with the tight-binding methodology (Section 6.2.3).

It is found in all cases that the atomic structure and the nearest neighbour distance are critical to determining the magnetic properties in clusters. Experimentally this information is sparsely available at present and has been determined in a limited number of cases by indirect methods such as the observation of magic numbers in the mass spectra of the clusters [79]. The situation is complicated by the existence in some cases of several isomers of clusters of a given size.

The development of very high-flux sources has enabled recently the measurement of X-ray absorption in a beam of free clusters [80] but up till now there is insufficient signal-to-noise for an XMCD experiment (see Chapter 9) on free particles. There have been many such studies on deposited clusters (as reviewed in Chapter 9), but in that case it is straightforward to accumulate enough material on the surface to obtain a low noise measurement. The magnetic measurements on free clusters have so far been obtained using exclusively the gradient-field experiment (described in the previous section) that measures the deflection and the modification of the beam profile of a beam of clusters by an inhomogenous magnetic field. In principle this can measure the total magnetic moment of a mass-selected cluster as a function of the field magnitude and the cluster temperature. As discussed earlier, initial controversies regarding the interpretation of data have been resolved. It is generally accepted that in transition metal clusters with their low magnetocrystalline anisotropy, at typical temperatures encountered in cluster sources, the moment is decoupled from the atomic lattice and the magnetisation follows the Langevin function Eq. (8.3) describing superparamagnetic behaviour. The magnetisation of rare-earth clusters on the other hand shows superparamagnetism only at some specific particle sizes. Clusters with different sizes show locked-moment behaviour. A review of results on free clusters of various types is presented below. For measurements on superparamagnetic clusters, the magnetisation measured at finite temperatures is used to extract the intrinsic value of μ_{clus} (usually expressed in μ_B/atom) using Eq. (8.3).

8.4.1. Ferromagnetic 3d transition metals

Figure 8.12(i) shows the total moments in free Fe_N, Co_N and Ni_N clusters for $N = 20$–700 atoms, measured by Billas et al. [71,81] using the gradient-field

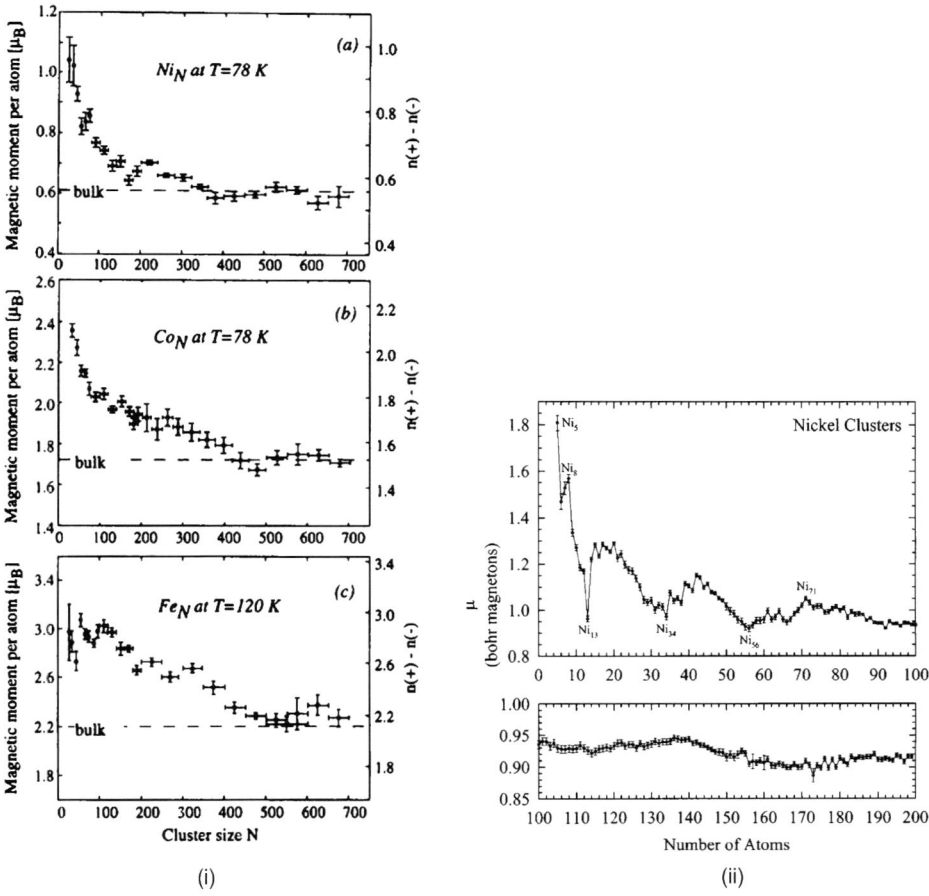

Figure 8.12. (i) Low temperature total moments (in μ_B/atom) as a function of cluster size N for (a) Ni_N at $T = 78$ K, (b) Co_N at $T = 78$ K, (c) Fe_N at $T = 120$ K. The right scale indicates the spin imbalance (see text). Reproduced with the permission of Elsevier Science from Billas et al. [71]. (ii) Nickel cluster moments as a function of cluster size at $T = 198$ K and below. Reproduced with the permission of the American Physical Society from Apsel et al. [82].

deflection technique. The scale on the right, called the spin imbalance by Billas et al. [71,81], is the spin moment M_{spin} obtained from the measured total moment by Eq. (8.16). Apsel et al. [82] performed similar measurements on Ni over the size range $N = 5$–740. Their data for sizes up to 200 atoms is shown in Figure 8.12(ii) and significant structure is apparent particularly at the lower end of the size range. Apsel et al. [82] quote a systematic uncertainty in their measurements of $\pm 0.05\,\mu_B$.

The Ni moment in the Billas et al. [71,81] data drops rapidly to the bulk value at about $N = 160$, and then shows a small increase until it reaches the bulk value finally at around $N = 350$. The fall in the moments of Fe and Co clusters is less rapid, reaching the bulk value for $N = 400$–500. The moments for Ni obtained by Billas et al. [71,81] are generally lower than those of Apsel et al. [82] who see enhancement over the whole size range studied. Knickelbein's [83] measurements on Ni ($N = 7$–25) are in general agreement with those of Apsel et al. [82]. Knickelbein [84] has also performed measurements on the moments of small Fe clusters ($N = 10$–25) and found a huge enhancement to 4.6–5.4 μ_B for $N = 10, 11, 12$, followed by a drop to 2.4 μ_B at $N = 13$, an increase 3.4 μ_B at $N = 14$, and then oscillations between 2.6 μ_B and 3.0 μ_B for the rest of the size range studied.

There has been a large amount of theoretical work attempting to understand the magnetic behaviour and to relate it to the structure of the clusters. Complementary chemical probe techniques (discussed in Section 8.3.1) and photoelectron spectroscopy (see Chapter 7) have also been employed to gain experimental insight into the structure and the electronic behaviour. We consider first the results of *ab initio* calculations on Ni [85–90], Fe [86,94–97] and Co [86,91–93] using DFT (either LSDA or GGA) and geometry optimisation. The results for structures, total spin S of a cluster and magnetic moment per atom ($\mu = 2S/N$) are summarised in Table 8.3 for $N \leq 13$ clusters. The generic structure is given; generally there is distortion from that, so that the optimised structure has lower symmetry and this is assumed in the subsequent discussion. Most authors find a large number of isomers with small differences in binding energy. We have listed just the lowest energy isomers.

It can be seen that the onset of nonplanar geometry occurs earlier than in the noble metals (Section 6.3.3). At $N = 4$, the majority of calculations predict a tetrahedral structure rather than a planar one; for Ni_4 Reuse and Khanna [85] find the square and tetrahedron energetically degenerate. Often the tetrahedron distorts considerably and can equally be regarded as a rhombus folded along a diagonal (butterfly structure). The trigonal bipyramid and the square pyramid compete at $N = 5$ and isomers are found for each; the trigonal bipyramid is clearly favoured as the ground state for Ni_5 and Fe_5 but the situation is less clear for Co_5. At $N = 7$ the pentagonal bipyramid is preferred for Fe_7 but for Ni_7 and Co_7 this structure and an octahedron with one face capped are in contention as the ground state. At $N = 8$ the bicapped octahedron (there are several isomers of this structure, one of which is known as bisdisphenoid) is the calculated ground state for Fe_8, but for Ni_8 and Co_8 this structure and a cube are in contention. At $N = 13$ the icosahedral structure is generally preferred for all the elements to one based on fcc and, at $N = 12$, the predicted ground state is an incomplete icosahedron. Grigoryan and Springborg [98] report extensive calculations of the structure of Ni clusters based on the embedded-atom method for the size range $2 \leq N \leq 150$.

Table 8.3. Calculated structure, spin (per cluster) and spin magnetic moment (μ_B per atom) for Ni_N, Co_N and Fe_N clusters for $2 \leq N \leq 13$.

N	Structure	Fe		Co		Ni	
		S	μ	S	μ	S	μ
2	Dimer	3 [86,94,95]	3.00	2 [86,91–93]	2.00	1 [85–88]	1.00
3	Triangle	5 [96]	3.33	5/2 [86,91]	1.67	1 [85–88]	0.67
		4 [86,94,95]	2.67	7/2 [92]	2.33		
	Linear			7/2 [93]	2.33		
4	Tetrahedron	6 [86,94,95]	3.00	5 [86,91,92]	2.50	3 [85]	1.50
		7 [96]	3.50			2 [86–88]	1.00
	Planar			4 [93]	2.00	3 [85]	1.50
5	Trigonal bipyramid	7 [94]	2.80	9/2 [86]	1.80	4 [85]	1.60
		8 [86,95]	3.20			2 [86–88]	0.80
	Square pyramid			11/2 [93]	2.20		
				13/2 [92]	2.60		
6	Octahedron	10 [94–96]	3.33	7 [92,93]	2.33	3 [85]	1.00
						4 [88]	1.33
7	Pentagonal bipyramid	11 [94–96]	3.14	15/2 [92]	2.14	3 [87]	0.86
	Capped Octahedron			15/2 [93]	2.14	4 [88,89]	1.14
8	Bicapped octahedron	12 [95,96]	3.00	8 [93]	2.00	4 [88,89]	1.00
	Cube			8 [92]	2.00	4 [85]	1.00
9		13 [95,96]	2.89	17/2 [93]	1.89	4 [88]	0.89
10		31/2 [96]	3.10	10 [93]	2.00	4 [88]	0.80
		14 [95]	2.80			2 [88]	0.40
11		17 [96]	3.09	21/2 [93]	1.91	4 [88]	0.73
		15 [95]	2.73				
12	Icosahedron	18 [96]	3.00	12 [93]	2.00	4 [88]	0.67
		16 [95]	2.67				
13	Icosahedron	17 [95]	2.62	23/2 [93]	1.77	4 [85,87,88,90]	0.62
		22 [96,97]	3.38				

Note: All calculations use DFT (either LSDA or GGA) and geometry optimisation. In two cases, Ni_4 [85] and Ni_{10} [88], two isomers have the same energy and both are listed. The $S = 2$ isomer of Ni_{10} [88] is ferrimagnetic; all others listed are ferromagnetic. Some of the structures are illustrated in Figure 6.1 (Section 6.3.1).

Experimental structure determinations for Ni [58,61] and Co [58,62] using the chemical probe method are in rather good agreement with the calculations (see Section 8.3.1). The triangle and tetrahedron are the definitive structures for N equal to 3 and 4 respectively. On balance the preferred structure at $N = 5$ is the trigonal bipyramid [58], but the results are not unambiguous and there are arguments [62] that would suggest that Co_5 forms a square pyramid. The octahedron is confirmed at $N = 6$, and at $N = 7$ the experiments favour the capped octahedron over the pentagonal bipyramid. The structure at $N = 8$ is the bicapped octahedron, with the bisdisphenoid isomer [58] being deduced for Ni_8 in agreement with calculations [88,89]. The icosahedral structure is indicated [58,61] in the chemical probe experiments on Ni_{12} and Ni_{13} in agreement with the calculations [85,87,88,90]. There is a disagreement in the case of Co_{12} and Co_{13} between theory and experiment with theory [93] favouring icosahedral while experiment [58,62] indicates an fcc or hcp structure.

We now turn to the spin and magnetic moments. In the calculations the total spin of the Ni clusters remains remarkably constant at about 4, for $N>4$, yielding a moment per atom that falls with cluster size as seen in the experiments. By contrast the Fe and Co clusters display a moment per atom that fluctuates around 3 and 2 μ_B respectively from the dimer to $N = 13$. Apart from a few sizes for Ni and for Fe_{13}, the calculated values fall significantly below the values determined experimentally. This is particularly true in iron, where as noted above, the experimental moments [84] for Fe_{10}, Fe_{11} and Fe_{12} are around 5 μ_B/atom.

The possibility of non-collinear spin configurations has been explored for a number of materials [7,99–101]. Oda et al. [99] considered Fe_N clusters for $N \leq 5$ and found that Fe_5 had a non-collinear ground state. The structure was a trigonal pyramid with the spins on the three atoms in the basal plane parallel; those at the apices are canted away from axis of the other spins in opposite directions by about 30°. The moment of the cluster, at about 2.9 μ_B/atom, is of similar magnitude to the parallel configurations. Fujima [7] predicted a pentagonal bipyramidal for Fe_7 clusters, and calculated a spin moment of 2.86 μ_B/atom using the bulk interatomic spacing; however a slight contraction would result in a coplanar (i.e. not collinear) arrangement of the moments.

There have been a number of attempts to extend the size range of calculations by various approximate methods. Andriotis and collaborators [102,103] have used tight-binding molecular dynamics (TBMD) to study the magnetic moments of geometry optimised clusters of Ni, Co and Fe for a selection of cluster sizes up to $N = 147$. Molecular dynamics simulations using the semi-empirical many-body Gupta potential have been employed to give an initial geometry optimisation followed by a tight-binding calculation to study the magnetisation for Ni [104,105] and Co [106].

The results for calculations on Ni_N ($5 \leq N \leq 60$) clusters by Aguilera-Granja et al. [104] are shown in Figure 8.13. They find that almost all the structures are based on an icosahedral growth pattern. The two exceptions are Ni_{38} for which an fcc

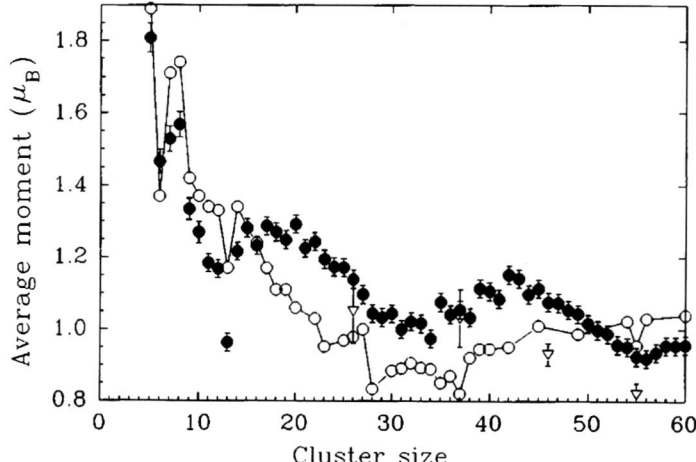

Figure 8.13. Calculated spin moments of Ni_N clusters (open circles) [104] compared with experimental results [81,82]. Reproduced with the permission of the American Physical Society from Aguilera-Granja et al. [104].

structure is preferred, and Ni_{36} where fcc and icosahedral binding energies are virtually the same. Interestingly the calculated moments [104,105] for the smaller clusters ($N \leq 13$) are significantly closer to experiment than those obtained from *ab initio* calculations (Table 8.3). Andriotis and Menon [102] and Lathiotakis *et al.* [103] compared the energies of fcc and icosahedral structures for $N = 19, 23, 24, 38, 43, 55$ and 147 and found that icosahedral structures were lower in energy for $N = 33, 55, 147$, while the fcc structure was more stable for the other sizes. Their results for the larger clusters agree rather well with experiments.

The chemical probe experiments indicate that the majority of Ni clusters with sizes from 13 atoms to several hundred atoms have some pentagonal symmetry [107], and thus have icosahedral characteristics. The pattern is more complex between Ni_{29} and Ni_{49} however. The atoms in Ni_{38} are found to pack as a truncated fcc octahedron [108], which is consistent with the calculations [102,104]. Adding one atom to form Ni_{39} results in a structure based on global pentagonal symmetry [109]. Ni_{46}, Ni_{47} and Ni_{48} are all deduced to have structures based on fcc packing [64], while the addition of one atom causes a change in structure to icosahedral at Ni_{49}.

Rodríguez-López *et al.* [106] use similar techniques to Aguilera-Granja *et al.* [104] to study Co_N clusters over the size range $4 \leq N \leq 60$. They obtain a moment that is strongly enhanced above the bulk value to $3.0\,\mu_B$/atom for Co_4 falling to $2.2\,\mu_B$/atom for Co_{20} and fluctuating around $2\,\mu_B$/atom for the rest of the size range examined. They also compare their results to other calculations [102,110,111]. The moments obtained by Rodríguez-López *et al.* [106] for $N \leq 13$ are significantly higher than those from *ab initio* calculations (Table 8.3) and from other calculations [102,110,111]. Unfortunately the size range covered by the Stern–Gerlach experiments ($25 \leq N \leq 700$) [71,81] does not allow a very critical comparison with theory. Andriotis and Menon [102] obtain an hcp structure for Co_{13}, icosahedral symmetry for Co_{19} and either an fcc or hcp structure for larger clusters. Rodríguez-López *et al.* [106] obtain icosahedra at Co_{13} and Co_{19}. Generally they find a tendency to polyicosahedral structures for the larger clusters with occasionally fcc or hcp distorted fragments. The other calculations [110,111] used fixed fcc or hcp geometry.

Calculations on Fe_N clusters over a large size range have been reported by a number of workers [102,110,112,113]. Guevara *et al.* [110] and Franco *et al.* [112] use fixed bcc geometry, while Andriotis and Menon [102] find bcc, fcc or icosahedral at different sizes. None of these calculations are able to explain the oscillations in the moment observed experimentally nor the very large moments found by Knickelbein [84] for $N < 13$.

Various attempts have been made to explain the oscillations in the magnetic moment as a function of size that are apparent in Figure 8.12(i) in terms of a shell model [65,81,114,115]. These have generally been phenomenological, and direct calculations of magnetisation over a range of sizes up to several hundred atoms generally show a rather smooth variation of the moment. However Tiago *et al.* [113], in DFT calculations on Fe_N, have shown that the moment can be lowered with surface faceting resulting in a non-monotonic decrease in moment with cluster size, and propose that the oscillatory structure could be explained by the variation of surface details with the size of cluster. It has been suggested [116,117] that an alternative explanation may lie in oscillations in the magnetocrystalline anisotropy.

There will certainly be oscillations in the anisotropy: high symmetry (magic number) clusters will have very low anisotropy and it will be significantly higher in lower symmetry clusters (between magic numbers). The Langevin function Eq. (8.3) is assumed to apply accurately enough to the moments of clusters to allow its use in the extraction of the moment from the experimental data. It has been shown [116,117] that reasonable assumptions about the magnitude of the cluster anisotropy could cause deviations from Langevin behaviour sufficient to account for the oscillations.

As note in Section 7.3.1, it is very difficult to make firm assignments of the structure of Fe clusters from experimental measurements and so most information comes from theoretical studies. Recent calculations [118] have extended the size range studied by DFT based methods up to 641 atoms. The energies and spin moments of icosahedral, cuboctahedral and bcc type structures were evaluated in calculations that included full geometry optimisation. Rollmann *et al.* [118] also studied magic number clusters with structures that comprised a close-packed core and an icosahedral surface, with intermediate shells that are partially transformed via the Mackay rotation between icosahedral and octahedral geometry (Section 2.3.3). The predicted moments as a function of cluster size for the various structures are shown in Figure 8.14, along with some earlier fixed geometry results from Tiago *et al.* [113] and the experimental data. For N greater than about 140, the bcc structure is preferred energetically, while for $N<100$ the Jahn–Teller distorted icosahedron or the Mackay transformed structure are favoured at certain sizes.

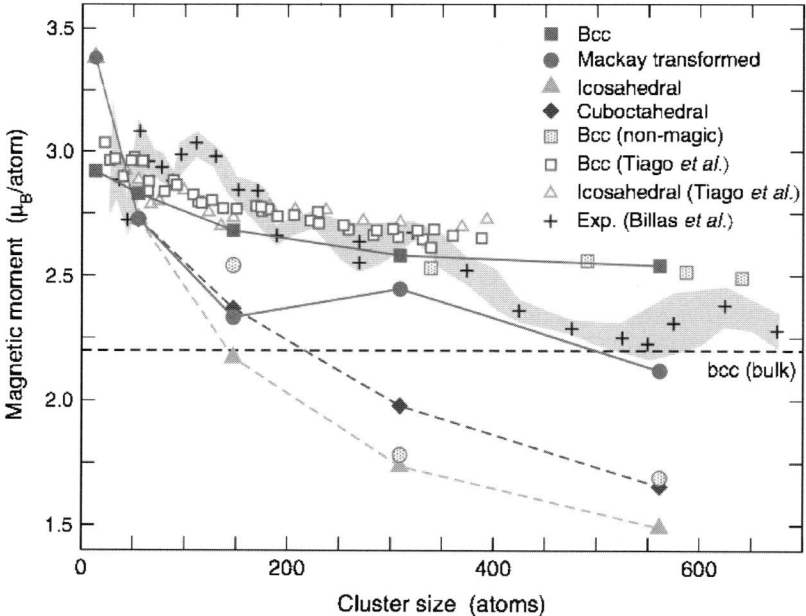

Figure 8.14. Calculated spin magnetic moments of Fe clusters for different structures and cluster sizes using relaxed geometry. The experimental results of Billas *et al.* [65,81] and the results from the calculations of Tiago *et al.* [113] using fixed geometry are also included in the figure. Reproduced with the permission of the American Physical Society from Rollmann *et al.* [118].

8.4.2. Orbital moment and magnetic anisotropy in ferromagnetic clusters

The calculations discussed so far have addressed just the spin moment and ignored the orbital moment completely. In bulk, of course, the orbital moment is virtually quenched, being typically of the order of $0.1\,\mu_B$/atom. However, it is well established by XMCD experiments [119,120] that can probe the orbital moment directly (see Section 9.2) that clusters deposited on surfaces exhibit a significantly enhanced orbital moment. Orbital moment enhancement has been explored theoretically, for example, in the context of $3d$ and $5d$ adatoms deposited on a Ag(0 0 1) surface [121] and transition metal wires [122,123].

Guirado-López et al. [124] have included the spin–orbit interaction within a tight-binding formalism and developed earlier work [125] to study the orbital magnetism in Ni clusters. They examined clusters based on fcc and icosahedral structures as well as bilayer clusters and showed that huge enhancements can be obtained. For Ni_3 and Ni_4 they obtain moments of 0.47 and $0.35\,\mu_B$/atom respectively which compares with a bulk value of $0.05\,\mu_B$/atom. Work on Ni clusters has been developed in further detail by Wan et al. [126], using cluster structures whose geometry has already been optimised using an embedded-atom method [127]. It can be seen from their plot in Figure 8.15 that the orbital contribution is a significant proportion of the total moment. The calculations [126] give agreement with experiments [83,82] that is as good as the agreement between the experiments themselves. It is possible that the large decrease in the magnetic moment [84] in going from Fe_{12} to Fe_{13} can be explained by a large change in orbital moment in going from a lower to a higher symmetry cluster.

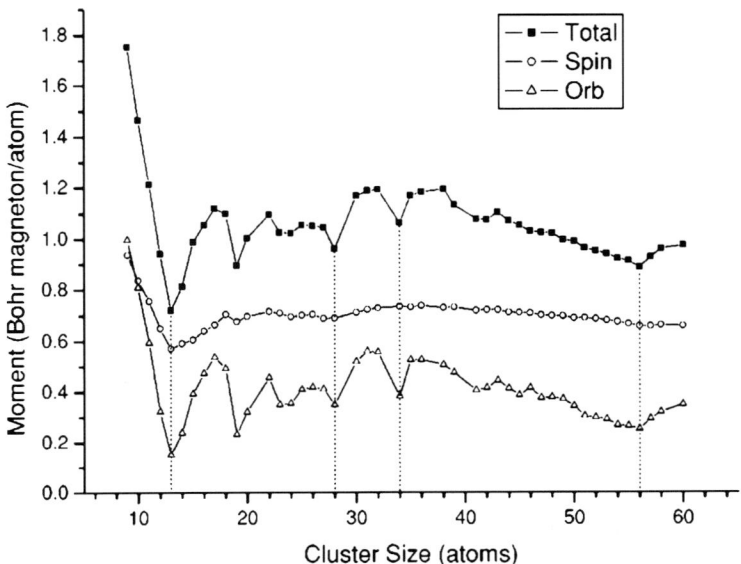

Figure 8.15. Calculated spin, orbital and total moments for Ni_N clusters for $9 \leq N \leq 60$. Reproduced with the permission of the American Physical Society from Wan et al. [126].

The orbital moment is closely related to the magnetic anisotropy through the spin–orbit coupling. It is well known that, like the orbital moment, the anisotropy in small clusters is enhanced above its value in the bulk material. Experimental data is often analysed by assuming that the measured anisotropy, K'_{eff}, can be expressed as the sum of volume and surface contributions

$$K'_{\text{eff}} V = K'_v V + K'_s S, \tag{8.40}$$

where V and S are respectively the volume and surface area of the cluster. Approximating to a spherical cluster,

$$K'_{\text{eff}} = \frac{K'_v + 6K'_s}{d}, \tag{8.41}$$

where d is the cluster radius. Bødker et al. [128] deduced values of K'_{eff} for Fe for a range of cluster sizes from Mössbauer experiments and found a reasonable fit to Eq. (8.41), with an extrapolation in the large d limit to a value of K'_v that was within a factor of 2 of the bulk anisotropy constant. However it is worth noting that the surface contribution dominates the anisotropy: the contribution from the second term in Eq. (8.40) was between 3 and 10 times larger than that from the first over the size range studied [128]. Since the measurements represent the average behaviour of an assembly of clusters and the division between a volume and surface contribution to the anisotropy is a phenomenological one, there is a strong motivation to obtain direct information about cluster anisotropy by study on an individual particle.

Single clusters of Co and Fe of about 1000 atoms have been studied recently by Jamet et al. [129,130] using a micro-superconducting quantum interference device (SQUID). The clusters are embedded in niobium, which forms the superconducting film of the micro-SQUID. The clusters are modelled according to the Wulff construction as a truncated octahedron for fcc Co and as a dodecahedron for bcc Fe (see Section 2.3.3). Surface atoms are added to the (0 0 1) and (1 1 1) facets of the truncated octahedron and to (1 1 0) and equivalent facets of the dodecahedron. This allows for a rather general expression for the anisotropy in terms of the direction cosines of the magnetisation vector that is neither constrained to cubic or uniaxial symmetry

$$E_{\text{anis}} = K_1 \alpha_z^2 + K_2 \alpha_y^2 + K_4(\alpha_{x'}^2 \alpha_{y'}^2 + \alpha_{y'}^2 \alpha_{z'}^2 + \alpha_{z'}^2 \alpha_{x'}^2) + \cdots, \tag{8.42}$$

where x', y', z' are the [1 0 0], [0 1 0] and [0 0 1] directions, and (x, y, z) is a coordinate system related to (x', y', z') by a rotation. Notice that K_4 here is equivalent to K_1 in Eq. (8.18) and Table 8.2 (for the cubic materials).

The following anisotropy constants were found for the Co clusters: $K_1 = -0.22$ MJ/m^3, $K_2 = 0.09$ MJ/m^3, $K_4 = -0.01$ MJ/m^3 for Co, with (x', y', z') related to (x, y, z) by a 45° rotation about the z $(=z')$ axis. For Fe the constants were $K_1 = -0.32$ MJ/m^3, $K_2 = 0.00$ MJ/m^3, $K_4 = 0.05$ MJ/m^3, with z as the easy axis lying in the (x', z') plane at 34.5° to z'. Interestingly the fourth order anisotropy constant for the fcc Co cluster is significantly smaller than the bulk value, whereas for Fe it is almost the same as in bulk (see Table 8.2, noting that K_4 in Eq. (8.42) is equivalent to K_1 in the table). The interpretation is complicated by the fact that Co and Nb are miscible elements and there is penetration [129] to almost two monolayers into the Co cluster resulting in two monolayers that are magnetically dead. Cobalt–silver clusters were also studied [130] in which the Ag atoms are expected to segregate at the cluster surface acting as a barrier against diffusion of Nb atoms into the Co.

Calculations of the magnetic anisotropy in small transition metal clusters ($N \leq 7$) have been performed by Pastor et al. [131], and for clusters ($19 \leq N \leq 79$) by Guirado-López [132]. In both calculations the anisotropy is found to be a complicated function of cluster size, bond length and d-band filling. Xie and Blackman [133] have performed calculations of the magnetic anisotropy in Co clusters of several hundred atoms in size with both cubic symmetry and reduced symmetry by adding surface atoms to facets. For the high symmetry clusters described by fourth order anisotropy they find that it is instructive to express the anisotropy as the sum of contributions from the interior and the surface of the cluster. The contributions have an opposite sign and there is a delicate balance between the two in determining the net anisotropy. For the majority of clusters, the energy difference between [1 1 1] and [0 0 1] magnetic orientations is equivalent to a value of K_4 between -0.4 and $-0.8\,\mu\text{eV/atom}$, although for some clusters it can be much smaller. This provides some insight into the small value of K_4 compared with bulk that was obtained by Jamet et al. [129,130]; their value of $-0.01\,\text{MJ/m}^3$ is equivalent to $-0.7\,\mu\text{eV/atom}$. Surface and interior contributions to the anisotropy that are of opposite signs has also been noted in calculations on thin film systems [134].

8.4.3. Antiferromagnetic 3d transition metals

8.4.3.1. Chromium

The small number of atoms in nanoclusters of metals that are antiferromagnetic in the bulk leads to the possibility of a significant imbalance of the spin sublattices and frustration. Thus ferrimagnetic or even ferromagnetic ground states may be expected. An initial experimental study of Cr clusters by Douglass et al. [135] using the gradient field deflection technique showed that Cr_N did not have a detectable net magnetic moment for N in the range 9–31 atoms. The sensitivity of the apparatus led them to conclude that the upper bounds for any net moment varied from $0.77\,\mu_B$/atom for Cr_9 to $0.42\,\mu_B$/atom for Cr_{31}. This result was in contrast to many early calculations that predicted large and easily measurable net moments for Cr_N clusters for N up to 51 [136–138], though the expected moments were shown to be highly sensitive to the structure [138]. Lee and Callaway [139] calculated the magnetic moment in Cr_9 and Cr_{15} clusters in a bcc structure as a function of lattice spacing and observed the disappearance of a net moment below a critical value. They found that, at the bulk Cr spacing, the magnetic moment for both Cr_9 and Cr_{15} clusters was below the detectable threshold of the experiment. A structure-optimised calculation by Cheng and Wang [140] showed a unique growth by dimers up to Cr_{11} after which the bcc structure is stabilised. They found that the clusters were always antiferromagnetic with a size-dependent atomic moment.

Reddy et al. [141] calculated the magnetic properties of Cr_8 and Cr_{13} clusters. In contrast to Cheng and Wang [140], they found no indication of dimerisation in Cr_8. They found that there are two degenerate antiferromagnetic isomers of Cr_8, while Cr_{13} has a ferromagnetic ground state with a net moment for the cluster of 14 μ_B. For both Cr_8 and Cr_{13} there are other isomers whose energy is just above that of the ground states. The finding of multiple moments at a given cluster size, however, is not necessarily due to isomers with a different atomic structure. An earlier calculation by Lee and Callaway [142] revealed that Cr_9 clusters with a bcc structure could exist in several magnetic states with different moments simultaneously. They found that for some atomic spacings four or five states could coexist.

Subsequent calculations have investigated the stability of non-collinear moments resulting from frustration [7,143,144]. Kohl and Bertsch [144] found that Cr_{13} clusters strongly favour non-collinear configurations, and Fujima [7] found that with decreasing interatomic spacing in Cr_7 clusters, with a pentagonal bipyramid structure, the spin configuration changes from collinear (anti-parallel) to coplanar.

More recently the field gradient deflection experiments on Cr clusters that earlier showed no detectable magnetic moment [135] were repeated by Payne et al. [145] using a more sensitive apparatus. They studied clusters in the size range $N = 20$–133 atoms at temperatures between 60 and 100 K. They find that some of their deflection profiles show a double peak structure that can be fitted to two Gaussians indicating that species with two distinct magnetic moments are present. Their results are reproduced in Figure 8.16.

Payne et al. [145] observe the double peak structure for all sizes from Cr_{34} upward. Below that, Cr_{30} is the only cluster for which it was possible to resolve two moments. There is a dramatic drop in the magnetisation at $N = 33$, on the borderline between the two regimes. Supporting evidence for this picture is rather ambiguous. There are indications from calculations [141,142] of several near degenerate isomers with different moments, but nearly all the theoretical work has been done on clusters with $N \leq 15$. Knickelbein [146] observed two different IPs for

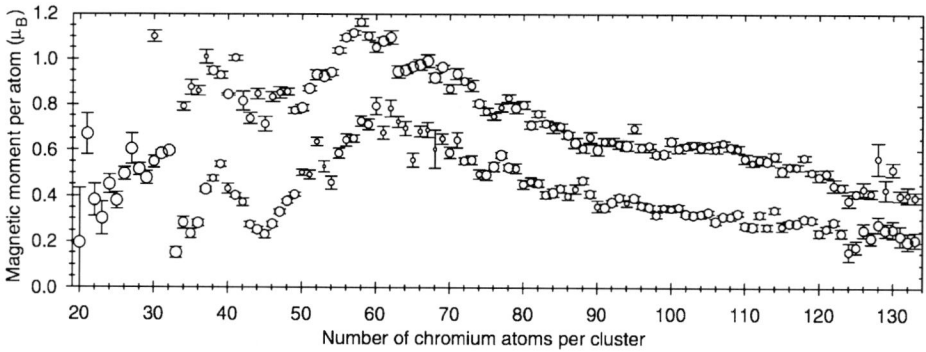

Figure 8.16. Magnetic moments per atom of Cr_N clusters for $20 \leq N \leq 133$. The clusters Cr_{30} and Cr_{34}–Cr_{133} each have two magnetically distinguishable isomers. The size of the circle indicates the intensity of the deflected beam. Reproduced with the permission of the American Physical Society from Payne et al. [145].

Cr$_8$ and Cr$_{18}$, but did not report on any clusters larger than Cr$_{25}$. An earlier photoelectron spectroscopy study of Cr clusters by Wang et al. [147] showed no evidence for distinguishable isomers for $N \leq 55$.

Payne et al. [145] offer a plausible explanation for their observed behaviour. Bulk Cr is a bcc structure. The magnetism is in the form of a spin-density wave (SDW). At a temperature of 78 K the wavelength of the SDW is about 21 bcc unit cells. The low-moment component of the cluster magnetism is near or below the peak moment per atom for the SDW in bulk Cr (0.62 μ_B). If the clusters were essentially fragments of bcc bulk Cr and the bulk behaviour persisted, the smaller clusters would be of a size to contain a single crest of the SDW. We could then anticipate a moment of $\sim 0.62\,\mu_B$. The gradual decrease with increasing cluster size could then be explained by the inclusion of more of the SDW within the cluster. The largest moment occurs at Cr$_{58}$ and, at 1.16 μ_B, is nearly twice the peak value of the SDW. Payne et al. [145] note that enhanced magnetism has been predicted in hcp or fcc Cr [148] and this may indicate the presence of isomers with a close-packed structure in the regime where the high moment occurs.

8.4.3.2. Manganese

Molecular beam deflection measurements on free Mn$_N$ clusters have been reported by Knickelbein [75] for $11 \leq N \leq 99$. His results are reproduced in Figure 8.17. Subsequently the behaviour of small clusters ($5 \leq N \leq 22$) was explored in greater detail [149]. There was some refinement in the values in the region common to the two experiments but the general behaviour is unchanged. There are clear maxima at Co$_{12}$ and Co$_{15}$ and deep minima at Co$_{13}$ and Co$_{19}$, followed by broader maxima for N around 24 and 55 and a minimum around 34. The minima at $N = 13$ and $N = 19$

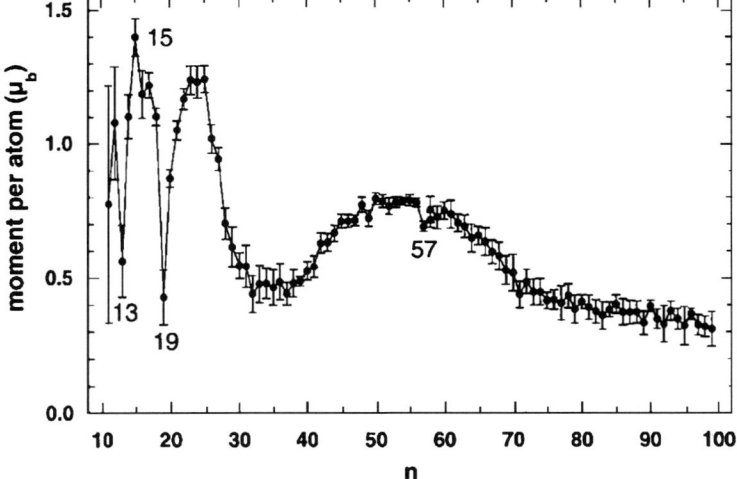

Figure 8.17. Magnetic moments (in μ_B/atom) of Mn$_N$ clusters plotted against size for $11 \leq N \leq 99$. The clusters were produced at 68 K. Reproduced with the permission of the American Physical Society from Knickelbein [75].

are suggestive of an icosahedral and double icosahedral structure at these cluster sizes.

There have been a number of theoretical studies of small clusters with $N \leq 8$ [7,150–155], including some that consider non-collinear spin configurations [7,87,154,155,160]. A number of authors [150,153] predict ferromagnetic ordering for small clusters with a moment of about $5\,\mu_B$/atom. The experiments, by contrast, find that the moments seldom exceed $1\,\mu_B$/atom in this small size range, indicating rather that ferrimagnetism occurs at least down to $N = 5$. Calculated moments of $0.87\,\mu_B$/atom [154] and $0.71\,\mu_B$/atom [151] respectively for Mn_6 and Mn_7 are consistent with this. A suggestion has also been made for Mn_5 and Mn_6 [151,155] that there is weak coupling between the localised atomic moments and that various isomers can be obtained by sequentially turning the local moments, and this leads to deviations from superparamagnetic behaviour. Fujima [7] examined clusters with a pentagonal bipyramidal structure and observed that, in the case of Mn_7, the spin configuration changes from collinear (anti-parallel) to disordered to coplanar. as the interatomic spacing relative to the bulk value varies between 0.8 and 1.

Bobadova-Parvanova et al. [156,157] have studied clusters up to $N = 13$ and find ferromagnetic and antiferromagnetic isomers almost degenerate in energy for $N \leq 6$, above which a ferrimagnetic state is clearly established. Other authors [158–161] have performed calculations on larger clusters as well in an attempt to address the structure observed in Figure 8.16. Briere et al. [158] performed a geometry optimisation on clusters containing 13, 15, 19 and 23 atoms, and found that Mn_{13}, Mn_{19} and Mn_{23} are icosahedral while Mn_{15} prefers a bcc structure. Their calculations were done with collinear spins and they found that in each case the spins tended to order ferromagnetically in layers (or approximate layers) and that the layers are antiferromagnetically coupled. This is reminiscent of the ordering in bulk γ-Mn and δ-Mn. Kabir et al. [159], also using a collinear theory, performed optimisations using a number of starting structures over the range $2 \leq N \leq 20$. Overall there is fair agreement between theory [158,159] and experiment [75,149].

However, the collinear calculations do appear to be consistent with each other in producing a minimum in the moment at $N = 15$ and a maximum at $N = 19$ which is contrary to experiment (see Figure 8.17). There have been attempts [160,161] at non-collinear calculations in this size range and it does appear that this can resolve the anomalies at $N = 15, 19$. The results of the non-collinear calculations of Xie and Blackman [160] for Mn_{15} and Mn_{19} are shown in Figure 8.18. They obtain results identical to Briere et al. [158] in a collinear calculation, but find marked changes when non-collinearity is allowed. It can be seen from Figure 8.18 that, with a collinear configuration, there is an excess of one atomic moment in the balance between up and down spins in the case of Mn_{15}, while for Mn_{19} the excess is five. This is the source of the minimum at $N = 15$ and the maximum at $N = 19$ in the collinear calculation. Allowing the spins to relax away from a single axis changes the behaviour considerably. For $N = 15$ the moment of the cluster is $0.20\,\mu_B$/atom in the collinear calculation and dramatically increases to $1.24\,\mu_B$/atom with non-collinearity. For $N = 19$, introducing non-collinearity reduces the moment from 1.21 to $0.79\,\mu_B$/atom. Although the values still do not

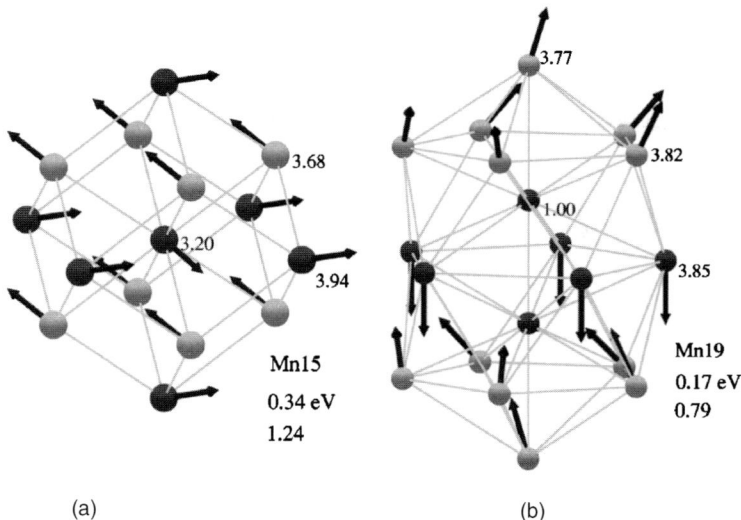

Figure 8.18. Calculated moments of Mn$_{15}$ and Mn$_{19}$ clusters. The energies shown are the difference between the ground state (non-collinear) and the lowest energy collinear state. The other numbers give the moment (in μ_B) on individual atoms and (below) the energies and the net moment of the cluster (in μ_B/atom). The dark and light shading represents spin-up and spin-down alignment in a collinear calculation. Reproduced with the permission of the American Physical Society from Xie and Blackman [160].

match the experimental moments precisely, the maxima and minima are now in the right places.

8.4.4. Non-magnetic 3d and 4d metals

The Stoner criterion [14] for itinerant magnetism is $N(\varepsilon_F)I > 1$, where $N(\varepsilon_F)$ is the density of states at the Fermi level and I is the exchange integral (see Eq. (8.8)). Both $N(\varepsilon_F)$ and I are modified at interfaces and in nanostructures leading to the possibility of stabilising magnetism in small clusters of elements that are paramagnetic in the bulk. Of the 3d transition elements, vanadium is a possible candidate for the stabilisation of magnetism in reduced dimensions and there have been several theoretical predictions for permanent magnetic moments in V clusters [136,162,163]. Gradient-field deflection measurements of V clusters did not reveal a magnetic moment [135] although there was some evidence presented in earlier work for the appearance of magnetism at the surface of V islands grown on graphite [164].

There is a general decrease in I with increasing atomic number [17] (Figure 8.1) so that magnetism is less favoured in 4d and 5d elements and is not found in any bulk elements from these series. Although many 4d and 5d metals have been predicted to show magnetism as monolayers [165] there are only two experimentally proven

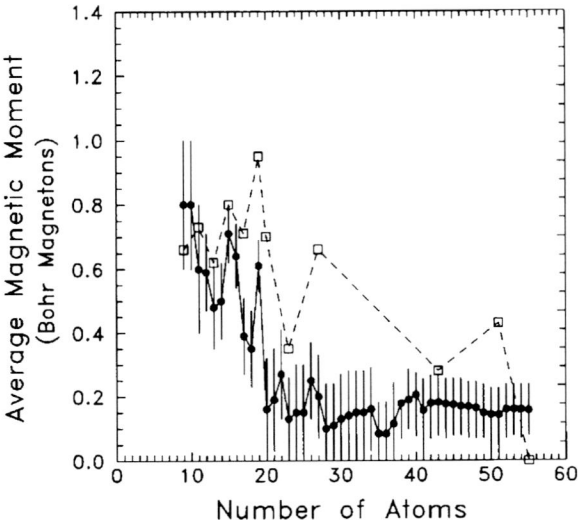

Figure 8.19. Total moments (in μ_B/atom) in free Rh_N clusters for $N<100$ measured by Cox et al. [69] (dots and error bars). The measured values are compared with calculations by Villaseñor-Gonzalez et al. [167] (open squares). The most stable relaxed structure for which the calculation was performed is shown at each cluster size. Reproduced with the permission of the American Physical Society from Villaseñor-Gonzalez et al. [167].

examples of 4d magnetism, that is, Ru monolayers on graphite [166] and free Rh nanoclusters [69].

The magnetic moment per atom as a function of Rh cluster size is shown in Figure 8.19 and compared with the unrestricted Hartree–Fock calculations by Villaseñor-Gonzalez et al. [167]. The structure was optimised at each size by determining the optimum bond length for icosahedral, fcc and bcc atomic arrangement and then observing which of these had the highest cohesive energy. The calculated moments for the most stable structures are shown in Figure 8.19 together with the experimental data. The structure that is most stable varies across the size range. The calculations reproduce the minima at $N=13$, $N=17$ and the maximum at $N=19$ that are observed experimentally. Interestingly the most stable structure at $N=13$ is found to be bcc-like instead of icosahedral. A 13-atom icosahedral Rh cluster is found to show a magnetisation maximum in contrast to the data. The double icosahedron at $N=19$ shows a maximum in agreement with the measurements. The variation in the magnetic moment per atom derives not only from changes in magnitude of the individual atomic moments but also from spin alignments that are sometimes antiferromagnetic. The calculated moments are systematically larger than experimental findings for the larger clusters, but obviously full geometry explorations are not possible at those sizes.

The neglect of the orbital contribution in this type of calculation may however change the agreement with the experiment, in which the total (orbital+spin) moment is measured. Cox et al. [69] also attempted to detect magnetism in Pd and Ru clusters in the size range 10–115 atoms but found no measurable moment and set upper limits of $0.32\,\mu_B$/atom for Ru_{10} clusters and $0.4\,\mu_B$/atom for Pd_{13} clusters. An *spd* tight-binding calculation of Pd_N and Rh_N clusters up to $N = 19$ showed either small or zero moments for Pd clusters and larger values for Rh clusters [168].

8.4.5. *Rare-earths*

The character of magnetism in rare-earths is very different to that of transition metals because the $4f$ orbitals, that produce most of the magnetic moment, are screened from the environment by the $6s$ electrons. The localisation of the $4f$ electrons within screening orbitals generally results in both the orbital and spin moments maintaining their full atomic values in the bulk. The large orbital moment and spin-orbit coupling is responsible for the very high values of the magnetic anisotropy energy observed in rare-earths.

Magnetic deflection experiments on free Gd clusters [66,79,169] at temperatures around 100 K have revealed two types of behaviour depending on the number of atoms in the cluster. At some cluster sizes, the deflection profiles show that the cluster magnetisation scales with H/T and these are assumed to be superparamagnetic as in the case of transition metal clusters. At other sizes the cluster moment, measured from the deflection profile, does not scale with H/T and in this regime the behaviour is ascribed to the moments being blocked so that they are locked to the crystal lattice and undergo the same rotation as the cluster. The simplest interpretation of these observations is that the magnetic anisotropy, which is higher in the rare-earths than the transition metals, varies with the number of atoms in the cluster producing a consequent variation in the cluster blocking temperature T_B. For example Gerion et al. [78] found that T_B is between 45 and 75 K for Gd_{13} and Gd_{21} but around 180 K for Gd_{22}.

The different experiments disagree however on the cluster sizes that show the different behaviours. Douglass et al. [169] found that all cluster sizes in the range Gd_{11}–Gd_{26} displayed locked-moment behaviour with the exception of Gd_{22}, which was superparamagnetic. In contrast, Gerion et al. [78] showed that Gd_{13} and Gd_{21} were superparamagnetic while Gd_{22} was blocked. Tb also shows the locked-moment or superparamagnetic behaviours described above depending on the cluster size [170].

For the superparamagnetic clusters it is possible to extract the total moment per atom from the magnetisation versus H/T using Eq. (8.3) as with the transition metals, and in all cases moments/atom significantly lower than the bulk values were found. For example Gd_{13} and Gd_{21} clusters have moments of 5.4 and $5\,\mu_B$/atom respectively, compared with a bulk value of $7.55\,\mu_B$/atom. This has been explained by a theoretical model that takes into account an RKKY-type interaction between the atoms in which the exchange force between nearest neighbours is ferromagnetic but switches to antiferromagnetic between second nearest-neighbours [59]. The behaviour varies according to the ratio of the ferromagnetic to antiferromagnetic coupling strengths The clusters are found to adopt a hexagonal structure and, within a range of values of this ratio, canting of individual atomic spins away from perfect ferromagnetic alignment is predicted. Thus the moment per atom measured along the applied field will be reduced. The reduction in moment relative to the bulk is also observed in Tb clusters [170], presumably for the same reason.

8.4.6. Magnetic ordering temperature

The thermal behaviour of the moment has so far been discussed in terms of superparamagnetism where the cluster super-moment is excited over the cluster anisotropy barrier. At sufficiently high temperatures the internal spin alignment will be disrupted leading eventually to normal paramagnetism. Gerion et al. [77] measured the specific heat of free $Fe_{250-290}$, $Co_{200-240}$ and $Ni_{200-240}$ clusters in the temperature range 80–900 K. The method involves using an experimental set-up similar to that used in measuring the magnetic moments, and the perturbation of the beam profile in response to a heating laser is used to determine the specific heat. The clearest results are for $Ni_{200-240}$, which shows a broad peak centred at 340 K on top of a nearly constant baseline of 6 cal/(mol K) – the classical Dulong and Petit value for bulk Ni. The peak is interpreted as due to the ferromagnetic to paramagnetic phase transition in the clusters, and its width is well described by the mean-field approximation. The temperature of the transition is significantly reduced relative to the bulk value of 627 K. The specific heat for the $Co_{200-240}$ clusters shows a monotonic rise from 5.5 cal/(mol K) at 300 K to 15 cal/(mol K) at 900 K implying that the phase transition is not reachable in the temperature range of the experiment. The behaviour of $Fe_{250-290}$ clusters is more enigmatic showing specific heat values significantly lower that the Dulong and Petit values. A peak at around $T=600$ K does occur but is poorly described by mean-field theory. One possibility discussed in the paper is that Fe clusters undergo a magnetic transition between a high-moment and low-moment state with different lattice parameters.

Billas et al. [65] observed the magnetic ordering temperatures in Fe_N and Ni_N clusters directly by measuring the magnetic moment as a function of temperature up to over 900 K. In all cases they found the magnetic moment was constant above some critical temperature $T_C(N)$, which was interpreted as the transition temperature for paramagnetism. The magnetisation in this state, μ_{para}, should attain a value $\mu_{T=0}/\sqrt{N}$ corresponding to N randomly aligned atomic moments so that in a cluster where $N \sim 100$, μ_{para} is significantly above zero. In fact it was found that μ_{para} was higher than $\mu_{T=0}/\sqrt{N}$. This was later explained by Pastor and Dorantes-Dávila [171] as due to short-range magnetic order (SRMO) above the transition temperature, which is also observed in bulk magnetic materials. The effect of this is to increase the paramagnetic moment to $\mu_{T=0}\sqrt{vN}$ where v is the average number of atoms in a SRMO domain.

The high temperature behaviour of rare-earth clusters contrasts with that of the transition metals. The measured ordering temperatures (420 K for Gd_{21} and over 500 K for Gd_{13}) are significantly higher than the bulk value of 293 K. In addition the moment above the ordering temperature tends asymptotically to $\mu_{T=0}/\sqrt{N}$ [78], indicating that there is no SRMO above the ordering temperature and the disorder is on the atomic scale. The decay in magnetic moment per atom with temperature found for Gd_{13} clusters is consistent with the canted spin model [78,172].

References

[1] R. Skomski, J. Phys. Condens. Matter **15** (2003) R841.
[2] E.C. Stoner and E.P. Wohlfarth, Philos. Trans. R. Soc. Lond. A **240** (1948) 599.
[3] L. Néel, Ann. Geophys. **5** (1949) 99.
[4] M. Respaud, J. Appl. Phys. **86** (1999) 556.
[5] A.V. Postnikov, P. Entel and J.M. Soler, Eur. J. Phys. D **25** (2003) 261.
[6] L. Theil Kuhn, A.K. Geim, J.G.S. Lok, P. Hedegård, K. Ylänen, J.B. Jensen, E. Johnson and P.E. Lindelof, Eur. Phys. J. D **10** (2000) 259.
[7] N. Fujima, Eur. Phys. J. D **16** (2001) 185.
[8] E.M. Chudnovsky and J. Tejada, Macroscopic Quantum Tunneling of the Magnetic Moment, Cambridge University Press, Cambridge, 1998.
[9] http://physics.nist.gov/PhysRefData
[10] M.A. Ruderman and C. Kittel, Phys. Rev. **96** (1954) 99.
[11] T. Kasuya, Progr. Theor. Phys. (Kyoto) **16** (1956) 45.
[12] K. Yosida, Phys. Rev. **106** (1957) 893.
[13] R.J. Elliott, Theory of magnetism in the rare earth metals, in: G.T. Rado and H. Suhl (Eds.), Magnetism, Vol. IIA, Academic Press, New York, 1965, pp. 385–424.
[14] E.C. Stoner, Proc. R. Soc. Lond. A **169** (1939) 339.
[15] S.H. Vosko and J.P. Perdew, Can. J. Phys. **53** (1975) 1385.
[16] O. Gunnarsson, J. Phys. F Metal Phys. **6** (1976) 587.
[17] J.F. Janak, Phys. Rev. B **16** (1977) 255.
[18] P.M. Marcus and V.L. Moruzzi, Phys. Rev. B **38** (1988) 6949.
[19] V.L. Moruzzi and P.M. Marcus, Energy band theory of metallic magnetism in the elements, in: H.J. Buschow (Ed.), Handbook of Magnetic Materials, Vol. 7, North Holland, Amsterdam, 1993, pp. 97–137.
[20] E. Fawcett, Rev. Mod. Phys. **60** (1988) 209.
[21] E. Fawcett, Rev. Mod. Phys. **66** (1994) 25.
[22] D. Hobbs, J. Hafner and D. Spišák, Phys. Rev. B **68** (2003) 014407.
[23] J. Hafner and D. Hobbs, Phys. Rev. B **68** (2003) 014408.
[24] M.B. Stearns, Magnetic properties of metals in: H.P.J. Wijn (Ed.), Landolt-Börnstein New Series, III/19aSpringer-Verlag, Berlin, 1986, pp. 24–141.
[25] J. Crangle, Philos. Mag. **46** (1955) 376.
[26] J. Trygg, B. Johansson, O. Eriksson and J.M. Wills, Phys. Rev. Lett. **75** (1995) 2871.
[27] V.A. Gubanov, A.I. Liechtenstein and A.V. Postnikov, Magnetism and the Electronic Structure of Crystals, Springer Series in Solid-State Sciences, Vol. 98, Springer-Verlag, Berlin, 1992.
[28] S.V. Beiden, W.M. Temmerman, Z. Szotek, G.A. Gehring, G.M. Stocks, Y. Wang, D.M.C. Nicholson, W.A. Shelton and H. Ebert, Phys. Rev. B **57** (1998) 14247.
[29] J. Kanamori, Anisotropy and magnetostriction of ferromagnetic and antiferromagnetic materials, in: G.T. Rado and H. Suhl (Eds.), Magnetism, Vol. I, Academic Press, New York, 1963, pp. 127–203.
[30] P. Bruno, Physical origins and theoretical models of magnetic anisotropy in: P.H. Dederichs, P. Grünberg and W. Zinn (Eds.), Magnetismus von Festkörpern und Grenzflächen, IFF-Ferienkurs Forschungszentrum Jülich, Jülich, 1993, pp. 24.1–24.27.
[31] D. Weller, G.R. Harp, R.F.C. Farrow, A. Cebollada and J. Sticht, Phys. Rev. Lett. **72** (1994) 2097.
[32] H. Brooks, Phys. Rev. **58** (1940) 909.
[33] H.J.F. Jansen, Phys. Rev. B **38** (1988) 8022.
[34] D. Craik, Magnetism, Principles and Applications, Wiley, Chichester, 1995.
[35] U. Pustogowa, J. Zabloudil, C. Uiberacker, C. Blaas, P. Weinberger, L. Szunyogh and C. Sommers, Phys. Rev. B **60** (1999) 414.
[36] P. Bruno, Phys. Rev. B **39** (1989) 865.

[37] A.R. Mackintosh and O.K. Andersen, The electronic structure of the transition metals, in: M. Springford (Ed.), Electrons at the Fermi Surface, Cambridge University Press, Cambridge, 1980, pp. 149–224.
[38] C. Antoniak, J. Lindner, M. Spasova, D. Sudfield, M. Acet, M. Farle, K. Fauth, U. Wiedwald, H.-G. Boyen, P. Ziemann, F. Wilhelm, A. Rogalev and S. Sun, Phys. Rev. Lett. **97** (2006) 117201.
[39] G.H.O. Daalderop, P.J. Kelly and M.F.H. Schuurmans, Phys. Rev. B **41** (1990) 11919.
[40] D.-S. Wang, R. Wu and A.J. Freeman, Phys. Rev. Lett. **70** (1993) 869.
[41] D.-S. Wang, R. Wu and A.J. Freeman, Phys. Rev. B **47** (1993) 14932.
[42] S.V. Halilov, A.Ya. Perlov, P.M. Oppeneer, A.N. Yaresko and V.N. Antonov, Phys. Rev. B **57** (1998) 9557.
[43] O. Eriksson, M.S.S. Brooks and B. Johansson, Phys. Rev. B **41** (1990) 7311.
[44] M.M. Steiner, R.C. Albers and L.J. Sham, Phys. Rev. B **45** (1992) 272.
[45] G.H.O. Daalderop, P.J. Kelly and M.F.H. Schuurmans, Phys. Rev. B **44** (1991) 12054.
[46] M.S.S. Brooks, Physica B **130** (1985) 6.
[47] I.V. Solovyev, A.I. Liechtenstein and K. Terakura, Phys. Rev. Lett. **80** (1998) 5758.
[48] A.I. Liechtenstein, V.I. Anisimov and J. Zaanen, Phys. Rev. B **52** (1995) R5467.
[49] V.I. Anisimov, F. Aryasetiawan and A.I. Liechtenstein, J. Phys. Condens. Matter **9** (1997) 767.
[50] I. Yang, S.Y. Savrasov and G. Kotliar, Phys. Rev. Lett. **87** (2001) 216405.
[51] Y. Xie and J.A. Blackman, Phys. Rev. B **69** (2004) 172407.
[52] J. Friedel, in: J.M. Ziman (Ed.), The Physics of Metals, Cambridge University Press, Cambridge, 1969, p. 340.
[53] A.P. Sutton, Electronic Structure of Materials, Oxford University Press, Oxford, 1993.
[54] P.J. Jensen and K.H. Bennemann, Z. Phys. D **35** (1995) 273.
[55] E.K. Parks, B.H. Weiller, P.S. Bechthold, W.F. Hoffman, G.C. Nieman, L.G. Pobo and S.J. Riley, J. Chem. Phys. **88** (1988) 1622.
[56] M.B. Knickelbein, Annu. Rev. Phys. Chem. **50** (1999) 79.
[57] E.K. Parks, G.C. Nieman, L.G. Pobo and S.J. Riley, J. Chem. Phys. **88** (1988) 6260.
[58] S.J. Riley, J. Non-Cryst. Solids **205–207** (1996) 781.
[59] R.L. Whetten, D.M. Cox, D.J. Trevor and A. Kaldor, Phys. Rev. Lett. **54** (1985) 1494.
[60] J. Conceicao, R.T. Laaksonen, L.-S. Wang, T. Guo, P. Nordlander and R.E. Smalley, Phys. Rev. B. **51** (1995) 4668.
[61] E.K. Parks, I. Zhu, J. Ho and S.J. Riley, J. Chem. Phys. **100** (1994) 7206.
[62] J. Ho, E.K. Parks, I. Zhu and S.J. Riley, Chem. Phys. **201** (1995) 245.
[63] T.D. Klots, B.J. Winter, E.K. Parks and S.J. Riley, J. Chem. Phys. **95** (1991) 8919.
[64] E.K. Parks, K.P. Kerns and S.J. Riley, J. Chem. Phys. **114** (2001) 2228.
[65] I.M.L. Billas, J.A. Becker, A. Châtelain and W.A. de Heer, Phys. Rev. Lett. **71** (1993) 4067.
[66] D.C. Douglass, A.J. Cox, J.P. Bucher and L.A. Bloomfield, Phys. Rev. B **47** (1993) 12874.
[67] D.M. Cox, D.J. Trevor, R.L. Whetton, E.A. Rolfing and A. Kaldor, Phys. Rev. B **32** (1985) 7290.
[68] W.A. de Heer, P. Milani and A. Châtelain, Phys. Rev. Lett. **65** (1990) 488.
[69] A.J. Cox, J.G. Louderback, S.E. Apsel and L.A. Bloomfield, Phys. Rev. B **49** (1994) 12295.
[70] W. Gerlach and O. Stern, Z. Physik **8** (1922) 110.
[71] I.M.L. Billas, A. Châtelain and W.A. de Heer, J. Magn. Magn. Mater. **168** (1997) 64.
[72] P. Milani and W.A. de Heer, Phys. Rev. B **44** (1991) 8346.
[73] I.M.L. Billas, J.A. Becker and W.A. de Heer, Z. Phys. D **26** (1993) 325.
[74] S.N. Khanna and S. Linderoth, Phys. Rev. Lett. **67** (1991) 742.
[75] M.B. Knickelbein, Phys. Rev. Lett. **86** (2001) 5255.
[76] D.P. Pappas, A.P. Popov, A.N. Anisimov, B.V. Reddy and S.N. Khanna, Phys. Rev. Lett. **76** (1996) 4332.
[77] D. Gerion, A. Hirt, I.M.L. Billas, A. Catelain and W.A. DeHeer, Phys. Rev. B **62** (2000) 7491.

[78] D. Gerion, A. Hirt and A. Chatelain, Phys. Rev. Lett. **83** (1999) 532.
[79] M. Sakurai, K. Watanabe, K. Sumiyama and K. Suzuki, J. Chem. Phys. **111** (1999) 235.
[80] V. Mazalova, A. Kravtsova, G. Yalovega, A. Soldatov, P. Piseri, M. Coreno, T. Mazza, C. Lenardi, G. Bongiorno and P. Milani, Nucl. Instrum. Methods A **575** (2007) 165.
[81] I.M.L. Billas, A. Châtelain and W.A. de Heer, Science **265** (1994) 1682.
[82] S.E. Apsel, J.W. Emmert, J. Deng and L.A. Bloomfield, Phys. Rev. Lett. **76** (1996) 1441.
[83] M.B. Knickelbein, J. Chem. Phys. **116** (2002) 9703.
[84] M.B. Knickelbein, Chem. Phys. Lett. **353** (2002) 221.
[85] F.A. Reuse and S.N. Khanna, Chem. Phys. Lett. **234** (1995) 77.
[86] M. Castro, C. Jamorski and D.R. Salahub, Chem. Phys. Lett. **271** (1997) 133.
[87] B.V. Reddy, S.K. Nayak, S.N. Khanna, B.K. Rao and P. Jena, J. Phys. Chem. A **102** (1998) 1748.
[88] T. Futschek, J. Hafner and M. Marsman, J. Phys. Condens. Matter **18** (2006) 9703.
[89] N. Desmarais, C. Jamorski, F.A. Reuse and S.N. Khanna, Chem. Phys. Lett. **294** (1998) 480.
[90] F.A. Reuse, S.N. Khanna and S. Bernel, Phys. Rev. B **52** (1995) R11650.
[91] C. Jamorski, A. Martinez, M. Castro and D.R. Salahub, Phys. Rev. B **35** (1997) 10905.
[92] H.-J. Fan, C.-W. Liu and M.-S. Liao, Chem. Phys. Lett. **273** (1997) 353.
[93] Q.-M. Ma, Z. Xie, J. Wang, Y. Liu and Y.-C. Li, Phys. Lett. A **358** (2006) 289.
[94] P. Ballone and R.O. Jones, Chem. Phys. Lett. **233** (1995) 632.
[95] O. Diéguez, M.M.G. Alemany, C. Rey, P. Ordejón and L.J. Gallego, Phys. Rev. B **63** (2001) 205407.
[96] Q.-M. Ma, Z. Xie, J. Wang, Y. Liu and Y.-C. Li, Solid State Commun. **142**.
[97] P. Bobadova-Parvanova, K.A. Jackson, S. Srinivas and M. Horoi, Phys. Rev. B **66** (2002) 195402.
[98] V.G. Grigoryan and M. Springborg, Phys. Rev. B **70** (2004) 205415.
[99] T. Oda, A. Pasquarello and R. Car, Phys. Rev. Lett. **80** (1998) 3622.
[100] M.A. Ojeda, J. Dorantes-Dávila and G.M. Pastor, Phys. Rev. B **60** (1999) 6121.
[101] N. Fujima, J. Phys. Soc. Jpn. **71** (2002) 1529.
[102] A.N. Andriotis and M. Menon, Phys. Rev. B **57** (1998) 10069.
[103] N.N. Lathiotakis, A.N. Andriotis, M. Menon and J. Connolly, J. Chem. Phys. **104** (1996) 992.
[104] F. Aguilera-Granja, S. Bouarab, M.J. López, A. Vega, J.M. Montejano-Carrizales, M.P. Iñiguez and J.A. Alonso, Phys. Rev. B **57** (1998) 12469.
[105] S. Bouarab, A. Vega, M.J. López, M.P. Iñiguez and J.A. Alonso, Phys. Rev. B **55** (1997) 13279.
[106] J.L. Rodríguez-López, F. Aguilera-Granja, K. Michaelian and A. Vega, Phys. Rev. B **67** (2003) 174413.
[107] E.K. Parks, B.J. Winter, T.D. Klots and S.J. Riley, J. Chem. Phys. **94** (1991) 1882.
[108] E.K. Parks, G.C. Nieman, K.P. Kerns and S.J. Riley, J. Chem. Phys **107** (1997) 1861.
[109] E.K. Parks, K.P. Kerns and S.J. Riley, J. Chem. Phys. **109** (1998) 10207.
[110] J. Guevara, F. Parisi, A.M. Llois and M. Weissmann, Phys. Rev. B **55** (1997) 13283.
[111] N. Fujima and S. Sakurai, J. Phys. Soc. Jpn. **68** (1999) 586.
[112] J.A. Franco, A. Vega and F. Aguilera-Granja, Phys. Rev. B **60** (1999) 434.
[113] M.L. Tiago, Y. Zhou, M.M.G. Alemany, Y. Saad and J.R. Chelikowsky, Phys. Rev. Lett. **97** (2006) 147201.
[114] P. Jensen and K.H. Bennemann, Z. Phys. D **35** (1995) 273.
[115] N. Fujima and T. Yamaguchi, Phys. Rev. B **56** (1996) 26.
[116] Y. Xie and J.A. Blackman, J. Phys. Condens. Matter **15** (2003) L615.
[117] Y. Xie and J.A. Blackman, J. Appl. Phys. **82** (2003) 1446.
[118] G. Rollmann, M.E. Gruner, A. Hucht, R. Meyer, P. Entel, M.L. Tiago and J.R. Chelikowsky, Phys. Rev. Lett. **99** (2007) 083402.
[119] J. Bansmann, S.H. Baker, C. Binns, J.A. Blackman, J.-P. Bucher, J. Dorantes-Dávila, V. Dupuis, L. Favrea, D. Kechrakos, A. Kleibert, K.-H. Meiwes-Broer, G.M. Pastor,

A. Xie, O. Toulemonde, K.N. Trohidou, V. Dupuis and Y. Xie, Surf. Sci. Rep. **56** (2005) 189.
[120] S.H. Baker, C. Binns, K.W. Edmonds, M.J. Mahea, S.C. Thornton, S. Louch and S.S. Dhesi, J. Magn. Magn. Mater. **247** (2002) 19.
[121] B. Nonas, I. Cabria, R. Zeller, P.H. Dederichs, T. Huhne and H. Ebert, Phys. Rev. Lett. **86** (2001) 2146.
[122] L. Zhou, D.S. Wang and Y. Kawazoe, Phys. Rev. B **60** (1999) 9545.
[123] M. Komelj, C. Ederer, J.W. Davenport and M. Fähnle, Phys. Rev. B **66** (2002) 140407.
[124] R.A. Guirado-López, J. Dorantes-Dávila and G.M. Pastor, Phys. Rev. Lett. **90** (2003) 226402.
[125] J. Dorantes-Dávila and G.M. Pastor, Phys. Rev. Lett. **81** (1998) 208.
[126] X. Wan, L. Zhou, J. Dong, T.L. Lee and D. Wang, Phys. Rev. B **69** (2004) 174414.
[127] J.M. Montejano-Carrizales, M.P. Iñiguez, J.A. Alonso and M.J. López, Phys. Rev. B **54** (1996) 5961.
[128] F. Bødker, S. Mørup and S. Linderoth, Phys. Rev. Lett. **72** (1994) 282.
[129] M. Jamet, W. Wernsdorfer, C. Thirion, D. Mailly, V. Dupuis, P. Mélinon and A. Pérez, Phys. Rev. Lett. **86** (2001) 4676.
[130] M. Jamet, W. Wernsdorfer, C. Thirion, V. Dupuis, P. Mélinon, A. Pérez and D. Mailly, Phys. Rev. B **69** (2004) 024401.
[131] G.M. Pastor, J. Dorantes-Dávila, S. Pick and H. Dreyssé, Phys. Rev. Lett. **75** (1995) 326.
[132] R. Guirado-López, Phys. Rev. B **63** (2001) 174420.
[133] Y. Xie and J.A. Blackman, J. Phys. Condens. Matter **16** (2004) 3163.
[134] J. Henk, A.M.N. Niklasson and B. Johansson, Phys. Rev. B **59** (1999) 9332.
[135] D.C. Douglass, J.P. Bucher and L.A. Bloomfield, Phys. Rev. B **45** (1992) 6341.
[136] D.R. Salahub and R.P. Messmer, Surf. Sci. **106** (1981) 415.
[137] V.L. Moruzzi, Phys. Rev. Lett. **57** (1986) 2211.
[138] G.M. Pastor, J. Dorantes-Dávila and K.H. Bennemann, Phys. Rev. B **40** (1989) 7642.
[139] K. Lee and J. Callaway, Phys. Rev. B **48** (1993) 15358.
[140] H. Cheng and L.-S. Wang, Phys. Rev. Lett. **77** (1996) 51.
[141] B.V. Reddy, S.N. Khanna and P. Jena, Phys. Rev. B **60** (1999) 15597.
[142] K. Lee and J. Callaway, Phys. Rev. B **49** (1994) 13906.
[143] M.A. Ojeda- López, Solid State Commun. **114** (2000) 301.
[144] C. Kohl and G.F. Bertsch, Phys. Rev. B **60** (1999) 4205.
[145] F.W. Payne, W. Jiang and L.A. Bloomfield, Phys. Rev. Lett. **97** (2006) 193401.
[146] M.B. Knickelbein, Phys. Rev. A **67** (2003) 013202.
[147] L.-S. Wang, H. Wu and H. Cheng, Phys. Rev. B **55** (1997) 12884.
[148] G.Y. Guo and H.H. Wang, Phys. Rev. B **62** (2000) 5136.
[149] M.B. Knickelbein, Phys. Rev. B **70** (2004) 014424.
[150] S.K. Nayak, B.K. Rao and P. Jena, J. Phys Condens. Matter **10** (1998) 10863.
[151] N.O. Jones, S.N. Khanna, T. Baruah and M.R. Pederson, Phys. Rev. B **70** (2004) 045416.
[152] S.N. Khanna, B.K. Rao, P. Jena and M. Knickelbein, Chem. Phys. Lett. **378** (2003) 374.
[153] M.R. Pederson, F. Reuse and S.N. Khanna, Phys. Rev. B **58** (1998) 5632.
[154] R.C. Longo, E.G. Noya and L.J. Gallego, J. Chem. Phys. **122** (2005) 226102.
[155] T. Morisato, S.N. Khanna and Y. Kawazoe, Phys. Rev. B **72** (2005) 014435.
[156] P. Bobadova-Parvanova, K.A. Jackson, S. Srinivas and M. Horoi, Phys. Rev. A **67** (2003) 061202(R).
[157] P. Bobadova-Parvanova, K.A. Jackson, S. Srinivas and M. Horoi, J. Chem. Phys. **122** (2005) 014310.
[158] T.M. Briere, M.H.F. Sluiter, V. Kumar and Y. Kawazoe, Phys. Rev. B **66** (2002) 064412.
[159] M. Kabir, A. Mookerjee and D.G. Kanhere, Phys. Rev. B **73** (2006) 224439.
[160] Y. Xie and J.A. Blackman, Phys. Rev. B **73** (2006) 214436.
[161] J. Mejía-López, A.H. Romero, M.E. Garcia and J.L. Morán- López, Phys. Rev. B **74** (2006) 140405(R).

[162] C. Rau, B. Xing and M. Robert, J. Vac. Sci. Technol. A **6** (1988) 579.
[163] F. Liu, S.N. Khanna and P. Jena, Phys. Rev. B **43** (1991) 8179.
[164] C. Binns, H.S. Derbyshire, S.C. Bayliss and C. Norris, Phys. Rev. B **45** (1992) 460.
[165] S. Blügel, Phys. Rev. Lett. **68** (1992) 851.
[166] R. Pfandzelter, G. Steierl and C. Rau, Phys. Rev. Lett. **74** (1995) 3467.
[167] P. Villaseñor-Gonzalez, J. Dorantes-Dávila, H. Dreyssé and G.M. Pastor, Phys. Rev. B **55** (1997) 15084.
[168] C. Barreteau, R. Guirado-López, D. Spanjaard, M.C. Desjonquères and A.M. Oleś, Phys. Rev. B **61** (2000) 7781.
[169] D.C. Douglass, J.P. Bucher and L.A. Bloomfield, Phys. Rev. Lett. **68** (1992) 1442.
[170] D.C. Douglass, J.P. Bucher, D.B. Haynes and L.A. Bloomfield, in: P. Jena, S.N. Khanna and B.K. Rao (Eds.), Physics and Chemistry of Finite Systems: From Clusters to Crystals, Kluwer NATO-ASI series, Dordrecht, 1992, p. 759.
[171] G.M. Pastor and J. Dorantes-Dávila, Phys. Rev. B **52** (1995) 13799.
[172] V.Z. Cerovski, S.D. Mahanti and S.N. Khanna, Eur. Phys. J. D **10** (2000) 119.

Chapter 9

Magnetism in Supported and Embedded Clusters

C. Binns[1] and J.A. Blackman[1,2]
[1]Department of Physics and Astronomy, University of Leicester, University Road, Leicester LE1 7RH, UK
[2]Department of Physics, University of Reading, Whiteknights, Reading RG6 6AF, UK

9.1.	Introduction	278
9.2.	Magnetic characterisation techniques	280
	9.2.1. VSM	280
	9.2.2. MOKE	280
	9.2.3. Micro-SQUID measurements	281
	9.2.4. Micro-Hall probes	282
	9.2.5. MLDAD and MCDAD	282
	9.2.6. XMCD	283
	9.2.7. X-ray magnetic linear dichroism (XMLD)	298
9.3.	Very small supported clusters ($N<10$) and 2D nanostructures	301
	9.3.1. Noble metal (Cu and Ag) surfaces	301
	9.3.1.1. $3d$ metal clusters	301
	9.3.1.2. $4d$ and $5d$ metal clusters	307
	9.3.1.3. Experimental results	308
	9.3.2. Ferromagnetic (Fe, Co, Ni) surfaces	308
	9.3.3. Pd and Pt surfaces	314
	9.3.4. Graphite surfaces	318
	9.3.5. The Kondo effect	321
	9.3.6. One-dimensional metal chains	327
9.4.	Supported clusters	330
9.5.	Clusters embedded in a matrix	333
	9.5.1. Isolated (non-interacting) clusters	334
	9.5.1.1. Co in Cu	334
	9.5.1.2. Fe and Co in Ag	335
	9.5.1.3. Fe in Cu	336
	9.5.1.4. Fe in Co	338
	9.5.2. Interacting clusters	340

HANDBOOK OF METAL PHYSICS
ISSN 1570-002X/DOI 10.1016/S1570-002X(08)00209-7

© 2009 ELSEVIER B.V.
ALL RIGHTS RESERVED

| 9.6. Exchange bias | 346 |
| References | 353 |

9.1. Introduction

The properties of free magnetic nanoparticles have been surveyed in Chapter 8, but of course, in most laboratory studies and certainly for device applications, the particles are deposited on a surface or embedded in another material. As we have seen with optical properties in Chapter 7, the embedding material can modify the behaviour considerably, and a similar effect occurs with magnetic particles with yet greater complexity.

The interest in magnetic particles has increased dramatically in the last few years because of their potential for applications in fields as diverse as ultrahigh-density recording and medicine [1–5]. There is also parallel interest in magnetic devices based on multilayer systems, generally known as spintronics, where there have been rapid developments particularly in the area of magnetoresistive effects [6,7].

Most nanoparticle applications depend on the magnetic order being stable with time and the superparamagnetic behaviour, introduced in Chapter 8, is a major factor limiting the miniaturisation of devices. The problem arises when the energy barrier due to the anisotropy, KV, becomes so small that the probability of switching due to the thermal energy becomes significant. The probability, P, of not switching (retaining the magnetisation) in time t is given by Boltzmann statistics and the Arrhenius relation

$$P = \exp\left(\frac{-t}{\tau}\right), \tag{9.1}$$

$$\frac{1}{\tau} = f_0 \exp\left(-\frac{KV}{kT}\right), \tag{9.2}$$

where f_0 is the attempt frequency [2]. The blocking temperature, which marks the onset of instability during an observation time τ, is then given by

$$T_B = \frac{KV}{k \ln(f_0\tau)}. \tag{9.3}$$

The attempt frequency is typically taken to be $f_0 = 10^9$ Hz, and a representative time constant for a laboratory experiment is $\tau = 100$ s, yielding $\ln(f_0\tau) = 25.3$. In the context of data storage, a time constant of $\tau \approx 10$ years would be more appropriate, which gives $\ln(f_0\tau) \approx 40$ and a lower blocking temperature. For a spherical particle of diameter D, Eq. (9.3) can be rewritten as

$$T_B = \frac{\pi K D^3}{6k \ln(f_0\tau)}, \tag{9.4}$$

emphasising the rapid drop in blocking temperature with decreasing particle size.

If we take $K = 0.2\,\mathrm{MJ/m^3}$ as a typical uniaxial anisotropy, Eq. (9.4) yields $T_B = 520\,\mathrm{K}$ as the 10 year blocking temperature for a nanoparticle with $D = 14\,\mathrm{nm}$, comfortably above room temperature or the operating temperature of a memory device, but the figure drops to $T_B = 65\,\mathrm{K}$ for a particle of half that diameter.

A number of strategies have been explored for enhancing the anisotropy, such as combining a high moment material like Co or Fe with an element with large spin–orbit interaction such as Pt. Possibilities include the deposition of ferromagnetic particles on a Pt surface [8,9] or the formation of alloys such CoPt or FePt [3,10]. Chemically ordered FePt, in the face-centred tetragonal structure $L1_0$ phase, has the ideal properties of high coercivity and large uniaxial anisotropy but, as nanoparticles, it forms in a disordered fcc structure which is magnetically soft. This is unfortunate as arrays of 3 nm particles in the $L1_0$ phase would have a sufficiently high anisotropy to be blocked at room temperature and thus the array would be capable of storing data at room temperature at a density greater that $10\,\mathrm{Tb/in^2}$. The fcc particles can be transformed to the $L1_0$ phase by high temperature annealing but this invariably destroys the ordered array. The FePt case will be discussed in more detail in Chapter 10. An alternative approach to enhanced effective anisotropy is through exploitation of the exchange bias phenomenon [11,12]. This relies on an interface between a ferromagnetic and an antiferromagnetic material, and the enhanced stability of the ferromagnetic component arises from exchange coupling at the interface and high anisotropy in the antiferromagnet.

We noted in Chapter 8 that isolated nanoparticles exhibit an enhancement in their spin and orbital moments as compared with the bulk materials, and clearly this effect is likely to be influenced by the presence of a substrate or an embedding material. At even a rather modest density of nanoparticles deposited on a surface, interactions between them can have a strong effect on the magnetic properties.

The focus in this chapter is mainly on the basic science of magnetic nanoparticles having an interface with a substrate or an embedding material. We begin in Section 9.2 with a survey of various techniques for probing the magnetic properties. Most of these are covered rather briefly except for one, X-ray magnetic circular dichroism (XMCD). XMCD is one of a number of powerful experimental techniques that employs radiation from a synchrotron source and yields element-specific information; another is extended X-ray absorption fine structure (EXAFS), which is used in structural studies. Generally the basis of XMCD is obscure to all except practitioners of the technique and thus it warrants a much fuller description.

Section 9.3 surveys the behaviour of clusters of just a few atoms deposited on various types of surface. The renewed interest in the Kondo effect is described. We also discuss the formation of one-dimensional chains at a step edge. This is one form of self-organisation. The reconstruction of a metal surface, such as the herringbone reconstruction of Au(1 1 1) described in Chapter 4, can also be used to engineer the self-organisation of clusters on a surface [13–15].

The behaviour of clusters of several hundred atoms deposited on a surface or embedded in a matrix is discussed in Sections 9.4 and 9.5, respectively. The chapter finishes in Section 9.6 with a description of exchange bias systems. Some applications of magnetic systems are surveyed in Chapter 10.

We note some reviews [11,12,15–18] that may be useful to readers who wish to explore aspects of this chapter in greater detail. Related to the topics of the current chapter are a number of emerging technologies that are still very much at the stage of fundamental science. Magnetic anisotropy at the level of a single atom [19,20] points to the prospect of single-atom data storage, while using a pulsed laser to flip a magnetic bit [21] has the potential for fast optical writing technology without the use of an external magnetic field. Finally we should note the extensive research activity in the field of single-molecule magnets [22,23]. The prototype material is Mn12acetate [24], with the chemical formula $Mn_{12}O_{12}(Ch_3COO)_{16}(H_2O)_4$, which is composed of 12 Mn atoms and bridging oxides, but many more molecules have been synthesised since this system was first studied. These systems are of particular interest because they exhibit the phenomenon of quantum tunnelling of the magnetisation, and thus have relevance to quantum computing as well as high-density data storage and magnetic refrigeration.

9.2. Magnetic characterisation techniques

9.2.1. VSM

UHV vibrating sample magnetometry (VSM), originally described by Foner [25] over 40 years ago, is a standard technique for measuring the magnetic moment in materials. The magnetic moment of a sample is determined by vibrating it between conducting coils and measuring the alternating voltage developed at the same frequency. The sensitivity of the technique has steadily improved and modern instruments can measure moments as low as $10^{-7}\,A\,m^2$, corresponding to $\sim 10^{16}$ Fe atoms. Recently, a system that allows *in situ* measurements on exposed clusters in UHV has been demonstrated [26]. A deposited cluster layer is deposited onto a substrate and then sealed, in its UHV environment, within an ampoule that can be withdrawn and inserted into the magnetometer without exposing the sample to atmosphere. The technique is not sensitive enough to measure significantly less than a single cluster layer, so it cannot be used to study isolated particles. However, the ability to measure exposed layers is important since even thick cluster films show a marked change in magnetic behaviour when coated with a protective non-magnetic capping layer to allow removal from the vacuum for *ex situ* measurements.

9.2.2. MOKE

In 1877, Kerr [27] observed that a ferromagnetic medium may affect the polarisation and intensity of polarised light when reflected by its surface. The phenomenon, known as the magneto-optical Kerr effect (MOKE), occurs by the same process as Faraday rotation but is seen in reflection rather than transmission. It is a widely used method for measuring magnetism in surfaces and ultrathin films [28,29] since it is relatively inexpensive to set up and is easily made compatible with UHV since the light source, usually a He–Ne laser, and all the polarisation detection equipment are external to the vacuum system with the incident and reflected beams transmitted through windows. The optical penetration depth of visible light is of the order of

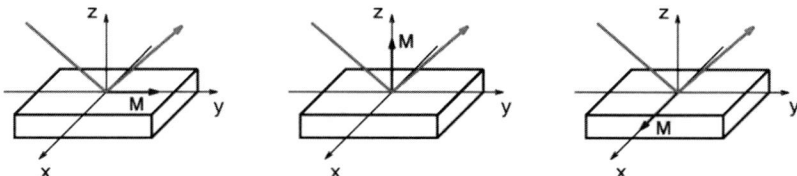

Figure 9.1. Different experimental configurations of the magneto-optical Kerr effect: longitudinal (left), polar (centre) and transverse (right). An external magnetic field is applied to orient the magnetisation vector **M** in the required direction. Reproduced with the permission of Elsevier Science from Bansmann et al. [15].

20 nm in metals and the signal from a deposited monolayer is a small fraction of the total signal, so the rotation of the polarisation vector is a fraction of a degree. Nevertheless, advanced experimental techniques are sufficiently sensitive to allow measurements on a few atomic monolayers or clusters embedded in non-magnetic matrices [30–32].

Three geometries are used: longitudinal, polar and transverse as illustrated in Figure 9.1. With careful design and good quality optics, it is possible to measure B–H loops in magnetic monolayers. The maximum signal is obtained when the ε vector of the light is perpendicular to the sample magnetisation.

With the advent of high intensity tuneable synchrotron radiation sources, MOKE has been extended from the visible light to the soft X-ray regime. Strongly enhanced effects at the element-specific core level resonances are observed [33–35]. The specular reflectivity of p-polarised light near the 2p absorption edges of the ferromagnetic $3d$ metals is measured with an external magnetic field perpendicular to the scattering plane (i.e. transverse MOKE), and large changes in reflectivity are observed upon reversal of the direction of the magnetic field. The advantage over transverse MOKE with visible light is the enhanced magnitude of the effect and the element specificity.

9.2.3. Micro-SQUID measurements

Superconducting quantum interference devices (SQUIDS) are the most sensitive magnetic detectors known and recent developments in lithographed devices of micrometre dimension have enabled the measurement of the switching field of individual nanoclusters [36]. The technique involves making the loop and Josephson junctions out of a cluster-assembled film consisting of nanoclusters embedded in a superconducting matrix. The film is deposited over a whole substrate that is then patterned using lithography into superconducting loops with weak links. Although every loop contains clusters over its entire surface, there is only significant flux coupling from those that are embedded within the bridge. In the devices reported, the junction regions have an area 30 nm × 50 nm and the clusters are embedded in a 20 nm thick Nb film; thus, if clusters with a volume ~ 20 nm^3 are embedded with a volume fraction (VFF) $\sim 0.1\%$, there will be on average 1 cluster per bridge region. One then makes measurements from a number of SQUIDS on the sample and it is clear from the switching field distribution which ones are detecting a single particle.

9.2.4. Micro-Hall probes

Using conventional photolithography and wet chemistry, it is possible to produce sub-micrometre size III–V semiconductor Hall bars in which the conducting layer is an ultrathin two-dimensional electron gas (2DEG) localised at the surface [37,38]. A magnetic particle deposited onto the central bar produces an intense local magnetic field that penetrates the 2DEG and generates a significant Hall voltage. Wirth and von Molnár [38] have shown how matching the size of the particle array to the active area of the Hall bar produces the optimum sensitivity, and recently Li et al. [39] measured the magnetisation curve of a single Fe nanoparticle with a diameter of 5 nm at temperatures up to 75 K. At present, the technique does not quite match the sensitivity of the micro-SQUID, but it is much more versatile since it can measure over a much wider temperature range (≤ 100 K) and is not restricted to particles embedded in superconducting materials.

9.2.5. MLDAD and MCDAD

A dichroism is observed in the 3p angle-resolved photoemission spectra of the transition metals taken with linearly or circularly polarised light with the geometries shown in Figures 9.2(a) and (b). There are two techniques. Magnetic linear dichroism in angular distributions (MLDAD) [40] measures the difference in the spectra in response to reversing the alignment between the in-plane sample magnetisation and the linear polarisation of the XUV light; in magnetic circular

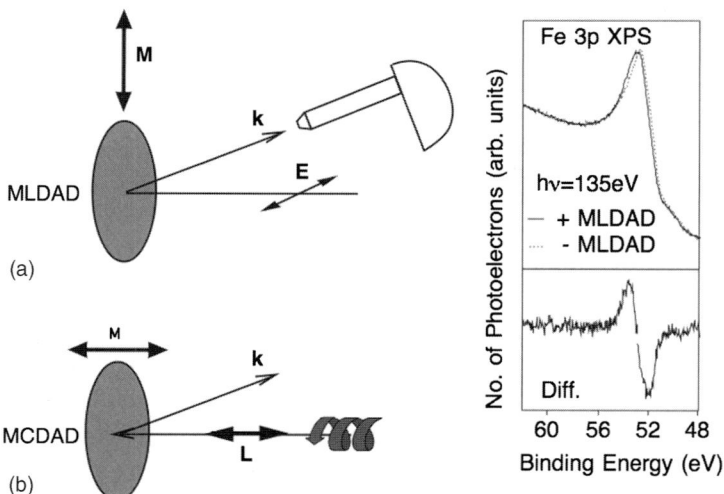

Figure 9.2. Experimental geometry and photoemission spectra from an Fe sample. (a) MLDAD measures the difference in the Fe 3p photoemission spectra taken with linearly polarised XUV light on reversing the magnetisation in the sample plane and perpendicular to the photon polarisation vector; (b) MCDAD measures the difference in the Fe 3p photoemission spectra taken with circularly polarised XUV light on reversing the magnetisation along the direction of the photon spin. All data shown were obtained from Fe nanoclusters on HOPG [43–45].

dichroism in angular distributions (MCDAD) [41], the difference is measured between the out-of-plane sample magnetisation and angular momentum of circularly polarised light. Since angle-resolved photoemission spectra must be collected in zero field, the techniques are, in general, restricted to measuring remanence but they are surface sensitive and capable of measuring sub-monolayer quantities of material. Schemes using magnetic substrates that expose the clusters to fields ∼ 1 T, but still allow photoemission measurements, have also been developed [42]. As with XMCD (see below), a synchrotron source must be used to provide the polarised XUV radiation. Endstations at which light of variable polarisation is available can be fitted with movable magnetising coils allowing both geometries to be combined in a single experiment. This enables the comparison of in-plane and out-of-plane remanent magnetisation.

9.2.6. XMCD

In the last 15 years, XMCD has emerged as a powerful technique to study the intra-atomic magnetism within magnetic materials. UV magnetic circular dichroism has been known since 1897 and X-ray dichroism involves essentially the same physics. The method involves measuring the X-ray absorption strength (XAS) as the wavelength of circularly polarised X-rays is tuned through an absorption edge. It is found experimentally, in the case of magnetic elements, that the spectral dependence of the XAS depends on the relative alignment of the sample magnetisation and the photon angular momentum. The effect is illustrated for thin films of Fe, Co and Ni in Figure 9.3 in which the XAS has been measured for wavelengths around the transition metal L edges arising from 2p to Fermi level transitions. The dichroism (i.e. the difference in the XAS for the two polarisations) is shown in the bottom panel of Figure 9.3 for the optimum case in which the photon angular momentum is parallel or antiparallel to the sample magnetisation.

The effect has been known for some time but the emergence of XMCD as a quantitative technique began with the formulation of the sum rules for the absorption of circularly polarised X-rays by magnetic materials [46,47]. Applying these sum rules to X-ray absorption data enables the evaluation, within certain restrictions, of the spin and orbital magnetic moments per atom in the material. The separation of the L edges in Figure 9.3 illustrates clearly that if a sample contained several magnetic elements, the spin and orbital moments for each element could be obtained independently.

The technique uses circularly polarised X-rays, that is, with the photons having well-defined angular momentum states $\pm\hbar$ in the direction of propagation. The most general definition of polarisation is given in terms of the electric fields, $E_x = E_x^0\, e^{i\omega t}$ and $E_y = E_y^0\, e^{i(\omega t + \phi)}$ of an electromagnetic wave. The polarisation, P, can be written:

$$P = \frac{E_y}{E_x} = \frac{E_y^0}{E_x^0}\, e^{i\phi}, \tag{9.5}$$

and thus as the wave oscillates, the polarisation remains constant. For circular polarisation, the phase factor $\phi = \pm\pi/2$ giving $P = \pm i$. The variation of the electric fields in the propagation direction for the two polarisations is illustrated in

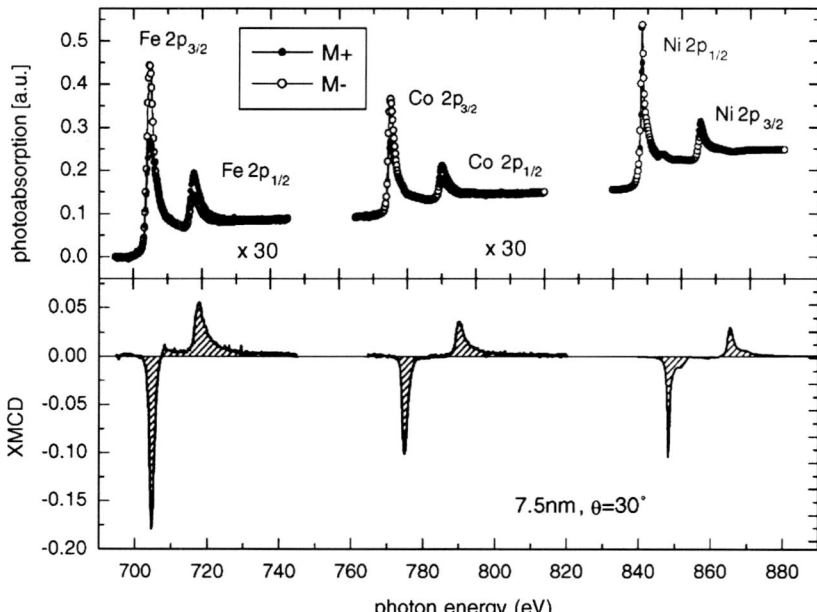

Figure 9.3. X-ray magnetic circular dichroism in the Fe, Co and Ni L edges for the optimum case when the photon angular momentum is aligned parallel or antiparallel to the sample magnetisation (denoted by M+ and M−). Reproduced with the permission of Elsevier Science from Bansmann et al. [15].

Figure 9.4. For photons of frequency ω, the direction of the ratio of electric field amplitudes in the x and y directions, $E_y/E_x = e^{i\phi}(E_y^0/E_x^0)$, spins at a rate ω clockwise or anticlockwise. For clockwise (anticlockwise) motion, the direction of the spiral is the same as for a right-handed (left-handed) screw.

The literature has unfortunately produced two naming conventions for the polarisation generally dividing between physics and optics literature since one defines polarisation as seen by the receiver and the other as seen by the source. The convention adopted here is that 'right circularly polarised' (RCP) corresponds to a right-handed screw in the direction of propagation. The important thing is that RCP imparts an angular momentum $\hbar$ into the sample while LCP imparts $-\hbar$ and it is this quantity that is involved in the optical selection rules and produces spin-polarised excitation from core levels. In the following, we will also present the polarisation in terms of the electric field unit vector, **e**, which is defined by:

$$\mathbf{e} = \frac{1}{\sqrt{2}}(\mathbf{e}_x + i\mathbf{e}_y) \quad \text{(RCP)} \tag{9.6}$$

and

$$\mathbf{e} = \frac{1}{\sqrt{2}}(\mathbf{e}_x - i\mathbf{e}_y) \quad \text{(LCP).} \tag{9.7}$$

Taking the example of transition metal L-edge XMCD, the X-ray absorption is by transitions from the 2p core level to empty states in the 3d valence band.

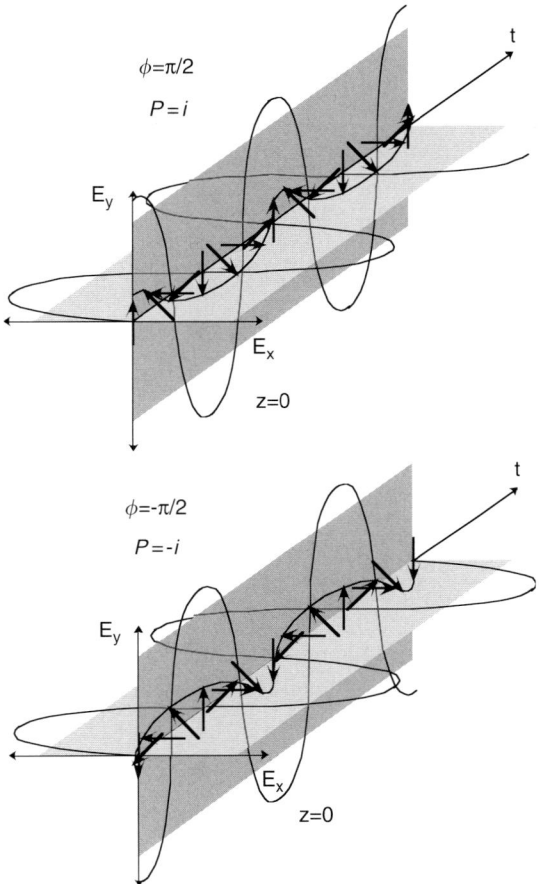

Figure 9.4. Circular polarisation specified by $\phi = \pm\pi/2$ in Eq. (9.5) giving $P = \pm i$. The direction of the ratio of electric field amplitudes in the x and y directions, $E_y/E_x = e^{i\phi}(E_y^0/E_x^0)$, spins at a rate ω clockwise or anticlockwise.

The angular momentum matrix elements for this transition for circularly polarised X-rays give the selection rules: $\Delta l = 1$, $\Delta m_l = 1$ (RCP) and $\Delta m_l = -1$ (LCP), where l is the angular momentum quantum number and m_l the magnetic quantum number of the electrons. Thus, the p ($l = 1$) to d ($l = 2$) transition satisfies the first selection rule and the second increments or decrements m_l by the absorption of the photon angular momentum. The X-ray dipole operator does not act on spin, so $\Delta m_s = 0$ for the initial and final states.

The mechanism of XMCD becomes clear when described in terms of an intuitive model presented by Stöhr and Wu [48]. They consider, in the first instance, a spin-only moment in a transition metal with a d^9 configuration, e.g. bulk Ni. Although a free Ni atom would have both a spin and an orbital magnetic moment, in the bulk the orbital component is quenched to close to zero. In fact, the small residual orbital moment in bulk transition metals is important in determining the magnetic anisotropy of the material but for the moment we will assume it is zero and examine the effect of the spin moment on the XAS. We will also assume that in the exchange-split band structure, the spin-up states are completely filled and there is one hole in the spin-down band. In a one-electron orbital picture, the angular part of the d wavefunctions is represented by the five spherical harmonics with angular momentum quantum number $l = 2$ and magnetic quantum numbers $m_l = -2, -1, 0, 1$ and 2, that is, $Y_2^{-2}, Y_2^{-1}, Y_2^0, Y_2^1$ and Y_2^2. In a non-relativistic treatment, these are multiplied by the spin functions α ($m_s = 1/2$) for spin-up electrons and β ($m_s = -1/2$) for spin-down electrons. Every member of the d-electron population is represented by a basis function $|l, m_l, s, m_s\rangle$, where in this case $d = 2$ and $s = 1/2$. The five states of a given spin are degenerate and there is an equal probability of finding the hole in any one of them. Thus, in order to evaluate the transition probability from the filled 2p states to the unfilled state in the d shell, we need to calculate the probability for all five m_l states and average them.

The strength of X-ray absorption can be evaluated from the so-called 'oscillator strength' arising from classical electrodynamics. It can be shown that a classical charge executing simple harmonic motion loses energy (or absorbs energy from an applied oscillating electric field) at a rate $\propto \langle \mathbf{r}\rangle^2$, where $\mathbf{r}$ is the displacement from its rest position. The equivalent statement in quantum mechanics is that during a transition from a stationary initial state i to a stationary final state f, the rate of absorption is proportional to the matrix element $\langle f|\mathbf{e}\cdot\mathbf{r}|i\rangle^2$. The symbols f and i contain all the quantum numbers associated with the two states and the relevant displacement is along the electric field unit vector. The dimensionless oscillator strength, obtained from standard time-dependent perturbation theory, is given by [49]:

$$I = \frac{2m(E_f - E_i)^2}{\hbar^2} \frac{1}{\hbar\omega} \langle f|\mathbf{e}\cdot\mathbf{r}|i\rangle^2. \tag{9.8}$$

From Eqs. (9.6) and (9.7), the following can be written:

$$I = \frac{2m(E_f - E_i)^2}{\hbar^2} \frac{1}{\hbar\omega} \langle f|x \pm iy|i\rangle^2, \tag{9.9}$$

where '+' represents RCP and '−' LCP. Care must be exercised in the standard notation used here to distinguish between the initial state i and the mathematical symbol i. The evaluation of the matrix elements is straightforward within the one-electron approximation where the initial state $\psi_i(\mathbf{r})$ (with quantum numbers n, l and m_l) and final state $\psi_f(\mathbf{r})$ (with quantum numbers n', l' and m'_l) are, in polar coordinates:

$$\psi_i(r) = R_{nl}(r) P_{lm_l}(\theta) e^{im_l\phi} \frac{1}{\sqrt{2\pi}}, \tag{9.10}$$

$$\psi_f(r) = R_{n'l'}(r) P_{lm'_l}(\theta) e^{im'_l\phi} \frac{1}{\sqrt{2\pi}}. \tag{9.11}$$

In a non-relativistic treatment, the electron spin is included by multiplying the space wavefunctions by α or β. The dipole operators $x \pm iy$ do not act on spin, so the spin quantum number and the spin function are constant during the X-ray absorption. Rewriting $x \pm iy$ as $r \sin\theta \, e^{\pm i\phi}$, the matrix element in Eq. (9.9) is given by:

$$\langle f|x \pm iy|i\rangle = \int_0^\infty R_{n'l'}(r)R_{nl}r^3 \, dr$$

$$\int_0^\pi P_{l'm_l'}(\theta)P_{lm}(\theta)\cos\theta \sin\theta \, d\theta \int_0^{2\pi} \frac{1}{2\pi} e^{i(m_l \pm 1 - m_l')\phi} \, d\phi, \quad (9.12)$$

where the $P_{lm}(\theta)$ are Legendre polynomials. It is evident that the integral over ϕ vanishes unless $m_l' - m_l = \pm 1$, which is the sum rule stated above. The integrals over θ give the sum rule $\Delta l = 1$. Thus, from an initial state n, l, m_l, the final state must be n', $l+1$, m_l+1 (RCP) or n', $l+1$, m_l-1 (LCP). Evaluating the integrals over the Legendre polynomials, it can be shown that they are zero unless [50]:

$$\langle n', l+1, m_l+1 | x+iy | n, l, m_l \rangle = -\sqrt{\frac{(l+m_l+2)(l+m_l+1)}{2(2l+3)(2l+1)}} R \quad \text{(RCP)}$$

(9.13)

and

$$\langle n', l+1, m_l+1 | x-iy | n, l, m_l \rangle = -\sqrt{\frac{(l-m_l+2)(l-m_l+1)}{2(2l+3)(2l+1)}} R \quad \text{(LCP)},$$

(9.14)

where the radial part of the matrix element is the same for absorption of either polarisation and has been replaced with R. Note that the matrix elements can be evaluated entirely from the n, l and m_l values of the initial state. In the case under consideration, this is the spin–orbit split 3p state, which separates into the degenerate doublet and degenerate quartet of states, each defined in terms of the total angular momentum m_j, shown in Table 9.1. The energy separation between the $^2P_{1/2}$ doublet (giving the L_2 edge) and the $^2P_{3/2}$ quartet (giving the L_3 edge) varies between 14 eV and 18 eV in the transition metals Fe, Co and Ni. The final column shows the spherical harmonic $(Y_l^{m_l}(\theta,\phi) = P_{lm_l}(\theta)e^{im_l\phi})$ times spin basis functions for the separate m_j states, which are a weighted mixture of spin-up and spin-down states.

For the moment, we will assume that the spin–orbit coupling in the 3d states is insignificant so that the five corresponding spherical harmonics ($l = 2$; $m_l = -2, -1, 0, 1, 2$) are pure spin states. Thus, returning to the special case of d^9 configuration and a single hole in the 3d spin-down states, only the spin-down (β) parts of the initial states in Table 9.1 need to be considered to evaluate the matrix elements in Eqs. (9.13) and (9.14). Since the transition can be from any initial state, the absorption strength at the L_3 edge for the two photon polarisations, from Eqs. (9.9), (9.13) and (9.14) and using the β coefficients for the four $^2P_{3/2}$ states in Table 9.1, is:

$$I_{L_3}^{RCP} = \left(\frac{0 + 1 \times 2 + (2/3) \times 6 + (1/3) \times 12}{30}\right) R^2 = \frac{1}{3} R^2, \quad (9.15)$$

Table 9.1. Spherical harmonics of m_j of $^2P_{1/2}$ doublet and the $^2P_{3/2}$ quartet.

Spectroscopic label ($^{2S+1}L_J$)	m_j	Corresponding spherical harmonic-spin basis functions, $Y_l^{m_l} \times (\alpha \text{ or } \beta)$
$^2P_{1/2}$	$\frac{1}{2}$	$\frac{1}{\sqrt{3}}\left(Y_1^0\alpha - \sqrt{2}Y_1^1\beta\right)$
	$-\frac{1}{2}$	$\frac{1}{\sqrt{3}}\left(\sqrt{2}Y_1^{-1}\alpha - Y_1^0\beta\right)$
$^2P_{3/2}$	$\frac{3}{2}$	$Y_1^1\alpha$
	$\frac{1}{2}$	$\frac{1}{\sqrt{3}}\left(\sqrt{2}Y_1^0\alpha + Y_1^1\beta\right)$
	$-\frac{1}{2}$	$\frac{1}{\sqrt{3}}\left(Y_1^{-1}\alpha + \sqrt{2}Y_1^0\beta\right)$
	$-\frac{3}{2}$	$Y_1^{-1}\beta$

$$I_{L_3}^{LCP} = \left(\frac{0 + 1 \times 12 + (2/3) \times 6 + (1/3) \times 2}{30}\right)R^2 = \frac{5}{9}R^2. \tag{9.16}$$

Thus, if we subtract the RCP XAS from the LCP XAS, we would get a difference between the two spectra (dichroism) of

$$\Delta I_{L_3} = I_{L_3}^{RCP} - I_{L_3}^{LCP} = -\frac{2}{9}R^2. \tag{9.17}$$

Repeating the process for the two $^2P_{1/2}$ states that produce the L_2 edge, we get:

$$\Delta I_{L_2} = I_{L_2}^{RCP} - I_{L_2}^{LCP} = \frac{2}{9}R^2. \tag{9.18}$$

Thus, the dichroism at the L_3 and L_2 edges is equal and opposite. It is evident that the dichroism arises directly from the matrix elements in Eqs. (9.13) and (9.14).

Another, more qualitative two-step explanation of X-ray dichroism was also presented by Stöhr and Wu [48]. Assuming, to begin with, there are an equal number of holes of each spin in the 3d band, the matrix elements for the transitions from the spin–orbit split 2p state reveal that for the $p_{3/2}$ initial state (L_3 edge), the LCP X-rays excite 62.5% electrons with spin down and 37.5% with spin up. The excitation from the $2p_{1/2}$ state has the opposite polarisation with 25% of the excited electrons with spin down and 75% with spin up. RCP X-rays produce the opposite spin polarisation in the excited electrons. The spin polarisation of photoelectrons from the 2p level excited by circularly polarised light has been measured in the case of Cu [51,52].

Thus, in the first step of the two-step model, the incident X-rays generate spin-polarised emission from the 2p core level. In a non-magnetic element in which there are equal numbers of empty spin-up and spin-down Fermi level states, this detail would be lost in an XAS measurement but in a magnetic material the imbalance of spins at the Fermi level produces different absorption strengths at the L_2 and L_3 edges for the two different light polarisations. In the second step, the spin-polarised valence band can be considered to act as a spin filter for the polarised electron emission. Figure 9.5 illustrates the two-step mechanism for XMCD.

The equal magnitude of the dichroism at each edge predicted by Eqs. (9.17) and (9.18) is a result of having a spin-only magnetic moment with no orbital component. The inclusion of an orbital moment and spin–orbit splitting in the $3d$ band results in unequal dichroism at the two edges. Within the one-electron model, it is easy to show that the orbital moment is proportional to the sum of dichroisms at the L_2 and L_3 edges [48]. This result is nicely illustrated in the data plotted in Figure 9.6, which compares the dichroism measured from very dilute and very dense films of 2 nm Fe nanoparticles deposited *in situ* onto graphite [53]. In the case of the very dilute (isolated) nanoparticles, the high proportion of surface atoms produces a

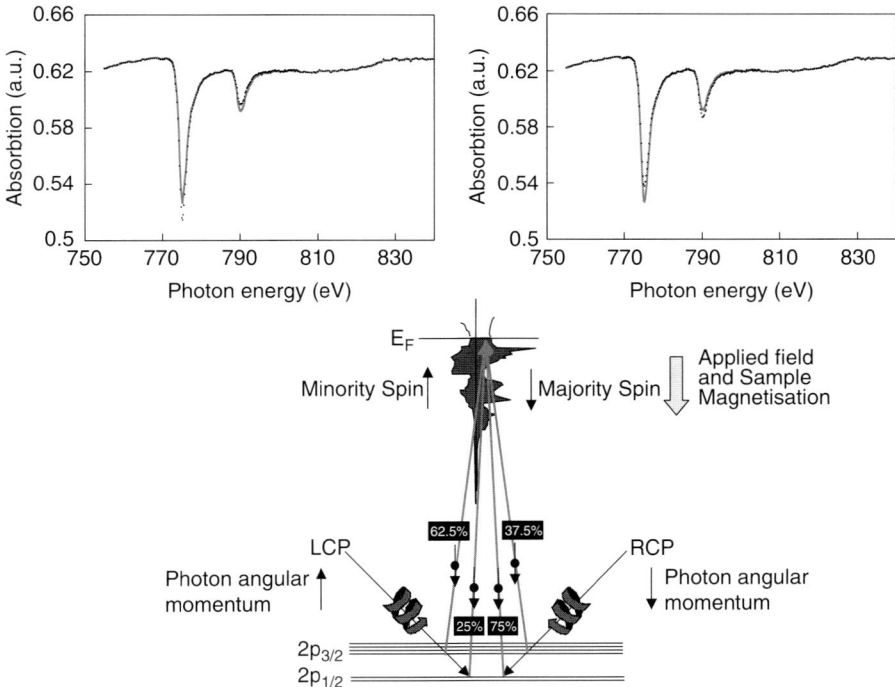

Figure 9.5. Illustration of the two-step model [48] for the mechanism producing XMCD. The sample is magnetised so that the majority spins are 'down'. The one-electron matrix elements for the transitions from the $2p_{3/2}$ initial state (L_3 edge) predict that LCP light excites 62.5% of the electrons with spin down and 37.5% with spin up. The excitation from the $2p_{1/2}$ state has the opposite polarisation with 25% of the excited electrons with spin down and 75% with spin up. The inverse polarisations are produced by RCP light. This with the excess of spin-down holes at the Fermi level XAS with LCP light will produce absorption at the L_3 and L_2 edges that is larger and smaller, respectively, than the averaged spectra with the opposite changes for RCP light. The data absorption data (dots) shown were taken by measuring the transmission through a thin film of Co nanoparticles. The grey line is the average of the two measured spectra.

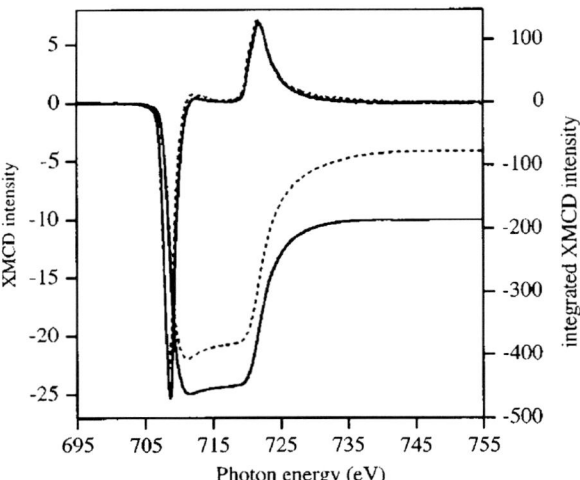

Figure 9.6. XMCD spectra and their integral across the Fe L_2 and L_3 edges from 400-atom Fe clusters deposited on HOPG [53]. The solid line is from a low coverage film (1.1 Å equivalent thickness) in which the clusters are isolated and the broken line is from a much higher coverage film (11.3 Å equivalent thickness) in which most of the clusters are in contact. The increased orbital moment in the isolated clusters is clearly observed in the increased value of the integral across the two edges.

significant enhancement in the orbital magnetic moment. In the dense film when the nanoparticles are in contact, the orbital moment is reduced to the (small) bulk value. This is revealed by integrating the dichroism across both edges and the magnitude of the integral beyond the edges is equal to the sum of dichroisms at the two edges. This is proportional to the orbital magnetic moment.

The above discussion presented the fundamental origin of dichroism in XAS but the technique has become rigorous and quantitative since the derivation of sum rules that connect total integrated intensities of the spectra to physical properties of the system [46,47]. Sum rule analysis is particularly powerful since it does not require a simulation of the absorption lineshape. This approach is familiar in chemistry with, for example, the Kuhn–Thomas sum rule [54] that connects the sum of oscillator strengths, f_{n0}, with the number of absorbing electrons, N_e, by the expression:

$$\sum_n f_{n0} = N_e. \qquad (9.19)$$

An analogous sum rule for the total X-ray absorption for linearly polarised light, $\mu_0(\omega)$, at the $L_{2,3}$ edge is [55]:

$$\int_{L_{2,3}} \mu_0(\omega) \propto n_h, \qquad (9.20)$$

where n_h is the number of holes in the $3d$ band. This is nicely illustrated for several $3d$ metals in the data shown in Figure 9.7 obtained by Stöhr and Nakajima [56].

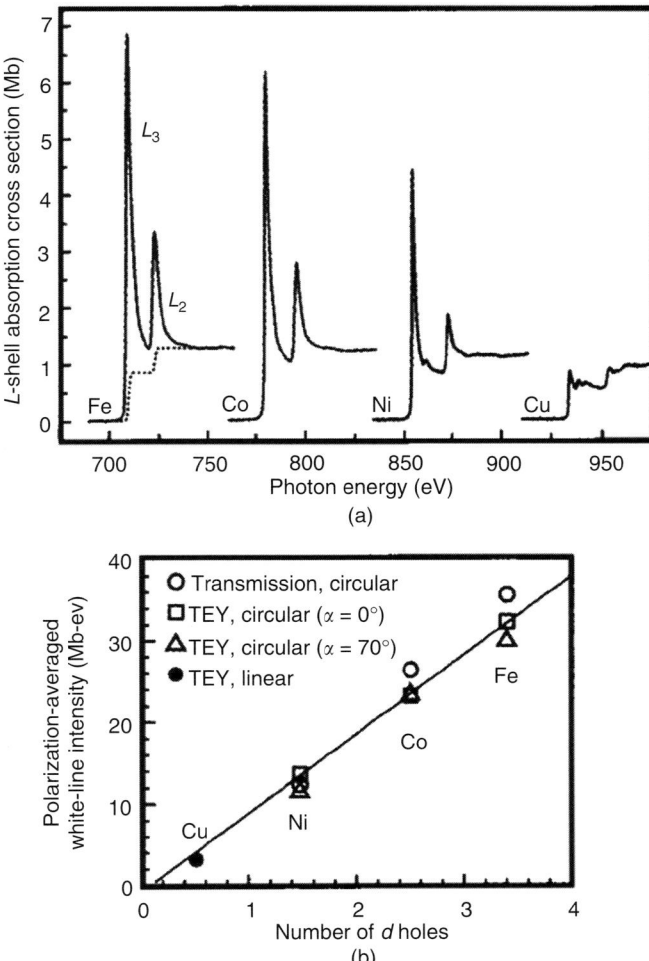

Figure 9.7. (a) X-ray absorption spectra for Fe, Co, Ni [57] and Cu [58] across the $L_{2,3}$ edges taken either with linearly polarised light or with circularly polarised light and averaged over the two polarisation directions. (b) White-line intensity, determined after subtraction of a double step function [57], shown dashed in the Fe spectrum in (a), versus the calculated number of d-orbital holes [59,60]. The plotted data were obtained by different detection methods, i.e. transmission and total electron yield (TEY) after correction for saturation effects (see below) and at different angles, α, relative to the surface normal. The proportionality between the total and $L_{2,3}$ absorption cross-section and n_h is clearly demonstrated. Reproduced with the permission of the *IBM Journal of Research and Development* from Stöhr and Nakajima [56].

The sum rule for the total X-ray absorption (Eq. (9.20)) is useful in determining the number of valence band holes. In nanoparticles, the 3d band narrows as a result of the reduced average coordination and the number of 3d band holes cannot be assumed to be the same as the bulk value. An experiment comparing the total normalised $L_{2,3}$ absorption of a nanoparticle sample with that of a bulk film grown *in situ* (for which n_h is well known) enables the determination of n_h for the particles.

This is important as the XMCD sum rules give the orbital and spin moment per valence band hole.

In order to analyse XMCD difference spectra such as shown in Figure 9.3, the most important sum rules relate the projection of the spin $\langle S_z \rangle$ and orbital $\langle L_z \rangle$ magnetic moments along the photon polarisation direction to partial differential absorption cross-sections at the L_2 and L_3 edges. Originally these were derived using a graphical angular momentum technique [47,48] but later the same sum rules were obtained within a Fermi golden rule formalism [61,62]. For transitions from core states with an angular momentum quantum number l_c to valence states with an angular momentum quantum number l_v, the orbital moment sum rule is given by [46]:

$$\langle L_z \rangle = 2 \frac{l_v(l_v+1)}{l_c(l_c+1) - l_v(l_v+1) - 2} \frac{\int_{\text{edge}} (\mu_{\uparrow\uparrow}(\omega) - \mu_{\uparrow\downarrow}(\omega)) d\omega}{\int_{\text{edge}} (\mu_{\uparrow\uparrow}(\omega) + \mu_{\uparrow\downarrow}(\omega) d\omega + \mu_0(\omega)) d\omega} n_h. \tag{9.21}$$

Here $\mu_{\uparrow\uparrow}(\omega)$ and $\mu_{\uparrow\downarrow}(\omega)$ are the X-ray absorption spectra measured with the photon angular momentum parallel and antiparallel to the applied (saturating) magnetic field, respectively, and $\mu_0(\omega)$ the average of the two, or, alternatively, the absorption spectrum measured with linearly polarised light. The integrals in this case are over the entire absorption edge. For absorption by a transition metal L edge, i.e. $2p$–$3d$ transitions, $l_c = 1$ and $l_v = 2$ and using

$$\mu_0(\omega) = \frac{1}{2} \mu_{\uparrow\uparrow}(\omega) + \mu_{\uparrow\downarrow}(\omega), \tag{9.22}$$

Eq. (9.21) becomes:

$$\langle L_z \rangle = \frac{4}{3} \frac{\int_{L_2+L_3} (\mu_{\uparrow\uparrow}(\omega) - \mu_{\uparrow\downarrow}(\omega)) d\omega}{\int_{L_2+L_3} (\mu_{\uparrow\uparrow}(\omega) + \mu_{\uparrow\downarrow}(\omega) d\omega) d\omega} n_h. \tag{9.23}$$

Thus, the orbital moment is proportional to the total area in the dichroism spectrum as illustrated in Figure 9.6.

The projection of the spin moment along the photon spin is given by [47]:

$$\frac{l_v(l_v+1) - 2 - l_c(l_c+1)}{3l_c} \langle S_z \rangle$$
$$+ \frac{l_v(l_v+1)[l_v(l_v+1) + 2l_c(l_c+1) + 4] - 3(l_c-1)^2(l_c+2)^2}{6l_v l_c(l_v+1)} \langle T_z \rangle$$
$$= \frac{\int_{j_+} (\mu_{\uparrow\uparrow}(\omega) - \mu_{\uparrow\downarrow}(\omega)) d\omega - ((l_c+1)/l_c) \int_{j_-} (\mu_{\uparrow\uparrow}(\omega) - \mu_{\uparrow\downarrow}(\omega)) d\omega}{\int_{\text{edge}} (\mu_{\uparrow\uparrow}(\omega) + \mu_{\uparrow\downarrow}(\omega) + \mu_0(\omega)) d\omega} n_h, \tag{9.24}$$

where $\langle T_z \rangle$ is the expectation value of the magnetic dipole operator along the photon spin and is a measure of the anisotropy of the spin distribution that emerges since the photon polarisation samples a directional cut through the atomic electron density. This term, discussed in more detail below, is small relative to $\langle S_z \rangle$ and for bulk samples has often been assumed to be zero but its significance increases in nanoscale particles and it must be taken into consideration. Note that the integrals in the numerator of Eq. (9.24) are taken over individual j components of the L edge.

In the case of transition metal L edges, Eq. (9.24) becomes

$$\langle S_z \rangle + \frac{7}{2} \langle T_z \rangle = \frac{\int_{L_3} (\mu_{\uparrow\uparrow}(\omega) - \mu_{\uparrow\downarrow}(\omega))d\omega - 2\int_{L_2} (\mu_{\uparrow\uparrow}(\omega) - \mu_{\uparrow\downarrow}(\omega))d\omega}{\int_{L_2+L_3} (\mu_{\uparrow\uparrow}(\omega) + \mu_{\uparrow\downarrow}(\omega))d\omega} n_h. \quad (9.25)$$

Figure 9.8(a) shows simulated dichroic XAS with a fitted integral background. Figures 9.8(b) and (c) show the background-subtracted μ_0 signal and the dichroism. The relevant integrals appearing in the $\langle L_z \rangle$ and $\langle S_z \rangle$ sum rules are also indicated in the figure and labelled:

$$I_{edge} = \frac{1}{2} \int_{L_2+L_3} (\mu_{\uparrow\uparrow}(\omega) + \mu_{\uparrow\downarrow}(\omega))d\omega, \quad (9.26)$$

$$I_{L_3} = \int_{L_3} (\mu_{\uparrow\uparrow}(\omega) - \mu_{\uparrow\downarrow}(\omega))d\omega, \quad (9.27)$$

$$I_{L_2} = \int_{L_2} (\mu_{\uparrow\uparrow}(\omega) - \mu_{\uparrow\downarrow}(\omega))d\omega. \quad (9.28)$$

In the simplest analysis for $3d$ transition metals, ignoring the $\langle T_z \rangle$ term, the spin and orbital sum rules reduce to:

$$\langle S_z \rangle = \frac{I_{L_3} - 2I_{L_2}}{I_{edge}} n_h \quad (9.29)$$

and

$$\langle L_z \rangle = \frac{4}{3} \frac{I_{L_2} + I_{L_3}}{I_{edge}} n_h. \quad (9.30)$$

The neglect of the $\langle T_z \rangle$ term is not valid in low-dimensional systems such as ultrathin films since the significant proportion of surface atoms introduces a significant anisotropy in the spin distribution and the spin term evaluated using Eq. (9.29) becomes dependent on the angle of the sample normal with respect to the photon incidence direction. (Note that we assume throughout that the photon direction and sample magnetisation are parallel or antiparallel.) In the case of nanoparticles on a surface, a random distribution of crystalline axes of the nanoparticles would average the $\langle T_z \rangle$ term to zero but it has been shown in the case of gas-phase nanoparticles deposited on surfaces that the measured spin moment is angle-dependent [63]. This was ascribed to a common asymmetry perpendicular to the sample surface, due, for example, to a degree of flattening of the clusters when adsorbed on the surface. Bruno has shown that the contribution

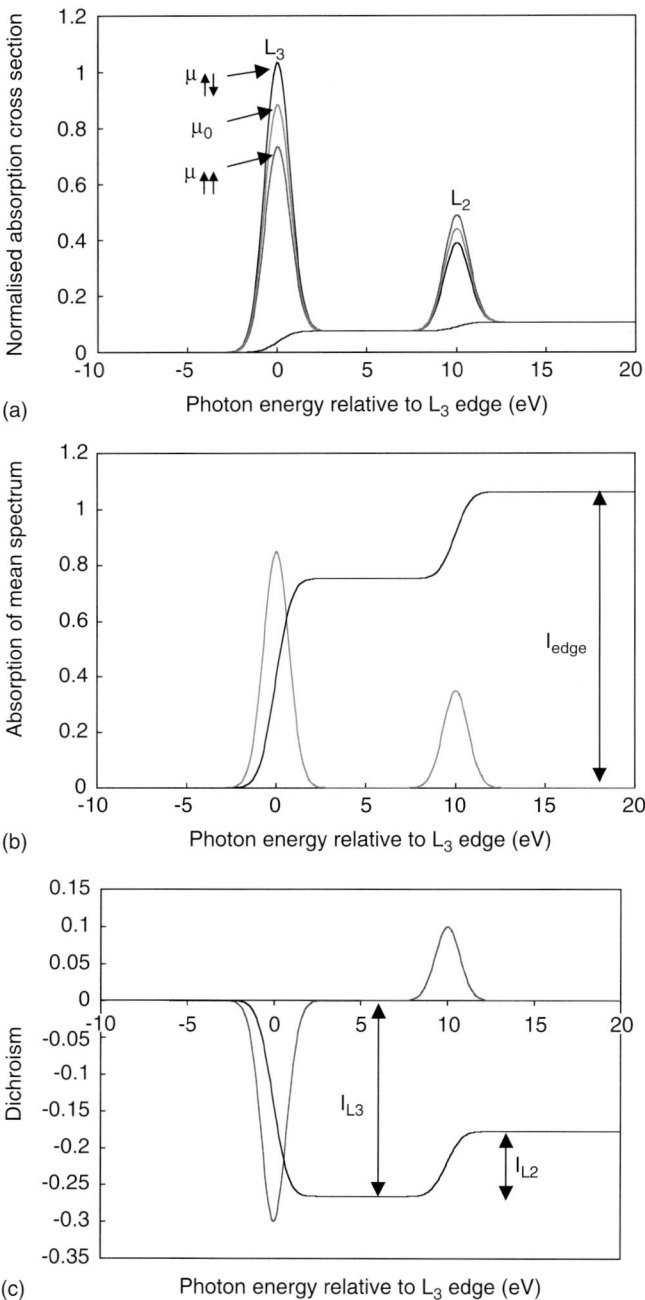

Figure 9.8. Simulated XAS and dichroism spectra for a transition metal L edge showing the relevant integrals used in the sum rules.

of $\langle T_Z \rangle$ to the measured spin moment with the sample normal at angle θ with the photon beam varies as $\sin^2 \theta$ [64]. The contribution of $\langle T_Z \rangle$ can be obtained by carrying out dichroism measurements as a function of angle and fitting to Bruno's model. For example, in the case of 690-atom Fe clusters deposited on highly oriented pyrolytic graphite (HOPG), it was found that the spin moment was 1.974 μ_B with a dipole moment, m_T ($= 7/2\langle T_Z \rangle$), of 0.012 μ_B [63], that is a 5% contribution. If one can assume that the sample has rotational symmetry parallel to the substrate surface (the normal situation), then a simpler analysis can be used. Separating the dipole moment into components parallel (m_T'') and perpendicular (m_T^z) to the sample surface, we want to find the measurement angle at which the dipole moment gives zero contribution, i.e. where:

$$m_T^x + m_T^y + m_T^z = 0, \qquad (9.31)$$

or, for the symmetry described

$$m_T^z + 2m_T'' = 0, \qquad (9.32)$$

where m_T'' is the component of the dipole moment parallel to the surface. The dipole moment at a general angle θ with respect to the sample normal is given by:

$$m_T^\theta = m_T^z \cos^2 \theta + m_T'' \sin^2 \theta, \qquad (9.33)$$

so from Eq. (9.32) this goes to zero when $\tan^2 \theta = 2$, i.e. $\theta = 54.7°$, the so-called 'magic angle'. Thus, a measurement at the magic angle will yield the pure spin moment without the dipole contribution and is the value to be compared with other measurement techniques, for example, magnetometry (after including the orbital moment). The relative importance of the dipole moment increases with the increasing proportion of surface atoms and this is illustrated in Figure 9.9, which shows the measured spin+dipole moment in isolated Fe clusters deposited on HOPG as a function of cluster size. Two sets of data taken with the photon beam along the sample normal and at the magic angle with respect to the sample normal are plotted. The difference between the two, plotted in the inset, is the dipole moment at normal incidence ($\theta = 0$). The dipole moment is observed to increase as the cluster size decreases and thus the proportion of surface atoms increases.

An accurate measurement of the absorption coefficient as a function of photon energy, $\mu(\omega)$, is essential to obtain reliable values for orbital and spin moments using XMCD and the three methods used are illustrated in Figure 9.10. One can measure the transmitted X-ray flux by, for example, measuring the drain current from photoemission grids before and after the sample. One can also determine the total electron yield (TEY), which is related to the absorption, either by a photoelectron detector near the sample surface or by measuring the sample drain current. Alternatively, the fluorescence following absorption can be detected. In general, the TEY or fluorescence has to be corrected for saturation effects (see below). The data in Figure 9.10 show a simultaneous measurement using transmission and TEY from a sample of dense 2 nm Co nanoparticles deposited on an amorphous carbon surface. There are advantages and disadvantages to both methods. The transmission measurement determines the sample absorption

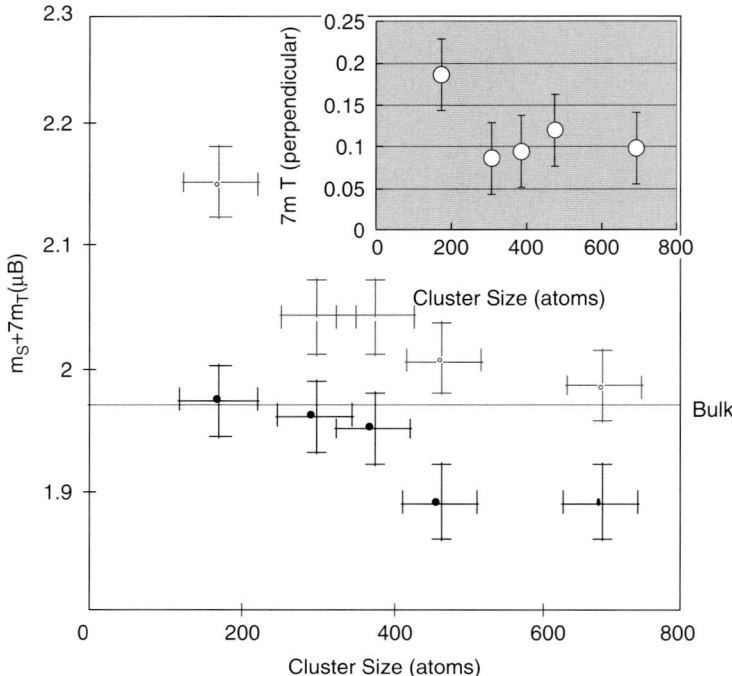

Figure 9.9. The measured spin+dipole moment in isolated Fe clusters deposited on HOPG as a function of cluster size. The grey data are taken with the photon beam at the magic angle (54.7°) with respect to the surface normal and the black data are with the photon beam along the surface normal. The difference between the two, plotted in the inset, is the dipole moment at normal incidence and is observed to increase as the cluster size decreases.

coefficient, $\mu(\omega)$, directly using:

$$\mu(\omega) = -\frac{1}{t}\log_e\left(\frac{I}{I_0}\right), \qquad (9.34)$$

where t is the sample thickness. It does however restrict samples plus their support to thicknesses that are reasonably transparent to soft X-rays, which is tens of nanometres for the transition metal L edges (600–800 eV). The sample must also be on a transparent support that is mechanically stable. The samples producing the data in Figure 9.10 were 50 nm equivalent thickness of 2 nm Co clusters deposited onto a 10 nm carbon film held on a Cu TEM grid, which has an approximately 80% transparency. The films were then coated with a further 10 nm carbon *in situ* to protect them from oxidation after removal from the UHV chamber in which the clusters were deposited.

Drain current or fluorescence measurements do not restrict the sample but are an indirect measurement of absorption and have to be corrected for saturation to get a reliable measurement of $\mu(\omega)$. The origin of the effect is that the spectral dependence of the incident X-ray beam changes as a function of depth due to absorption by the edge being probed, which is illustrated nicely in Figure 9.11,

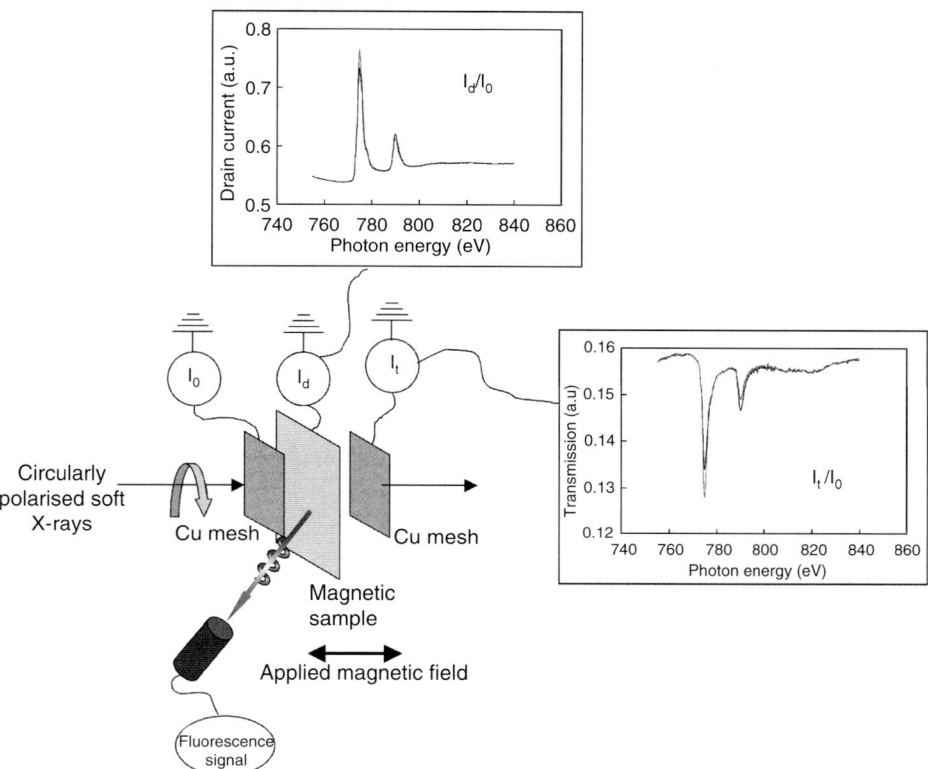

Figure 9.10. Various methods used to determine X-ray absorption in XMCD experiments, including total electron yield (TEY), fluorescence and transmission measurements. The raw data shown are a simultaneous measurement of transmission and TEY of a dense deposit of 2 nm Co clusters on an amorphous carbon surface.

reproduced from Nakajima et al. [65]. The photoelectrons that produce the measured signal originate from a region near the surface with a characteristic escape depth, λ_e, of ~ 20 Å. Because of the strong absorption by, for example, the transition metal L edge, the photon absorption at the resonant energy is significant after penetrating to a depth λ_e so that the layer at this depth produces a smaller drain current at the peak of the L absorption edges than the layer at the surface. In the limit of strong absorption, the deeper layers will produce no photoemission current at the edges, hence the term 'saturation'. Clearly any saturation correction is a function of the photon incidence angle as this determines the effective absorption length of the photons at a given vertical depth.

An expression for the electron yield, Y_e, from a surface of a thick sample after correcting for saturation was given by Nakajima et al. [65]:

$$Y_e = C\left(\frac{1}{1 + \lambda_e/\lambda_x \cos\theta}\right)\mu, \tag{9.35}$$

where C is a constant, λ_x the photon penetration depth ($= 1/\mu$) and θ the photon incidence angle. In the case, $\lambda_x \gg \lambda_e$, $Y_e \propto \mu$ as required. The corresponding

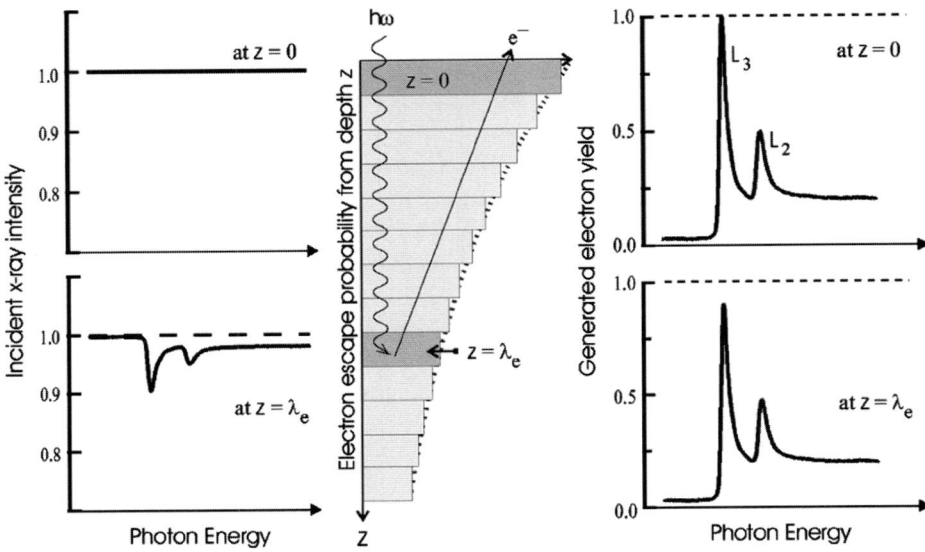

Figure 9.11. Illustration of saturation effect for Fe. *Left*: The difference in the X-ray intensity arriving at depth $z = \lambda_e$ is shown, compared to the incident X-ray intensity at $z = 0$ ($I_0 = 1$). *Centre*: Probability of electron escape, plotted in a horizontal bar graph as a function of depth z. The probability of escape is unity at the sample surface ($z = 0$) and drops to $1/e$ at $z = \lambda_e$. *Right*: Hypothetical electron yield spectra generated from the absorbed photons in layers at a depth $z = 0$ and λ_e. The measured total electron yield spectrum outside the surface consists of contributions from the various layers weighted by the probability of escape from the surface (shown in centre). Reproduced with permission of the American Physical Society from Nakajima et al. [65].

expression for a thin film of thickness t is:

$$Y_e = C\left(\frac{1}{1 + \lambda_e/\lambda_x \cos\theta}\right)(1 - e^{-t(1/\lambda_e + 1/\lambda_x \cos\theta)})\mu. \tag{9.36}$$

The different probing depths of TEY and transmission measurements can be used to good effect to distinguish between surface and bulk magnetisations. For example, closer inspection of the data in Figure 9.10 reveals a much smaller dichroism in the TEY spectra than the transmission data. This is due to the surface of the film being oxidised and the presence of antiferromagnetic CoO.

9.2.7. X-ray magnetic linear dichroism (XMLD)

An X-ray dichroism is also observed with linearly polarised light and shares some of the powerful characteristics of XMCD, including extreme sensitivity and element specificity. The electric field in linearly polarised light is along an axis and the absorption strength will be related to overlap of initial state and final state orbitals along that axis. Thus, any charge asymmetry in the unit cell will produce an absorption that depends on the direction of polarisation. In non-magnetic cubic

9.2. *Magnetic characterisation techniques*

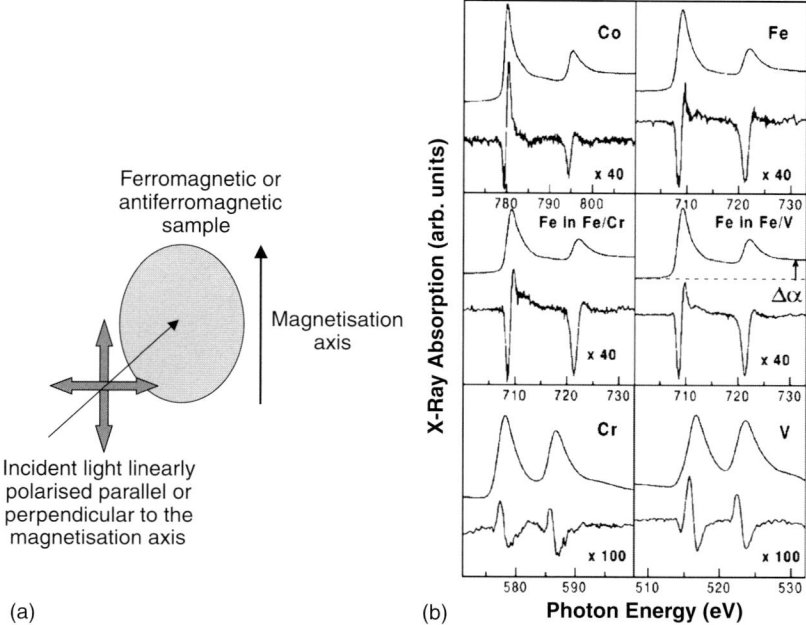

Figure 9.12. (a) Geometry to obtain the maximum linear dichroism from magnetic samples. (b) X-ray absorption spectra at the L edges for Co and Fe in sputtered films and Fe, Cr and V in multilayer films. The difference in absorption for the two polarisations is displayed in the bottom panel of each spectrum and a dichroism in the region ∼ 1–5% is observed. Reproduced with permission of the American Physical Society from Schwickert *et al.* [68].

metals, the charge asymmetry is very small, so they show no linear dichroism, but in a magnetic material magnetised along an axis, spin–orbit coupling of the valence electrons will induce a charge asymmetry about the magnetic axis, so magnetic samples show linear dichroism. The effect was first predicted in 1985 by Thole *et al.* [66] and demonstrated experimentally by the same group a year later [67].

The maximum dichroism is obtained by aligning the linear polarisation parallel and perpendicular to the magnetisation axis as shown in Figure 9.12(a). The spin–orbit coupling in the 3*d* band of the magnetic transition metals is small and thus so is the induced charge asymmetry and resulting linear dichroism. This is not the case in rare-earths, which can demonstrate much stronger linear dichroism [67]. Figure 9.12(b) shows the dichroism at the L edges for sputtered Fe and Co films and also for Fe, V and Cr in FeV and FeCr multilayer films produced by sputtering [68].

A quantitative measure of the maximum linear dichroism was defined by [68]:

$$I_{\text{XMLD}} = \max \left| \frac{\mu^\perp - \mu^\|}{\Delta \alpha} \right|, \tag{9.37}$$

where $\mu^\perp$ and $\mu^\|$ are the absorptions measured with the light polarisation perpendicular and parallel to the magnetisation axis and $\Delta\alpha$ the step height well past

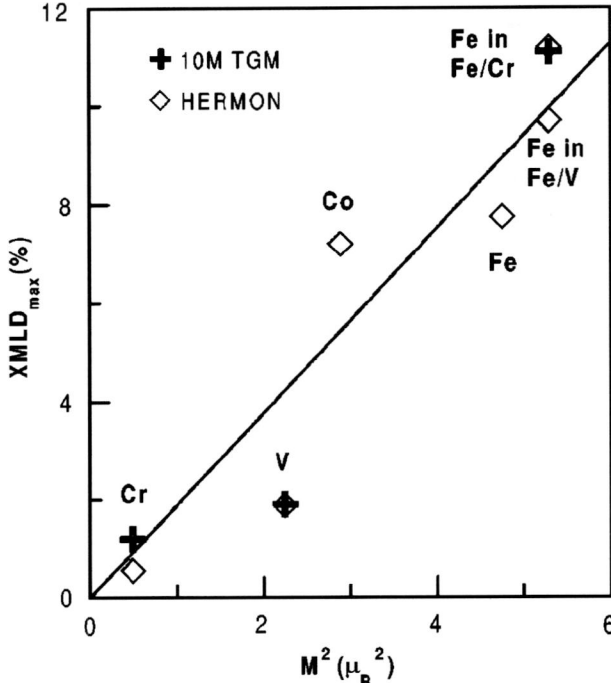

Figure 9.13. I_{XMLD} plotted against $\langle M^2 \rangle$ for the four elements Fe, Co, V and Cr. The two kinds of symbols represent two sets of measurements taken on different beamlines with different photon energy resolutions. The line fit follows the trend $I_{\text{XMLD}} = 1.88 \langle M^2 \rangle$. While Cr and V are not normally magnetic at room temperature, they have been shown to acquire the displayed magnetic moments in the multilayers used in this study [69,70]. Reproduced with permission of the American Physical Society from Schwickert et al. [68].

the edge as illustrated for the Fe L-edge data in the panel showing the FeV data. The authors then related this quantity to the known magnetic moments for the four elements. As shown in Figure 9.13, it is found experimentally that

$$I_{\text{XMLD}} \propto \langle M^2 \rangle. \tag{9.38}$$

The $\langle M^2 \rangle$ dependence of XMLD was predicted in the pioneering paper by Thole et al. [66] in 1985 and shows an important fundamental difference to XMCD. The maximum circular dichroism measured in an XMCD experiment is related to $\langle M \rangle$ making it a *unidirectional* technique and sensitive only to the ferromagnetic alignment in a sample. On the other hand, XMLD has a *uniaxial* sensitivity and is thus sensitive to any alignment with respect to an axis including antiferromagnetic alignment. The technique has thus focused on antiferromagnetic materials. Many antiferromagnets are oxides and in these systems, with a significant localisation of charge, the linear dichroism can be much larger than in metals.

9.3. Very small supported clusters ($N<10$) and 2D nanostructures

We review first theoretical and experimental studies of very small clusters (mostly smaller than 10 atoms) of $3d$, $4d$ and $5d$ metals on a variety of surfaces. Calculations based on various techniques form the bulk of work in this field, because an experimental study on very small clusters is far from straightforward. Nevertheless, over the last decade, a number of important experimental investigations have been reported. Where useful in elucidating the behaviour of the clusters, reference is made to results from studies of monolayers, but a full description of monolayers and multilayers *per se* is outside the scope of this book.

The initial sub-sectioning of this part of the chapter is by surfaces. Copper and silver represent the noble metal surfaces; gold has additional interest because of its surface reconstruction, which is relevant to self-organisation [13–15]. The magnetic behaviour of clusters of $3d$, $4d$ and $5d$ metals is discussed, but it is clear that, despite the anticipation of novel behaviour from the $4d$ and $5d$ elements, the main interest will remain on the traditional magnetic $3d$-based systems and alloys. The ferromagnetic surfaces (Fe, Co and Ni) provide added interest because of the interplay between their magnetism and that of the deposited clusters. Pd and Pt have a large spin–orbit coupling constant (see Figure 8.2 in Chapter 8) and have the potential to enhance the magnetocrystalline anisotropy of a deposited cluster. Other surfaces such as graphite are also considered. We discuss the Kondo effect, first explained in 1964, but receiving renewed interest in the context of isolated magnetic impurities on a metallic substrate, and finally one-dimensional chains of atoms formed at a step edge.

9.3.1. Noble metal (Cu and Ag) surfaces

9.3.1.1. 3d metal clusters

Stepanyuk *et al.* perform calculations based on density functional theory and the Korringa–Kohn–Rostoker (KKR) method. A Green's function technique is used to treat the perturbation of introducing a surface into the ideal crystal and then placing an adatom or cluster on the surface [71–73]. Their results [72] for small $3d$ metal clusters on a Cu(0 0 1) surface are shown in Figure 9.14. The atoms are fixed at the relevant positions in the ideal fcc crystal lattice without any lattice relaxation. The plots show the behaviour for single adatoms, dimers, trimers and square islands. The results for monolayers (ferromagnetic configurations) are also displayed. A similar trend is shown by monomers [74] and dimers [75] on an Ag(0 0 1) surface, but with somewhat larger moments because the lattice constant of Ag is 12% larger than that of Cu.

Vanadium is of interest, being non-magnetic in its bulk bcc form, but potentially magnetic as clusters or in layer structures, but both theoretically and experimentally the results have been contradictory. It would appear, from Figure 9.14, that magnetism in V disappears at an island size somewhere between the tetramer and the complete monolayer, whereas with a Ag(0 0 1) substrate V remains magnetic through to monolayer coverage [75]. Silver seems more promising therefore as a

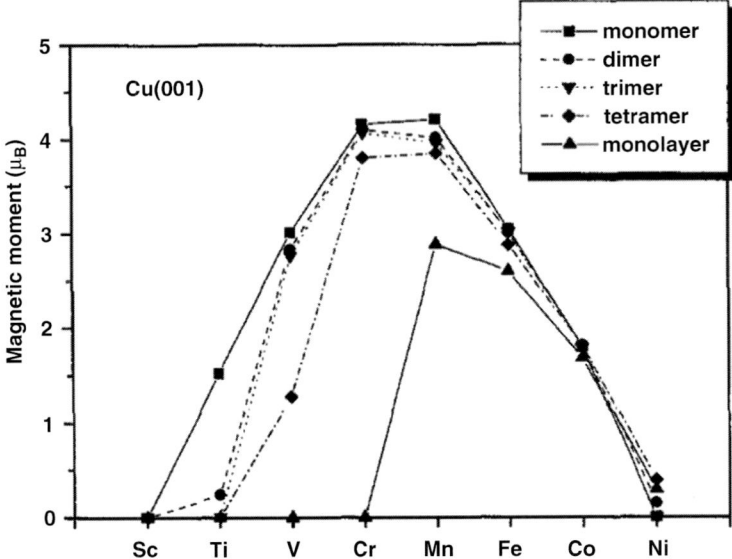

Figure 9.14. Spin magnetic moment of 3d metal clusters and monolayer on Cu(0 0 1). Reproduced with the permission of Elsevier Science from Stepanyuk et al. [72].

substrate for magnetism in vanadium but, even for the monolayer, results are in conflict. Theoretical studies [76–78] predict a magnetic moment of about $2\,\mu_B$ and ferromagnetic ordering, while other work [79] suggests that V, as well as Cr and Mn, favours an antiferromagnetic $c(2 \times 2)^1$ configuration on noble metal and Pd and Pt surfaces, while only Fe, Co and Ni prefer ferromagnetic ordering. There has also been some controversy in the interpretation of experimental measurements on V monolayers on Ag(0 0 1) ([80] and references therein), but recent XMCD results [80] appear to rule out ferromagnetic ordering. However, scanning tunnelling microscopy (STM) images show [80] that the growth mode of V on Ag(0 0 1) is of the Volmer–Weber type with the formation of 3D islands rather than layers, which is consequent on the lattice mismatch of 5%, and so the model monolayer calculations do not reflect the actual growth process. The situation is clearer with a Cu(0 0 1) substrate with no evidence of magnetic order in a V monolayer either theoretically [81] or experimentally [82].

Small V islands on Cu(0 0 1) also appear to be non-magnetic. Although an adsorbed dimer fixed to lattice sites exhibits a large magnetic moment [72,83], it is found that if geometry optimisation is included in the calculation, the Cu atoms near the adsorbed V dimer undergo significant relaxations which results in complete quenching of the magnetic moment of the vanadium [83]. Calculations [84,85] on small V islands on an Ag(0 0 1) surface indicate a magnetic moment on the V atoms

---

[1] In Wood's notation: a p(2 × 2) overlayer has a unit cell twice the size of the unit cell of the surface layer; the notation c(2 × 2) denotes an additional atom at the centre of the primitive (2 × 2) adsorbate unit cell.

with a tendency to antiferromagnetic ordering, but the effect of the relaxation of the positions of the atoms was not included.

The moments on the Co and Fe atoms plotted in Figure 9.14 are relatively insensitive to island size but, before discussing these elements further, it is useful to recall work on the deposition of monolayers onto noble metal surfaces. Heteroepitaxy of Co on Cu(0 0 1) represents a model system since the lattice constant of magnetic fcc Co is only 2% smaller than that of fcc Cu; also Co in bulk Cu has a low miscibility. In principle, these are ideal conditions for a high quality, sharp interface, and there are numerous studies of this system. Calculations ([86] and references therein) for a monolayer on a Cu(0 0 1) surface show an enhanced spin moment compared with the bulk for both Co and Fe, but a reduction in the case of Ni. The reduced spin moment for the Ni monolayer is ascribed to Ni-d–Cu-d hybridisation. The orbital moment is enhanced in all three cases, sometimes by as much as a factor of 2 compared with the bulk value [86]. This enhancement to the orbital moment has been observed in XMCD experiments [87]. However, despite the fact that the Co on Cu(0 0 1) system has the characteristics to be a prototype for layer-by-layer growth, experimental studies show evidence [88–95] of island formation, a bilayer growth mode and surface alloying. Nouvertné et al. [94] have analysed the growth behaviour by STM using CO titration and density functional theory total energy calculations. They find that initially substitutional adsorption with a Co atom exchanging with a Cu atom in the top substrate layer is energetically favoured over on-surface adsorption (see Figure 9.15). The adsorbed Co atoms then act as pinning centres for subsequent island nucleation. There are thus both Cu and Co atoms diffusing on the surface with the former having a higher mobility. The result at a fraction of a monolayer coverage is a bimodal island size distribution with smaller islands that are predominantly Co and larger islands that are mainly Cu with some adsorption of Co around their edges. This brief review illustrates the interest in both spin and orbital moments, and indicates that

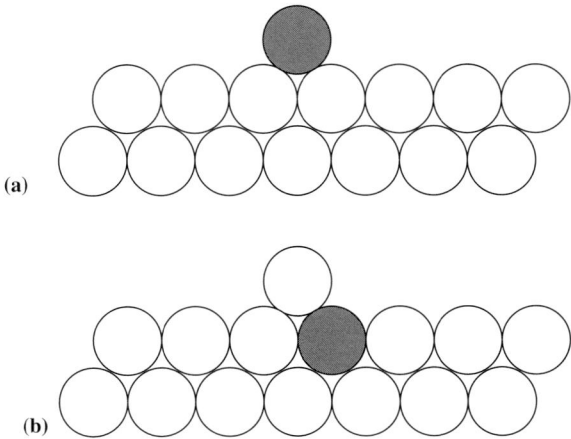

Figure 9.15. Adsorption of a Co atom into the surface layer of Cu(0 0 1) substrate with exchange of positions of Co and Cu atoms (b) is energetically favoured over surface adsorption (a).

structural modifications, including the incorporation of adatoms into the substrate, are important. We now turn our focus back to small supported islands.

The initial adsorption and relaxed geometry of small islands of Co on Cu(0 0 1) has been studied [96–100] using molecular dynamics with many-body potentials of the type discussed in Section 6.2.4, and by total energy calculations on repeated slabs separated by a vacuum region [101]. Molecular dynamics studies [102] for Fe islands on Cu(0 0 1) have also been performed. The fact that the macroscopic misfit between Co and Cu is small ($\sim 2\%$) would suggest a small tensile strain in Co nanostructures on Cu(0 0 1). However, atoms at the edge of an island have fewer nearest neighbours than bulk Co causing a lower binding energy and a relaxation in the direction of the centre of the island. The result is a shortening of the average Co–Co bond length, an effect which is most pronounced for the smallest islands. Embedding of the cluster is favoured with a typical relaxed geometry illustrated in Figure 9.16.

Island morphology will depend on experimental conditions, and clearly there is evidence both for islands on the surface and for the burrowing of clusters into the substrate. A transition from two- to three-dimensional structures has also been explored. Izquierdo et al. [98] show that a 2D–3D transition occurs for Co islands on Cu(0 0 1), and that the inclusion of magnetism in the calculation drives the transition at a smaller size than when it is excluded. Gómez et al. [104] study Co islands on Cu(1 1 1) and show that at a certain size, the islands exhibit a 2D–3D transition and form mixed Co–Cu clusters.

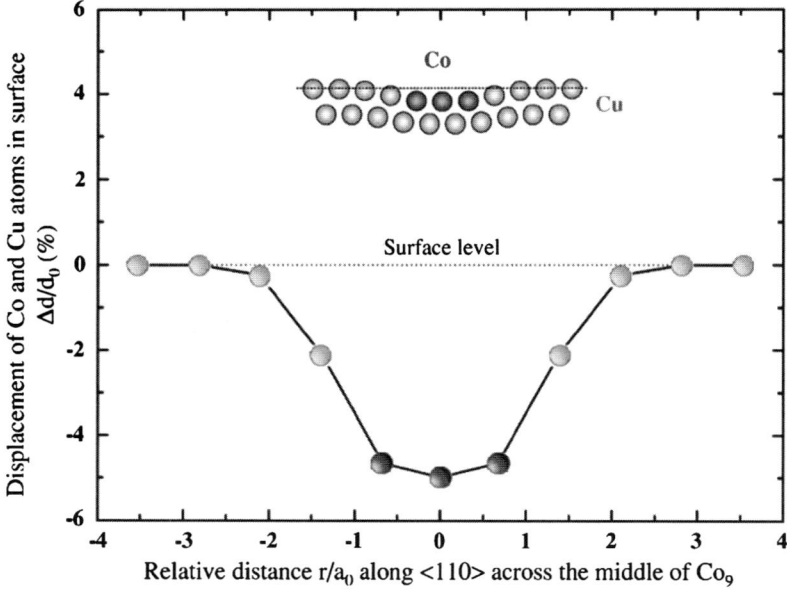

Figure 9.16. The shape of an embedded 3×3 square Co island and the Cu(0 0 1) substrate along the (1 1 0) direction. The displacement of atoms relative to the interlayer distance d_0 is shown. The surface level corresponds to the atomic positions of the surface without the Co cluster. Reproduced with the permission of the American Physical Society from Pick et al. [103].

9.3. Very small supported clusters (N < 10) and 2D nanostructures

A number of studies have been performed of the magnetic properties of small 3d islands under various scenarios. Enhancement of the orbital moment above its bulk value is expected. Calculations on small islands on the surface, that neglect geometry relaxation, are fairly consistent in their predictions of the spin moment, and find its value enhanced with respect to bulk but relatively insensitive to cluster size: Co on Cu (1.7–2.0 μ_B) [72,109], Co on Ag (2.0–2.1 μ_B) [106–109] and Fe on Ag (3.2–3.4 μ_B) [106–109]. The bulk values for Co and Fe are, respectively, 1.58 μ_B and 2.12 μ_B (see Table 6.1 in Chapter 6).

The situation with the orbital moment is much less definitive however. We recall (see Section 8.2.2) that in bulk systems, the orbital moment is underestimated and the orbital polarisation term [105] was introduced to try to bring the calculated values into closer agreement with experiment. The bulk orbital moments for bcc Fe and hcp Co are 0.10 μ_B and 0.14 μ_B, respectively; that of fcc Co is somewhat less (see Table 8.1 in Chapter 8).

Calculations have been performed on small clusters at fixed lattice positions on the surfaces of Cu and Ag both with [106,107,109] and without [106–109] orbital polarisation. The predicted orbital moments for Co and Fe islands on Ag(0 0 1) from calculations not using orbital polarisation are shown in Figure 9.17. Calculations that include orbital polarisation [106,107,109] give a huge enhancement to the orbital moment with values of about 2.2 μ_B and 2.5 μ_B, respectively, for Fe and Co on Ag(0 0 1). From calculations on larger clusters with which comparison with experiment is easier, it is probably true to say that orbital polarisation overestimates the orbital moment while its exclusion yields an underestimate.

Pick et al. [103,110] study the orbital moments of Co islands both on the surface of Cu(0 0 1) allowing for lattice relaxation [110] and as embedded clusters as in

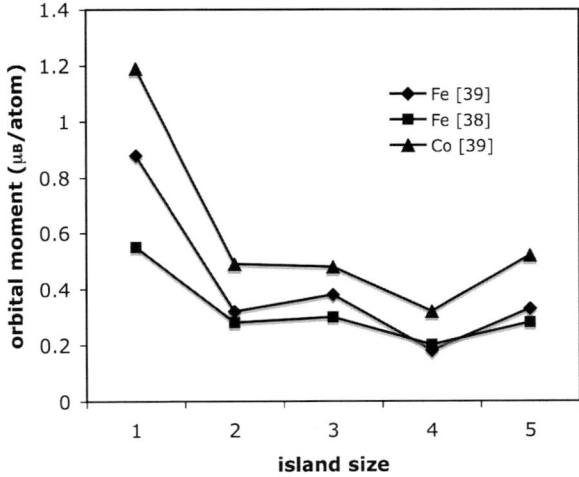

Figure 9.17. Calculated orbital moments for Fe and Co islands on an Ag(0 0 1) substrate with no lattice relaxation or orbital polarisation. The data have been taken from Cabria et al. [107] (squares) and Lazarovits et al. [108] (diamonds and triangles). The results of Klautau and Frota-Pessôa [109] for Fe and Co adatoms are very close to those of Ref. [108].

Figure 9.16 [103]. The magnetic anisotropy energy (MAE) is calculated; this is the energy difference between configurations with the magnetic vector perpendicular and parallel to the surface. Without atomic relaxations, the adatom shows a tendency to perpendicular magnetisation, but this is totally suppressed giving in-plane magnetisation when relaxation is included [110]. A major contribution to the MAE comes from the edge atoms of a supported cluster, but this is significantly reduced if the cluster is embedded in the surface layer [103]. For all embedded clusters, in-plane magnetisation is preferred over the perpendicular one. A simple approximate relation connecting the MAE and the change in the orbital moment when the magnetisation direction is switched [64] has been tested in calculations [103,110].

The orbital moments of Cr and Mn adatoms are extremely small [109], as might be expected given the ground state configurations of the individual atoms (see Section 8.2.1). Dimers of Cr are antiferromagnetic, while ferromagnetic and antiferromagnetic configurations of the Mn dimer are virtually degenerate in energy [111]. Several small Mn islands on Ag(0 0 1) have been shown to exhibit this degeneracy [112]. Such bistability is of interest for quantum tunnelling devices. Non-collinear spin arrangements have also been investigated [113–116] and are likely to be particularly relevant for Cr and Mn clusters. Results from first principles calculations of the ground states of some small Cr clusters on a Cu(1 1 1) substrate are reproduced in Figure 9.18.

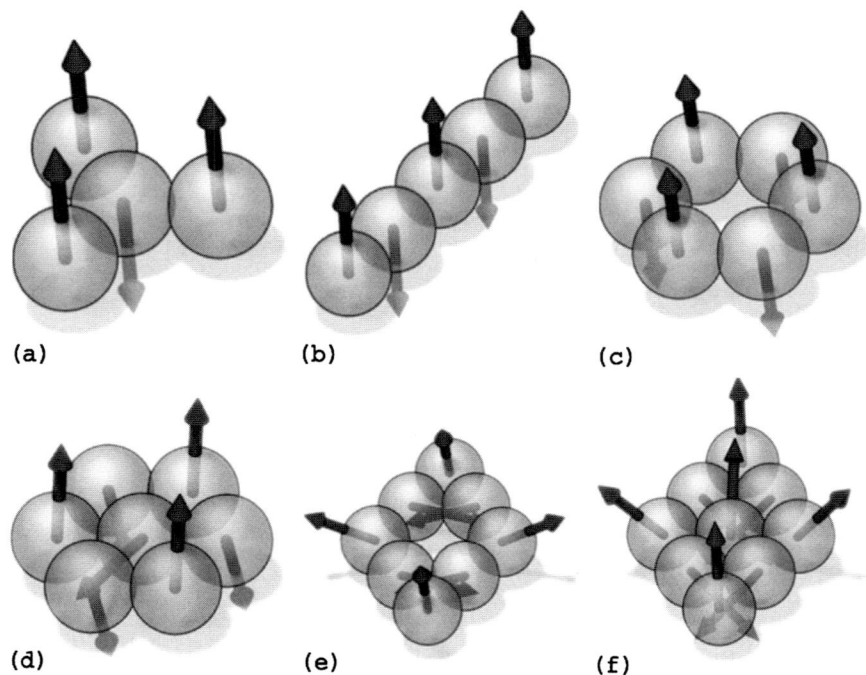

(a)   (b)   (c)

(d)   (e)   (f)

Figure 9.18. The calculated magnetic ground state for Cr clusters on a Cu(1 1 1) surface. Reproduced with the permission of the American Physical Society from Bergman et al. [116].

9.3.1.2. 4d and 5d metal clusters

There has been considerable interest in the possibility of magnetism in clusters of the 4d and 5d metals, and a number of studies have examined the spin [71,72,75,106,107,117,118] and orbital moments [106,117] on adatoms and small clusters. Representative plots are reproduced in Figures 9.19 and 9.20.

Although most adatoms are predicted to have a local moment, this rapidly disappears with increasing cluster size, and Ru and Rh in the 4d series and Os and Ir in the 5d series would appear to be the only candidates for displaying magnetic properties. There are also predictions of ferromagnetism in overlayers of Ru and Rh when placed on an Ag(0 0 1) substrate [119,78].

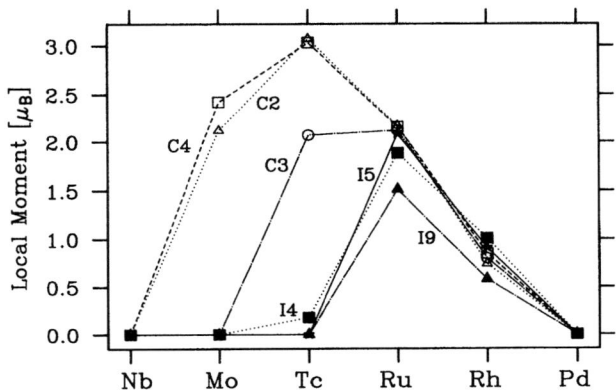

Figure 9.19. Spin magnetic moment of 4d metal clusters on Ag(0 0 1): C2 dimer, C3 trimer, C4 tetramer, I4 2 × 2 square, I5 cross (centred square) and I9 3 × 3 square. Reproduced with the permission of the American Physical Society from Wildberger et al. [117].

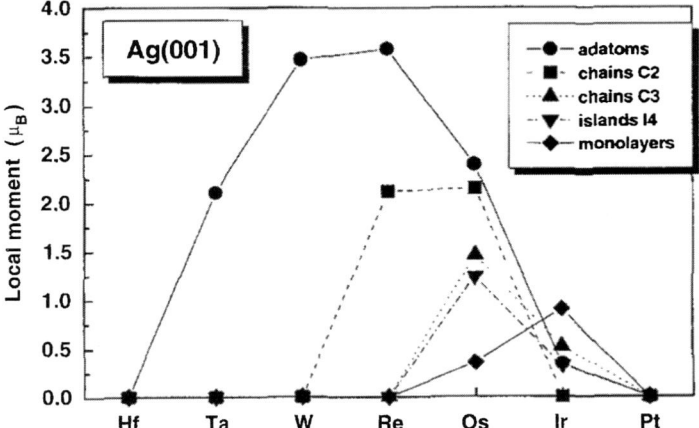

Figure 9.20. Spin magnetic moment of 5d metal clusters on Ag(0 0 1). Reproduced with the permission of Elsevier Science from Stepanyuk et al. [72].

9.3.1.3. Experimental results

The magnetic properties of 3d clusters (100–10000 atoms) on various substrates have been well documented (see Section 9.4), but experimental data on very small clusters are rather sparse. Single Fe, Co and Ni atoms on alkali metal surfaces have been studied by means of the anomalous Hall effect [120] and by XAS and XMCD [121]. Gambardella *et al.* [121] study films with coverages as low as 0.002 of a monolayer, at which concentration the film comprises mostly individual atoms. XAS is used to determine the d-state configurations, which are found to be localised (i.e. atomic-like with little d-hybridisation with the free electrons of the substrate). When the atoms are on a potassium surface, the configurations are identified as d^7, d^8 and d^9 for Fe, Co and Ni, respectively. Assuming these valence states and atomic-like behaviour, Hund's rules were used to obtain spin and orbital moments and thus a theoretical prediction of the ratio $R = \langle L_z \rangle / (2\langle S_z \rangle + 7\langle T_z \rangle)$ (from Eqs. (9.23) and (9.25)), with which XMCD results could be compared [121]. There is excellent agreement between theory and experiment for Ni confirming a huge total moment of about 3.5 μ_B. Large moments for Fe and Co are also indicated, but with some discrepancy between predicted and observed values of R, implying that some quenching of the orbital moment is taking place. Gambardella *et al.* [121] find similar behaviour for Fe and Co on a sodium substrate. Ni is found to be non-magnetic however; it is concluded that the d electrons are still well localised but that the valence states have a strongly increased d^{10} character.

The search for magnetism in the 4d and 5d systems has been far less fruitful. Stern–Gerlach experiments show that isolated Rh clusters smaller than about 60 atoms do exhibit a magnetic moment [122,123], as discussed in Section 8.4.4, and in-plane magnetisation has been observed [124] in a Ru monolayer on C(0 0 0 1) below a surface Curie temperature of about 250 K. However, so far, no experimental evidence for magnetism in 4d or 5d nanostructures on noble metal surfaces has been found. MOKE experiments on Rh on Ag [125] and Rh on Au [126,127] failed to confirm the presence of ferromagnetism. More recently, Honolka *et al.* [128] used XMCD to study Ru on Ag(1 0 0) and Rh on Ag(1 0 0) and Pt(9 9 7) for a range of coverages, and concluded that the magnetic moments of single impurities, small clusters and monolayers of Rh and Ru are below the detection limit of 0.04 μ_B/atom. Since Ru and Rh are the most likely candidates for 4d magnetism (see Figure 9.19), these experiments would appear to rule out a non-zero magnetic moment for all 4d nanostructures on most substrates. Experiments then are in conflict with theoretical predictions. Alloying of the 4d atoms with the substrate, lattice relaxation effects or many-body effects such as the Kondo phenomenon or local spin fluctuations have been discussed as possible sources of this discrepancy [128], but as yet the issue has not been resolved.

9.3.2. Ferromagnetic (Fe, Co, Ni) surfaces

The deposition of atoms of the 3d elements onto a ferromagnetic surface provides additional fascination: the magnetic moments of the adatoms may align parallel or antiparallel to the direction of magnetisation of the substrate. With 2D clusters or

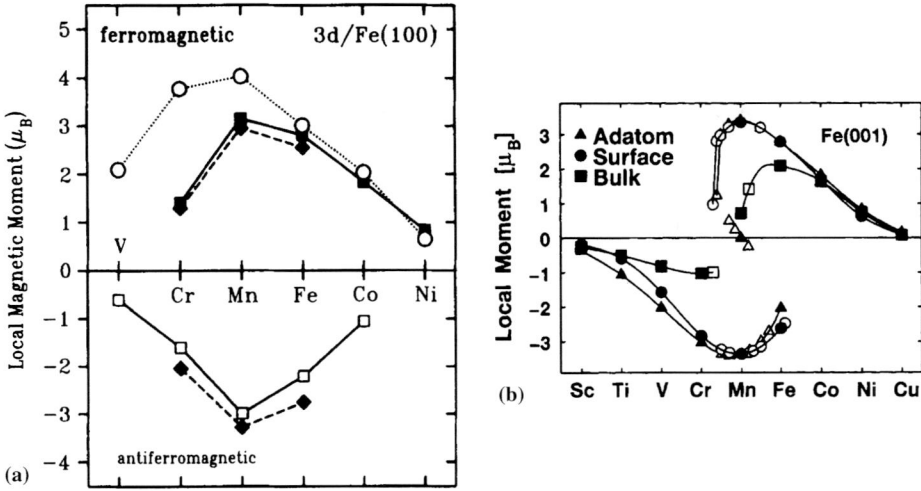

Figure 9.21. Calculated magnetic moments of monolayers and single impurities on or in bcc Fe(0 0 1). (a) Monolayer on surface: p(1 × 1) ferromagnetic coupling (solid squares/solid line), layered antiferromagnetically (open squares), c(2 × 2) ferrimagnetic (diamonds/dashed line), 3d monolayers on Ag(0 0 1) for comparison. (b) Adatoms on surface (triangle), impurity in surface layer (circle), impurity in bulk (square); unfilled symbols represent calculations for non-integral nuclear charges. Reproduced with the permission of Elsevier Science from (a) Handschuh and Blügel [130] and (b) Nonas et al. [129].

monolayers, there is further complexity if there is a tendency for antiferromagnetic ordering within the nanostructure itself. Surface alloying is also a factor that has to be considered.

The spin magnetic moments obtained from calculations on monolayers [179] and adatoms [129] on bcc Fe(0 0 1) are shown in Figure 9.21. Nonas et al. [129] use a KKR Green's function method with exchange and correlation effects included in the local spin density approximation. They compare the behaviour of adatoms on top of the Fe(0 0 1) surface, substitutional atoms in the surface layer and also impurities in the bulk. Atoms are fixed at lattice positions and relaxation is neglected. In Figure 9.21(b), a positive (negative) moment means ferromagnetic (antiferromagnetic) coupling to the host. In some cases, both types of configuration are found. The moments of Fe, Mn and Cr are considerably enhanced compared with the bulk. For the adatom, elements from Mn to Ni prefer ferromagnetic coupling, while those earlier in the series to Cr show antiferromagnetic coupling. A similar situation occurs for substitutional atoms in the surface layer or in bulk; the one exception is Mn which couples antiferromagnetically when in the surface layer.

Broadly similar behaviour occurs with an fcc Ni(0 0 1) substrate, apart from Mn, for which total energy calculations indicate ferromagnetic coupling for all positions [129]. The enhancement of the impurity moments is larger with Ni(0 0 1) than with Fe(0 0 1), because the Ni wavefunction is less extended and so hybridisation is weaker. Dimers on next nearest neighbour sites on top of Fe(0 0 1) were also examined [129]. The ground states of all dimers except Mn couple to the surface with the same configuration as the adatoms, with the moments of the dimer atoms parallel to each other. For the Mn dimer, two states are energetically degenerate: one with the moments on both atoms ferromagnetically coupled to the substrate; the other has the moments of the dimer atoms antiparallel.

Handschuh and Blügel [130] use the local spin density functional approximation and the full-potential linearised augmented-plane-wave method (FLAPW) to study monolayers of the 3d metals on Fe(0 0 1) and allow the interlayer spacing to relax to a minimum energy. They consider two configurations with parallel moment alignment within the monolayer, one coupled ferromagnetically and the other antiferromagnetically to the substrate. A third arrangement has a c(2 × 2) structure in the monolayer with up and down moments of different magnitudes (ferrimagnetic). The results are shown in Figure 9.21(a) for the structurally unrelaxed monolayers. For Cr, Mn, Fe and Co, all three configurations exist and are stable. The lowest energy states display antiferromagnetic coupling for V and Cr, and ferromagnetic coupling for Fe, Co and Ni, although for Cr the ferrimagnetic arrangement is almost degenerate in energy; Mn couples ferrimagnetically. The assignments are not changed with interlayer relaxation.

Surface alloying was discussed in Section 9.3.2 in the context of noble metal substrates. The possibility of similar behaviour has to be considered here. Nonas et al. [131] have examined the exchange of 3d adatoms on the surface of Fe(0 0 1) with ones in the surface layer, processes of the type illustrated in Figure 9.15, and calculated the relative energies by the KKR Green's function method. They find that in all cases except Co, the surface position is preferred to the adatom one, while for Co the difference in energies for the two positions is negligible. They also find a repulsion with pairs of substituted atoms indicating a tendency to surface segregation. The results are consistent with experimental observations of alloying at a Cr–Fe(0 0 1) interface [132,133].

Perhaps the most interesting case is Mn which, as we have seen from calculations, shows c(2 × 2) ferrimagnetic ordering as an overlayer and, for a dimer, one ground state with the moments on the Mn atoms antiparallel. The work on overlayers is most instructive because initial discrepancies between theory and experiment motivated detailed studies that have led to a more complete understanding of the behaviour of deposited Mn. The c(2 × 2) monolayer configuration on a Fe(0 0 1) surface [130] has been corroborated in a number of other calculations both for Fe(0 0 1) [134–136] and for Co(0 0 1) [137] surfaces. It is also established experimentally [138] that Mn on Fe(0 0 1) forms epitaxially in the bcc phase.

There is a small difference between the local moments of the up and down sublattices of the c(2 × 2) configuration of the monolayer, and thus a very small net magnetic moment is predicted. This is in stark contrast to the results of XMCD measurements on Mn monolayers on Co(0 0 1) [139], which indicated that the Mn was in a high spin state and ferromagnetically ordered. A similar conclusion was reached in experiments [140] on Mn layers on Fe(0 0 1).

Some of the discrepancy between theory and experiment has been resolved with evidence that Mn forms an alloy at the Co(0 0 1) surface. M'Passi-Mabiala et al. [141,142] performed calculations using gradient corrected density functional theory and studied several different configurations. They found that the ground state is an interfacial alloy in the surface layer with ferromagnetic coupling both between the Mn and the Co atoms in the surface and with the magnetisation of the underlying

Co(0 0 1) substrate. The calculations were done on an ordered $Co_{0.5}Mn_{0.5}$ surface layer with a c(2 × 2) unit cell. The moments on the Co and Mn atoms are shown in the row labelled 0.5 ML in Table 9.2. They also studied the ground state magnetisation when one and two complete Mn layers were added on top of the alloyed surface layer. The results are shown in Table 9.2 in the rows labelled 1.5 ML and 2.5 ML, respectively. The single added layer orders with a two sub-lattice structure of up and down spins; with two added layers, the top layer orders in the same way, while the middle layer is ferromagnetic but coupled antiferromagnetically to the underlying surface.

The net magnetic moment per atom on the Mn as a function of coverage is shown in Figure 9.22. This is relevant to XMCD experiments, which are element specific and probe the total moment (see Section 9.2.6); some XMCD results are also reproduced. A large ferromagnetic moment occurs at sub-monolayer coverage but drops rapidly with increasing coverage; this is consistent with experimental data

Table 9.2. Summary of results of the calculations of M'Passi-Mabiala et al. [142].

	Surface layer		Layer 1		Layer 2	
	Co	Mn	Mn	Mn	Mn	Mn
0.5 ML	1.73	3.67				
1.5 ML	1.01	2.35	−3.37	3.13		
2.5 ML	1.14	2.48	−1.49	−1.49	−2.94	3.25

Magnetic moments in μ_B of atoms in c(2 × 2) layer unit cells for the surface layer ($Co_{0.5}Mn_{0.5}$ alloy) and two pure Mn overlayers for Mn coverages of 0.5 ML, 1.5 ML and 2.5 ML. The Co substrate is ferromagnetic. A plus (minus) sign indicates a moment parallel (antiparallel) with the substrate magnetisation.

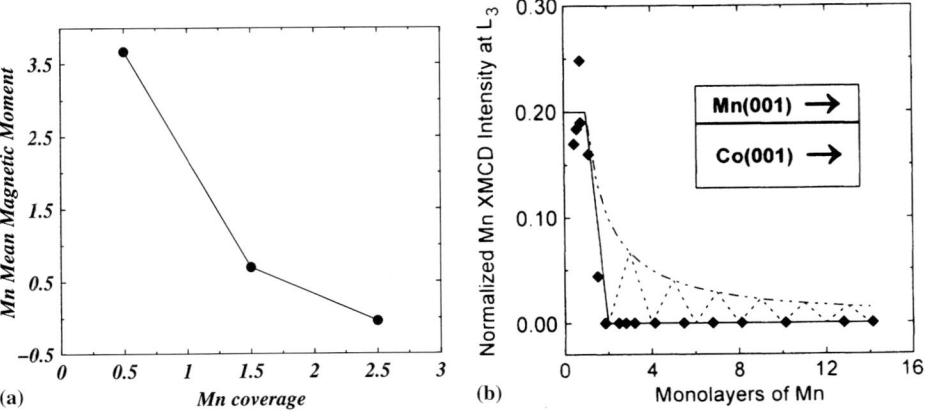

Figure 9.22. (a) Calculated net Mn magnetic moment (μ_B/atom) for three Mn coverages. Reproduced with the permission of Elsevier Science from M'Passi-Mabiala et al. [142]. (b) Normalised Mn $L_{2,3}$ XMCD intensity versus film thickness (data points); the lines are from various theoretical models. Reproduced with the permission of the American Physical Society from O'Brien and Tonner [139].

[139]. Choi et al. [143,144] have confirmed the stabilisation of a two-dimensional magnetic c(2×2) Mn–Co(0 0 1) surface alloy using MOKE and low-energy electron diffraction (LEED) at room temperature, and have shown that the Mn couples ferromagnetically to the Co(0 0 1) substrate. Surface alloying of Mn layers on a Fe(0 0 1) whisker at 370 K has been studied by Yamada et al. [145] using scanning spectroscopy. Intermixing of Fe into the Mn layers was observed at 14, 4 and 2% for the first, second and third Mn layers, respectively.

The analysis of results is further complicated by oxidation, which appears to be an issue in many of the experimental measurements. Pick and Demangeat [146] have shown that the presence of oxygen destabilises the ferromagnetic coupling between Co and Mn, which may explain the experimental results of O'Brien and Tonner [147] and Yonamoto et al. [148]. Oxidation also modifies the coupling in the Mn on Fe(0 0 1) system [149]. An extensive review of the behaviour of Mn nanostructures on a range of substrates is given by Demangeat and Parlebas [150].

The competition between antiferromagnetic exchange coupling within a Cr or Mn cluster and the coupling (whether ferromagnetic or antiferromagnetic) between the atoms of the cluster and the substrate leads to frustration. The system can respond by adopting non-collinear moment configurations to reduce the frustration. Calculations have been performed to explore non-collinear ordering in Cr and Mn dimers and trimers on Ni(0 0 1) [151], Cr and Mn dimers, trimers and tetramers on Ni(1 1 1) [152], and small Cr clusters on Fe(0 0 1) bcc and fcc surfaces [153].

We turn now to one of the relatively few experiments on magnetism in very small deposited metal clusters. Lau et al. [154,155] produce Fe clusters with a sputter source. These are soft-landed on a Ni film, 20–40 monolayers thick, that has been deposited onto a Cu(0 0 1) substrate. The coverage of the Fe clusters on the surface is less than 0.03 monolayers and the experiments are carried out at temperatures of less than 20 K so as to avoid cluster agglomeration. Spin and orbital moments of the Fe_N ($N = 2$–9) clusters were extracted from XMCD measurements by use of the sum rules. Their results are reproduced in Figures 9.23 and 9.24.

The XMCD asymmetries for iron and nickel at their respective L_3 and L_2 edges have the same sign, from which it can be inferred that the Fe clusters couple ferromagnetically to the Ni/Cu(0 0 1) substrate. In order to obtain the spin and orbital moments per atom, one needs to estimate the number of unoccupied $3d$ states per atom, n_h. Lau et al. [155] note that the theoretical values for bulk bcc Fe, fcc Fe and a $NiFe_3$ alloy are in the range 3.39–3.66. The spin moment per $3d$ hole m_S cannot exceed 1 μ_B and so the dipolar term $7m_T$ is contributing significantly to the sum rule. Lau et al. [155] estimate $7m_T$ to be 40–50% of m_S for Fe_2 and 25–35% for the other clusters. Their derived values of the spin and orbital moments per atom are shown in Figure 9.25.

The spin moment and particularly the orbital moment are enhanced compared with bulk bcc Fe (see Table 8.1 in Chapter 8), and there is significant variation across the size range studied. Two theoretical calculations [156,157], employing different methods, have studied the spin moment of $N = 2$–9 Fe clusters on Ni(0 0 1). Both find that all clusters adopt a planar geometry on top of the Ni(0 0 1) surface, but obtain spin moments that vary smoothly from about 3.15 μ_B to 2.85 μ_B in going from the smallest to the largest cluster, without any evidence of the oscillations appearing in the experimental data. The results of Martínez et al. [156]

9.3. Very small supported clusters (N<10) and 2D nanostructures

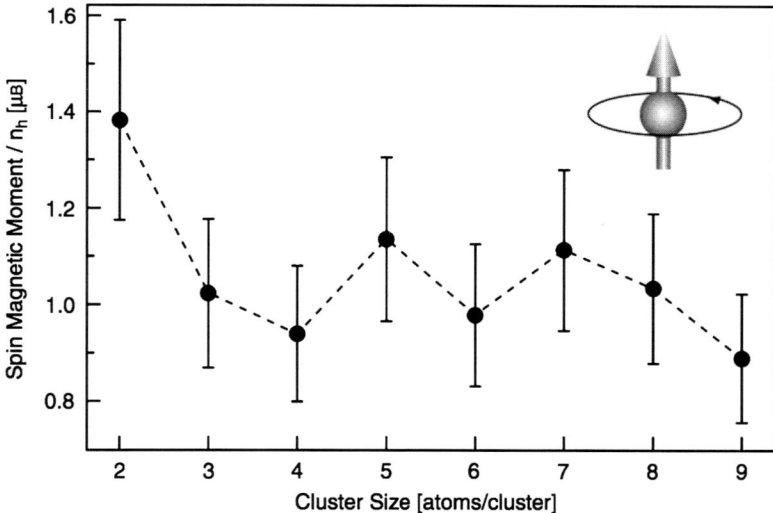

Figure 9.23. Normalised spin magnetic moments $m_S + 7m_T$ per unoccupied $3d$ state for deposited Fe clusters on a Ni/Cu(0 0 1) substrate. Reproduced with the permission of the Institute of Physics from Lau et al. [155].

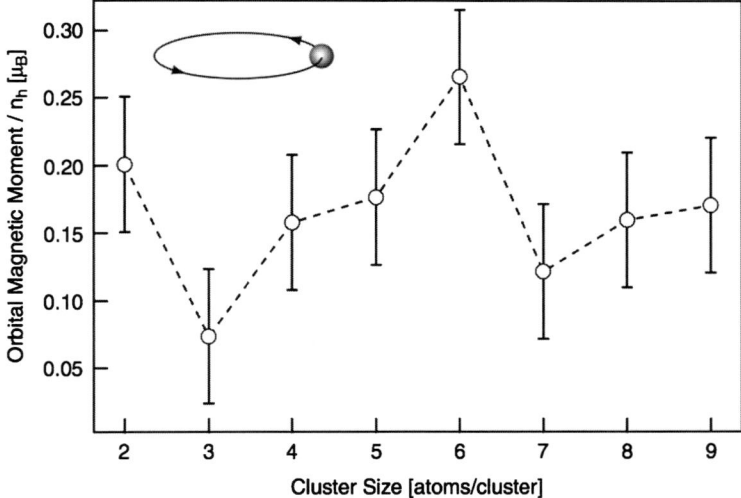

Figure 9.24. Normalised magnetic orbital moments per unoccupied $3d$ state for deposited Fe clusters on a Ni/Cu(0 0 1) substrate. Reproduced with the permission of the Institute of Physics from Lau et al. [155].

are plotted in Figure 9.25. Those of Mavropoulos et al. [157] are very similar. It is possible that some size dependence to the dipolar corrections is required to extract the spin moments from the experimental data and this could account for the discrepancy between theory and experiment. Martínez et al. [156] also report their calculated values of n_h, the number of $3d$ holes; these range from 3.37 to 3.49 over the various clusters.

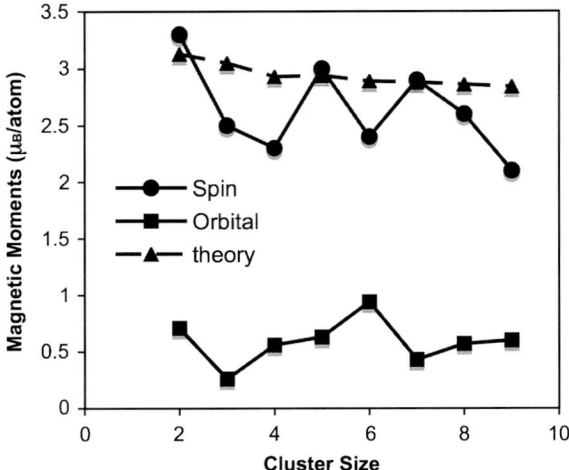

Figure 9.25. Spin and orbital moments (in μ_B/atom) for deposited Fe clusters on a Ni/Cu(0 0 1) substrate as estimated by Lau et al. [155]; the error bars on spin and orbital moments are $\pm 0.6\,\mu_B$/atom and $\pm 0.15\,\mu_B$/atom, respectively. The theoretical spin moments are from a modified embedded atom model calculation of Martínez et al. [156].

Yayon et al. [158] have studied individual magnetic adatoms on a ferromagnetic surface. They used low temperature spin-polarised scanning tunnelling spectroscopy to observe the spin polarisation of Fe and Cr adatoms adsorbed onto Co nanoislands deposited on a Cu(1 1 1) substrate. They find that the Fe adatoms exhibit parallel spin polarisation to the Co surface while the Cr adatoms exhibit antiparallel spin polarisation.

9.3.3. Pd and Pt surfaces

A key property that determines whether a magnetic system will be useful in technological applications is the magnetic anisotropy energy. The sign of the MAE determines the direction of alignment of the magnetisation (the easy axis); in nanostructures, the magnitude of the MAE determines the level of miniaturisation that is possible while still maintaining stability against thermal fluctuations. Of the elemental bulk materials, hcp Co has the largest MAE at about 0.053 meV/atom, an order of magnitude larger than that of bcc Fe at 0.0038 meV/atom (see Table 8.2 in Chapter 8). Various factors are significant in the search for high anisotropy materials: the symmetry of the environment of a magnetic atom and the spin–orbit interaction, which determines the magnetocrystalline anisotropy and whose effect can be augmented with reduced coordination through d-band narrowing.

Co/Pt and Fe/Pt are systems that have come under intense scrutiny because they offer the opportunity of combining a high moment material (the $3d$ atoms) with Pt, which has a very high spin–orbit coupling constant (see Figure 8.2 in Chapter 8). In bulk, CoPt and FePt form layer-ordered L1$_0$ face-centred tetragonal structures with a tendency for the magnetisation to align perpendicular to the layer stacking. MAEs have been reported [1,159,160] as high as 0.8 meV/ formula unit (fu) for CoPt and 1.2 meV/fu for FePt. Apart from Co_5Sm, with an MAE of 1.8 meV/Co atom [1], these are the highest values among bulk hard magnetic materials (see Figure 9 of Ref. [2]). The spin and orbital moments and the MAE of bulk CoPt and FePt (and CoPd and FePd which have a similar structure) have been the subject of numerical theoretical studies [161–168]. Most calculations estimate the MAE to be a factor of 2–3 higher than that observed experimentally. As Burkert et al. [167] note, this discrepancy could be due to a lack of chemical order in the actual samples and/or the decrease in the MAE between absolute zero and room temperature [166,168]. Perez et al. [169] discuss the formation of nanostructures of these materials from nanoclusters preformed in the gas phase, a topic that will be reviewed further in Section 9.4.

The deposition of 3d nanoparticles onto a Pt surface offers considerable scope for tuning the MAE through modifications of the size and shape of the particles. Work on the epitaxial growth of layers of Co or Fe onto a Pt substrate has yielded contradictory results, particularly with regard to the sign of the anisotropy: whether in-plane or out-of-plane magnetisation is preferred. However, studies by Gambardella et al. [8,170] on deposited clusters as small as a single atom perhaps yield a clearer understanding of the behaviour.

Isolated adatoms [8,170] were obtained by depositing minute amounts of Co (less than 0.03 monolayers) onto a Pt(1 1 1) surface at a temperature of 5.5 K to inhibit diffusion. Larger islands were obtained through diffusion-controlled aggregation at various substrate temperatures and deposition fluxes, generating samples with different narrow size distributions so that islands over a size range 1–40 atoms could be studied. XMCD was used to measure the orbital moment; the anisotropy was obtained by measuring the magnetic field dependence of the magnetisation, with the field applied along different directions with respect to the surface normal. The results are reproduced in Figure 9.26.

The orbital moment for a single adatom is 1.1 μ_B, which is large compared with that of the bulk system, and the anisotropy is huge at 9.3 meV/atom. A comparison of the two plots in Figure 9.26 demonstrates the correlation between the orbital

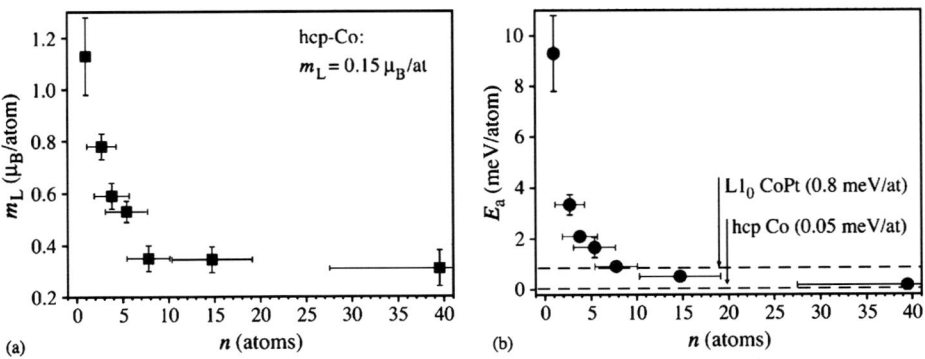

Figure 9.26. (a) Orbital moment m_L and (b) anisotropy energy E_a as a function of average island size n for Co/Pt(1 1 1). The orbital moment for hcp Co, and anisotropy energies of the $L1_0$ CoPt alloy and hcp Co are indicated for comparison. Reproduced with the permission of the AAAS from Gambardella et al. [8].

moment and the anisotropy energy (see Section 8.2.2). The MAE drops quite rapidly as the island size increases falling to the L1$_0$ CoPt value at $n = 8$ and approaching the hcp Co value at the top of the size range. The spin and orbital moments and the MAE of small Co islands on Pt(1 1 1) have been studied theoretically using a number of techniques [170–175]. The dependence of MAE on cluster size is reproduced quite well in the calculations [170,172]. Other combinations studied include Fe on Co [174] and Fe on Pd [176].

The enhanced MAE depends on the 3d–5d hybridisation, which is clearly sensitive to the island morphology, and the achievement of a high value clearly requires a low coordination. Rusponi et al. [9,170,177] have studied larger islands with varying geometries with the purpose of identifying the factors that contribute to the MAE. They studied ensembles of monolayer islands with a mean island size of 1200 ± 1000 atoms, with magnetisation measurements and used MOKE to obtain the susceptibility. Both ramified and compact shapes were examined. It was proposed that the anisotropy, K, of an island could be represented by separate surface and perimeter contributions, $K = sE_{as} + pE_{ap}$, where E_{as} and E_{ap} denote, respectively, the contributions per atom from the surface and perimeter, s the number of surface atoms and p the number of atoms on the perimeter of the island. Using this representation for K in the fit of a theoretical expression for the susceptibility to experimental data, they obtained the following values: $E_{ap} = 0.9 \pm 0.1$ meV/atom and $E_{as} = -0.03 \pm 0.01$ meV/atom. The surface contribution, E_{as}, was further split into a magnetocrystalline contribution, 0.06 ± 0.01 meV/atom, and a shape (dipole–dipole) contribution, -0.090 meV/atom. This implies that the demagnetising energy (shape contribution) dominates over the surface magnetocrystalline anisotropy and, without the anisotropy from the perimeter atoms, the islands would be magnetised in-plane.

Rusponi et al. [9,170] provide further confirmation for their model by producing bimetallic one monolayer high islands with a non-magnetic Pt core and a 2–3 atom wide Co rim. The measured susceptibility agrees well with the theoretical expression using the values of the anisotropy constants noted above obtained from the analysis of the pure Co islands. This emphasises further the importance of low coordinated atoms for obtaining enhanced anisotropy. Calculations [172] show that the MAE of Co islands on Pt(1 1 1) can be enhanced further by a factor of 2 or more with some capping of the islands with Pt or Os atoms. The presence of contamination can reduce the MAE, however, with oxidation being the commonest cause of this effect [9,170]. We have already noted, in the context of Mn on a ferromagnetic surface, that the magnetic behaviour at a surface is very sensitive to the presence of oxygen.

The studies [8,9,170] of small Co islands deposited on Pt(1 1 1) possibly can shed light on conflicting results from experimental and theoretical work on monolayers and multilayers of Co or Fe grown epitaxially on a Pt (or a Pd) surface. There have been a number of experiments [178–186] that have attempted to determine the easy axis of Co and Fe films. Many of these report results show out-of-plane anisotropy (i.e. the magnetisation aligns perpendicular to the surface) at monolayer coverage and a transition to in-plane anisotropy as the film thickness is increased. For a Pt substrate, the reorientation transition has been reported for a variety of film thicknesses ranging from about 4 to 12 monolayers, while for Pd it tends to take place somewhat earlier. This behaviour conflicts with the results of calculations

9.3. Very small supported clusters (N<10) and 2D nanostructures

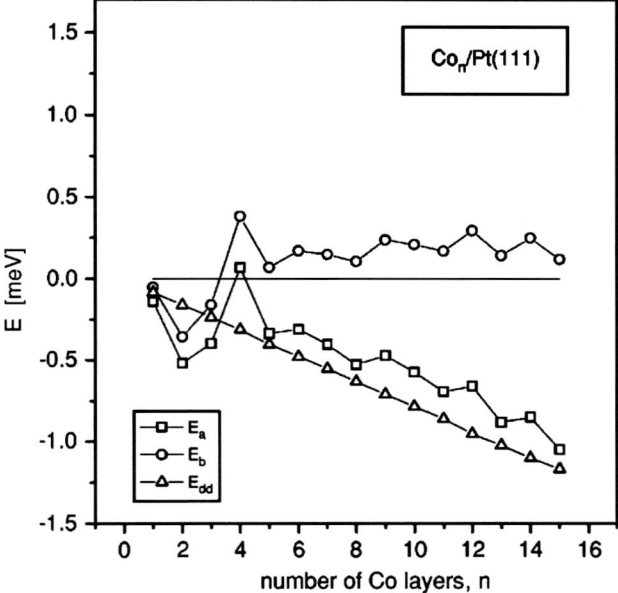

Figure 9.27. Magnetic anisotropy energy for n Co layers on Pt(1 1 1). E_a is the total energy and E_b and E_{dd} are the band and dipole–dipole contributions, respectively. Out-of-plane magnetisation is favoured with $E_a>0$, and in-plane magnetisation with $E_a<0$. Reproduced with the permission of the American Physical Society from Pustogowa et al. [187].

for Co on Pt [187], Fe on Pt [188,189] and Co on Pd [190,191]. In each case, in-plane anisotropy is generally found even at a film thickness of a single monolayer. Figure 9.27 shows results from a relativistic spin-polarised KKR calculation of Co monolayers deposited onto a Pt(1 1 1) surface. The anisotropy is plotted against the number of Co layers, and the band (magnetocrystalline) and dipole–dipole contributions to the total anisotropy energy are shown ($E_a = E_b+E_{dd}$).

The dipole–dipole interaction always favours in-plane magnetisation while the band contribution favours out-of-plane (in-plane) anisotropy for $n \geq 4$ ($n \leq 3$). The prediction from the total anisotropy in Figure 9.27 is for in-plane magnetisation except at $n = 4$. For the Pt(0 0 1) surface, perpendicular magnetic anisotropy is predicted for $n = 2$, 4 and 6 [187], but otherwise there is in-plane orientation. Gambardella et al. [170] suggest that the resolution of this conflict lies in the details of the interface between the substrate and the deposited layers. The theoretical calculations are done for a perfect monolayer with an abrupt interface with the substrate. There is strong evidence from the experiments, however, of surface roughness, interfacial alloying or the occurrence of several simultaneous growth modes. Any of these will increase the number of low coordinated sites or the number of 3d–5d bonds and the consequent tendency for perpendicular magnetocrystalline anisotropy, which is necessary to overcome the dipole–dipole anisotropy that favours in-plane magnetisation. Nakajima et al. [180], for example, observe a structural transition from fcc to hcp in Co/Pt, while Ferrer et al. [179] report imperfect layer-by-layer growth, Lundgren et al. [192] observe 3D growth of

Co at room temperature and Robach et al. [181] note Co–Pt exchange at the interface under certain temperature conditions. The tendency to perpendicular anisotropy can also be increased by capping the $3d$ islands or layers with $5d$ atoms to create a second interface [172,187,190,193–196]. Finally it should be noted that, although it is appealing to separate the anisotropy of an island into a weakly in-plane contribution from the surface and a strongly out-of-plane contribution from the perimeter [9], the model has not gone unchallenged. Meier et al. [197] have performed spin-resolved scanning tunnelling measurements on Co nanostructures on Pt(1 1 1) at low temperatures, and found no evidence of any departure from perpendicular anisotropy as the island size is increased.

9.3.4. Graphite surfaces

Amorphous carbon and HOPG are widely used as substrates for the study of supported metal clusters. Often use is being made of the relatively weak bonding at the surface, which results in high adatom mobility. In addition, HOPG can serve as an ideal atomically flat conducting surface for use in scanning probe investigations. A top view of an ideal graphite surface is shown in Figure 9.28. Possible adatom positions are above the α- and β-sites (defined in the figure caption). Other high

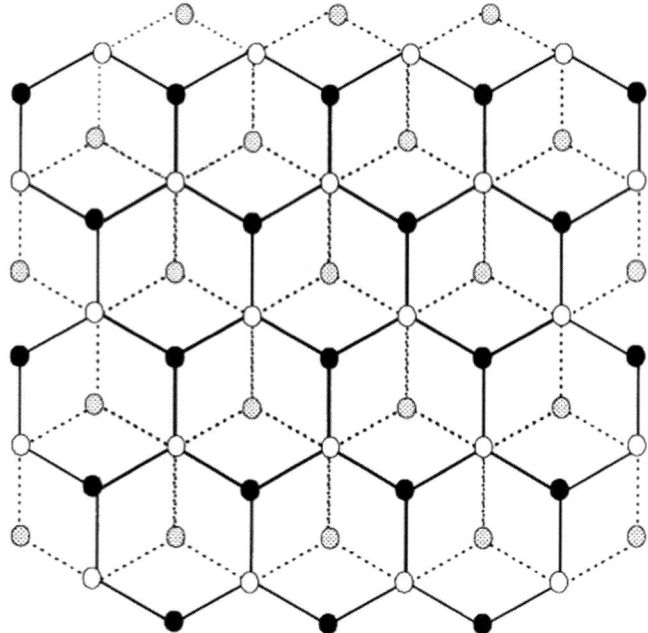

Figure 9.28. The top view of a graphite surface. The solid lines join sites in the top plane and the broken lines join sites in the second plane. Subsequent planes repeat in an ABABAB⋯ pattern. White circles represent α-sites (sites in the top layer with a C atom in the plane below) and black circles represent β-sites (sites in the top layer without a C atom in the plane below). Grey circles are sites in the second plane.

symmetry sites that are potential locations for an adatom are above the centre of a hexagonal ring (hole sites) and above a bond linking neighbouring surface atoms.

The growth, electronic, magnetic and spectroscopic properties of transition metals on graphite have been reviewed by Binns et al. [198], but we begin with a discussion of the noble metals. Ganz et al. [199,200] have studied two-dimensional islands of Cu, Ag, Au and Al and found them weakly bonded and incommensurate with the graphite substrate, while Francis et al. [201,202] have investigated the growth mechanisms of Ag clusters on terraces and steps. Some of these growth mechanisms have been modelled [203–205] using methodology of the sort described in Chapter 4.

The preferred position of an adatom on the surface appears to depend on the material. A Car–Parinello *ab initio* molecular dynamics calculation [206] on Al found an over-bond site to be marginally favoured over the α- and β-sites, for which the energies are degenerate, while the hole site is least favoured. Duffy and Blackman [207] found a complete contrast in behaviour in an *ab initio* study on Ag adatoms and dimers. In this case, an adatom position over the β-site was preferred, with the hole, α-site and bond positions in decreasing order of binding energy. The C atom at a β-site is not bonded to an atom in the plane below and therefore the π-orbital is free to hybridise with the Ag atom. Conversely, the C atom at an α-site is weakly bound to the C atom in the plane below and it binds less strongly with the Ag atom. In fact, the binding energy at the β-site is very close to the energy obtained if just a single graphite layer is used in the calculation. The preference for β-site adsorption accords with the STM measurements of Ganz et al. [200] on Ag adatoms. Dimers were studied [207] at various positions on the surface; the variation in binding energy was negligible and the Ag–Ag separation was essentially that of the free dimer. This suggests that dimers will be very mobile on a graphite surface and that clusters will tend to adopt a geometry similar to that of a free cluster. This is consistent with the observation in STM experiments of islands incommensurate with the underlying substrate. A preference for β-sites has also been found by Wang et al. [208] in density functional calculations on Au_n ($n = 3$–5) clusters on graphite; in addition, they obtained both parallel and perpendicular orientations of the clusters with respect to the surface.

The adsorption of $3d$ transition metal adatoms and dimers on a graphite surface has been studied by Duffy and Blackman [209]; Krüger et al. [210] have investigated monolayers. A section of a single layer of graphite was used in the calculation on adatoms and dimers [209]. As noted in the calculation on Ag [207], the results were essentially the same whether one or two graphite layers were used (a similar observation was made in the Al work [206]). The geometry of the graphite was fixed and the atoms and dimers were allowed to find their lowest energy positions. The monolayer calculation [210] employed the TB-LMTO-ASA method and was based on a graphite trilayer. To obtain accurate band structures, it was necessary to add empty spheres because of the open structure of graphite. The monolayer calculation positioned the adatoms 1.9 Å above the centre of the graphite hexagon (hole position). This was done following a non-spin-polarised full-potential LMTO calculation [211] that showed this arrangement to be a stable one for a Fe monolayer.

The spin magnetic moments from the calculations [209,210] are displayed in Figure 9.29. Interestingly the adatoms behave rather differently in the upper and lower parts of the $3d$ series [209]. The metals Sc–Mn prefer positions about 2.1 Å

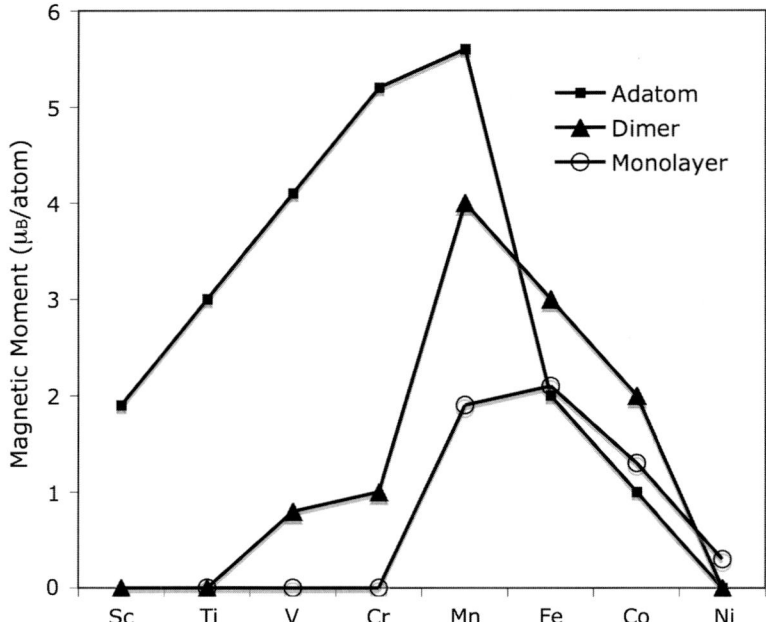

Figure 9.29. Composite showing the results of calculations of the spin moments of the $3d$ transition metal adatoms, dimers by Duffy and Blackman [209] and monolayers by Krüger et al. [210] on graphite. The Mn monolayer has antiferromagnetic and ferrimagnetic ground states that are degenerate [210]; the average magnitudes of the sub-lattice moments are plotted.

above a C atom (presumably above a β-site had additional graphite layers been used in calculations), whereas Fe, Co and Ni are in a hole position about 1.5 Å above the surface. Minimum energy positions for the dimers are with their centres either above C–C bonds (V, Cr, Fe and Co) or over holes (Sc, Ti, Mn and Ni). The Fe dimers lie 1.92 Å above the surface, which is close to that predicted by the Peng et al. calculation [211], although the interatomic separation of the dimer atoms at 2.09 Å is somewhat less than the distance between the centres of neighbouring graphite hexagons (2.46 Å).

The moments of the Sc–V adatoms are large (atomic like), but fall significantly for the dimers and collapse completely for the monolayers. Fe and Co as dimers and monolayers and Ni as a monolayer order ferromagnetically. The rather low moments of the Fe and Co adatoms are due to the hybridisation with the C π-orbital, an effect that is reduced on dimer formation. Although the ground states of Cr_2 and Fe_2 as free dimers were antiferromagnetic, only ferromagnetic configurations were found [209] for the deposited supported dimers. The monolayers of Cr and Mn experience topological frustration because the atoms lie on a triangular lattice. Cr responds to this with a non-magnetic ground state, while for Mn there are two degenerate ground states with arrangements of up and down spins (one, for example, orders in ferromagnetic rows, with neighbouring rows coupling antiferromagnetically) [210]. Non-collinearity is likely to be relevant for Cr and Mn, as we have noted earlier in the context of a noble metal substrate (see Figure 9.18).

There is rather little experimental work on very small 3d metal particles on graphite. However, results for Fe_N particles in the range $7 < N < 100$ from Fauth et al. [212] are intriguing and serve to highlight a strong influence of the graphite substrate on the magnetic properties. They cover a HOPG substrate with an argon film and then deposit the Fe clusters at a low coverage on the argon and a temperature of about 14 K to prevent aggregation. Oxygen contamination is carefully avoided. XMCD measurements are then performed. The argon cover effectively shields the Fe particles from the substrate. The sample is then warmed for a short period to remove the argon and the particles come into contact with the graphite surface. The XMCD measurements are then repeated. The results are shown in Figure 9.30.

Figure 9.30(a) shows the spectra for the particles shielded from the substrate by the argon film. The spectra are taken in the presence of a magnetic field oriented parallel or antiparallel to the direction of light propagation, and strong magnetic contrast is observed. The authors claim that a sum rule analysis yields moments in rather good agreement with those obtained for free particles by gradient-field deflection techniques (as discussed in Section 8.4). By contrast, we see a complete absence of magnetic contrast when the same clusters are in contact with the graphite surface. The implication is that the Fe clusters are driven into a non-magnetic state by the interaction with the graphite. It is interesting to note that Edmonds et al. [53] used a similar technique to study 300-atom Fe particles deposited on graphite, without any evidence of quenching the Fe moment at this larger size. In fact, they observed a strongly enhanced orbital moment, although the overall enhancement was much less than that found in free clusters. Fauth et al. [213,214] have also found a strong influence on the electronic structure of Pt particles when they are deposited on graphite. A full understanding of these observations has yet to be reached.

Finally we note the lively interest in the magnetic properties of carbon nanotubes doped with 3d transition metals [215–222], which obviously relates to the above discussion.

9.3.5. *The Kondo effect*

The ability to image single impurity atoms on a surface by STM has led to a revival of interest in the Kondo effect [223], a phenomenon discovered in the 1930s and first explained in the 1960s. In experiments on gold (probably containing a small amount of iron impurities), de Haas and van den Berg [224] in 1936 observed that the electric resistivity dropped as the temperature decreased until about 8 K, and below that it increased. With normal metals (i.e. non-superconductors), one expects the resistance to fall with temperature as the amplitude of atomic vibrations decreases and, below about 10 K, to remain constant; this normal low temperature residual resistivity is due to scattering by defects in the material.

Since the 1930s, there have been many observations of the anomalous resistance minimum, yet it took until 1964 for a satisfactory explanation to appear [225]. Kondo [225] used perturbation theory to calculate the scattering of the conduction electrons of a noble metal by magnetic defects, using a Heisenberg model to describe the interaction between the conduction electron spins and the spin

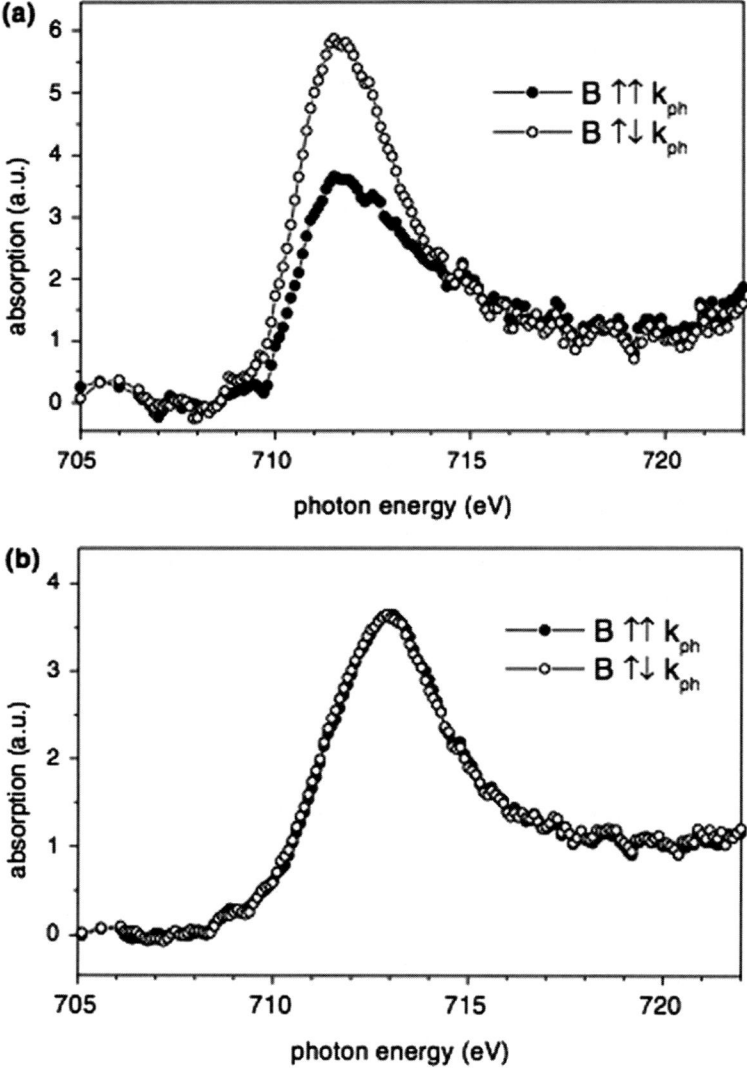

Figure 9.30. (a) L_3 absorption spectra from Ar matrix isolated Fe clusters, the applied field being oriented parallel (filled circles) and antiparallel (empty circles) with respect to the light propagation direction; (b) same spectra as before, but after removal of the Ar layer. Reproduced with the permission of Elsevier Science from Fauth et al. [212].

moments of the defects. To first order nothing dramatic happens, but Kondo took the perturbation theory to second order when spin flip processes can occur, and found that these resulted in a resistance that increased logarithmically as the temperature is lowered. Kondo's result is valid above a temperature T_K, which has become known as the Kondo temperature, but for $T < T_K$ perturbation theory breaks down. The following year, Nagaoka [226] published a self-consistent treatment; this gave solutions applicable both above and below the Kondo temperature, and reproduced the resistance minimum.

However, it took a number of years for a proper understanding of the Kondo problem to emerge and much of the analysis was based on a highly simplified model due to Anderson [227] for the Hamiltonian of a single magnetic impurity in a Fermi sea of electrons

$$H = \sum_{\vec{k}\sigma} \varepsilon_k c^\dagger_{\vec{k}\sigma} c_{\vec{k}\sigma} + \varepsilon_d \sum_\sigma d^\dagger_\sigma d_\sigma + \sum_{\vec{k}\sigma} (V_{dk} d^\dagger_\sigma c_{\vec{k}\sigma} + V^*_{dk} c^\dagger_{\vec{k}\sigma} d_\sigma) + U d^\dagger_\uparrow d_\uparrow d^\dagger_\downarrow d_\downarrow.$$

(9.39)

The first term is just the kinetic energy of the electrons, with $c^\dagger_{\vec{k}\sigma}$ and $c_{\vec{k}\sigma}$ the creation and annihilation operators for electrons with spin σ in state $\vec{k}$. The magnetic impurity is represented by a single state with energy ε_d and electron operators $d^\dagger_\sigma$ and d_σ. The third term represents the hybridisation of the d-level and the conduction band (the s–d interaction), and the fourth term takes account of the cost, U, in Coulomb energy of double occupancy of the d-level. The model assumes U is large and $\varepsilon_d < \varepsilon_F$, where ε_F is the Fermi energy, so in the absence of the hybridisation term we have a singly occupied (magnetic) d-state. Obviously, with a single d-level, the model is not attempting to describe the details of the atomic 3d orbitals, but it does capture the essential physics of the Kondo problem.

The Hamiltonian allows for the so-called spin exchange processes. Via the V term, an electron can tunnel from the localised impurity state to an unoccupied state just above the Fermi energy and then another from the Fermi sea replaces it. If the electrons departing and arriving at the impurity site have opposite spins, we have a spin flip process. An alternative spin flip process involves double occupancy of the impurity state, but that is with a cost in Coulomb energy. When many such processes are taken together, a new many-body state known as the Kondo resonance is generated. The Kondo resonance, which is a spin singlet state, has width kT_K and is pinned at the Fermi level as illustrated in Figure 9.31. The impurity state at ε_d is shifted and broadened through hybridisation with electron states of the Fermi sea (Γ is derived from the Anderson model); we will call that a d-resonance. We can picture the net effect as an alignment of the spins of the conduction electrons near to the impurity atom so as to screen the spin on the local moment. A clearer understanding of the physics below T_K began to emerge in the late 1960s from Anderson's derivation of scaling laws for the Kondo problem [228–232]. Using scaling ideas, the Kondo temperature can be expressed [233] in terms of the parameters of the Anderson model by $kT_K = \frac{1}{2}(\Gamma U)^{1/2} \exp[\pi \varepsilon_d(\varepsilon_d + U)/\Gamma U]$. Due to the exponential dependence on the parameters, the Kondo temperature can vary, in practice, from 1 K to 100 K.

The Kondo resonance forms at temperatures below the Kondo temperature. Because the resonance is at the Fermi level, it is very effective at scattering electrons with energies close to ε_F. It is the increased scattering as the Kondo resonance appears below T_K that is responsible for the resistance minimum. Remarkably, the resistance R at temperature T can be expressed as a function of the ratio of T to the Kondo temperature, i.e. $R = R_0 f(T/T_K)$, where R_0 is the resistance at absolute zero. This is a further consequence of the scaling properties.

Previously the role of the Kondo effect could be inferred only through measurements of electric resistivity or magnetic susceptibility. It would be nice to

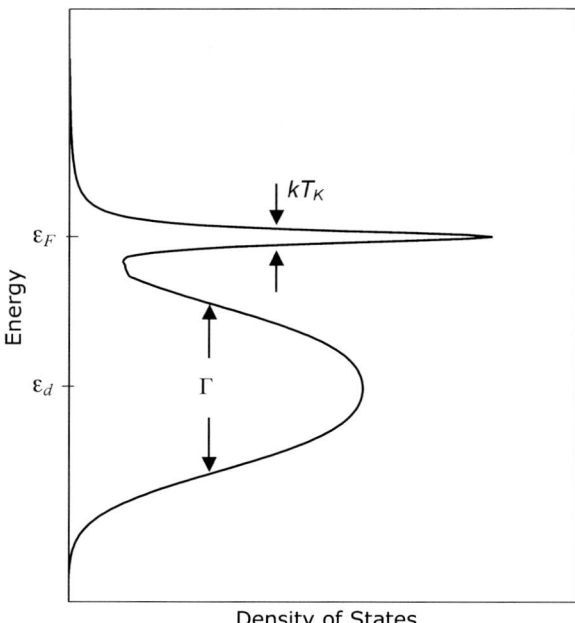

Figure 9.31. The Kondo resonance of width kT_K at the Fermi level and the impurity state broadened to width Γ by hybridisation with the conduction electrons.

have a more direct method of detecting the resonance. In principle, photoemission should be able to detect the enhanced density of states near the Fermi surface, but unfortunately the resolution of the method is not good enough to resolve the feature. However, the advent of the STM has opened the possibility of imaging magnetic atoms on the surface of a metal and measuring the Kondo resonance from the current versus voltage characteristics. The first measurements of this kind were performed in 1998 on cobalt atoms on an Au(1 1 1) surface [234] and cerium atoms on Ag(1 1 1) [235].

STM experiments measure the bias voltage dependence of the differential conductance dI/dV. Electrons tunnel between states at ε_F on the tip and states at $\varepsilon_F + eV$ on the surface and so structure in the measured differential conductance reflects the surface density of states at that energy. Both Madhavan et al. [234] and Li et al. [235] note that the STM is probing a localised state immersed in a continuum (the Fermi sea) and this bears a resemblance to the spectroscopy of a discrete autoionised state, which was discussed by Fano [236]. The Kondo resonance is analysed as a type of Fano resonance.

Fano's original analysis [236] concerned a discrete state and a continuum, and spin was ignored. The Hamiltonian describing his system can be written

$$H = \sum_{\vec{k}\sigma} \varepsilon_k c^\dagger_{\vec{k}} c_{\vec{k}} + \varepsilon_a a^\dagger a + \sum_{\vec{k}} (V_{ak} a^\dagger c_{\vec{k}} + V^*_{ak} c^\dagger_{\vec{k}} a), \qquad (9.40)$$

where $a^\dagger$, a and ε_0 are the electron operators and energy of the discrete state. It is essentially the Anderson Hamiltonian of Eq. (9.39) without the Coulomb term.

The discrete level will be broadened like the d-resonance above. Before discussing the Kondo problem, it is useful to review the analysis of Fano lineshapes based on Eq. (9.40) in the context of STM [237].

Matrix elements M_{at} and $M_{\vec{k}t}$ are introduced; the first describes the tunnelling between the tip and the discrete atomic state, while the latter describes tunnelling between the tip and the continuum states. The differential conductance can be written as [237]

$$\frac{dI}{dV} = \frac{2\pi e^2}{\hbar} \frac{(\varepsilon' + q)^2}{1 + \varepsilon'^2} \rho_{tip} \sum_{\vec{k}} |M_{\vec{k}t}|^2 \delta(\varepsilon - \varepsilon_k) + \text{constant}, \tag{9.41}$$

where ρ_{tip} is the tip density of states. The factor involving q determines the shape of the signal; q itself is the ratio of two terms, $q = A/B$, where

$$A = M_{at} + \sum_{\vec{k}} M_{\vec{k}t} V_{a\vec{k}} P\left(\frac{1}{\varepsilon - \varepsilon_k}\right), \tag{9.42}$$

$$B = \pi \sum_{\vec{k}} M_{\vec{k}t} V_{a\vec{k}} \delta(\varepsilon - \varepsilon_k), \tag{9.43}$$

and the quantity ε' is defined as

$$\varepsilon' = \frac{\varepsilon - \varepsilon_0 - \text{Re}\Sigma(\varepsilon)}{\text{Im}\Sigma(\varepsilon)}. \tag{9.44}$$

Here $\Sigma(\varepsilon)$ is the self-energy of the discrete state Green's function

$$G_{aa}(\varepsilon) = \frac{1}{\varepsilon - \varepsilon_0 - \Sigma(\varepsilon)}, \tag{9.45}$$

where the real and imaginary parts of $\Sigma(\varepsilon)$ are given by

$$\text{Re}\Sigma(\varepsilon) = \sum_{\vec{k}} |V_{a\vec{k}}|^2 P\left(\frac{1}{\varepsilon - \varepsilon_k}\right), \tag{9.46}$$

$$\text{Im}\Sigma(\varepsilon) = \pi \sum_{\vec{k}} |V_{a\vec{k}}|^2 \delta(\varepsilon - \varepsilon_k). \tag{9.47}$$

P denotes the principal part in the summations in Eqs. (9.42) and (9.46).

The real and imaginary parts of the self-energy give the shift and half line-width ($\Gamma/2$) of the resonant state formed from the discrete energy level through hybridisation with the continuum states. The quantities A and B in Eqs. (9.42) and (9.43) have the following interpretation. A is an amplitude for tunnelling to the discrete state modified (second term on right-hand side of Eq. (9.42)) by hybridisation with the continuum. B is an amplitude for tunnelling into a set of continuum states contained in a range of energies of the order of the width of the resonance. If $q \gg 1$ tunnelling via the resonance dominates, but if $q \ll 1$ it is the continuum states that dominate the tunnelling. In the context of STM, the energy ε in the above equations is set equal to eV.

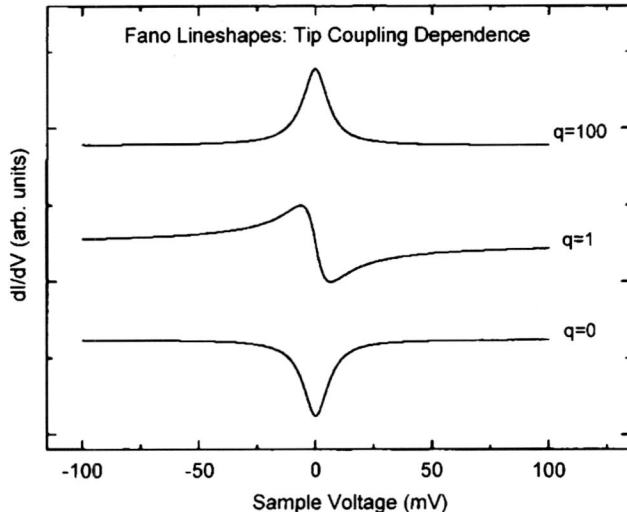

Figure 9.32. Theoretical Fano resonance lineshapes for three values of q using the model of Eq. (9.40). Reproduced with the permission of the American Physical Society from Madhavan et al. [237].

Madahavan et al. [237] performed a model calculation to illustrate the behaviour of the differential conductance if a resonance occurred at the Fermi energy. Their results are shown in Figure 9.32 and one can see that the shape of the signal depends strongly on the value of q. If $q \gg 1$, the tunnelling is mostly into the localised state and a Lorentzian lineshape is observed, while if $q \to 1$ an antiresonant dip occurs. This is due to a depletion in the continuum density of states around the energy of the resonance, which arises because the continuum states are, so to speak, 'pushed away' to make room for the new state when the defect is introduced into the system. At $q = 1$, we get interference between the two effects.

Madhavan et al. [234,237] and Schiller and Hershfield [238] have discussed the adaptation of the Fano analysis outlined above for the Kondo problem, and Újsághy et al. [239], Lin et al. [240] and Plihal and Gadzuk [241] have performed calculations with a more realistic treatment of a $3d$ metal than that provided by the Anderson model. However, a Kondo resonance is expected to manifest itself in a signal like one of the generic forms shown in Figure 9.32 at $V = 0$.

Experimental results for Co atoms on an Au(1 1 1) substrate are reproduced in Figure 9.33. It shows how the dI/dV spectra vary as the STM tip is moved with a clear feature of the intermediate q type appearing when the tip is in the vicinity of the Co atom indicating the Kondo resonance. Since the first experiments, the Kondo resonance has been observed in number of $3d$ atomic species on various surfaces [242–246,19], and dimers [247,248] and three atom clusters [249–251] have also been studied. Co dimers on Au(1 1 1) show an abrupt disappearance of the Kondo resonance for Co–Co less than 6 Å [247].

A very pretty experiment, dubbed a quantum mirage [252–254], has been reported from IBM's Almaden Research Center. Manoharan et al. [252] built an ellipse of Co atoms on a Cu(1 1 1) surface. A Co atom was then placed at one focus of the ellipse and the dI/dV spectrum measured. A Fano lineshape (low q type) at

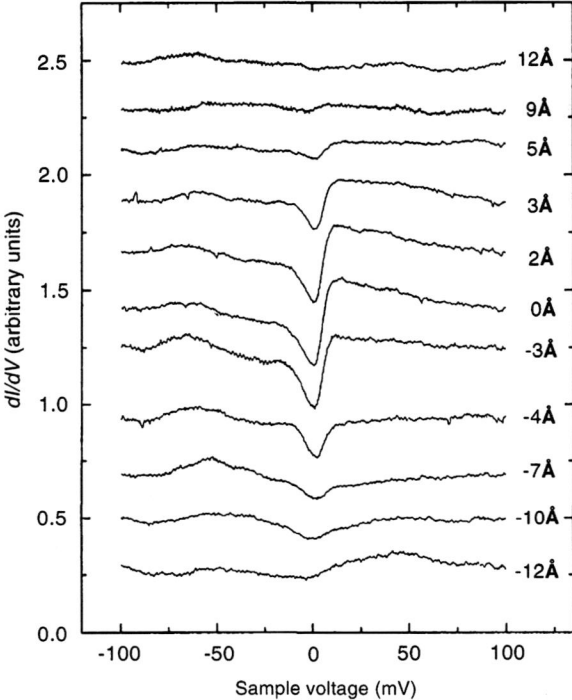

Figure 9.33. A series of dI/dV spectra taken with the STM tip held at various lateral distances away from the centre of a single Co atom on Au(1 1 1). Reproduced with the permission of the AAAS from Madhavan et al. [234].

zero bias voltage signalled the presence of a Kondo resonance. The term 'quantum mirage' arises because a similar measurement made at the other focus of the ellipse yields a signal with a spectrum of similar character even though there is no Co atom at that position. The ellipse of atoms is a quantum corral and this acts as a focusing device or quantum mechanical resonator. Electrons in surface states are scattered by the atom at the first focus of the ellipse and are refocused creating a mirror image of the Kondo resonance at the empty position.

9.3.6. *One-dimensional metal chains*

We have seen that reducing the dimensionality of a physical system from the bulk can lead to an enhancement in the spin and orbital moments and, importantly for stability against fluctuations, an enhancement to the magnetic anisotropy energy as well. What are the implications for one-dimensional chains of 3d metal atoms? It is well known from statistical mechanics that long-range magnetic order does not occur in simple models (Ising or Heisenberg) in one-dimension because of fluctuations, but it was not clear what to expect if a chain is on a substrate in a high anisotropy environment. There is a sizeable body of theoretical work [255–272] that has extended studies of islands and monolayers to 1D chains. The self-organisation of atoms diffusing on terraces on vicinal surfaces results in the formation of chains at the steps and, since 2000, a number of measurements on the magnetism of these systems have been reported.

A crystal surface cut at a small angle to a symmetry direction is called a vicinal surface. If the Miller indices of the plane of the cut are $(h\,h\,l)$, the resulting surface for an fcc crystal comprises (1 1 1) terraces separated by evenly spaced parallel steps. The number, n, of rows of atoms on a terrace parallel to a step is given by $n = (h+l)/(h-l)$. A side view of a (9 9 7) crystal surface is shown in Figure 9.34.

Gambardella et al. [273–275] have used a Pt(9 9 7) surface to deposit Co atoms and grow ordered arrays of Co nanowires. Their stepped surface had a high degree of uniformity with terraces of average width 20.1 Å and standard deviation 2.9 Å. Diffusion and aggregation processes and island formation have been discussed in Chapter 4. In this case, the essential feature is preferential nucleation at the steps due to the increased coordination with respect to the terrace sites. Temperatures higher than 250 K are required. The mean free path of the Co adatoms is larger than the terrace width and, at lower temperatures, wire formation is kinetically hindered by slow edge- and corner-diffusion processes. In this way, virtually ideal row-by-row growth can be achieved with the formation of an array of 1D Co chains. If the growth process is allowed to continue, additional chains parallel to the first can be formed with eventually complete coverage of the terraces.

In later work, Gambardella et al. [276–278] explored the magnetic properties of the nanowires using photoemission and XMCD experiments. To determine the easy axis and the magnetic anisotropy energy, the XMCD was measured along different crystal orientations. Several different coverages were studied, providing data for both single chains and for multiple chains adjacent to each other. The results are reproduced in Figure 9.35. The rows a, b and c are for 1, 2 and 4 chains, respectively, while d shows the behaviour for 1.3 ML coverage.

The maximum in the magnetisation determines the easy axis. An abrupt change occurs from $+46°$ (step up direction) for a single wire to $-60°$ (step down) for two wires. The easy axis remains step down as more wires are grown parallel to the first two and tends towards perpendicular as the terrace coverage increases. The temperature dependence of the magnetisation was also measured. The results for a single chain array are shown in Figure 9.36. Superparamagnetic behaviour is observed at $T = 45$ K, but as temperature is decreased hysteretic behaviour sets in, with an estimated blocking temperature $T_B = 15 \pm 5$ K. The XMCD sum rules were

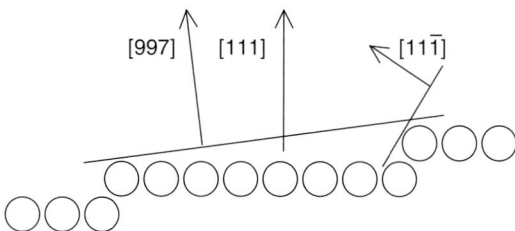

Figure 9.34. Side view of the top layers of a (9 9 7) crystal surface. The normals to the planes of the cut, terrace and step are shown. The step edges lie along the [1 1 0] direction into the page, and a terrace comprises eight rows of atoms.

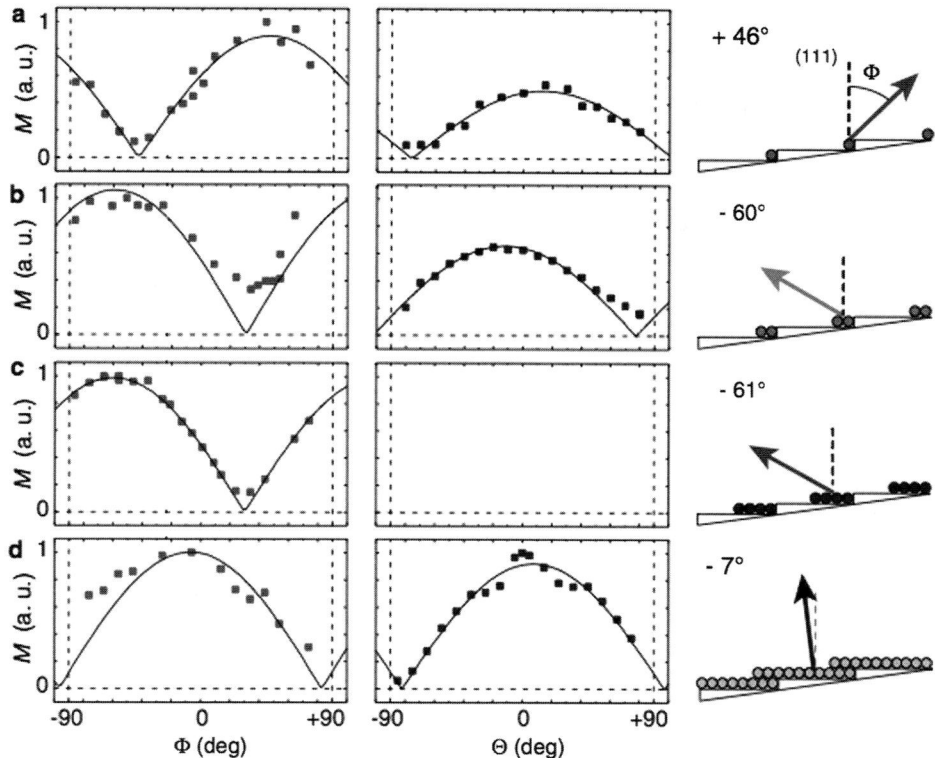

Figure 9.35. XMCD data for Co on Pt(9 9 7). The data points represent the XMCD signal at the Co L_3 edge normalised by the total absorption yield. The first three rows are for (a) one-wire, (b) two-wires and (c) four-wires, all at $T = 10$ K; (d) is for 1.3 ML coverage and $T = 45$ K. The magnetic fields used were sufficient for 50% saturation or above. The first and second columns show the magnetisation at different angles and the third column indicates the direction of the easy axis. Φ is an angle with respect to the [1 1 1] direction for a rotation axis along the axis of the chain. Θ is an angle with respect to the [1 1 1] direction for a rotation axis along the terrace and perpendicular to the chain. Reproduced with the permission of the American Physical Society from Gambardella et al. [278].

used to obtain values of the orbital momentum. High values (in excess of $0.6\,\mu_B$/atom) were found for the single chains, dropping to between $0.3\,\mu_B$/atom and $0.4\,\mu_B$/atom as the coverage was increased.

The observation of hysteresis below 15 K raises the question of long-range ferromagnetic order in one dimension. The Heisenberg model is particularly susceptible to fluctuations and even the Ising model exhibits long-range order only at absolute zero. With high uniaxial anisotropy, it could be argued that these systems are Ising like in the sense that spin alignment is constrained to lie along the easy axis and can take just an up or down orientation. In that case, one can use an argument due to Landau and Lifshitz [279] to suggest that ferromagnetic correlations do exist but over a limited length scale. Gambardella and Kern [280] estimate that this could be as high as 50 atoms.

Theoretical calculations of the magnetic anisotropy have shown the easy axis canting away from perpendicular as is seen in the experiments. Earlier

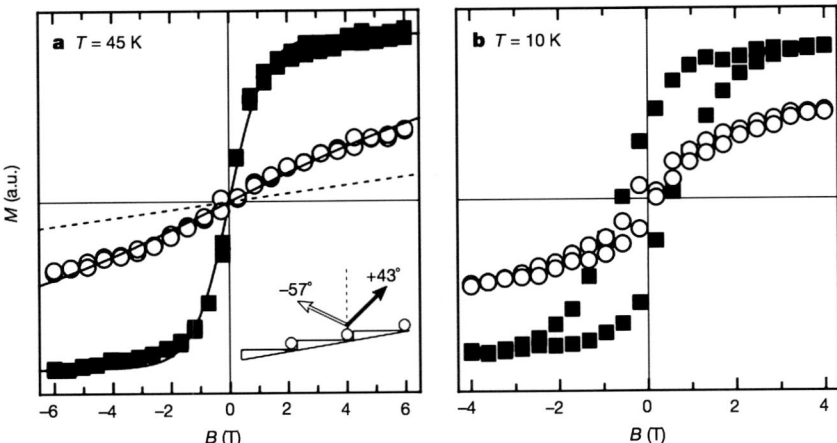

Figure 9.36. Magnetisation of array of single chains in easy axis direction (filled squares) and away from easy axis (open circles). Hysteretic behaviour sets in at $T = 10$ K (b), but is absent at $T = 45$ K. The dashed curve in (a) represents the magnetisation expected for an isolated Co atom on Pt(997). Reproduced with the permission of Macmillan Publishers Ltd. (*Nature*) from Gambardella et al. [276].

calculations [255,258] reported similar behaviour for Co wires on a Pd surface, and more recent work on Co wires on Pt is in general accord with Lazarovits et al. [266] predicting an angle very close to that observed. There have been experiments [279] on Ni chains on Pt(997), but alloying seems to be a major factor in this case.

9.4. Supported clusters

In the previous section, we considered small islands formed on a surface by atom diffusion and aggregation. The clusters were generally two-dimensional and various models for the island–surface interface were explored. An alternative experimental procedure is the deposition of pre-formed clusters. Clearly, in this case, the nature of the interface is much less well defined and will depend, for example, on the shape of the particle and on whether the deposited clusters undergo a soft or hard landing; for the latter, burrowing into the surface is likely to be important. In the experimental work that follows, the clusters are on the surface. As far as magnetism is concerned, the size range up to about 500 atoms is of most interest since that is where deviations from bulk behaviour are most apparent in free clusters (see Section 8.4).

Figure 9.37 shows the spin and orbital moments in mass-selected Fe clusters for sizes below 700 atoms deposited *in situ* onto HOPG substrates held at 6 K to a coverage of 0.03 cluster monolayers [63]. The spin m_S+7m_T values are shown for data taken at normal incidence ($\theta = 0°$) and with the sample rotated to the 'magic angle' of 55° incidence. As mentioned in Section 9.2.6, the contribution from the magnetic dipole term m_T vanishes at this angle revealing the pure spin moment. The spin moment shown in Figure 9.37(a) increases with decreasing cluster size and is

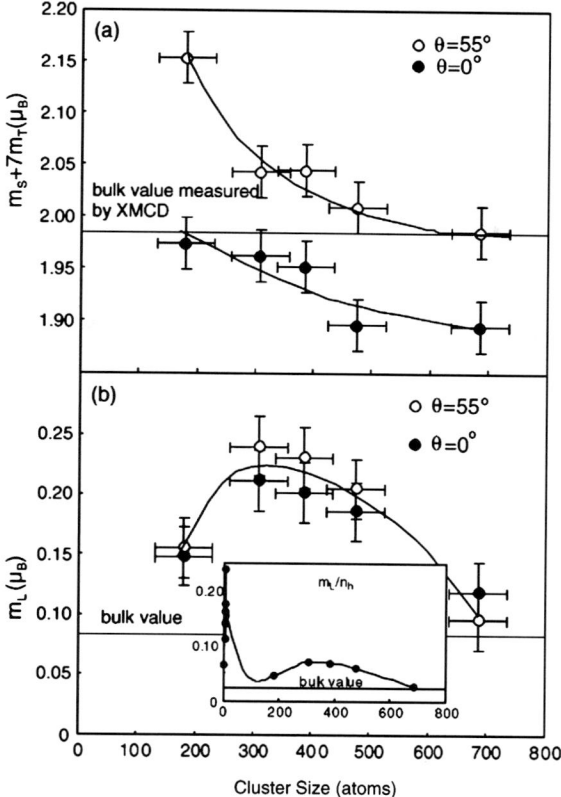

Figure 9.37. (a) m_S+7m_T values obtained by XMCD from isolated Fe clusters on HOPG as a function of size. Measurements at $\theta = 0°$ (filled circles) and 55° (open circles) are displayed and $m_T = 0$ in the data at $\theta = 55°$ (the 'magic angle') revealing the pure spin moment. The measured bulk value was from a 250 Å thick conventional Fe film. (b) m_L values obtained by XMCD showing a peak at ~ 300 atoms. There is no observable difference between the measurements at $\theta = 0°$ (filled circles) and 55° (open circles) indicating that the orbital moment is isotropic. The inset compares the present data with the measurements of Lau et al. [154,155] for very small Fe clusters ($n = 2$–9) on Ni substrates (see Figure 9.25). The combined data reveal a remarkable change in m_L with cluster size [63]. Reproduced with the permission of Elsevier Science from Bansmann et al. [15].

enhanced by 10% relative to the bulk value at the smallest size measured (181 atoms). Note how the data taken at 55° and normal incidence diverge revealing the increasing importance of the dipole contribution as the cluster size decreases. The bulk value, for comparison, was obtained by XMCD from a 250 Å thick conventional Fe film deposited *in situ*. The orbital moment shown in Figure 9.37(b) also increases with decreasing cluster size and is enhanced by a factor of 3 relative to the bulk value for cluster sizes around 300 atoms. In the smallest clusters, however, it diminishes again, which may be due to the formation of a closed shell system, for example, the 147-atom icosahedron. The inset in Figure 9.37(b) compares the measurement of m_L/n_h from Fe clusters on HOPG obtained by Baker et al. [63] with those of Lau et al. [154,155] for very small Fe clusters in the size range 2–9 atoms deposited on Ni substrates. The moments of Lau et al. [154,155] are shown in more

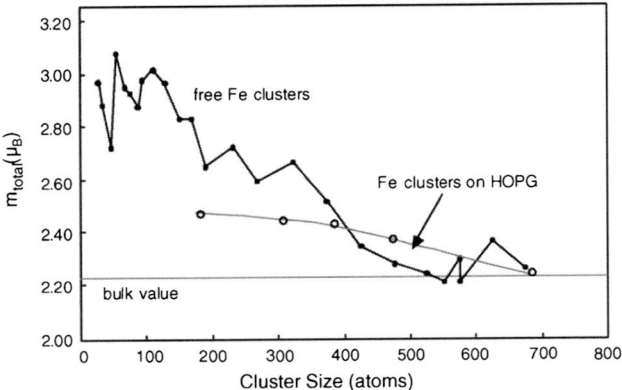

Figure 9.38. Total magnetic moment m_{total} in isolated Fe clusters on HOPG as a function of size (open circles) [63] in comparison to the total moments in free Fe clusters (filled circles, taken from Ref. [281]). It should be mentioned that the XMCD data published in Ref. [63] have additionally been corrected as outlined in Section 9.2. Reproduced with the permission of Elsevier Science from Bansmann et al. [15].

detail in Figure 9.25. Thus, the orbital moments of Fe clusters shown in the lower part of Figure 9.37 support the strong influence of the particle geometry (especially open and closed shells) on the orbital moment.

The total magnetic moments of the Fe clusters (spin+orbital from Figure 9.37) are shown in Figure 9.38 together with gradient-field deflection measurements of free Fe clusters. It is evident that the supported clusters have a significantly enhanced moment relative to the bulk (about half of which comes from the increased orbital moment). The total moment is smaller than in the free clusters.

A more systematic overview on the size dependence of m_L/m_S is given in Figure 9.39, where the results from the exposed Fe clusters (up to 2.3 nm) on HOPG [63] are combined with data from XMCD measurements by Bansmann and Kleibert [282] on Fe clusters between 6 nm and 12 nm. These larger clusters were deposited on an epitaxially ordered Co(0 0 0 1) film on W(1 1 0). A data point from very small Fe clusters on Ni/Cu(1 0 0) [155] has also been included in the left part of the figure. This represents an average of the moments over the $n = 2$–9 clusters [155]. The experimental data for large Fe clusters (filled circle) exhibit m_L/m_S values from 0.07 at a cluster size of 12 nm up to 0.095 for Fe clusters with 6 nm. All these values are clearly above the corresponding bulk value of 0.043. The enhanced ratio is related to an increase of the orbital moment in the outer two shells, which exhibit a large number of surface atoms compared to the total number of atoms in the cluster. When assuming a spherical shape, the ratio of surface to volume atoms amounts to 43% in case of a 6 nm Fe cluster and still 23% for Fe clusters with a size of 12 nm. The X-ray absorption in all cases was determined by measuring the electron yield as a function of photon energy. For the larger clusters, systematic errors arising from saturation (Section 9.2.6) will become more significant and although there is a well-defined algorithm for treating this, the uncertainty inevitably grows with the level of correction required. Here, the uncertainties are included in error bars in Figure 9.39.

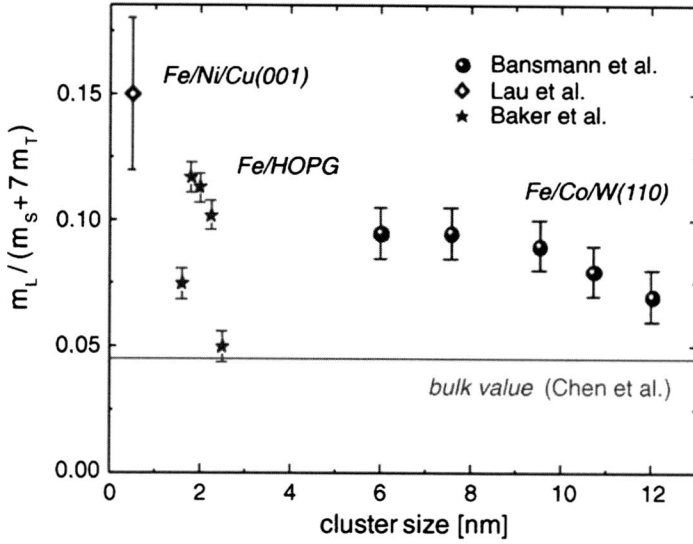

Figure 9.39. Ratio of orbital (m_L) to spin moment ($m_S + 7m_T$) for Fe clusters on surfaces: Fe clusters on Ni/Cu(1 0 0) (from Ref. [154]) (left part), Fe clusters on HOPG (from Ref. [63]) (middle) and large Fe clusters on Co/W(1 1 0) (from Ref. [282]) (right part). The bulk value (solid line) has been taken from Ref. [57]. Reproduced with the permission of Elsevier Science from Bansmann et al. [15].

9.5. Clusters embedded in a matrix

The properties of a metallic cluster are modified when it is embedded in a matrix of another material. We have already encountered this situation in the context of optical properties (see Section 7.3), in which the position of the plasmon resonance depends on the dielectric function of both the particle and the embedding medium. In discussing the magnetic properties of embedded particles, there are two issues to address. The enhancement in the magnetic moment of free particles over the bulk value is due largely to the reduced coordination of atoms at the surface of the particle. However, an embedded particle has the coordination of its surface atoms restored to something like the number in bulk (at least in the ideal case of perfect epitaxy). Whether the magnetic moment of the particle reverts back to its bulk value will then depend on the difference in the electronic band structures of the particle and the host matrix.

The second issue concerns the geometric structure. If the particle and the matrix have the same bulk crystal structure with similar lattice spacings, then an embedding with reasonable good epitaxy is likely to take place. If the two materials have different structures, however, the particle structure could remain robust or adapt to that of the host matrix.

Some results from measurements of the structures of Fe and Co particles embedded in various matrices are shown in Table 9.3. The samples were prepared by co-depositing pre-formed clusters and atomic vapour of the embedding material. The relative proportion of the two determines the VFF of the sample. The structures were determined by EXAFS measurements [17]. This is an

Table 9.3. Structures of Fe and Co clusters embedded in various matrices as determined by EXAFS [17].

Cluster	Matrix	Volume fraction (%)	Structure of clusters
Fe	Ag	5, 40	bcc
Fe	Cu	6	fcc
Fe	Co	9	bcc
Fe	Amorphous carbon	4, 40	Mixed bcc and fcc
Co	Ag	5	hcp or fcc
Co	Fe	7	bcc

In two cases, measurements were made at different volume fractions. The structure of the clusters was independent of the volume fractions, but there were small differences in the lattice parameters.

element-specific technique, which yields values of the neighbour distances of a particular atomic species and, from this information, a crystal structure can be assigned. In two cases, measurements were made at low and high VFFs, but no differences were found in the structures of the embedded clusters, although there were small variations in lattice spacing [17]. It was clear that Co in Ag adopts a close packed structure, but it was not possible to make a definite assignment to fcc or hcp.

Fe clusters in a Co matrix maintain their bulk bcc structure, whereas Co clusters in Fe modify their close packed structure to match the bcc structure of the Fe matrix. The bcc structure of Fe is also robust when embedded in Ag, but shifts to fcc when in a Cu matrix.

There are two scenarios to consider when discussing embedded clusters. At very low concentrations, the particles are essentially isolated from each other and, when modelling their properties, we can treat them as single particles embedded in a medium. At higher concentrations, the particles interact with each other, either through the dipole–dipole interaction or indirectly via the electron sea in the embedding medium. Of course, at higher concentrations, there is also an increasing probability that particles will be in contact and form aggregates, and for VFFs above the percolation threshold we are essentially dealing with a granular material.

9.5.1. Isolated (non-interacting) clusters

9.5.1.1. Co in Cu

There have been a number of calculations [283–291] covering several cluster materials and different host matrices. The simplest case perhaps is Co embedded in Cu [284,286,288,291], where the Co moment is found to be little changed from that of the bulk material. There is some variation in the local moment within the cluster, with generally a lower moment on atoms near the interface and a higher one on atoms near the centre of the cluster. For the smallest Co clusters the average moment is slightly below the bulk value, but for clusters larger than about 100 atoms the mean moment is close to that of the bulk. In contrast to the spin moment, the local orbital moment tends to increase away from the centre of the cluster. The lattice parameters of Cu and fcc Co are similar and lattice relaxation has little effect on the results. Co in Ag does show a moment somewhat above the bulk value, however, probably due to the larger lattice spacing of the host material [288].

9.5.1.2. Fe and Co in Ag

Magnetometry measurements have been reported on dilute (1–2% VFFs) assemblies of Fe and Co clusters in a Ag matrix prepared by co-depositing the clusters and the matrix [292]. Figures 9.40(a) and (c) show, respectively, in-plane magnetic isotherms from an assembly of Fe clusters in Ag with a VFF of 1% (Fe_1Ag_{99}) and Co clusters in Ag with a VFF of 2% (Co_2Ag_{98}). The Fe_1Ag_{99} assembly displays ideal superparamagnetism over the temperature range 50–300 K, as can be seen by the absence of hysteresis. The size distribution of the clusters can be obtained by fitting a set of Langevin functions with different moment values where the amplitude of each Langevin function is a fitting variable. A log-normal distribution was assumed and the ones giving a best fit to the magnetisation data are

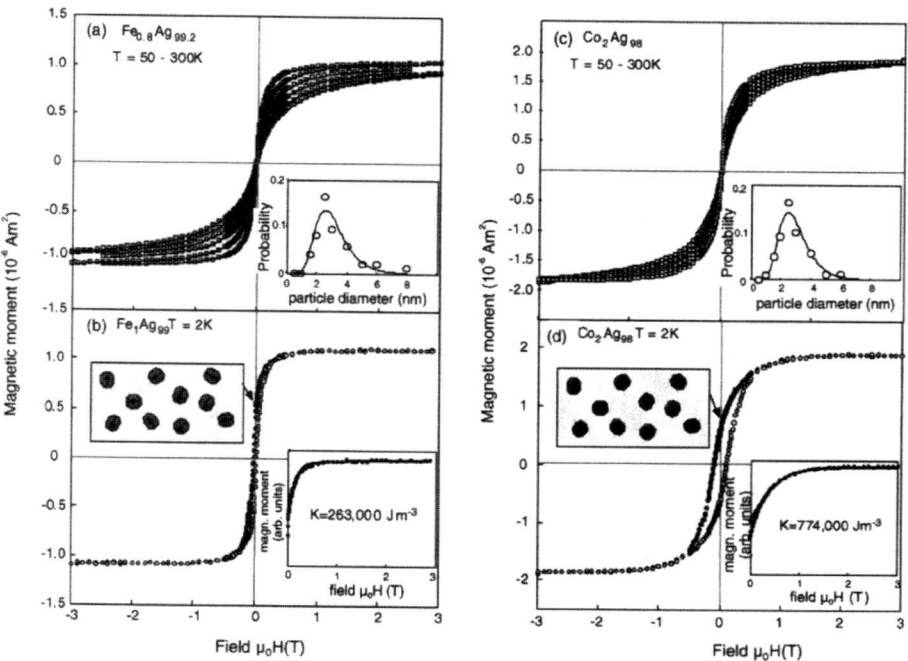

Figure 9.40. Magnetisation isotherms in the range 50–300 K of (a) Fe_1Ag_{99} and (c) Co_2Ag_{98} cluster-assembled film (open squares) compared to fits by Langevin functions (line) with a size distribution represented by 10 size bins in the range 0.5–8 nm. The insets show the average probability of each bin for the optimum fit to curves at temperatures $T > 50$ K (open circles) and the corresponding log-normal distribution (line). (b) In-plane magnetisation isotherms at 2 K of Fe_1Ag_{99} sample and (d) out-of-plane magnetisation isotherms at 2 K of Co_2Ag_{98} sample: (filled circles) field sweeping down; (open circles) field sweeping up. The insets show the decay from saturation (filled circles) compared to a calculation (line) assuming a random distribution of uniaxial anisotropy axes. The best fit anisotropy constants are displayed in the insets [293]. Reproduced with the permission of Elsevier Science from Bansmann et al. [15].

shown as an inset. This distribution is similar to those obtained by direct STM imaging of deposited cluster films [292] confirming that the clusters are isolated in the film. In the case of the Co_2Ag_{98}, there was some departure from ideal superparamagnetism below 150 K and so the fitting procedure was confined to higher temperatures.

At 2 K, most of the clusters in both the Fe_1Ag_{99} and $Co_2Ag_{0.88}$ samples are below the blocking temperature and as shown in Figures 9.40(b) and (d) the magnetic isotherms develop hysteresis. The remanence (magnetisation remaining after the magnetic field is removed), M_r, gives information about the anisotropy through the ratio M_r/M_S, where M_S is the saturation magnetisation [15,17]. It was deduced that the clusters display uniaxial anisotropy with axes distributed randomly over three dimensions, i.e. there is no preferred direction of the axes with respect to the substrate. Once this is established, the anisotropy constants, K_1 (see Section 8.2.2), can be obtained. For the Fe clusters, an optimised fit to the data yielded $K_1 = 2.63 \times 10^5$ J/m^3 (~ 3 meV/atom), about five times as large as the bulk value and very similar to the results from SQUID measurements on similar samples [294]. The anisotropy constant that fits the Co data is two to three times larger than this.

9.5.1.3. Fe in Cu

As noted in Table 9.3, Fe clusters in Cu at low concentration adopt an fcc structure. In bulk, fcc γ-Fe can only be stabilised above 1200 K, although it is well established experimentally that fcc and fct Fe films can be grown epitaxially at lower temperatures on Cu [295–300]. Baker et al. [301] prepared samples of Fe clusters (~ 250 atoms) embedded in Cu for a range of Fe VFFs by co-deposition of pre-formed Fe clusters and Cu. Magnetometry and EXAFS measurements determined the average moment on the Fe atoms and the structures, including an estimate of the lattice parameters. The fcc structure in the Fe clusters persisted up to a VFF of about 20%. At higher VFFs when the samples were essentially granular Fe, the bcc structure reappeared. The results are summarised in Figure 9.41.

The low VFF is of most interest. There have been a number of calculations [302–304] on bulk fcc γ-Fe investigating the dependence of the magnetic moment on the lattice spacing. For an fcc unit cell with lattice parameter larger than about 3.6 Å, the magnetic moment is very similar to that of bcc Fe, but as the lattice spacing is decreased there is a transition from a ferromagnetic to an antiferromagnetic phase (ferromagnetic planes of spins coupled antiferromagnetically). Eventually, at a little below 3.5 Å, a non-magnetic phase becomes energetically favourable.

The lattice parameter for the embedded Fe clusters, as measured by EXAFS, is 3.58 Å and is therefore within what, in bulk, would be a transition regime. The magnetometry measurements yield low moments in the range 0.5–0.8 μ_B/atom. Calculations have been performed [301] on fcc Fe particles covered with shells of Cu atoms to give a rough simulation of the experimental system and results are shown in Figure 9.42. The magnetisation of bulk fcc Fe is plotted for comparison. The systems studied were core–shell particles comprising an Fe_{147} or an Fe_{309} core embedded in shells of Cu of a number of thicknesses. The Fe clusters investigated

9.5. *Clusters embedded in a matrix* 337

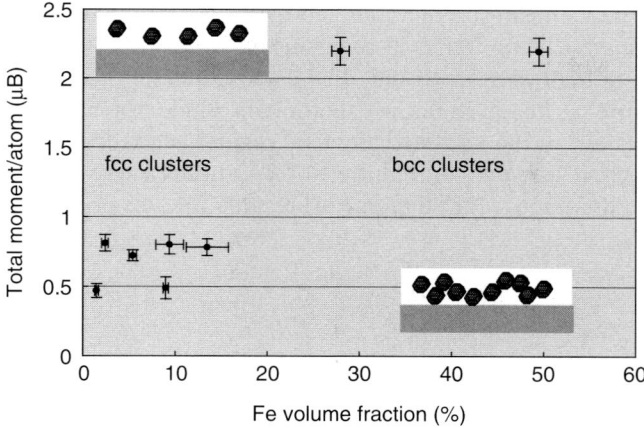

Figure 9.41. Magnetic moments of Fe clusters (~250 atoms) embedded in a Cu matrix as a function of the volume fraction of Fe [301].

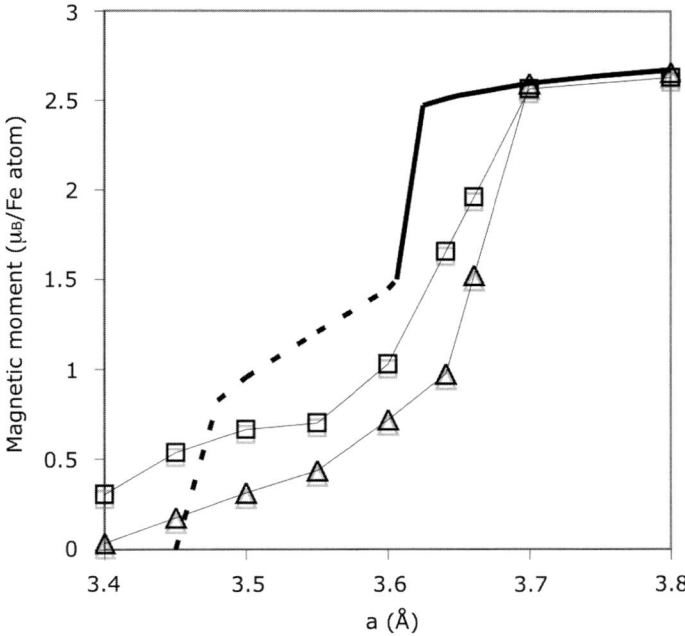

Figure 9.42. Calculated magnetic moments for bulk fcc Fe and for Fe clusters covered by two shells of Cu atoms all at fixed fcc lattice spacings a [301]. The solid and broken lines represent, respectively, the ferromagnetic and antiferromagnetic states of bulk fcc Fe. Note, for the antiferromagnetic state (broken line), it is the sub-lattice magnetisation that is plotted; the net moment is zero. The data points refer to Fe_{147} (squares) and Fe_{309} clusters embedded in Cu shells.

experimentally had a mean size of about 260 atoms. A ferrimagnetic regime replaces the antiferromagnetic state found in bulk since, for the cluster, we no longer have the possibility of equivalent up and down sub-lattices. There is a reduced net moment therefore, as observed in the experiments, unlike the behaviour that would be observed in bulk in the antiferromagnetic regime where the up and down sub-lattices give an equal and opposite contribution and thus a zero net moment. As can be seen from the plots, the theoretical results are in good agreement with experiments.

9.5.1.4. Fe in Co

A particularly interesting system is Fe embedded in Co. As noted in Table 9.3, EXAFS measurements show that the Fe clusters maintain their bulk bcc structure. Figure 9.43 shows the results of calculations on this system by Xie and Blackman [290]. Since there is no experimental knowledge of the details of the Fe/Co interface, the procedure in the calculation [290] was simply to add a bcc shell of Co to the bcc Fe core. The figure shows the average spin magnetic moment per atom on the Fe (see plot labelled Fe_nCo_{1021-n}-I) as a function of cluster size for a total system size of 1021 atoms. The results are compared with those of free Fe clusters and for Fe clusters with a Cu coating. It can be seen that, while a Cu coating reduces the free Fe enhanced cluster moment back to virtually the bulk Fe value, a Co coating leaves the moment essentially unchanged from the free cluster value. The effect of some alloying at the Fe/Co interface was also studied and, surprisingly, there was even enhancement to the moment above the free cluster values (see plot Fe_nCo_{1021-n}-II).

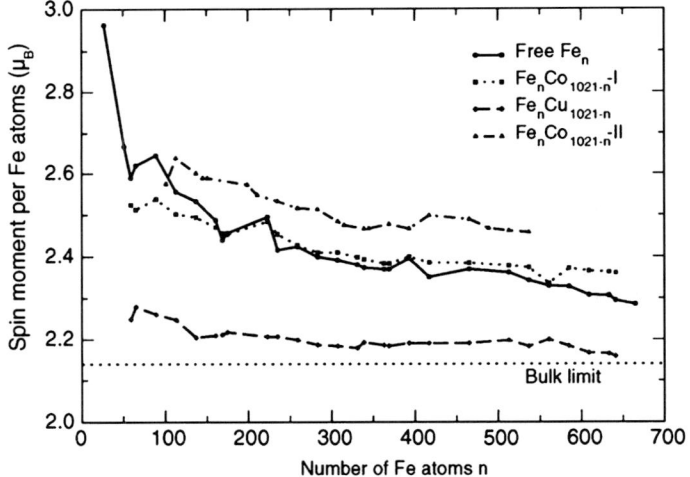

Figure 9.43. Averaged local spin magnetic moments of Fe atoms for free clusters (free Fe_n), Cu-coated Fe clusters (Fe_nCu_{1021-n}) and Co-coated Fe clusters (Fe_nCo_{1021-n}-I). Fe_nCo_{1021-n}-II are clusters with some Co atoms substituted for Fe atoms inside the interface of the Co-coated Fe clusters [290]. Reproduced with the permission of Elsevier Science from Bansmann et al. [15].

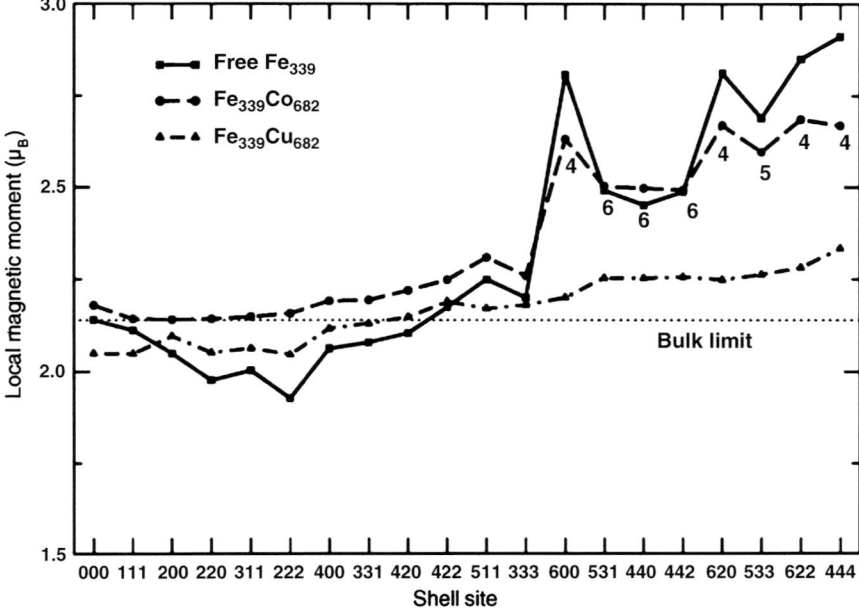

Figure 9.44. Local moment of Fe atoms for free and for Cu- or Co-coated Fe$_{339}$ clusters as a function to positions on bcc lattice (in units of one-half the lattice constant). The figures on the plot are the numbers of Fe neighbours of surface (or interface) Fe atoms [290]. Reproduced with the permission of Elsevier Science from Bansmann et al. [15].

The local moments on the Fe atoms of a Fe$_{339}$ cluster are shown in Figure 9.44 as a function of bcc lattice positions [290]. The enhancement to the moment comes largely from the interface (surface for free cluster) atoms. The enhancement of the local moment of atoms on the surface of free clusters is due to the reduced number of neighbours compared to bulk. When these atoms are at the interface with Co, the cause of enhancement is somewhat different. It has been pointed out before that Fe is magnetically weak because of an insufficient electron–electron interaction to bandwidth ratio, while in Fe–Co alloys this interaction for the Co atoms could increase the exchange splitting of the Fe 3d bands [305–307]. This comment applies equally to the Fe–Co interface and to Fe–Co alloys.

Edmonds et al. [308] investigated the effect of coating assemblies of exposed Fe cluster films on HOPG with Co films in situ and they showed that the combined film consisted of Co islands with embedded Fe clusters. The Fe clusters were exchange-coupled to the Co film and the magnetisation showed a strong in-plane shape anisotropy due to the Co islands. The fundamental orbital and spin moments localised in the Fe clusters could be extracted due to the chemical specificity of XMCD. Figure 9.45 shows the measured size-dependent spin moment, after correcting for the systematic error discussed earlier, compared with the calculation by Xie and Blackman [290] for free Co-coated Fe clusters (see Figure 9.43).

The agreement is good and demonstrates an increase in the spin moment of about 0.2 μ_B relative to the exposed clusters on HOPG. Over a significant fraction of

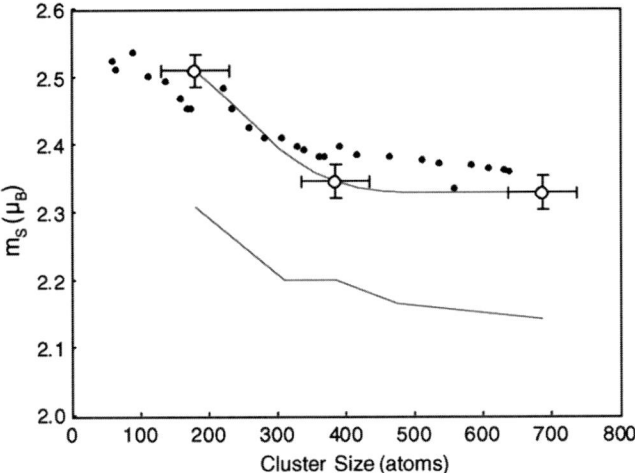

Figure 9.45. Spin moment as a function of Fe cluster size measured by XMCD in Co-coated Fe clusters on HOPG (open circles, from Ref. [63]) compared with the calculated spin moment in free Co-coated Fe clusters (from $Fe_N Co_{1021-N}$-I plot in Figure 9.29; filled circles) [290]. The lower grey line shows the measured spin moments in the exposed clusters prior to coating [63]. Note that the experimental values from Ref. [63] have been corrected analogously to Figure 9.38. Reproduced with the permission of Elsevier Science from Bansmann et al. [15].

the size range, the total magnetic moment per atom measured in the supported Co-coated Fe clusters is as large as in free Fe clusters.

9.5.2. Interacting clusters

In the previous sub-section, we examined the behaviour of embedded clusters at low volume filling fraction, for which the particles behave as essentially independent entities. As the VFF is increased, a number of new factors come into play. We have already referred to Fe nanoclusters embedded in Ag in Section 9.5.1, and this system provides a convenient case study for exploring over a range of VFFs [17,292].

The situation at VFFs of 1 and 10% is shown schematically in Figure 9.46. It is assumed that the particles (represented by elongated spheres) have uniaxial anisotropy with axes that are oriented in random directions. At 1% (upper panel), most of the particles are isolated from each other and the moments are aligned close to the anisotropy axes, with small deviations due to the dipole–dipole interactions between the moments. As the VFF is increased, the effects of the dipolar interactions become stronger. The effect of dipolar interactions on the magnetic properties of an assembly of single domain particles has been studied by Monte Carlo modelling [309]. In addition, the chances of particles being in contact with others increase so that, by about 10% (lower panel), significant aggregation has taken place. There is now strong exchange coupling between particles within an aggregate and this competes with the intra-particle anisotropy and dipolar coupling. We can consider an aggregate as a distinct entity with a net magnetic moment.

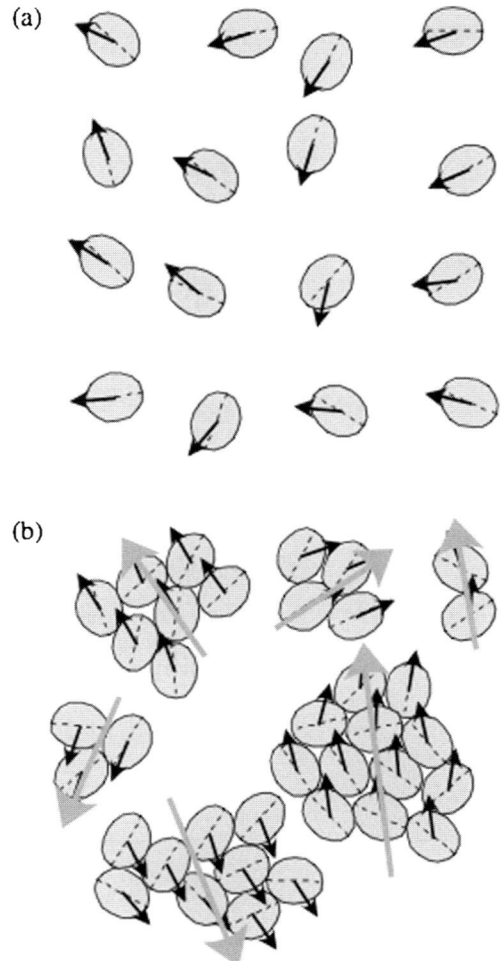

Figure 9.46. Schematic of Fe nanoparticles dispersed in Ag with a uniaxial anisotropy (axes indicated by broken lines) and random orientation of the anisotropy axes at $T = 0$ and zero field. (a) 1% volume fraction: all particle moments (black arrows) oriented close to anisotropy axes; (b) 10% volume fraction: deposited nanoparticles form aggregates containing on average seven particles that are strongly exchange coupled to form a single entity with aggregate moment indicated by grey arrows. Reproduced with the permission of the Institute of Physics from Binns et al. [17].

If we could neglect the dipolar interaction between aggregates, we would expect the aggregates themselves to behave as large superparamagnetic particles with effective moments and sizes that increased with the VFF. The magnetisation isotherms for various VFFs are shown in Figure 9.47, and the most noticeable feature is the increase in low field susceptibility (dM/dH at $H = 0$) as the cluster density increases. This is consistent with larger particle sizes at higher densities and indeed it is possible to get a good fit using a Langevin function with a higher supermoment than in the dilute limit.

What emerges, if we follow this fitting procedure, is a particle size that depends on temperature, which is clearly nonsensical. This is illustrated in Figure 9.48 and

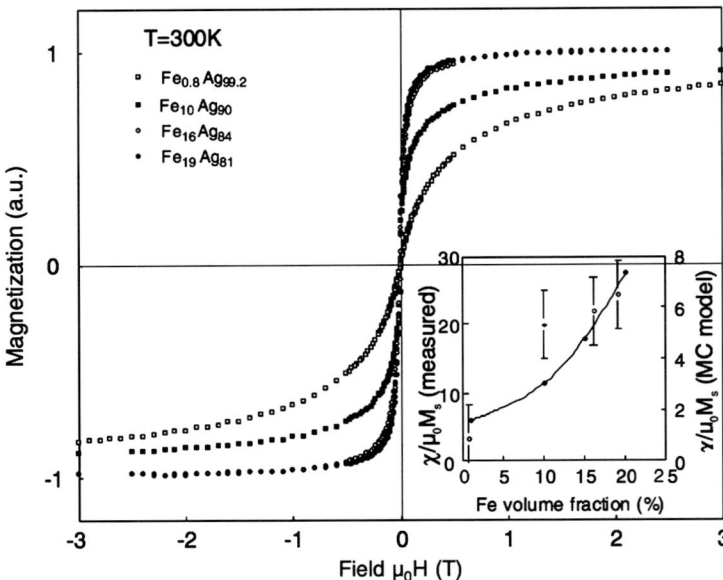

Figure 9.47. Magnetisation isotherms at 300 K for four VFFs. The plots of the two largest VFFs are virtually indistinguishable. The inset compares the initial susceptibility with the results of Monte Carlo calculations (note different scales). Reproduced with the permission of the Institute of Physics from Binns et al. [17].

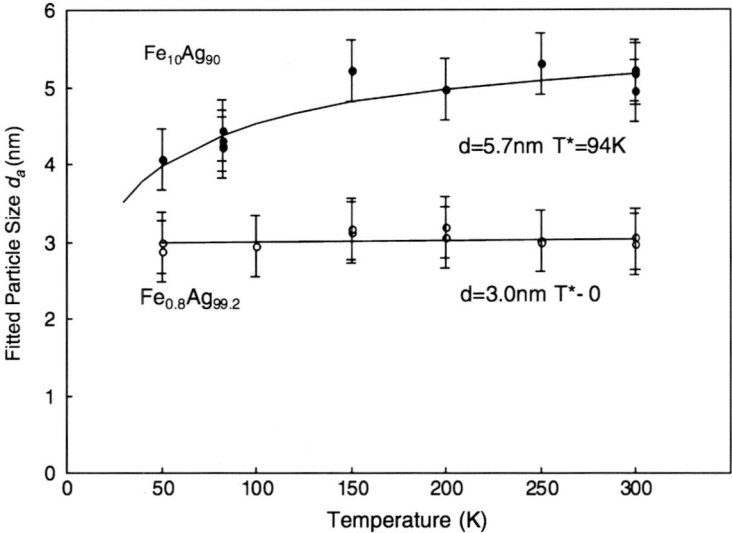

Figure 9.48. Median particle diameter d_a obtained by fitting unmodified Langevin functions to the magnetisation curves taken at 50–300 K of the samples $Fe_{0.8}Ag_{99.2}$ (open symbols) and $Fe_{10}Ag_{90}$ (solid symbols). Reproduced with the permission of the Institute of Physics from Binns et al. [17].

compared with the sensible behaviour obtained at low particle density. The origin of the anomalous behaviour is the neglect of the inter-aggregate dipolar interactions. Allia et al. [310] showed that the dipolar interactions could be incorporated into the analysis by fitting the experimental data to a modified Langevin function that has a form similar to the Curie–Weiss law

$$M = N\mu \mathcal{L}\left(\frac{\mu H}{k(T+T^*)}\right). \tag{9.48}$$

The parameter T^* is proportional to the dipolar energy, and μ is the average aggregate moment. The true particle diameter d is related to the apparent diameter d_a (the value obtained by using the simple Langevin function in the fitting) by

$$d_a = \left(\frac{T}{T+T^*}\right)^{1/3} d. \tag{9.49}$$

The apparent diameter for $Fe_{10}Ag_{90}$ shown in Figure 9.48 can be fitted to Eq. (9.49) with a true diameter $d = 5.7$ nm, which corresponds to aggregates containing on average six to seven particles.

The scenario we have described operates at VFFs up to the percolation threshold, which is 24.88% according to theory [311]. Above the percolation threshold, it is no longer appropriate to think in terms of an assembly of aggregates, since the nanoparticles now form an interconnected network. The random anisotropy model [312–315] provides a useful description in this density regime. The random anisotropy model was originally developed for granular material, and considered single domain grains with uniaxial anisotropy and randomly oriented anisotropy axes. The other component in the model was exchange coupling between neighbouring grains. The model is expressed in terms of two parameters, a random anisotropy field

$$H_r = \frac{2K_r}{M_S}, \tag{9.50}$$

and an exchange field

$$H_{ex} = \frac{2A}{M_S R_a^2}. \tag{9.51}$$

Here, K_r is the (randomly oriented) anisotropy of the grains, M_S their saturation magnetisation, A the exchange constant for the interaction between the grains and R_a the nanometre-scale region over which the local anisotropy axis is correlated, i.e. the characteristic grain size. The relative strength of the fields is given by the dimensionless parameter

$$\lambda_r = \frac{H_r}{H_{ex}}. \tag{9.52}$$

The model was originally developed to describe amorphous films in which a local, randomly oriented, anisotropy is due to local atomic disorder. It is even better suited to providing a description of the magnetisation in cluster-assembled films

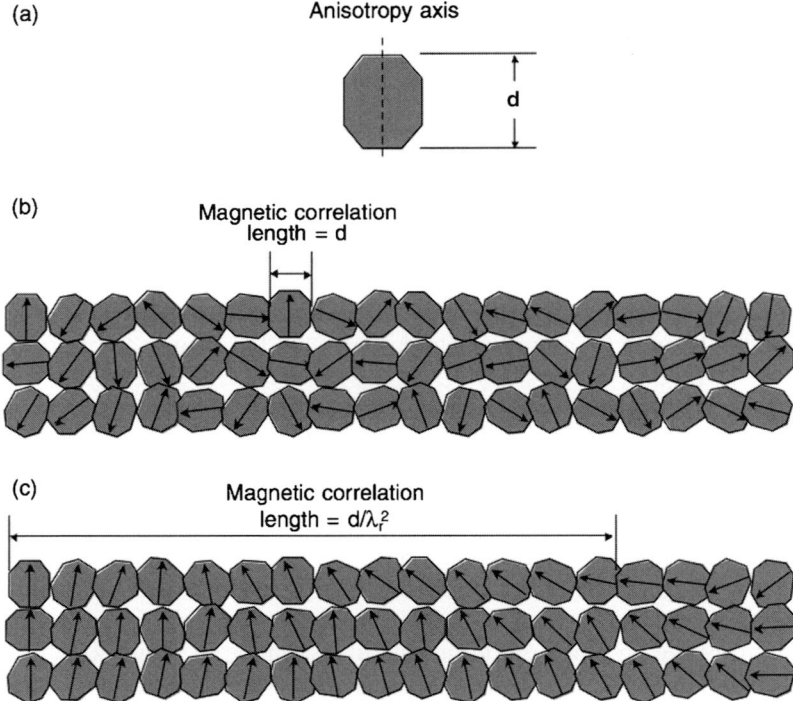

Figure 9.49. (a) Schematic representation of a magnetic nanoparticle with a uniaxial anisotropy axis (represented by a slight elongation); (b and c) stack of particles with randomly oriented anisotropy axes. In (b) $\lambda_r \geq 1$ and the magnetisation vector points along the local anisotropy axis, so the magnetic correlation length is a single particle diameter. In (c) $\lambda_r < 1$ and the magnetic vectors are nearly aligned. The random perturbation from perfect alignment results in a finite magnetic correlation length that is a factor $1/\lambda_r^2$ larger than a single particle diameter. Reproduced with the permission of the Institute of Physics from Binns et al. [17].

[17,292,316–318] in which the grains are now the nanoparticles themselves, and the distance R_a over which an anisotropy axis is correlated is well defined (i.e. the particle diameter). For $\lambda_r \geq 1$, the magnetic correlation length at zero field is R_a, and the magnetic vector in each particle points along the local intra-particle anisotropy axis. Note that in an arrow representation, this state would be identical to that in isolated non-interacting particles at absolute zero. With increasing inter-particle exchange (or decreasing intra-particle anisotropy), the configuration becomes a correlated superspin glass (CSSG) in which the magnetisation vector in neighbouring particles is nearly aligned, but the random deviation of the moments from perfect alignment produces a smooth rotation of the magnetisation throughout the system with a magnetic correlation length that is a factor $1/\lambda_r^2$ larger than the particle diameter. The difference between the two states is illustrated in Figure 9.49. An important characteristic of the CSSG state is that it excludes magnetic domains, which require long-range order. The randomly oriented magnetisation within the sample produces no external magnetic field, thus minimising the materials dipolar energy without domains. This has technological implications since a magnetisation reversal will not occur by the normal process involving the motion of domain walls. Time-resolved magnetisation experiments on materials assembled from Fe clusters indicated that the magnetisation reversal could be significantly faster than a conventional film [319].

Evidence of a CSSG state at high VFFs is obtained from the field and temperature (field cooled (FC) and zero field cooled (ZFC)) dependence of the magnetisation and a comparison with predictions of the random anisotropy model. The details of the procedure are given in Refs. [17,293]. Further evidence emerged from a recent experiment using X-ray photoelectron microscopy (XPEEM) that showed that thin films of Fe nanoparticles had no domain structure larger than the spatial resolution limit of 100 nm [320].

The global picture over the full concentration range is summarised in Figure 9.50. Above a blocking temperature, ideal superparamagnetism occurs at the lowest concentrations, and as the VFF increases dipolar interactions and aggregation determine the behaviour. At still higher concentrations, there is firm evidence for a CSSG state. At low temperatures, blocking occurs with the moments of individual particles aligned close to their own anisotropy axes. The diagram distinguishes between single particle blocking in isolated clusters where the intra-particle anisotropy is the sole influence stabilising the magnetic moment and collective blocking, which occurs in the presence of strong interparticle interactions. The latter will occur at higher temperatures since, for example, a particle that would normally be thermally activated if it was isolated has its magnetisation stabilised by a more robust neighbour, which is slightly larger or more anisotropic.

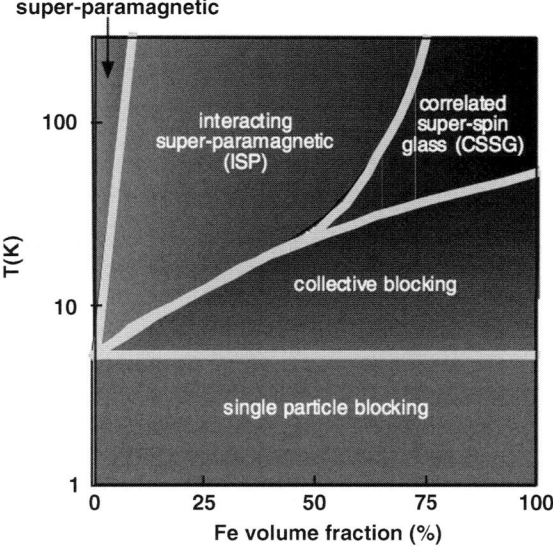

Figure 9.50. Magnetic phase diagram for films of deposited 3 nm diameter Fe nanoparticles embedded in Ag matrices as a function of volume fraction and temperature. Reproduced with the permission of the Institute of Physics from Binns *et al.* [17].

9.6. Exchange bias

One of the most interesting interfaces between two materials is that between a ferromagnet and an antiferromagnet. The first system studied [321] of this type consisted of small particles of Co, about 20 nm in size, with their surfaces oxidised to form a cobalt oxide shell. Cobalt oxide is an antiferromagnet with a Néel temperature of 293 K. Meiklejohn and Bean [322–323] cooled the material from the paramagnetic state of the oxide to the antiferromagnetic state in a saturating field and measured the hysteresis curves. Remarkably they found that the loops were displaced along the magnetic field axis (as in Figure 9.51) rather than being centred about zero field as in the familiar behaviour of a ferromagnet. The phenomenon is called exchange bias.

The particulate system is rather hard to analyse, so Meiklejohn and Bean [322] proposed a simple model to explain how the effect could arise in a simpler system, namely an antiferromagnetic substrate covered with a ferromagnetic layer as in Figure 9.52. We assume strong exchange coupling in the ferromagnet so that an external field will rotate the moments coherently, while the antiferromagnet remains unaffected by the field due to very strong uniaxial anisotropy.

Consider first Figure 9.52(a) where the AFM component orders as ferromagnetic layers that are antiferromagnetically coupled. The energy of the system is written as

$$E = -HM \cos\theta - J \cos\theta + K \sin^2\theta, \qquad (9.53)$$

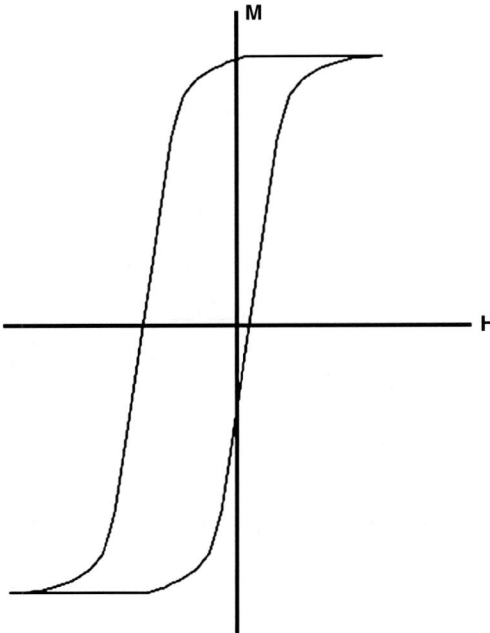

Figure 9.51. Schematic representation of shift in the hysteresis loop in an exchange bias system.

where θ is the angle the ferromagnetic moment makes with the interface. Considering just zero temperature, M is the saturation magnetisation of the ferromagnetic component, and the first and third terms in Eq. (9.53) are, respectively, the Zeeman energy and the energy due to the anisotropy of the ferromagnet. The second term represents exchange coupling between the FM and AFM components across the interface. We assume ferromagnetic coupling across the interface is preferred so that the configuration depicted in the upper panel has a lower energy than that shown in the lower panel.

To illustrate the behaviour, the energy at zero field according to Eq. (9.53) is plotted in Figure 9.53 for three values of the exchange coupling. The $J=0$ case is the familiar one with only uniaxial anisotropy and two equal energy minima at $\theta=0$ and π. The other two plots include the exchange term. For $J<2K$, there are again two energy minima, but the $\theta=0$ one has lower energy than that at $\theta=\pi$. For $J\geq 2K$, there is a single minima and the configuration shown in the lower panel of Figure 9.52(a) is unstable unless forced by an applied field.

The hysteresis behaviour is obtained straightforwardly from Eq. (9.53), now including the magnetic field. The energy always has maxima or minima at $\theta=0$ and π. To reverse the direction of magnetisation from $\theta=0$ to π requires a maximum at $\theta=0$ and a minimum at $\theta=\pi$; the opposite is required for a reversal from $\theta=\pi$ to 0. The second derivative of the energy immediately yields the conditions for the respective cases: $HM+J+2K<0$ and $-(HM+J)+2K<0$, and the two coercive fields

$$H_{c1} = -\frac{2K+J}{M} \qquad (9.54)$$

and

$$H_{c2} = \frac{2K-J}{M}. \qquad (9.55)$$

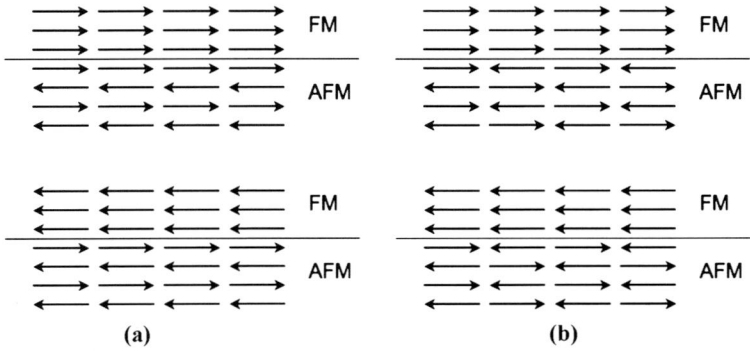

Figure 9.52. Uniaxial anisotropy aligns both ferromagnetic (FM) and antiferromagnetic (AFM) moments parallel to the interface, with the ferromagnetic moments pointing to the right and left, respectively, in the upper and lower panels. Two different types of antiferromagnetic ordering are shown in (a) and (b).

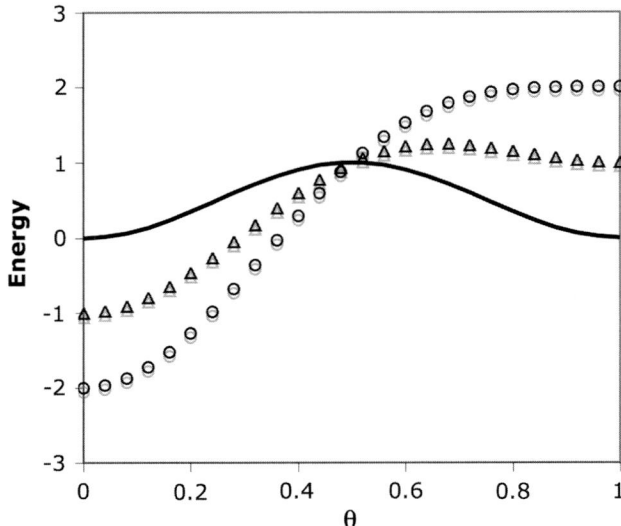

Figure 9.53. Plot of energy (in units of K) against θ (in units of π) from Eq. (9.53) for $H = 0$. The plots are for three values of the exchange constant J: full line ($J = 0$), triangles ($J = K$) and circles ($J = 2K$).

The centre of the hysteresis loop is shifted from zero applied magnetic field by an amount H_E, known as the exchange bias field, and is given in this model by

$$H_E = \frac{J}{M}. \qquad (9.56)$$

Schematic hysteresis loops are shown in Figure 9.54. In a layer system, J will scale with the area of the interface, while M is proportional to the volume (i.e. area of interface times layer thickness) of the ferromagnetic component. Writing $M = M't$, where M' is the layer magnetisation and t the film thickness (or number of layers)

$$H_E = \frac{J}{M't}, \qquad (9.57)$$

and it is clear that the exchange bias field is expected to decrease with increasing thickness.

The simple picture just outlined offers a plausible explanation for the exchange bias phenomenon, but it relies on the rather special antiferromagnetic configuration shown in Figure 9.52(a). An alternative antiferromagnetic arrangement is illustrated in Figure 9.52(b). Clearly, in this case, there is no difference in energy between the two ferromagnetic alignments shown in the upper and lower panels and no shift in the hysteresis loops would be predicted. The term 'compensated' is used to describe an antiferromagnetic interface with an equal number of spins in opposite alignment in the first AFM layer as in Figure 9.52(b). The interface in Figure 9.52(a) where the net magnetisation of the first AFM layer is non-zero is uncompensated.

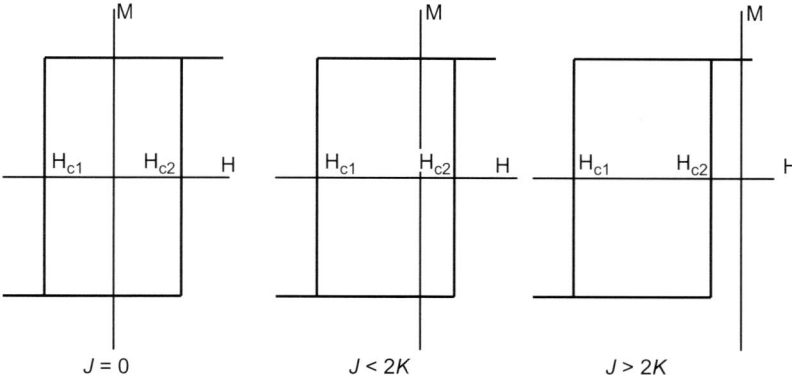

Figure 9.54. Schematic hysteresis loops for three values of J. The coercive fields are given by Eqs. (9.16) and (9.17).

However, even for the uncompensated case, estimates of the loop shift from the model are very much higher than the values obtained in experiments on layer systems. A resolution of this discrepancy came with the recognition [324,325] that a perfectly rigid antiferromagnet and a perfectly uniform ferromagnet are unlikely to describe the lowest energy configuration near the interface, and the formation of a partial or complete domain wall will allow the reversal of the magnetisation to take place at a much reduced field. The formation of domain walls in bulk systems between regions magnetised in different directions is, of course, well understood [326]. In a layer system, however, the wall can form in the antiferromagnet or the ferromagnet, or through both, depending on the exchange interactions and uniaxial anisotropies within the two components themselves. For example, if the domain wall is entirely within the antiferromagnet, the bias field is reduced from Eq. (9.57) to $H_E \propto \sqrt{K_{\mathrm{AFM}} A_{\mathrm{AFM}}}/M't$, where K_{AFM} and A_{AFM} are, respectively, the anisotropy and exchange stiffness within the antiferromagnet. In addition, the layer thickness must be big enough to support wall formation. Strong evidence for a domain wall has been found in experiments on single crystal Co/NiO(0 0 1) [327], and a dependence of bias field on the thickness of the antiferromagnetic layer has been observed [328–330].

However, the model does not allow for the possibility of exchange bias effects in the case of a compensated interface and does not explain the increased width of hysteresis loops (coercivity) that is commonly observed when a ferromagnet interfaces with an antiferromagnet. A number of developments have been proposed to address one or both of these issues. Koon [331] introduced a model based on the canting of spins to reduce the frustration at the interface. It was later shown that this does not give rise to exchange bias but can account for coercivity enhancement [332]. Kiwi et al. [333,334] also present a model based on canted spins. A number of approaches have been based on grains within a polycrystalline structure in the antiferromagnet, with various coupling schemes between the grains themselves and between the grains and the ferromagnet [335–338]. Alternative approaches that focus on the microstructure of the antiferromagnet but are not based on polycrystallinity propose the formation of magnetic domains. It is suggested that these can result either from roughness at the interface [339,340] or by the presence of non-magnetic [341,342] or magnetic [343] defects in the antiferromagnet. The behaviour of the domains under field cooling can result in a net magnetic moment coupling to the ferromagnet even in the compensated system with an exchange bias shift as a consequence.

The literature on the mechanisms for exchange bias is vast, and concentrates mainly on layer structures. We refer to a number of comprehensive reviews [11,12,344–347] for more details and turn now to systems based on ferromagnetic nanoparticles in an antiferromagnetic environment.

The commonest way to prepare ferromagnetic nanoparticles with an antiferromagnetic coating is via oxidation of the particles. This results in a core–shell structure. Depending on the method used, the particles can be prepared with a reasonably well-defined diameter, but the oxide shell then grows at the expense of ferromagnetic core. The thickness of the shell is rather difficult to control and often the formation of the oxide on the surface inhibits further oxidation (passivation). Various chemical reactions can be used to produce alternatives to an oxide shell, such as nitride or sulphides. The core–shell particles are deposited on a surface or embedded in some matrix to facilitate detailed measurements. An alternative approach is the co-evaporation of ferromagnetic and antiferromagnetic materials or the co-deposition of pre-formed ferromagnetic clusters with the antiferromagnetic component. The object in this case is to form ferromagnetic clusters embedded in an antiferromagnetic matrix. Yet a third approach is the mechanical milling (ball milling) of the two components, which increases the range of materials that can be studied. For a detailed discussion of these and other methods of production, see Ref. [12].

Inevitably the nature of the interface between core and shell (or matrix) is less well understood than for the corresponding interface in layer structures. Nevertheless, the nanoparticle systems do show exchange bias effects and, of course, the effects were first observed in such systems [321]. The main characteristics of exchange bias systems, as we have noted, are a shift in and an increased width of the hysteresis loops. A further property is known as the 'training effect', in which repeated traces of the hysteresis loop yield different values of the bias field. It should be emphasised that exchange bias effects appear only at temperatures below the Néel temperature of the antiferromagnetic material.

Interest in magnetic nanoparticles has been sharpened because of their potential for applications in fields such as high-density recording and medicine. Most applications rely on magnetic order being stable with time, but with decreasing size one is confronted by the superparamagnetic limit. The possibility of using an antiferromagnetic interface to effectively enhance the anisotropy of a ferromagnetic particle is therefore an attractive proposition. For reference, some properties of important antiferromagnets (and ferrimagnets) are listed in Table 9.4.

Co–CoO [321–323,348–365] and Fe–FeO or Fe–γ-Fe$_2$O$_3$ [366–377] are systems that have been particularly well studied. In the case of Fe, the shell often consists of a mixture of oxides. Examples of other systems that have been studied using various preparation techniques include Ni–NiO [366,378–381], Co–CoN [349], Fe–FeCl$_2$ [382], Co–Cr$_2$O$_3$ [383], Fe–Cr$_2$O$_3$ [384], Fe–MnF$_2$ [385], Fe–MnO$_2$ [386] and Co–Mn [387].

9.6. Exchange bias

Table 9.4. Structure, magnetic ordering and Néel (Curie for ferrimagnets) temperature of some bulk transition metal compounds.

Material	Structure	Ordering	T_N (K)
FeO	fcc	AFM	198
CoO	fcc	AFM	291
NiO	fcc	AFM	523
α-Fe$_2$O$_3$	Hexagonal	AFM	948
FeF$_2$	bc tetragonal	AFM	79
FeCl$_2$	Hexagonal layer	AFM	24
Cr$_2$O$_3$	Complex	AFM	307
α-Mn	Complex	AFM	~ 100
MnF$_2$	bc tetragonal	AFM	67
MnO$_2$	bc tetragonal	AFM	84
Fe$_3$O$_4$	Spinel	FIM	858
γ-Fe$_2$O$_3$	Spinel	FIM	~ 870

Abbreviations: AFM, antiferromagnetic; FIM, ferrimagnetic. Note the structures refer to $T > T_N$. For $T < T_N$, there is generally a deformation due to magnetostrictive effects with, for example, CoO adopting tetragonal symmetry and CoO and NiO undergoing rhombohedral distortion. α-Fe$_2$O$_3$ (hematite) is an antiferromagnet below its Morin temperature (260 K) and a canted antiferromagnet (weakly ferromagnetic) between the Morin and Néel temperatures.

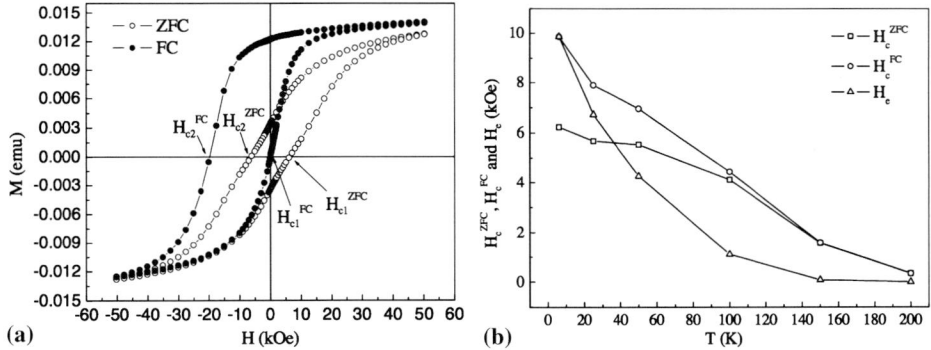

Figure 9.55. (a) Hysteresis loops for Co–CoO particles (20 nm core with 3 nm shell) measured at 6 K after ZFC and FC; (b) the temperature dependence of loop shift (triangles), FC coercivity (circles) and ZFC coercivity (squares). Reproduced with the permission of Elsevier Science from Wen *et al.* [360].

There is a large variation in the effect of the exchange coupling reported in the experiments, reflecting differences in the microstructures of the systems. In core–shell structures, the behaviour is sensitive to the shell thickness. In many cases, the shell is probably amorphous or composed of small crystallites, as is evidenced by a change in the exchange bias effects on annealing.

An example of results from a study that showed strong exchange bias effects is illustrated in Figure 9.55. Wen *et al.* [360] used a particle beam deposition system to produce ~ 20 nm Co clusters with 3 nm CoO shells. From the FC plots at low temperature, they obtained an exchange bias field of about 9.85 kOe and a coercivity (half the loop width at $M = 0$) of 9.90 kOe. This compared with the ZFC coercivity of 6.25 kOe. The exchange bias field and the FC and ZFC coercivities fall with temperature and at ~ 150 K the loop shift has dropped to zero and the two coercivities are indistinguishable. This is significantly below the Néel temperature of the antiferromagnetic shell (see Table 9.4), which the authors attribute to superparamagnetic behaviour of the oxide shell, which is composed of small crystallites. The blocking temperature (where the hysteresis disappears completely) is ~ 300 K.

Peng et al. [353] studied a similar cluster-assembled system, but with Co cores of smaller diameter. With 6 nm particles, they found a similar loop shift (10.2 kOe) falling essentially to zero at ~ 150 K, but with a smaller blocking temperature of ~ 200 K.

Particles embedded in a matrix generally show exchange bias effects to higher temperatures than deposited core–shell particles. The blocking temperature is usually higher as well. Skumryev et al. [355] produced Co particles with diameters 3–4 nm and oxidised them to passivation producing a thin (~ 1 nm) shell of CoO. The core–shell particles were then embedded in a matrix of CoO and the behaviour compared with a reference embedding in Al_2O_3. A blocking temperature of ~ 10 K was found for the Al_2O_3 embedded particles, which increased to 200 K if 7 nm particles were used. Below the blocking temperature, the systems showed a small coercivity of ~ 0.02 T ($= 0.2$ kOe) and no shift in the hysteresis loop. The behaviour is very different when CoO is the embedding material, with a loop shift of 0.74 T and an enhanced coercivity of 0.76 T at a temperature of 4.2 K.

An alternative technique for particle embedding was taken by Sort et al. [384] facilitating a different metal in the ferromagnetic and antiferromagnetic components. They used selective high temperature reduction under a hydrogen atmosphere of $Cr_{1.8}Fe_{0.2}O_3$ powders to produce a microstructure consisting of Fe nanoparticles embedded in an antiferromagnetic Cr_2O_3 matrix. The results of their magnetisation measurements are reproduced in Figure 9.56. A loop shift is seen to temperatures up to the Néel temperature of the oxide (see Table 9.4) and the Fe nanoparticles remain ferromagnetic at least up to 350 K.

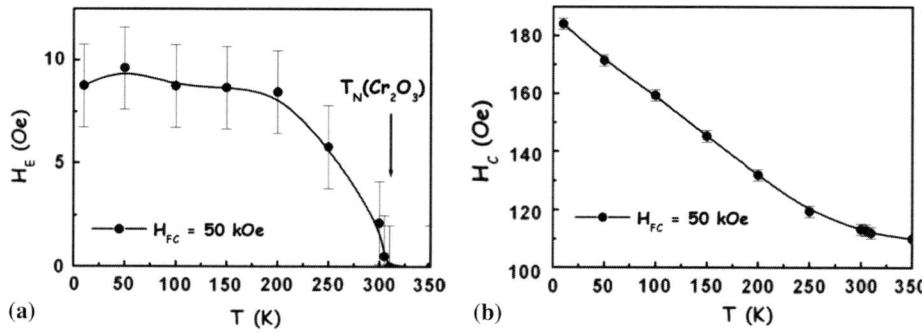

Figure 9.56. Temperature dependence of loop shift (a) and coercivity (b) for Fe particles embedded in Cr_2O_3. Reproduced with the permission of the Institute of Physics from Sort et al. [384].

Of course, the effect on exchange bias and superparamagnetism of dipolar interactions between the particles cannot be discounted, and indeed it is most likely to a significant influence. To get a rough estimate of interparticle separation, imagine that particles of diameter d are positioned regularly on an fcc lattice. The distance apart of their centres, d_{cc}, is given by $d_{cc}/d = [\pi/(3\sqrt{2}\phi)]^{1/3}$, where ϕ is the volume filling fraction. For a VFF of 10%, $d_{cc}/d = 1.9$ and, even at 1% with $d_{cc}/d = 4.2$, the particles can hardly be regarded as isolated. In addition, aggregation of the particles will influence the magnetic behaviour as discussed in Section 9.5.2. The influence of interparticle interactions is explicit in the study of a Co–CoO system by Koch et al. [363], where the results have been interpreted in terms of a CSSG.

Finally we note that there have been a number of attempts to model the behaviour of core–shell systems [388–390]. Inevitably it is difficult to make a quantitative comparison with experiments because of lack of knowledge about the detailed microstructure. However, with the assumption of uncompensated spins and random anisotropy at the interface, the trends observed experimentally are reproduced in the modelling.

References

[1] D. Weller and A. Moser, IEEE Trans. Magn. **35** (1999) 4423.
[2] D. Weller and M.F. Doerner, Annu. Rev. Mater. Sci. **30** (2000) 611.
[3] S.H. Sun, C.B. Murray, D. Weller, L. Folks and A. Moser, Science **287** (2000) 1989.
[4] U. Hafeli, W. Schütt, J. Teller and M. Zborowski (Eds.), Scientific and Clinical Applications of Magnetic Materials, Plenum, New York, 1997.
[5] J.I. Martín, J. Nogués, K. Liu, J.L. Vincent and I.K. Schuller, J. Magn. Magn. Mater. **256** (2003) 449.
[6] F.J. Himpsel, J.E. Ortega, G.J. Mankey and R.F. Willis, Adv. Phys. **47** (1998) 511.
[7] I. Žutić, J. Fabian and S. Das Sarma, Rev. Mod. Phys. **76** (2004) 323.
[8] P. Gambardella, S. Rusponi, M. Veronese, S.S. Dhesi, C. Grazioli, A. Dallmeyer, I. Cabria, R. Zeller, P.H. Dederichs, K. Kern, C. Carbone and H. Brune, Science **300** (2003) 1130.
[9] S. Rusponi, T. Cren, N. Weiss, M. Epple, P. Buluschek, L. Claude and H. Brune, Nat. Mater. **2** (2003) 546.
[10] B. Stahl, J. Ellrich, R. Theissmann, et al. Phys. Rev. B **67** (2003) 14422.
[11] R.L. Stamps, J. Phys. D: Appl. Phys. **33** (2000) R247.
[12] J. Nogués, J. Sort, V. Langlais, V. Skumryev, S. Suriñach, J.S. Muñoz and M.D. Baró, Phys. Rep. **422** (2005) 65.
[13] S. Padovani, F. Chado, F. Scheurer and J.P. Bucher, Phys. Rev. B **59** (1999) 11887.
[14] S. Padovani, P. Scheurer and J.P. Bucher, Europhys. Lett. **45** (1999) 327.
[15] J. Bansmann, S.H. Baker, C. Binns, J.A. Blackman, J.-P. Bucher, J. Dorantes-Dávila, V. Dupiuis, L. Favre, D. Kechrakos, A. Kleibert, K.-H. Meiwes-Broer, G.M. Pastor, A. Perez, O. Toulemonde, K.N. Trohidou, J. Tuaillon and Y. Xie, Surf. Sci. Rep. **56** (2005) 189.
[16] C. Binns, Surf. Sci. Rep. **44** (2001) 1.
[17] C. Binns, K.N. Trohidou, J. Bansmann, S.H. Baker, J.A. Blackman, J.-P. Bucher, D. Kechrakos, A. Kleibert, S. Louch, K.-H. Meiwes-Broer, G.M. Pastor, A. Perez and Y. Xie, J. Phys. D: Appl. Phys. **38** (2005) R357.
[18] K.-H. Meiwes-Broer (Ed.) Metal Clusters at Surfaces, Springer-Verlag, Berlin, 2000.
[19] A.J. Heinrich, J.A. Gupta, C.P. Lutz and D.M. Eigler, Science **306** (2004) 466.

[20] C.F. Hirjibehedin, C.-Y. Lin, A.F. Otte, M. Ternes, C.P. Lutz, B.A. Jones and A.J. Heinrich, Science **317** (2007) 1199.
[21] C.D. Stanciu, F. Hansteen, A.V. Kimel, A. Kirilyuk, A. Tsukamoto, A. Itoh and T. Rasing, Phys. Rev. Lett. **99** (2007) 047601.
[22] W. Wernsdorfer, Adv. Chem. Phys. **118** (2001) 99.
[23] D. Gatteschi, R. Sessoli and J. Villain, Molecular Nanomagnets, Oxford University Press, Oxford, 2006.
[24] R. Sessoli, D. Gatteschi, A. Caneschi and M.A. Novak, Nature **365** (1993) 141.
[25] S. Foner, Rev. Sci. Instrum. **30** (1959) 548.
[26] C. Binns, M.J. Maher, Q.A. Pankhurst, D. Kechrakos and K.N. Trohidou, Phys. Rev. B **66** (2002) 184413.
[27] J. Kerr, Philos. Mag. **3** (1877) 321.
[28] E.R. Moog and S.D. Bader, Superlatt. Microstruct. **1** (1985) 543.
[29] S.D. Bader, Proc. Inst. Elect. Electron. Eng. **78** (1990) 909.
[30] Z.J. Yang and M.R. Scheinfein, J. Appl. Phys. **74** (1993) 6810.
[31] Z.Q. Qiu and S.D. Bader, J. Magn. Magn. Mater. **200** (1999) 664.
[32] J. Bansmann, V. Senz, R.P. Methling, R. Röhlsberger and K.-H. Meiwes-Broer, Mater. Res. Eng. C **19** (2002) 305.
[33] C. Kao, J.B. Hastings, E.D. Johnson, D.P. Siddons, G.C. Smith and G.A. Prinz, Phys. Rev. Lett. **65** (1990) 373.
[34] J.B. Kortright and S.-K. Kim, Phys. Rev. B **62** (2000) 12216.
[35] H.C. Mertins, D. Abramsohn, A. Gaupp, F. Schäfers, W. Gudat, O. Zaharko, H. Grimmer and P.M. Oppeneer, Phys. Rev. B **66** (2002) 184404.
[36] M. Jamet, W. Wernsdorfer, C. Thirion, D. Mailly, V. Dupuis, P. Mélinon and A. Perez, Phys. Rev. Lett. **86** (2001) 4676.
[37] A.D. Kent, S. von Molnár, S. Gider and D.D. Awschalom, J. Appl. Phys. **76** (1994) 6656.
[38] S. Wirth and S. von Molnár, Appl. Phys. Lett. **76** (2000) 3282.
[39] Y. Li, P. Xiong, S. von Molnár, S. Wirth, Y. Ohno and H. Ohno, Appl. Phys. Lett. **80** (2002) 4644.
[40] G. Panaccione, F. Sirotti and G. Rossi, Solid State Commun. **113** (2000) 373.
[41] G. van der Laan, Phys. Rev. B **51** (1995) 240.
[42] F. Sirotti, S. Girlando, P. Prieto, L. Floreano, G. Panaccione and G. Rossi, Phys. Rev. B **61** (2000) R9221.
[43] K.W. Edmonds, C. Binns, S.H. Baker, M.J. Maher, S.C. Thornton, O. Tjernberg and N.B. Brookes, J. Magn. Magn. Mater. **231** (2001) 113.
[44] K.W. Edmonds, C. Binns, S.H. Baker, S.C. Thornton and P. Finetti, J. Appl. Phys. **88** (2000) 3414.
[45] C. Binns, J. Nanosci. Nanotechnol. **1** (2001) 243.
[46] B.T. Thole, P. Carra, F. Sette and G. van der Laan, Phys. Rev. B **68** (1992) 1943.
[47] P. Carra, B.T. Thole, M. Altarelli and X. Wang, Phys. Rev. Lett. **70** (1993) 694.
[48] J. Stöhr and Y. Wu in A.S. Schlachter and F.J. Wuilleumier (Eds.), New Directions in Research with Third Generation Synchrotron Radiation Sources, NATO ASI Series E: Applied Sciences, Kluwer, Dordrecht, 1994, p. 221.
[49] J. Stöhr, NEXAFS Spectroscopy, Springer Series in Surface Science, Vol. 25, Springer, Heidelberg, 1992.
[50] H.A. Bethe and E.E. Salpeter, Quantum Mechanics of One- and Two-Electron Atoms, Springer-Verlag, Berlin, 1957, p. 254.
[51] L.H. Tjeng, B. Sinkovic, N.B. Brookes, J.B. Goedkoop, R. Hesper, E. Pellegrin, F.M.F. de Groot, S. Altieri, S.L. Hulbert, E. Shekel and G.A. Sawatzky, Phys. Rev. Lett. **78** (1997) 1126.
[52] N.B. Brookes, B. Sinkovic, L.H. Tjeng, J.B. Goedkoop, R. Hesper, E. Pellegrin, F.M.F. de Groot, S. Altieri, S.L. Hulbert, E. Shekel and G.A. Sawatzky, J. Electron Spectrosc. Relat. Phenom. **92** (1998) 11.
[53] K.W. Edmonds, C. Binns, S.H. Baker, S.C. Thornton, C. Norris, J.B. Goedkoop, M. Finazzi and N.B. Brookes, Phys. Rev. B **60** (1999) 472.

[54] P.W. Atkins and R.S. Friedman, Molecular Quantum Mechanics, 3rd edn. Oxford University Press, Oxford, 1997.
[55] J. Stöhr and H. Konig, Phys. Rev. Lett. **75** (1995) 3748.
[56] J. Stöhr and R. Nakajima, IBM J. Res. Dev. **42** (1998) 73.
[57] C.T. Chen, Y.U. Idzerda, H.-J. Lin, N.V. Smith, G. Meigs, G.H. Ho, E. Pellegrin and F. Sette, Phys. Rev. Lett. **75** (1995) 152.
[58] H. Ebert, J. Stöhr, S.S.P. Parkin, M. Samant and A. Nilsson, Phys. Rev. B **53** (1996) 16067.
[59] O. Eriksson, B. Johansson, R.C. Albers, A.M. Boring and M.S.S. Brooks, Phys. Rev. B **42** (1990) 2707.
[60] P. Soderlind, O. Eriksson, B. Johansson, R.C. Albers and A.M. Boring, Phys. Rev. B **45** (1992) 12911.
[61] M. Altarelli, Phys. Rev. B **47** (1993) 597.
[62] A. Ankudinov and J.J. Rehr, Phys. Rev. B **51** (1995) 1282.
[63] S.H. Baker, C. Binns, K.W. Edmonds, M.J. Maher, S.C. Thornton, S. Louch and S.S. Dhesi, J. Magn. Magn. Mater. **247** (2002) 19.
[64] P. Bruno, Phys. Rev. B **39** (1989) 865.
[65] R. Nakajima, J. Stohr and Y.U. Idzerda, Phys. Rev. B **59** (1999) 6421.
[66] B.T. Thole, G. van der Laan and G.A. Sawatzky, Phys. Rev. Lett. **55** (1985) 2086.
[67] G. van der Laan, B.T. Thole, G.A. Sawatzky, J.B. Goedkoop, J.C. Fuggle, J.-M. Esteva, R. Karnatak, J.P. Remeika and H.A. Dablowska, Phys. Rev. B **34** (1986) 6529.
[68] M.M. Schwickert, G.Y. Guo, M.A. Tomaz, W.L. O'Brien and G.R. Harp, Phys. Rev. B **58** (1999) R4289.
[69] M.A. Tomaz, W.J. Antel, W.L. O'Brien and G.R. Harp, Phys. Rev. B **55** (1997) 3716.
[70] M.A. Tomaz, W.J. Antel, W.L. O'Brien and G.R. Harp, J. Phys.: Condens. Matter **9** (1997) L179.
[71] P. Lang, V.S. Stepanyuk, K. Wildberger, R. Zeller and P.H. Dederichs, Solid State Commun. **92** (1994) 755.
[72] V.S. Stepanyuk, W. Hergert, P. Rennert, K. Wildberger, R. Zeller and P.H. Dederichs, Surf. Sci. **377–379** (1997) 495.
[73] V.S. Stepanyuk, P. Lang, K. Wildberger, R. Zeller and P.H. Dederichs, Surf. Rev. Lett. **1** (1994) 477.
[74] V.S. Stepanyuk, W. Hergert, K. Wildberger, R. Zeller and P.H. Dederichs, Phys. Rev. B **53** (1996) 2121.
[75] V.S. Stepanyuk, W. Hergert, P. Rennert, K. Wildberger, R. Zeller and P.H. Dederichs, Phys. Rev. B **54** (1996) 14121.
[76] C.L. Fu, A.J. Freeman and T. Oguchi, Phys. Rev. Lett. **54** (1985) 2700.
[77] I.G. Kim and J.I. Lee, J. Magn. Magn. Mater. **272–276** (2004) 1188.
[78] S. Blügel, Phys. Rev. Lett. **68** (1992) 851.
[79] S. Blügel, M. Weinert and P.H. Dederichs, Phys. Rev. Lett. **60** (1988) 1077.
[80] P. Benčok, S. Stanescu, J. Cezar and N.B. Brookes, Thin Solid Films **515** (2006) 724.
[81] B.-S. Kang, S.-K. Oh, J.-S. Chung and K.-S. Sohn, Phys. B **304** (2001) 67.
[82] D.P. Moore, O. Ozturk, F.O. Schumann, S.A. Morton and G.D. Waddill, Surf. Sci. **449** (2000) 31.
[83] B.V. Reddy, M.R. Pederson and S.N. Khanna, Phys. Rev. B **55** (1997) R7414.
[84] H. Nait-Laziz, C. Demangeat and A. Mokrani, J. Magn. Magn. Mater. **121** (1993) 123.
[85] S.E. Weber, B.K. Rao, P. Jena, V.S. Stepanyuk, W. Hergert, K. Wildberger, R. Zeller and P.H. Dederichs, J. Phys.: Condens. Matter **9** (1997) 10739.
[86] O. Hjortstam, J. Trygg, J.M. Wills, B. Johansson and O. Eriksson, Phys. Rev. B **53** (1996) 9204.
[87] M. Tischer, O. Hjortstam, D. Arvanitis, J. Hunter Dunn, F. May, K. Baberschke, J. Trygg, J.M. Wills, B. Johansson and O. Eriksson, Phys. Rev. Lett. **75** (1995) 1602.
[88] H. Li and B.P. Tonner, Surf. Sci. **237** (1990) 141.
[89] A.K. Schmid and J. Kirschner, Ultramicroscopy **42–44** (1992) 483.
[90] M.T. Kief and E.F. Egelhoff Jr., Phys. Rev. B **47** (1993) 10785.

[91] J.R. Cerdá, P.L. Andres, A. Cebollada, R. Miranda, E. Navas, P. Schuster, C.M. Schneider and J. Kirschner, J. Phys.: Condens. Matter **5** (1993) 2055.
[92] W. Weber, A. Bischof, R. Allerspach and C.H. Back, Phys. Rev. B **54** (1996) 4075.
[93] J. Fassbender, R. Allenspach and U. Dürig, Surf. Sci. **383** (1997) L748.
[94] F. Nouvertné, U. May, M. Bamming, A. Rampe, U. Korte, G. Güntherodt, R. Pentcheva and M. Scheffler, Phys. Rev. B **60** (1999) 14382.
[95] S.-K. Kim, J.-S. Kim, J.Y. Han, J.M. Seo, C.K. Lee and S.C. Hong, Surf. Sci. **453** (2000) 47.
[96] N.A. Levanov, V.S. Stepanyuk, W. Hergert, D.I. Bazhanov, P.H. Dederichs, A. Katsnelson and C. Massobrio, Phys. Rev. B **61** (2000) 2230.
[97] V.S. Stepanyuk, D.I. Bazhanov, A.N. Baranov, W. Hergert, P.H. Dederichs and J. Kirschner, Phys. Rev. B **62** (2000) 15398.
[98] J. Izquierdo, D.I. Bazhanov, A. Vega, V.S. Stepanyuk and W. Hergert, Phys. Rev. B **63** (2001) 140413.
[99] V.S. Stepanyuk, D.V. Tsivline, D.I. Bazhanov, W. Hergert and A. Katsnelson, Phys. Rev. B **63** (2001) 235406.
[100] R. Robles, R.C. Longo, A. Vega, C. Rey, V. Stepanyuk and L.J. Gallego, Phys. Rev. B **66** (2002) 064410.
[101] R. Pentcheva and M. Scheffler, Phys. Rev. B **65** (2002) 155418.
[102] R.C. Longo, V.S. Stepanyuk, W. Hergert, A. Vega, L.J. Gallego and J. Kirschner, Phys. Rev. B **69** (2004) 073406.
[103] Š. Pick, V.S. Stepanyuk, A.V. Klavsyuk, L. Niebergall, W. Hergert, J. Kirschner and P. Bruno, Phys. Rev. B **70** (2004) 224419.
[104] L. Gómez, C. Slutzky, J. Férron, J. de la Figuera, J. Camarero, A.J. Vázquez de Parga, J.J. de Miguel and R. Miranda, Phys. Rev. Lett. **84** (2000) 4397.
[105] M.S.S. Brookes, Physica B **130** (1985) 6.
[106] B. Nonas, I. Cabria, R. Zeller, P.H. Dederichs, T. Huhne and H. Ebert, Phys. Rev. Lett. **86** (2001) 2146.
[107] I. Cabria, B. Nonas, R. Zeller and P.H. Dederichs, Phys. Rev. B **65** (2002) 054414.
[108] B. Lazarovits, L. Szunyogh and P. Weinberger, Phys. Rev. B **65** (2002) 104441.
[109] A.B. Klautau and S. Frota-Pessôa, Surf. Sci. **579** (2005) 27.
[110] Š. Pick, V.S. Stepanyuk, A.N. Baranov, W. Hergert and P. Bruno, Phys. Rev. B **68** (2003) 104410.
[111] V.S. Stepanyuk, A.N. Baranov, W. Hergert and P. Bruno, Phys. Rev. B **68** (2003) 205442.
[112] V.S. Stepanyuk, W. Hergert, K. Wildberger, S.K. Nayak and P. Jena, Surf. Sci. **384** (1997) L892.
[113] S. Uzdin, V. Uzdin and C. Demangeat, Europhys. Lett. **47** (1999) 556.
[114] S. Uzdin, V. Uzdin and C. Demangeat, Surf. Sci. **482–485** (2001) 965.
[115] A. Bergman, L. Nordström, A.B. Klautau, S. Frota-Pessôa and O. Eriksson, Phys. Rev. B **73** (2006) 174434.
[116] A. Bergman, L. Nordström, A.B. Klautau, S. Frota-Pessôa and O. Eriksson, Phys. Rev. B **75** (2007) 224455.
[117] K. Wildberger, V.S. Stepanyuk, P. Lang, R. Zeller and P.H. Dederichs, Phys. Rev. Lett. **75** (1995) 509.
[118] V.S. Stepanyuk, W. Hergert, P. Rennert, J. Izquierdo, A. Vega and L.C. Balbás, Phys. Rev. B **57** (1998) 14020.
[119] O. Eriksson, R.C. Albers and A.M. Boring, Phys. Rev. Lett. **66** (1991) 1350.
[120] G. Bergmann and M. Hossain, Phys. Rev. Lett. **86** (2000) 2138.
[121] P. Gambardella, S.S. Dhesi, S. Gardonio, C. Grazioli, P. Ohresser and C. Carbone, Phys. Rev. Lett. **88** (2002) 047202.
[122] A.J. Cox, J.G. Louderback and L.A. Bloomfield, Phys. Rev. Lett. **71** (1993) 923.
[123] A.J. Cox, J.G. Louderback, S.E. Apsel and L.A. Bloomfield, Phys. Rev. B **49** (1994) 12295.
[124] R. Pfandzelter, G. Steierl and C. Rau, Phys. Rev. Lett. **74** (1995) 3467.
[125] G.A. Mulhollan, R.L. Fink and J.L. Erskine, Phys. Rev. B **44** (1991) 2393.

[126] C. Liu and S.D. Bader, Phys. Rev. B **44** (1991) 12062.
[127] I. Chado, F. Scheurer and J.P. Bucher, Phys. Rev. B **64** (2001) 094410.
[128] J. Honolka, K. Kuhnke, L. Vitali, A. Enders, K. Kern, S. Gardonio, C. Carbone, S.R. Krishnakumar, P. Bencok, S. Stepanow and P. Garbardella, Phys. Rev. B **76** (2007) 144412.
[129] B. Nonas, K. Wildberger, R. Zeller and P.H. Dederichs, J. Magn. Magn. Mater. **165** (1997) 137.
[130] S. Handschuh and S. Blügel, Solid State Commun. **105** (1998) 633.
[131] B. Nonas, K. Wildberger, R. Zeller and P.H. Dederichs, Phys. Rev. Lett. **80** (1998) 4574.
[132] A. Davies, J.A. Stroscio, D.T. Pierce and R.J. Celotta, Phys. Rev. Lett. **76** (1996) 4175.
[133] D. Venus and B. Heinrich, Phys. Rev. B **53** (1996) R1733.
[134] R.Q. Wu and A.J. Freeman, Phys. Rev. B **51** (1995) 17131.
[135] R. Wu and A.J. Freeman, J. Magn. Magn. Mater. **161** (1996) 89.
[136] O. Elmouhssine, G. Moraïtis, C. Demangeat and J.C. Parlebas, Phys. Rev. B **55** (1997) R7410.
[137] A. Noguera, S. Bouarab, A. Mokrani, C. Demangeat and H. Dreyssé, J. Magn. Magn. Mater. **156** (1996) 21.
[138] B. Heinrich, A.S. Arrott, C. Liu and S.T. Purcell, J. Vac. Sci. Technol. A **5** (1987) 1935.
[139] W.L. O'Brien and B.P. Tonner, Phys. Rev. B **50** (1994) 2963.
[140] T.G. Walker and H. Hopster, Phys. Rev. B **48** (1993) 563.
[141] B. M'Passi-Mabiala, S. Meza-Aguilar and C. Demangeat, Phys. Rev. B **65** (2001) 012414.
[142] B. M'Passi-Mabiala, S. Meza-Aguilar and C. Demangeat, Surf. Sci. **518** (2002) 104.
[143] B.-Ch. Choi, P.J. Bode and J.A.C. Bland, Phys. Rev. B **58** (1998) 5166.
[144] B.-Ch. Choi, P.J. Bode and J.A.C. Bland, Phys. Rev. B **59** (1999) 7029.
[145] T.K. Yamada, M.M.J. Bischoff, T. Mizoguchi and H. van Kempen, Surf. Sci. **516** (2002) 179.
[146] Š. Pick and C. Demangeat, Surf. Sci. **584** (2005) 146.
[147] W.L. O'Brien and B.P. Tonner, Phys. Rev. B **58** (1998) 3191.
[148] Y. Yonamoto, T. Yokoyam, K. Amemiya, K. Matsumara and T. Ohta, Phys. Rev. B **63** (2001) 214406.
[149] H. Zenia, S. Bourab, J. Ferrer and C. Demangeat, Surf. Sci. **564** (2004) 12.
[150] C. Demangeat and J.C. Parlebas, Rep. Prog. Phys. **65** (2002) 1679.
[151] S. Lounis, P. Mavropoulos, R. Zeller, P.H. Dederichs and S. Blügel, Phys. Rev. B **72** (2005) 224437.
[152] S. Lounis, P. Mavropoulos, R. Zeller, P.H. Dederichs and S. Blügel, Phys. Rev. B **75** (2007) 174436.
[153] R. Robles and L. Nordström, Phys. Rev. B **74** (2006) 094403.
[154] J.T. Lau, A. Föhlisch, R. Nietubyć, M. Reif and W. Wurth, Phys. Rev. Lett. **89** (2002) 057201.
[155] J.T. Lau, A. Föhlisch, M. Martins, R. Nietubyć, M. Reif and W. Wurth, New J. Phys. **4** (2002) 98.
[156] E. Martínez, R.C. Longo, R. Robles, A. Vega and L.J. Gallego, Phys. Rev. B **71** (2005) 165425.
[157] P. Mavropoulos, S. Lounis, R. Zeller and S. Blügel, Appl. Phys. A **82** (2006) 103.
[158] Y. Yayon, V.W. Brar, L. Senapati, S.C. Erwin and M.F. Crommie, Phys. Rev. Lett. **99** (2007) 067202.
[159] D. Weller, A. Moser, L. Folks, M.E. Best, W. Lee, M. Toney, M. Schwickert, J.-U. Thiele and M. Doerner, IEEE Trans. Magn. **36** (2000) 10.
[160] O.A. Ivanov, L.V. Solina, V.A. Demshina and L.M. Magat, Phys. Met. Metallogr. **35** (1973) 92.
[161] I.V. Solovyev, P.H. Dederichs and I. Mertig, Phys. Rev. B **52** (1995) 13419.
[162] I. Galanakis, M. Alouani and H. Dreyssé, Phys. Rev. B **62** (2000) 6475.
[163] P. Ravindran, A. Kjekshus, H. Fjellvåg, P. James, L. Nordström, B. Johansson and O. Eriksson, Phys. Rev. B **63** (2001) 144409.

[164] A.B. Shick and O.N. Mryasov, Phys. Rev. B **67** (2003) 172407.
[165] J.B. Staunton, S. Ostanin, S.S.A. Razee, B.L. Gyorffy, L. Szunyogh, B. Ginatempo and E. Bruno, J. Phys.: Condens. Matter **16** (2004) S5623.
[166] J.B. Staunton, S. Ostanin, S.S.A. Razee, B.L. Gyorffy, L. Szunyogh, B. Ginatempo and E. Bruno, Phys. Rev. Lett. **93** (2004) 257204.
[167] T. Burkert, O. Eriksson, S.I. Simak, A.V. Ruban, B. Sanyal, L. Nordström and J.M. Wills, Phys. Rev. B **71** (2005) 134411.
[168] J.B. Staunton, L. Szunyogh, A. Buruzs, B.L. Gyorffy, S. Ostanin and L. Udvardi, Phys. Rev. B **74** (2006) 144411.
[169] A. Perez, V. Dupuis, J. Tuaillon-Combes, L. Bardotti, B. Prevel, E. Bernstein, P. Mélinon, L. Favre, A. Hannour and M. Jamet, Adv. Eng. Mater. **7** (2005) 475.
[170] P. Gambardella, S. Rusponi, T. Cren, N. Weiss and H. Brune, C.R. Phys. **6** (2005) 75.
[171] S. Bornemann, J. Minár, S. Polesya, S. Mankovsky, H. Ebert and O. Šipr, Phase Trans. **78** (2005) 701.
[172] Y. Xie and J.A. Blackman, Phys. Rev. B **74** (2006) 054401.
[173] J. Minár, S. Bornemann, O. Šipr, S. Polesya and H. Ebert, Appl. Phys. A **82** (2006) 139.
[174] S. Bornemann, J. Minár, J.B. Staunton, J. Honolka, A. Enders, K. Kern and H. Ebert, Eur. Phys. J. D **45** (2007) 529.
[175] O. Šipr, S. Bornemann, J. Minár, S. Polesya, V. Popescu, A. Šimůnek and H. Ebert, J. Phys.: Condens. Matter **19** (2007) 096203.
[176] J. Hafner and D. Spišák, Phys. Rev. B **76** (2007) 094420.
[177] W. Kuch, Nat. Mater. **2** (2003) 505.
[178] J. Thiele, C. Boeglin, K. Hricovini and F. Chevrier, Phys. Rev. B **53** (1996) R11934.
[179] S. Ferrer, J. Alvarez, E. Lundgren, X. Torelles, P. Fajardo and F. Boscherini, Phys. Rev. B **56** (1997) 9848.
[180] N. Nakajima, T. Koide, T. Shidara, H. Miyauchi, H. Fukutani, A. Fujimori, K. Ilio, T. Katayama, M. Nývlt and Y. Suzuki, Phys. Rev. Lett. **81** (1998) 5229.
[181] O. Robach, C. Quiros, P. Steadman, K.F. Peters, E. Lundgren, J. Alvarez, H. Isern and S. Ferrer, Phys. Rev. B **65** (2002) 054423.
[182] J. Kim, J.-W. Lee, J.-R. Jeong, S.-C. Shin, Y.H. Ha, Y. Park and D.W. Moon, Phys. Rev. B **65** (2002) 104428.
[183] J.-W. Lee, J.-R. Jeong, S.-C. Shin, J. Kim and S.-K. Kim, Phys. Rev. B **66** (2002) 172409.
[184] R. Cheng, K.Y. Guslineko, F.Y. Fradin, J.E. Pearson, H.F. Ding, D. Li and S.D. Bader, Phys. Rev. B **72** (2005) 014409.
[185] D. Repetto, T.Y. Lee, S. Rusponi, J. Honolka, K. Kuhnke, V. Sessi, U. Starke, H. Brune, P. Gambardella, C. Carbone, A. Enders and K. Kern, Phys. Rev. B **74** (2006) 054408.
[186] R. Cheng, J. Pearson, D. Li and F.Y. Fradin, J. Appl. Phys. **100** (2006) 073911.
[187] U. Pustogowa, J. Zabloudil, C. Uiberacker, C. Blaas, P. Weinberger, L. Szunyogh and C. Sommers, Phys. Rev. B **60** (1999) 414.
[188] M. Tsujikawa, A. Hosokawa and T. Oda, J. Magn. Magn. Mater. **310** (2007) 2189.
[189] M. Tsujikawa, A. Hosokawa and T. Oda, J. Phys.: Condens. Matter **19** (2007) 365208.
[190] J. Dorantes-Dávila, H. Dreyssé and G.M. Pastor, Phys. Rev. Lett. **91** (2003) 197206.
[191] R. Félix-Medina, J. Dorantes-Dávila and G.M. Pastor, Phys. Rev. B **67** (2003) 094430.
[192] E. Lundgren, B. Stanka, M. Schmid and P. Varga, Phys. Rev. B **62** (2000) 2843.
[193] P. Beauvillain, A. Bounouh, C. Chappert, R. Mégy, S. Ould-Mahfoud, J.P. Renard, P. Veillet, D. Weller and J. Corno, J. Appl. Phys. **76** (1994) 6078.
[194] J. Kohlhepp and U. Gradmann, J. Magn. Magn. Mater. **139** (1995) 347.
[195] H. Fritzsche, J. Kohlhepp and U. Gradmann, J. Magn. Magn. Mater. **148** (1995) 154.
[196] B.N. Engel, M.H. Wiedmann, R.A. von Leeuwen and C.M. Falco, Phys. Rev. B **48** (1993) 9894.
[197] F. Meier, K. von Bergmann, P. Ferriani, J. Wiebe, M. Bode, K. Hashimoto, S. Heinze and R. Wiesendanger, Phys. Rev. B **74** (2006) 195411.
[198] C. Binns, S.H. Baker, C. Demangeat and J.C. Parlebas, Surf. Sci. Rep. **34** (1999) 10.
[199] E. Ganz, K. Sattler and J. Clarke, Phys. Rev. Lett. **60** (1988) 1856.

[200] E. Ganz, K. Sattler and J. Clarke, Surf. Sci. **219** (1989) 33.
[201] G.M. Francis, I.M. Goldby, L. Kuipers, B. von Issendorff and R.E. Palmer, J. Chem. Soc., Dalton Trans. **5** (1996) 665.
[202] G.M. Francis, L. Kuipers, J.R.A. Cleaver and R.E. Palmer, J. Appl. Phys. **79** (1996) 2942.
[203] P.A. Mulheran and J.A. Blackman, Surf. Sci. **376** (1997) 403.
[204] S.J. Carroll, R.E. Palmer, P.A. Mulheran, S. Hobday and R. Smith, Appl. Phys. A **67** (1998) 613.
[205] C.D. Pownall and P.A. Mulheran, Phys. Rev. B **60** (1999) 9037.
[206] I. Moullet, Surf. Sci. **331–333** (1995) 697.
[207] D. Duffy and J.A. Blackman, Surf. Sci. **415** (1998) L1016.
[208] G.M. Wang, J.J. BelBruno, S.D. Kenny and R. Smith, Surf. Sci. **576** (2005) 107.
[209] D. Duffy and J.A. Blackman, Phys. Rev. B **58** (1998) 7443.
[210] P. Krüger, A. Rakotomahevitra, J.C. Parlebas and C. Demangeat, Phys. Rev. B **57** (1998) 5276.
[211] S.S. Peng, B.R. Cooper and Y.G. Hao, Philos. Mag. B **73** (1996) 611.
[212] K. Fauth, S. Gold, M. Heßler, N. Schneider and G. Schütz, Chem. Phys. Lett. **392** (2004) 498.
[213] K. Fauth, M. Heßler, D. Batchelor and G. Schütz, Surf. Sci. **529** (2003) 397.
[214] K. Fauth, N. Schneider, M. Heßler and G. Schütz, Eur. Phys. J. D **29** (2004) 57.
[215] A.N. Andriotis, M. Menon and G.E. Froudakis, Phys. Rev. B **62** (2000) 9867.
[216] M. Menon, A.N. Andriotis and G.E. Froudakis, Chem. Phys. Lett. **320** (2000) 425.
[217] T.W. Odom, J.-L. Huang, C.L. Cheung and C.M. Lieber, Science **290** (2000) 1549.
[218] F. Banhart, N. Grobert, M. Terrones, J.C. Charlier and P.M. Ajayan, Int. J. Mod. Phys. B **15** (2001) 4037.
[219] S.B. Fagan, R. Mota, A.J.R. da Silva and A. Fazzio, Phys. Rev. B **67** (2003) 205414.
[220] Y. Yagi, T.M. Briere, M.H.F. Sluiter, V. Kumar, A.A. Farajian and Y. Kawazoe, Phys. Rev. B **69** (2004) 075414.
[221] N.Y. Jin-Phillipp and M. Rüle, Phys. Rev. B **70** (2004) 245421.
[222] N. Fujima and T. Oda, Phys. Rev. B **71** (2005) 115412.
[223] L. Kouwenhoven and L. Glazman, Phys. World **14** (2001) 33.
[224] W.J. de Haas and G.J. van den Berg, Physica **3** (1936) 440.
[225] J. Kondo, Prog. Theor. Phys. **32** (1964) 37.
[226] Y. Nagaoka, Phys. Rev. **138** (1965) A1112.
[227] P.W. Anderson, Phys. Rev. **124** (1961) 41.
[228] P.W. Anderson, G. Yuval and D.R. Hamann, Phys. Rev. B **1** (1970) 4664.
[229] P.W. Anderson, G. Yuval and D.R. Hamann, Solid State Commun. **8** (1970) 1033.
[230] P.W. Anderson, J. Phys. C: Solid State Phys. **3** (1970) 2436.
[231] K.G. Wilson, Rev. Mod. Phys. **47** (1975) 773.
[232] A.C. Hewson, The Kondo Problem to Heavy Fermions, Cambridge University Press, Cambridge, 1997.
[233] F.D.M. Haldane, Phys. Rev. Lett. **40** (1978) 416.
[234] V. Madhavan, W. Chen, T. Jamneala, M.F. Crommie and N.S. Wingreen, Science **280** (1998) 567.
[235] J. Li, W.-D. Schneider, R. Berndt and B. Delley, Phys. Rev. Lett. **80** (1998) 2893.
[236] U. Fano, Phys. Rev. **124** (1961) 1866.
[237] V. Madhavan, W. Chen, T. Jamneala, M.F. Crommie and N.S. Wingreen, Phys. Rev. B **64** (2001) 165412.
[238] A. Schiller and S. Hershfield, Phys. Rev. B **61** (2000) 9036.
[239] O. Újsághy, J. Kroha, L. Szunyogh and A. Zawadowski, Phys. Rev. Lett. **85** (2000) 2557.
[240] C.Y. Lin, A.H. Castro Neto and B.A. Jones, Phys. Rev. Lett. **97** (2006) 156102.
[241] M. Plihal and J.W. Gadzuk, Phys. Rev. B **63** (2001) 085404.
[242] T. Jamneala, V. Madhavan, W. Chen and M.F. Crommie, Phys. Rev. B **61** (2000) 9990.

[243] K. Nagaoka, T. Jamneala, M. Grobis and M.F. Crommie, Phys. Rev. Lett. **88** (2002) 077205.
[244] N. Knorr, M.A. Schneider, L. Diekhöner, P. Wahl and K. Kern, Phys. Rev. Lett. **88** (2002) 096804.
[245] M.A. Schneider, L. Vitali, N. Knorr and K. Kern, Phys. Rev. B **65** (2002) 121406.
[246] P. Wahl, L. Diekhöner, M.A. Schneider, L. Vitali, G. Wittich and K. Kern, Phys. Rev. Lett. **93** (2004) 176603.
[247] W. Chen, T. Jamneala, V. Madhavan and M.F. Crommie, Phys. Rev. B **60** (1999) R8529.
[248] P. Wahl, P. Simon, L. Diekhöner, V.S. Stepanyuk, P. Bruno, M.A. Schneider and K. Kern, Phys. Rev. Lett. **98** (2007) 056601.
[249] T. Jamneala, V. Madhavan and M.F. Crommie, Phys. Rev. Lett. **87** (2001) 256804.
[250] Yu.B. Kudasov and V.M. Uzdin, Phys. Rev. Lett. **89** (2002) 276802.
[251] K. Ingersent, A.W.W. Ludwig and I. Affleck, Phys. Rev. Lett. **95** (2005) 257204.
[252] H.C. Manoharan, C.P. Lutz and D.M. Eigler, Nature **403** (2000) 512.
[253] G.A. Fiete and E.J. Heller, Rev. Mod. Phys. **75** (2003) 933.
[254] A.A. Aligia and A.M. Lobos, J. Phys.: Condens. Matter **17** (2005) S1095.
[255] J. Dorantes-Dávila and G.M. Pastor, Phys. Rev. Lett. **81** (1998) 208.
[256] M. Eisenbach, B.L. Györffy, G.M. Stocks and B. Újfalussy, Phys. Rev. B **65** (2002) 144424.
[257] M. Komelj, C. Ederer, J.W. Davenport and M. Fähnle, Phys. Rev. B **66** (2002) 140407.
[258] R. Félix-Medina, J. Dorantes-Dávila and G.M. Pastor, New J. Phys. **4** (2002) 100.
[259] B. Lazarovits, L. Szunyogh and P. Weinberger, Phys. Rev. B **67** (2003) 024415.
[260] B. Lazarovits, L. Szunyogh, P. Weinberger and B. Újfalussy, Phys. Rev. B **68** (2003) 024433.
[261] J. Hong and R.Q. Wu, Phys. Rev. B **67** (2003) 020406.
[262] A.B. Shick, F. Máca and P.M. Oppeneer, Phys. Rev. B **69** (2004) 212410.
[263] J. Hong and R.Q. Wu, Phys. Rev. B **70** (2004) 060406.
[264] B. Újfalussy, B. Lazarovits, L. Szunyogh, G.M. Stocks and P. Weinberger, Phys. Rev. B **70** (2004) 100404.
[265] A.B. Klautau and S. Frota-Pessôa, Phys. Rev. B **70** (2004) 193407.
[266] B. Lazarovits, B. Újfalussy, L. Szunyogh, G.M. Stocks and P. Weinberger, J. Phys, Condens. Matter **16** (2004) S5833.
[267] A.B. Shick, F. Máca and P.M. Oppeneer, J. Magn. Magn. Mater. **290–291** (2005) 257.
[268] M. Komelj, D. Steiauf and M. Fähnle, Phys. Rev. B **73** (2006) 134428.
[269] Y. Li and B.-G. Liu, Phys. Rev. B **73** (2006) 174418.
[270] S. Baud, G. Bihlmayer, S. Blügel and Ch. Ramseyer, Surf. Sci. **600** (2006) 4301.
[271] Š. Pick, P.A. Ignatiev, A.V. Klavsyuk, W. Hergert, V.S. Stepanyuk and P. Bruno, J. Phys.: Condens. Matter **19** (2007) 446001.
[272] M.C. Desjonquères, C. Barreteau, G. Autès and D. Spanjaard, Eur. Phys. J. B **55** (2007) 23.
[273] P. Gambardella, M. Blanc, H. Brune, K. Kuhnke and K. Kern, Phys. Rev. B **61** (2000) 2254.
[274] P. Gambardella, M. Blanc, L. Bürgi, K. Kuknke and K. Kern, Surf. Sci. **449** (2000) 93.
[275] A. Dallmeyer, C. Carbone, W. Eberhardt, C. Pampuch, O. Rader, W. Gudat, P. Gambardella and K. Kern, Phys. Rev. B **61** (2000) 5133.
[276] P. Gambardella, A. Dallmeyer, K. Maiti, M.C. Malagoli, W. Eberhardt, K. Kern and C. Carbone, Nature **416** (2002) 301.
[277] P. Gambardella, J. Phys.: Condens. Matter **15** (2003) S2533.
[278] P. Gambardella, A. Dallmeyer, K. Maiti, M.C. Malagoli, S. Rusponi, P. Ohresser, W. Eberhardt, C. Carbone and K. Kern, Phys. Rev. Lett. **93** (2004) 077203.
[279] L. Landau and E. Lifshitz, Statistical Physics, Vol. 5, Pergamon, London, 1959.
[280] P. Gambardella and K. Kern, Surf. Sci. **475** (2001) L229.
[281] I.M.L. Billas, J.A. Becker, A. Châtelain and W.A. de Heer, Phys. Rev. Lett. **71** (1993) 4067.
[282] J. Bansmann and A. Kleibert, Appl. Phys. A **80** (2005) 957.

[283] A. Vega, L.C. Balbás, J. Dorantes-Dávila and G.M. Pastor, Phys. Rev. B **50** (1994) 3899.
[284] C. Xiao, J. Yang, K. Deng and K. Wang, Phys. Rev. B **55** (1997) 3677.
[285] P. Alvarado, J. Dorantes-Dávila and G.M. Pastor, Phys. Rev. B **58** (1998) 12216.
[286] J. Guevara, A.M. Llois and M. Weissmann, Phys. Rev. Lett. **81** (1998) 5306.
[287] R.N. Nogueira and H.M. Petrilli, Phys. Rev. B **60** (1999) 4120.
[288] R.N. Nogueira and H.M. Petrilli, Phys. Rev. B **63** (2000) 012405.
[289] S. Krüger, M. Stener and N. Rösch, J. Chem. Phys. **114** (2001) 5207.
[290] Y. Xie and J.A. Blackman, Phys. Rev. B **66** (2002) 085410.
[291] *ibid*, p. 155417.
[292] C. Binns, M.J. Maher, Q.A. Pankhurst, D. Kechrakos and K.N. Trohidou, Phys. Rev. B **66** (2002) 184413.
[293] C. Binns and M.J. Maher, New J. Phys. **4** (2002) 85.
[294] T.J. Jackson, C. Binns, E.M. Forgan, E. Morenzoni, C. Niedermayer, H. Glückler, A. Hofer, H. Luetkens, T. Prokscha, T.M. Riseman, A. Schatz, M. Birke, J. Litterst, G. Schatz and H.P. Weber, J. Phys.: Condens. Matter **12** (2000) 1399.
[295] H. Magnan, D. Chandesris, B. Villette, O. Heckmann and J. Lecante, Phys. Rev. Lett. **67** (1991) 859.
[296] S. Müller, P. Bayer, C. Reischl, K. Heinz, B. Feldmann, H. Zillgen and M. Wuttig, Phys. Rev. Lett. **74** (1995) 765.
[297] R.D. Ellerbrock, A. Fuest, A. Schatz, W. Keune and R.A. Brand, Phys. Rev. Lett. **74** (1995) 3053.
[298] D. Schmitz, C. Charton, A. Scholl, C. Carbone and W. Eberhardt, Phys. Rev. B **59** (1999) 4327.
[299] J. Shen, M. Klaua, P. Ohresser, H. Jenniches, J. Barthel, Ch.V. Mohan and J. Kirschner, Phys. Rev. B **56** (1997) 11134.
[300] J. Shen, P. Ohresser, Ch.V. Mohan, M. Klaua, J. Barthel and J. Kirschner, Phys. Rev. Lett. **80** (1998) 1980.
[301] S.H. Baker, A.M. Asaduzzaman, M. Roy, S.J. Gurman, C. Binns, J.A. Blackman and Y. Xie, Phys. Rev. B **78** (2008) 014422.
[302] V.L. Moruzzi, P.M. Marcus, K. Schwarz and P. Mohn, Phys. Rev. B **34** (1986) 1784.
[303] V.L. Moruzzi, P.M. Marcus and J. Kübler, Phys. Rev. B **39** (1989) 6957.
[304] D. Guenzburger and D.E. Ellis, Phys. Rev. B **51** (1995) 1519.
[305] C. Paduani and J.C. Krause, J. Appl. Phys. **86** (1999).
[306] R.H. Victora and L.M. Falicov, Phys. Rev. B **30** (1984) 259.
[307] R.H. Victora, L.M. Falicov and S. Ishida, Phys. Rev. B **30** (1984) 3896.
[308] K. Edmonds, C. Binns, S.H. Baker, M.J. Maher, S.C. Thornton, O. Tjernberg and N.B. Brookes, J. Magn. Magn. Mater. **231** (2001) 113.
[309] D. Kechrakos and K.N. Trohidou, Phys. Rev. B **58** (1998) 12169.
[310] P. Allia, M. Coisson, P. Tiberto, F. Vinai, M. Knobel, M.A. Novak and W.C. Nunes, Phys. Rev. B **64** (2001) 144420.
[311] D. Stauffer and A. Aharony, Introduction to Percolation Theory, Taylor and Francis, London, 1994.
[312] E. Chudnovsky, J. Magn. Magn. Mater. **40** (1983) 21.
[313] E. Chudnovsky, W.M. Saslow and R.A. Serota, Phys. Rev. B **33** (1986) 251.
[314] W.M. Saslow, Phys. Rev. B **35** (1987) 5770.
[315] E. Chudnovsky, J. Appl. Phys. **64** (1988) 5770.
[316] L. Thomas, J. Tuaillon, J.P. Perez, V. Dupuis, A. Perez and B. Barbara, J. Magn. Magn. Mater. **140** (1995) 437.
[317] J.P. Perez, V. Dupuis, J. Tuaillon, A. Perez, V. Paillard, P. Mélinon, M. Treilleux, L. Thomas, B. Barbara and B. Bouchet-Fabre, J. Magn. Magn. Mater. **145** (1995) 74.
[318] J.F. Löffler, H.B. Braun and W. Wagner, Phys. Rev. Lett. **85** (2000) 1990.
[319] C. Binns, F. Sirotti, H. Gruegel, P. Prieto, S.H. Baker and S.C. Thornton, J. Phys: Condens. Matter **15** (2003) 4287.
[320] C. Binns, P.B. Howes, S.H. Baker, H. Marchetto, A. Potenza, P. Steadman, S.S. Dhesi, M. Roy, M.J. Everard and A. Rushforth, J. Phys.: Condens. Matter **20** (2008) 055213.

[321] W.H. Meiklejohn and C.P. Bean, Phys. Rev. **102** (1956) 1413.
[322] W.H. Meiklejohn and C.P. Bean, Phys. Rev. **105** (1957) 904.
[323] W.H. Meiklejohn, J. Appl. Phys. **33** (1962) 1328.
[324] L. Néel, Ann. Phys. Paris **2** (1967) 61.
[325] C. Mauri, H.C. Siegmann, P.S. Bagus and E. Kay, J. Appl. Phys. **62** (1987) 3047.
[326] C. Kittel, Introduction to Solid State Physics, Wiley, New York, 1971.
[327] A. Scholl, M. Liberati, E. Arenholz, H. Ohldag and J. Stohr, Phys. Rev. Lett. **92** (2004) 247201.
[328] J. van Driel, F.R. de Boer, K.-M.H. Lenssen and R. Coehoorn, J. Appl. Phys. **8** (2000) 975.
[329] H. Xi and R.M. White, Phys. Rev. B **61** (2000) 80.
[330] M. Ali, C.H. Marrows, M. Al-Jawad, B.J. Hickey, A. Misra, U. Nowak and K.D. Usadel, Phys. Rev. B **68** (2003) 214420.
[331] N.C. Koon, Phys. Rev. Lett. **78** (1997) 4865.
[332] T.C. Schulthess and W.H. Butler, Phys. Rev. Lett. **81** (1998) 4516.
[333] M. Kiwi, J. Mejía-López, R.D. Portugal and R. Ramírez, Europhys. Lett. **48** (1999) 573.
[334] M. Kiwi, J. Mejía-López, R.D. Portugal and R. Ramírez, Appl. Phys. Lett. **75** (1999) 3995.
[335] E. Fulcomer and S.H. Charap, J. Appl. Phys. **43** (1972) 4190.
[336] H. Fujiwara, K. Zhang, T. Kai and T. Zhao, J. Magn. Magn. Mater. **235** (2001) 319.
[337] M.D. Stiles and R.D. McMichael, Phys. Rev. B **63** (2001) 064405.
[338] D. Suess, M. Kirschner, T. Schrefl, J. Fidler, R.L. Stamps and J.-V. Kim, Phys. Rev. B **67** (2003) 054419.
[339] A.P. Malozemoff, Phys. Rev. B **35** (1987) 3679.
[340] A.P. Malozemoff, Phys. Rev. B **37** (1988) 7673.
[341] P. Miltényi, M. Gierlings, J. Keller, B. Beschoten, G. Güntherodt, U. Nowak and K.D. Usadel, Phys. Rev. Lett. **84** (2000) 4224.
[342] U. Nowak, K.D. Usadel, J. Keller, P. Miltényi, B. Beschoten and G. Güntherodt, Phys. Rev. B **66** (2002) 014430.
[343] J.-V. Kim and R.L. Stamps, Phys. Rev. B **71** (2005) 094405.
[344] J. Nogués and I.K. Schuller, J. Magn. Magn. Mater. **192** (1999) 203.
[345] A.E. Berkowitz and K. Takano, J. Magn. Magn. Mater. **200** (1999) 552.
[346] M. Kiwi, J. Magn. Magn. Mater. **234** (2001) 584.
[347] M. Finazzi, L. Duò and F. Ciccacci, Surf. Sci. Rep. **62** (2007) 337.
[348] S. Gangopadhyay, G.G. Hadjipanayis, C.M. Sorensen and K.J. Klabunde, J. Appl. Phys. **73** (1993) 6964.
[349] H.M. Lin, C.M. Hsu, Y.D. Yao, Y.Y. Chen, T.T. Kuan, F.A. Yang and C.Y. Tung, Nanostruct. Mater. **6** (1995) 977.
[350] D.L. Peng, K. Sumiyama, T.J. Konno, T. Hihara and S. Yamamuro, Phys. Rev. B **60** (1999) 2093.
[351] R.H. Kodama and A.S. Edelstein, J. Appl. Phys. **85** (1999) 4316.
[352] S. Yamamuro, K. Sumiyama, T. Kamiyama and K. Suzuki, J. Appl. Phys. **86** (1999) 5726.
[353] D.L. Peng, K. Sumiyama, T. Hihara, S. Yamamuro and T.J. Konno, Phys. Rev. B **61** (2000) 3103.
[354] J.M. Meldrim, Y. Qiang, Y. Liu, H. haberland and D.J. Sellmyer, J. Appl. Phys. **87** (2000) 7013.
[355] V. Skumryev, S. Stoyanov, Y. Zhang, G. Hadjipanayis, D. Givord and J. Nogués, Nature **423** (2003) 850.
[356] J. van Lierop, H.S. Isaacs, D.H. Ryan, A. Beath and E. McCalla, Phys. Rev. B **67** (2003) 134430.
[357] U. Wiedwald, M. Spasova, E.L. Salabas, M. Ulmeanu, M. Farle, Z. Frait, A. Fraile Rodriguez, D. Arvanitis, N.S. Sobal, M. Hilgendorff and M. Giersig, Phys. Rev. B **68** (2003) 064424.
[358] R. Morel, A. Brenac and C. Portemont, J. Appl. Phys. **95** (2004) 3757.

[359] J. Connolly, T.G. St. Pierre, M. Rutnakornpituk and J.S. Riffle, J. Phys. D: Appl. Phys. **37** (2004) 2475.
[360] G.H. Wen, R.K. Zheng, K.K. Fung and X.X. Zhang, J. Magn. Magn. Mater. **270** (2004) 407.
[361] J.M. Rivero, J.A. De Toro, J.P. Andrés, J.A. Gonzaléz, T. Muñoz and J.P. Goff, Appl. Phys. Lett. **86** (2005) 172503.
[362] A.N. Dobrynin, D.N. Ievlev, K. Temst, P. Lievens, J. Margueritat, J. Gonzalo, C.N. Afonso, S.Q. Zhou, A. Vantomme, E. Piscopiello and G. Van Tendeloo, Appl. Phys. Lett. **87** (2005) 012501.
[363] S.A. Koch, G. Palasantzas, T. Vystavel, J.T.M. De Hosson, C. Binns and S. Louch, Phys. Rev. B **71** (2005) 085410.
[364] D. Givord, V. Skumryev and J. Nogués, J. Magn. Magn. Mater. **294** (2005) 111.
[365] G. Palasantzas, S.A. Koch, T. Vystavel and J.T.M. De Hosson, J. Alloys Compd. **449** (2008) 237.
[366] J.F. Löffler, J.P. Meier, B. Doudin, J.-A. Ansermet and W. Wagner, Phys. Rev. B **57** (1998) 2915.
[367] C. Prados, M. Multigner, A. Hernando, J.C. Sánchez, A. Fernández, C.F. Conde and A. Conde, J. Appl. Phys. **85** (1999) 6118.
[368] L. Del Bianco, D. Fiorani, A.M. Testa, E. Bonetti, L. Savini and S. Signoretti, Phys. Rev. B **66** (2002) 174418.
[369] L. Del Bianco, D. Fiorani, A.M. Testa, E. Bonetti and L. Signorini, Phys. Rev. B **70** (2004) 052401.
[370] R.K. Zheng, G.H. Wen, K.K. Fung and X.X. Zhang, J. Appl. Phys. **95** (2004) 5244.
[371] R.K. Zheng, G.H. Weng, K.K. Fung and X.X. Zhang, Phys. Rev. B **69** (2004) 214431.
[372] C. Baker, S.K. Hasanain and S.I. Shah, J. Appl. Phys. **96** (2004) 6657.
[373] J. Nogués, V. Skumryev, J. Sort, S. Stoyanov and D. Givord, Phys. Rev. Lett. **97** (2006) 157203.
[374] D. Fiorani, L. Del Bianco and A.M. Testa, J. Magn. Magn. Mater. **300** (2006) 179.
[375] D. Fiorani, L. Del Bianco, A.M. Testa and K.N. Trohidou, Phys. Rev. B **73** (2006) 092403.
[376] D. Fiorani, L. Del Bianco, A.M. Testa and K.N. Trohidou, J. Phys.: Condens. Matter **19** (2007) 225007.
[377] F. Jiménez-Villacorta and C. Prieto, J. Phys.: Condens. Matter **20** (2008) 085216.
[378] Y.D. Yao, Y.Y. Chen, C.M. Hsu, H.M. Lin, C.Y. Tung, M.F. Tai, D.H. Wang and C.T. Suo, Nanostruct. Mater. **6** (1995) 933.
[379] S. Sako, K. Ohshima, M. Sakai and S. Bandow, J. Vac. Sci. Technol. B **15** (1997) 1338.
[380] T. Seto, H. Akinaga, F. Takano, K. Koga, T. Orii and M. Hirasawa, J. Phys. Chem. B **109** (2005) 13403.
[381] L. Del Bianco, F. Boscherini, A.L. Fiorini, M. Tamisari, F. Spizzo, M.V. Antisari and E. Piscopiello, Phys. Rev. B **77** (2008) 094408.
[382] S. Sahoo, C. Binek and W. Kleemann, Phys. Rev. B **68** (2003) 174431.
[383] E. Winkler, R.D. Zysler, H.E. Troiani and D. Fiorani, Physica B **384** (2006) 268.
[384] J. Sort, V. Langlais, S. Doppiu, B. Dieny, S. Surinach, J.S. Munoz, M.D. Baro, C.X. Laurent and J. Nogues, Nanotechnology **15** (2004) S211.
[385] T. Furubayashi and H. Mamiya, IEEE Trans. Magn. **41** (2005) 3418.
[386] E.C. Passamani, C. Larica, C. Marques, J.R. Proveti, A.Y. Takeuchi and F.H. Sanchez, J. Magn. Magn. Mater. **299** (2006) 11.
[387] N. Domingo, A.M. Testa, D. Fiorani, C. Binns, S. Baker and J. Tejada, J. Magn. Magn. Mater. **316** (2007) 155.
[388] E. Eftaxias and K.N. Trohidou, Phys. Rev. B **71** (2005) 134406.
[389] E. Eftaxias, K.N. Trohidou and C. Binns, Phys. Status Solidi C **1** (2004) 3361.
[390] M.H. Wu, Q.C. Li and J.-M. Liu, J. Phys.: Condens. Matter **19** (2007) 186202.

Chapter 10

Some Applications of Nanoparticles

C.-H. Yu[1], W. Oduro[1], Kin Tam[2] and Edman S.C. Tsang[1]
[1]*Wolfson Catalysis Centre, Inorganic Chemistry Laboratory, University of Oxford, Oxford OX1 3QR, UK*
[2]*AstraZeneca, Mereside, Alderley Park, Macclesfield, Cheshire SK10 4TG, UK*

10.1.	Introduction	365
10.2.	Functionalisation of nanoparticles with biomolecules	366
	10.2.1. Non-covalent biomolecule–nanoparticle conjugation	366
	10.2.2. Covalent biomolecule–nanoparticle conjugation	367
10.3.	Magnetic nanoparticles	370
	10.3.1. Some biological applications of magnetic nanoparticles	372
	10.3.2. Bio-separation using magnetic nanoparticles	372
	10.3.3. High-gradient magnetic separation (HGMS) using magnetic colloid crystals	373
10.4.	Fundamentals in heterogeneous nanocatalysis	374
	10.4.1. The effect of size and shape on heterogeneous catalysts	375
	10.4.2. Nanoparticles in catalysis	376
	10.4.3. Engineering of nanocatalysts	377
	References	378

10.1. Introduction

It is evident that nanoparticles are going to have significant impacts on our daily life, ranging from fabrication of new construction materials to applications in bio-medicine and pharmaceuticals. In this chapter, two areas of applications, namely biotechnology and industrial catalysis, are particularly highlighted.

In the biotechnology area, nanoparticles with magnetic properties have been applied not only in electrical, mechanical and optical areas but also for drug screening, contrast enhancement agents for magnetic resonance imaging, medicine and bio-separation, etc. Some of these applications will be discussed in Sections 10.2 and 10.3. The background to magnetism and the properties of magnetic nanoparticles have been discussed in Chapters 8 and 9.

HANDBOOK OF METAL PHYSICS
ISSN 1570-002X/DOI 10.1016/S1570-002X(08)00210-3

© 2009 ELSEVIER B.V.
ALL RIGHTS RESERVED

Catalysis plays an important role in our society. Presently, over 90% of all industrial chemical productions (with an annual market value of between $40 and $70 billion) involve the use of at least one catalytic step. A catalyst in general is a substance that increases the rate at which a thermodynamically feasible system attains chemical equilibrium without itself undergoing change. In a given reaction, there may be more than one permissible reaction paths. However, the type of catalyst used may determine the path taken, thus exerting a specific directing influence and hence promoting the synthesis of specific products. The challenges of modern industrial processes require the development of novel catalysts with a unique preferential reaction for only one of two or more different chemical functional groups (chemo-selectivity), for one direction of bond formation or breaking over all other possible directions (regio-selectivity) or for formation of one stereoisomer over the other (stereo-selectivity). Chemoselective catalysts, for example, are essential in the industrial synthesis of flavours and fragrances. Thus, selectivity is one key consideration in the successful design of a catalytic process in order to minimise undesirable side products and, most importantly, to enable the development of environmentally friendly processes. It is noted that the use of tailored nanoparticles may offer exciting possibilities for engineering selectivity of catalytic systems of interest at the molecular level. In Section 10.4, we will summarise some successful attempts in this area.

10.2. Functionalisation of nanoparticles with biomolecules

The characteristic small size of nanoparticles associated with high surface area, high reactivity and unique physicochemical phenomena such as optical, thermal, electrical and magnetic properties render nanoparticles interesting candidates for many potential applications [1]. However, for most applications, raw nanoparticles may have to be physically or chemically modified before they can be used. In this section, emphasis will be placed on the conjugate formation between nanoparticles and biomolecules, which is a crucial step for most biological applications of nanoparticles.

10.2.1. Non-covalent biomolecule–nanoparticle conjugation

Non-covalent binding is a simple technique based on the electrostatic interactions between biomolecule and nanoparticle. The biomolecule is electrostatically attracted to the oppositely charged shell of the nanoparticles. At a pH above the isoelectric point (IEP), the biomolecule is negatively charged, while at pH below the IEP, it is positively charged. The advantage of this method is that it is not necessary to expose the biomolecule to any drastic chemical environment. Hence, the biomolecule is likely to be in its native form and remain active, and could be used as a bio-sensor, bio-detector or in bio-diagnostics. As shown in Figure 10.1, modification of the nanoparticles' surface with an adsorbed species can alter the surface charge on the nanoparticles, and this offers flexibility for binding positively or negatively charged biomolecules [2].

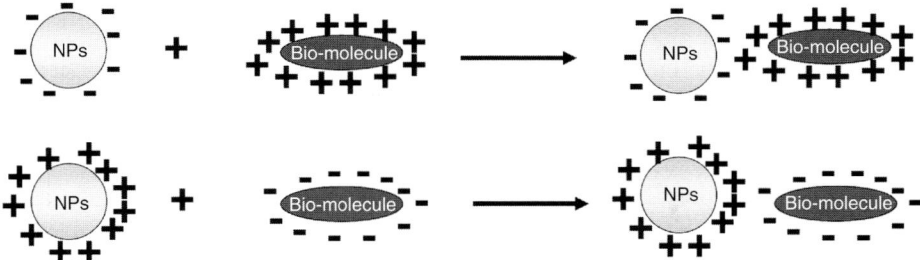

Figure 10.1. Formation of non-covalent biomolecule–nanoparticle (NP) conjugates.

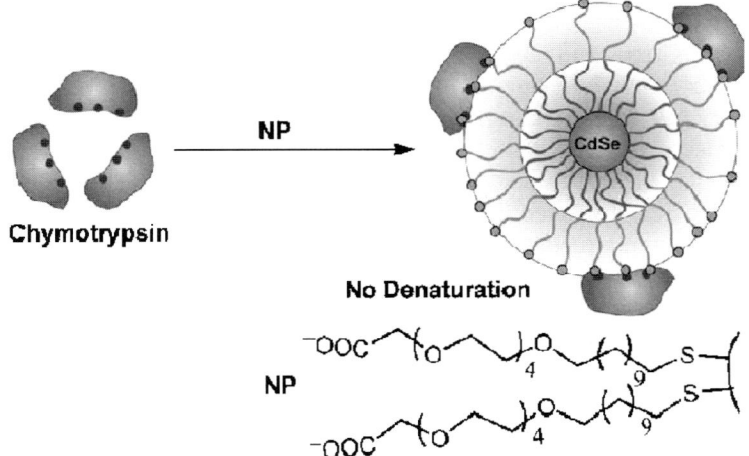

Figure 10.2. Binding of a nanoparticle (NP) with chymotrypsin protein through electrostatic interaction. Reproduced with the permission of Wiley-VCH from Mirkin [2].

Fisher et al. used the electrostatic principle to carry out surface-binding interactions between α-chymotrypsin (ChT) and nanoparticles. The cationic side chains surrounding the active site of ChT and the CdSe clusters terminated with anionic functional group (COO^-) can be held together by electrostatic forces (Figure 10.2) [2]. In parallel with this finding, binding of a gold nanoparticle to the ChT protein system can also be accomplished by electrostatic interactions using anionic surfactants on the gold surface [40]. Following the formation of biomolecular conjugates, tryptophan fluorescence and circular dichroism (CD) can be used to confirm the integrity of the secondary structure. This spectroscopic data provide a clear diagnostic examination to ascertain whether the protein is in its native form for eliciting desirable activity [3,4].

10.2.2. Covalent biomolecule–nanoparticle conjugation

Covalent protein–nanoparticle (e.g. Fe_3O_4, CdSe, Au, SiO_2 and ZnS) conjugates can provide strong binding affinity and site specificity for specific applications.

The nanoparticles can provide unique properties such as magnetic, optical, and electronic properties that can be exploited as bio-sensors or in bio-diagnoses. General methods for covalent protein–nanoparticle bio-conjugation can couple with either an active group (e.g. $-NH_2$, $-SH$ and $-COOH$) on the functionalised surface of nanoparticles or functionalised group on the site of protein. The last decades of research indicate that the most significant problem encountered in the conjugation formation is that the protein or enzyme on a nanoparticle surface, after immobilisation, suffers from leaching or denature. These issues may be resolved by building surface-activated group derivatives (e.g. thiol or amine groups) or cross-linkers [e.g. sulphosuccinimidyl 4-(N-maleimidomethyl) cyclohexane-1-carboxylate or 1-ethyl-3-(3-dimethylaminopropyl) carbodiimide hydrochloride] for the protein, enzyme, antigens or antibodies to create specific bio-conjugates via one- or two-step cross-linking reactions (Figure 10.3).

The use of semiconductor nanoparticles as quantum dots with intrinsic optical properties for labelling biological species or medicine has been extensively reported [6]. As most capping agents, for example, tri-n-octylphosphine oxide (TOPO) molecules, involved in the synthesis of these quantum dots are strongly coordinated to their surface, the non-polar nature of TOPO enables the nanomaterials well dispersed in organic solvents. However, this prevents the use of the technology in aqueous solution where most biotechnology applications take place. Recent progress has demonstrated that, by displacing the TOPO molecules with hydrophilic groups to render the quantum dot conjugates water soluble, many important applications in biology and medicine may be realised (Figure 10.4). As shown in Figure 10.4(a), the surface of quantum dots can be decorated (via a condensation process under alkaline conditions) *first* by using mercaptopropyl-trimethoxysilane (MPS) and then by a trihydroxysilylpropyl methylphosphonate monosodium salt (TMP). MPS provides the quantum dot surface with thiol groups for subsequent bio-conjugation reaction, while the TMP groups help the solubilisation in aqueous media [7]. Moreover, the surface functional groups can be converted to amine or carboxyl groups by using bifunctional linkers, which offers diversified functional groups for the formation of covalent bonds with the biomolecule of interest [8]. For instance, Chan used highly luminescent

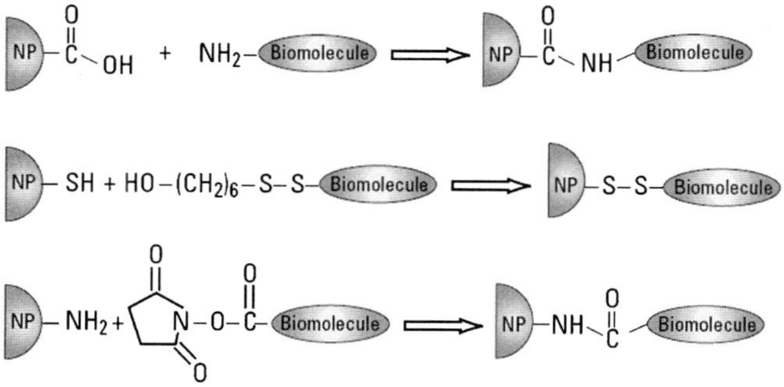

Figure 10.3. Representative bio-conjugation protocols for the attachment of biomolecule on dye-doped silica nanoparticles. Reproduced with the permission of the American Chemical Society from Wang et al. [5].

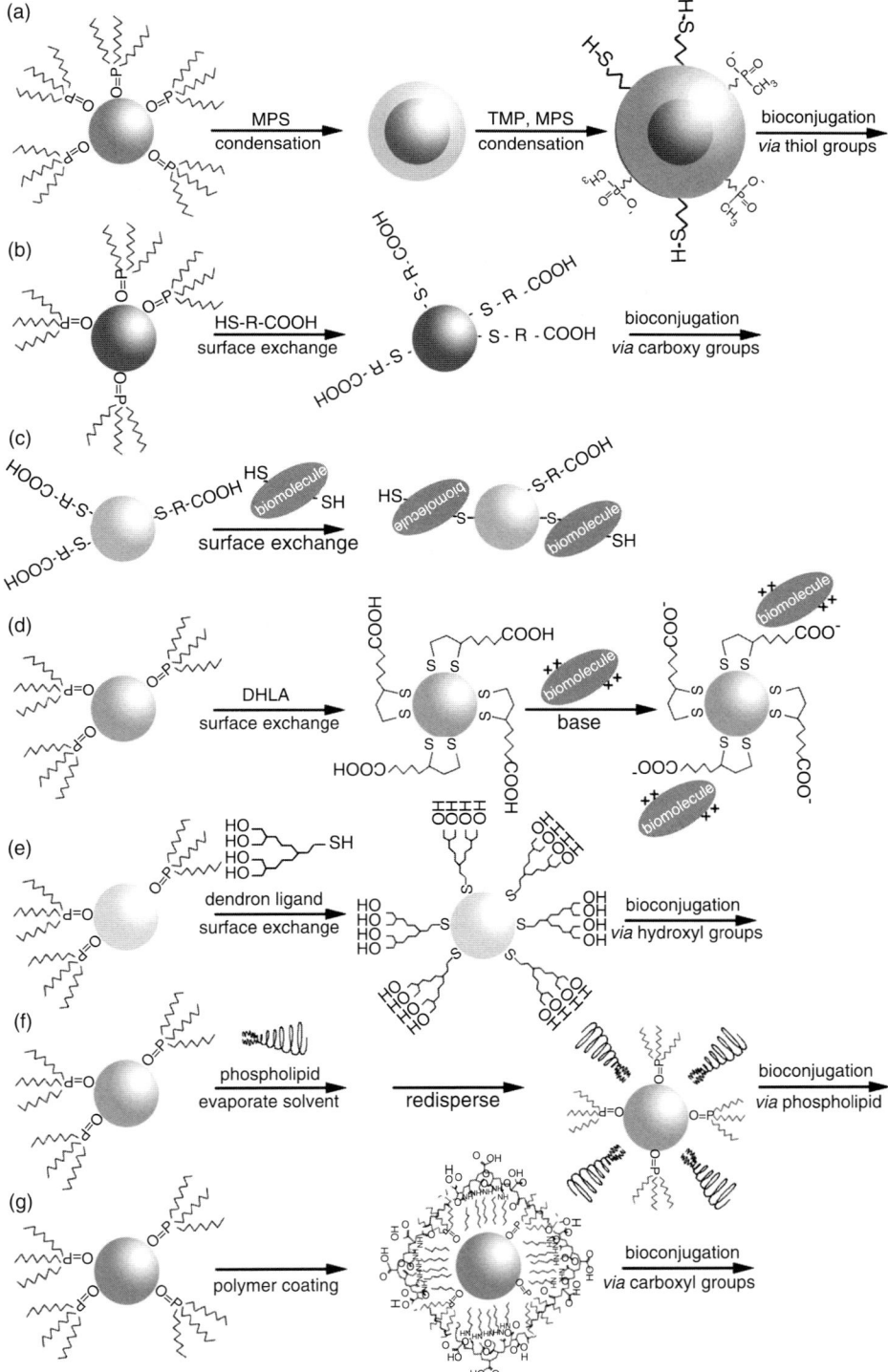

Figure 10.4. Representatives for quantum dot surface modification and bio-conjugation: (a) coating quantum dots with a thiol–silica shell, (b) direct covalently coupled to biomolecules by using bifunctional linkers, (c) directly coupled with thiolated biomolecules, (d) electrostatically attracted to the oppositely charged shell of nanoparticles using positively and negatively charged biomolecules, (e) stabilisation and bio-conjugation using organic dendrimers, (f) block copolymer encapsulation and (g) polymer coating. Reproduced with the permission of Wiley-VCH from Rao et al. [6].

semiconductor quantum dots that are covalently attached to proteins by mercaptoacetic acid for use in ultra-sensitive biological detection [Figures 10.4(b) and (c). Excellent water solubility and bio-compatibility have been demonstrated [9]. Other bio-conjugation strategies including the use of dihydrolipoic acid (DHLA) to attach the quantum dot surface followed by electrostatic interaction with biomolecules have been reported [Figure 10.4(d) [10]. Peng *et al.* employed simple hydrophilic organic dendron ligands onto the quantum dot surface to form bio-conjugated groups [Figure 10.4(e) [11]. Dubertret *et al.* used phospholipids as block copolymers to encapsulate a quantum dot [Figure 10.4(f) [12]. Wu used streptavidin to coat quantum dots for application in cellular imaging [13].

10.3. Magnetic nanoparticles

The engineering of nanomaterials generally refers to the synthesis/fabrication of nanomaterials by design to suit a particular application. This definition implies the ability to assemble basic components into the desired nanoarchitecture by the fabrication of a target network so that the assembled materials exhibit the sought properties. A good example to illustrate this point is the development of magnetic nanomaterials.

Magnetic nanoparticles can be widely used for electrical, optical, magnetic and other new technological applications that include magnetic-field-assisted bioseparation and biocatalysis [14–18]. A brief account on various synthetic methodologies for the production of magnetic nanoparticles can be found in Chapter 5. Although a majority of current research is focussed on the application of magnetic nanoparticles for recording purpose, magnetic nanoparticles with superparamagnetism are actually required in biotechnology areas [15]. To achieve super-paramagnetism, individual magnetic bits should be aligned under external magnetic field to give a strong and coherent magnetic response, while thermal energy is sufficient to totally disrupt the alignment upon the removal of the external field. A nanometre-sized magnetic particle of lesser magnetic domain size can display super-paramagnetic properties. In bio-medicine, the super-paramagnetic nanoparticles can be used as labelling, sensor or imaging reagents when tagged with biological entities of compatible size. Such approaches are very attractive to industry as the magnetically tagged biomolecules with super-paramagnetic properties can be isolated and recycled easily from solution using magnetic means (thus minimising the waste production through regeneration) [17–21].

Typically, single-phase FePt nanoalloy particles are important magnetic nanomaterial candidates with a good chemical stability. However, control synthesis of these nanoparticles with tailored size, morphology and chemical composition is still a major challenge. As a result, a large effort has recently been devoted to exploring a wide variety of methods in order to prepare this stable alloy nanoparticle of tailored size and composition. Chemical routes using high-temperature solution-phase conditions are the most practical and widely chosen fabrication methods [22,23]. Thus, FePt nanoparticles can be synthesised by

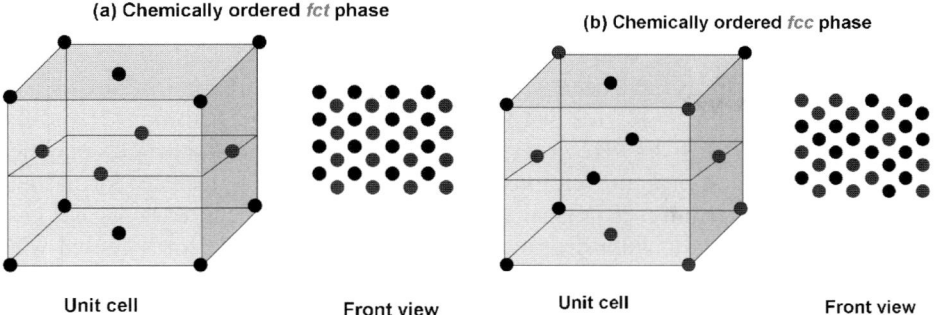

Figure 10.5. The cartoon illustrates the structure of FePt in (a) face-centred tetragonal (*fct*) and (b) face-centred cubic (*fcc*) phase.

co-reduction of platinum and decomposition of iron precursors, in the presence of polyol reducing agents, as the 'polyol' process [23] (see also Chapter 5). It is noted that the size distribution prepared from the polyol method can be extremely narrow with standard deviations usually within 10% [24,25]. Most as-synthesised FePt nanoparticles are reported to have a chemically disordered face-centred cubic (fcc) structure (Figure 10.5) and are super-paramagnetic due to their small crystal size [26]. Annealing the material at temperatures above 500°C can convert this phase to a face-centred tetragonal (fct) structure (also known as $L1_0$ phase), which shows higher magneto-crystalline anisotropy [27–29]. It has recently been demonstrated that by placing FePt in close vicinity to Fe_3O_4, the ferromagnetic Fe_3O_4 can dramatically enhance the magnetisation of the composite material by magnetic exchange coupling between the components [30]. It is also shown that reductive annealing of the FePt-Fe_3O_4 in hydrogen gas produces a magnetic nanocomposite FePt-Fe_3Pt with more effective exchange coupling between the FePt hard phase and the Fe_3Pt soft phase, giving a further enhanced energy product. Such systems offer the major advantage of combining the permanent magnetic field with magnetisation, hence rendering the mixture with a higher magnetisation compared to traditional single-phase nanomagnets. However, phase segregation and sintering were clearly observed in the previous work [31], forming large islands or domains of FePt and Fe_3Pt during reduction. Despite the enhancement in magnetisation, the sample is also undoubtedly associated with a significantly high coercivity due to the cooperative coupling between the domains and, hence, may not be suitable for use as a biomolecule carrier [22]. On the other hand, it has been reported that using sequential nanochemistry preparation techniques, FePt-Fe_3O_4 and FePt-Fe_3Pt nanocrystallites of tailored size can be fabricated and then encapsulated in silica with a high precision [26]. The silica coating offers the composite nanoparticle a physical barrier against sintering by modifying its diffusion rate and surface energy. Moreover, the silica encapsulation also offers provision of anchored sites for biomolecule immobilisation and isolation of the nanomagnets from magnetic interference from other domains. As a result, the silica-encapsulated FePt–Fe_3O_4 or FePt–Fe_3Pt display enhanced magnetisation while maintaining low coercivity, which could provide interesting candidates as magnetic bio-carriers [26].

10.3.1. Some biological applications of magnetic nanoparticles

Magnetic nanoparticles have been widely studied for biological applications such as for binding BSA[1] [17,32–34], for drug delivery [35–38], for bio-sensing [17] and for bio-separations [5,14,39–41]. However, naked magnetic alloy nanoparticles may not be suitable for wide-range applications because of their high propensity for particle aggregation. Furthermore, biological species may not be stable in the presence of a metallic phase and may cause degradation. Therefore, magnetic cores coated with inorganic shells have been synthesised to tag onto biological systems through modification of their surface groups. These magnetically tagged biological species could then be separated by using an external magnetic field. Silica is generally believed to be chemically compatible to most biological systems, and hence many hybrid conjugates based on silica have been developed (Sections 10.3.1 and 10.3.2). These materials can be characterised by various techniques including X-ray diffraction (XRD), transmission electron microscopy (TEM), scanning electron microscopy (SEM), Brunauer–Emmett–Teller (BET) adsorption isotherm, Fourier transform infrared spectroscopy (FTIR), thermogravimetric analysis (TGA) and UV-visible spectroscopy (UV/vis) [18]. Very recently, the encapsulation of protein molecules by silica has also been reported [40,42]. Mesoporous silica nanorods capped with magnetic nanoparticles for controlled-release delivery have recently been developed [38]. A novel method using a magnetic nanoabsorbent for the determination of the partition coefficient of a drug candidate between aqueous buffer and n-octanol has been suggested [14]. Supercritical carbon dioxide has been applied as a carrier to deposit the species of interest into the porous overlayers of these magnetic nanoabsorbents, which provides an elegant way to prepare a uniform distribution of nanomaterials for high-throughput screening applications [43].

10.3.2. Bio-separation using magnetic nanoparticles

Magnetic nanoparticles have infiltrated a variety of fields of physics, chemistry and bio-medicine, particularly imaging (e.g. as imaging contrast enhancer) and transdermal drug delivery. However, their application to the field of biomolecule separation has been very limited to date, as the fabrication of nanoparticles of many materials has been difficult until recently, and the options to regenerate the retained biomolecules after separation also remain limited. In contrast, commercial suspension of magnetic beads of submicrometre or micrometre size in solution for magnetic isolation and separation of biomolecules (e.g. DNA, proteins) has been applied for a number of years. An area of notable success has been in the field of fine chemical manufacture, where free forms of paramagnetic nanomaterials coupled to chemical catalysts have been shown to be stable in harsh chemical conditions and have been successfully captured for reuse following dispersal by magnetic retrieval. Promising results were obtained using such magnetised

---

[1] Bovine serum albumin (BSA) is a kind of protein; it is also used in bio-applications such as enzyme-linked immunosorbent assay (ELISA).

nanomaterials as bio-catalysts, where they were shown to successfully carry β-lactamase via chemical linkage to the silica overlayer [18]. Enzyme activities were as good as the free enzyme, and recovery and reusability upon application of magnetic separation was achieved. It should be noted that the current technology of magnetic separation using a nanomagnet is entirely dependent upon a two-step process that involves (i) tagging or labelling of the desired biological/chemical entity on colloid magnetic nanoparticles for recognition of complementary species in solution, and (ii) separation of the resulting solid entities via a fluid-based magnetic separation followed by regeneration of the species from the particles. Although this technique is now widely adopted in protein purification, immunoassays, pre-processing in polymerase chain reactions and pre-concentration of biological entities, the combination of highly specific tagging and recognition chemistry and generally weak magnetic response of the nanoparticle–biomolecule conjugates towards external magnetic field in solution does not allow efficient separation of closely related species.

10.3.3. *High-gradient magnetic separation (HGMS) using magnetic colloid crystals*

As discussed above, the potential for applying magnetic nanoparticles in bio-separation is immense. In combination with a variable external magnetic field, the system could be useful in magnetic gradient separation of nanoscaled chemical or biochemical entities such as proteins. To enable sufficiently strong magnetisation, it is essential to assemble the magnetic nanoparticles as larger sized entities, for example via the bottom-up approach, to form magnetic colloidal crystals. This section seeks to outline various possibilities for developing these kinds of magnetic colloidal crystals, which may be suitable for use in magnetic gradient separation. These colloidal crystals, which are three-dimensional (3D) regular arrays of super-paramagnetic nanoparticles encapsulated in a protective shell with controllable thickness, may provide strong localised magnetic forces at well-defined interstitial sites to attract and retain nanosized paramagnetic entities when they percolate through the arrays. This new approach may allow both the efficiency and the selectivity of magnetic separation to be enhanced simultaneously as the size of the interstitial sites can be optimised for given paramagnetic entities of interest, while the local magnetic field gradients can be controlled by magnetising the particles using an applied field. Thus, the porous colloidal crystals are expected to have immense technological impacts in view of the increasing demand for novel methods for pre-concentration, separation or purification of valuable chemicals or biomolecules such as proteins.

In fact, the use of a high magnetic field gradient for the separation of paramagnetic entities could date back to the studies of Kolm *et al.* [44]; its applications in chromatography were reported later by Vickrey and Garcia-Ramirez in 1980 [45] and by Mori in 1986 [46]. In a non-uniform magnetic field H, the magnetic force F acting on a paramagnetic particle in a medium is proportional to the particle magnetisation M. It has been demonstrated that magnetic separation of paramagnetic chemical or biochemical entities can be successfully achieved by this technique, although the retardation in elution time for species was very small due to the small local field gradient and hence weak magnetic interactions with the entities [45].

Reducing the length scale of the magnetic field source offers an avenue to enhancing the field gradient so that the technique can be applied to separate smaller paramagnetic entities such as proteins or DNA, which are tens of nanometres in size. Arrays of super-paramagnetic nanoparticles with a large magnetic moment can, in principle, be magnetised under externally applied fields to provide localised field gradients for this purpose.

It is surprising to find that very little work, if any, has been devoted to exploring the potential use of assembled magnetic nanoparticles as 3D crystals for magnetic separations, despite the fact that magnetic nanoparticles offer a distinctive advantage: the extremely high–surface area to volume ratio of magnetic nanoparticles allows for enhanced magnetic interactions with the chemical or biochemical species of interest so that they can be pre-concentrated in a very small volume for separation or purification purposes. On the other hand, there have been intense efforts made over the past decade to fabricate nanoparticle arrays as 3D regular structures for use in optical filters [47], switches [48], sensors [49] and waveguides [50]. Several techniques have been developed for the synthesis of such 3D nanoparticle arrays, including colloidal self-assembly [51,52], 3D holography using multiple laser beams [53] and photolithography [54]. It is anticipated that the development of magnetic colloid crystals from an assembly of magnetic nanoparticles for HGMS applications will be fuelled in the near future.

10.4. Fundamentals in heterogeneous nanocatalysis

Catalysis is a surface phenomenon. The environment of a surface atom is quite different from that of an atom situated in the bulk of a crystal. Their location on the surface makes them considerably reactive towards foreign atoms or molecules because they do not have a full complement of neighbours and hence have a lower coordination number and also are of relatively high energy. They therefore exhibit the tendency to saturate missing bonds and reconstruct, thereby re-establishing the symmetry of the field of force to which they would be subject were they in the interior of the crystal. The ability to form new bonds is highly dependent on the crystallographic orientation of the surface, and each crystalline plane of a metal may be considered as a real chemical entity with its own specific properties. This influence of the crystallographic orientation, which affects chemical behaviour, is also revealed in significant differences in physical properties such as optical properties, surface energy or the work function of the material [55]. The role of the metal surface is to provide an energetically favourable pathway for the reaction. The synthesis of heterogeneous catalysts therefore, of necessity, must take into consideration methods aimed at optimising surface characteristics such as particle morphology.

Catalyst synthesis takes various forms and approaches depending on the type of reaction being considered. In the literature, a number of engineering processes have been conducted on the surface of metals in order to enhance their activity or selectivity. Some of the popular approaches include: metals on inert supports, metals modified by organic moieties, co-metals in bimetallic catalysts and immobilised enzymes on metal surfaces.

10.4.1. *The effect of size and shape on heterogeneous catalysts*

For efficient catalysis, a high surface area is required that ensures high atom efficiency for a chemical reaction. The relevance of nanotechnology on catalyst development can therefore not be overemphasised as it affords a largely exposed surface. Metals that have shown quite inert catalytic effects in a larger crystalline form, such as coinage metals of group 11 of the periodic table, have been found to exhibit exemplary catalytic effects both as pure metals and as alloys in the nanoscale. Catalytic properties are therefore very much particle size dependent. Considering that metal atoms located in a typical fcc crystal structure tend to have a coordination number of 12, atoms situated in the bulk of a crystal are quite different from those situated on the surface. This is because atoms of the crystal particle situated in the bulk tend to have a full complement of their coordination number, which is of course higher than for those at the surface. Thus, some surface atoms are in high-index planes such as atomic steps, kinks, corners or adatoms as shown in Figures 10.6 and 10.7 respectively. In order to minimise the surface energy of the crystal, crystallographic defects tend to make these surface atoms the favourable reactive sites.

Thus, particular atoms or groups of atoms on the surface are the active sites responsible for the catalytic activity and selectivity [57,58]. As a result, once the active sites for a particular reaction are identified, one can, in principle, design and prepare an optimal catalyst to maximise the active sites to meet the needs of the reaction. Similarly, the particle shape has also been found to be an important parameter in the activity and, more so, in the selectivity of a catalyst. This is because reaction molecules have different sorption structures on the various surface planes

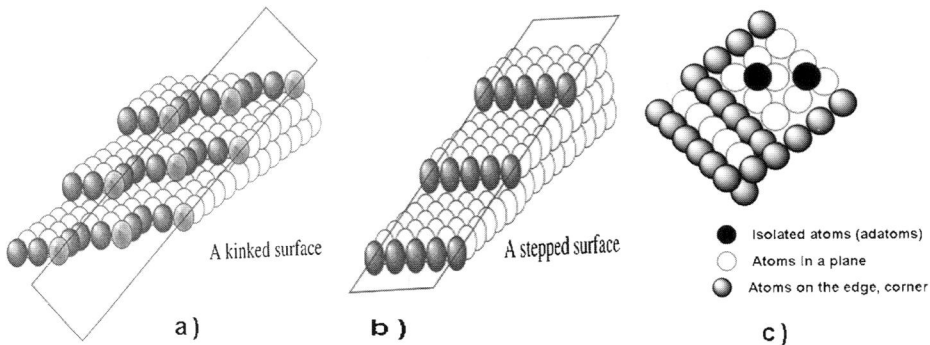

Figure 10.6. Typical defects on a crystal surface that combine to minimise the surface energy: (a) kinked surface, (b) stepped surface and (c) isolated atoms on the surface of a plane. Reproduced with the permission of Oxford University Press from McCash [56].

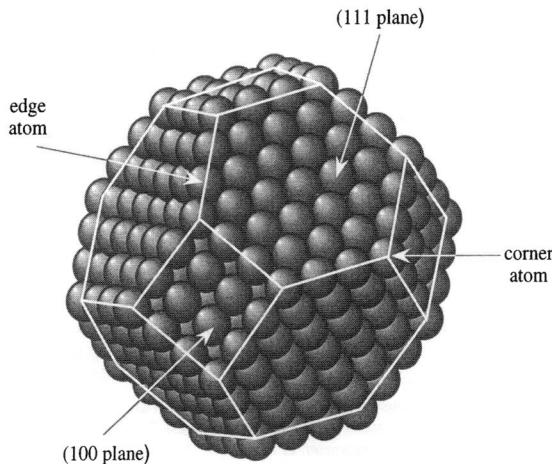

Figure 10.7. A crystal surface showing the various morphologies of a cubo-octahedral metal particle. Facets of the low-indexed faces are shown with the presence of step edges and corner atoms. Reproduced with the permission of Oxford University Press from McCash [56].

and on highly exposed atomic structures such as those in steps and kinks. Restricted adsorption may have a significant impact on selectivity. For instance, Pt(1 1 1) has been shown to favour selectivity in catalytic hydrogenation of α,β-unsaturated aldehydes to unsaturated alcohols as compared to a Pt(1 0 0) surface that allows unrestricted sorption of C=C and C=O bonds [59].

10.4.2. Nanoparticles in catalysis

Haruta and his co-workers recently discovered that supported gold nanoparticles are extremely catalytically active for carbon monoxide (CO) oxidation at temperatures much lower than room temperature when the gold particle size is below 5 nm. This finding has fuelled the recent searches for more efficient nanocatalysts for many chemical reactions [60,61]. The interfaces created between the gold nanoparticles with transition metal oxide supports, such as zirconia, iron oxide, titania, cerium oxide and others, are attributed to account for the ultra-high catalytic activity and selectivity. However, the activity of these metal clusters (in particular gold nanoparticles) in a reaction is very much dependent on the method of preparation, particle size, shape, dispersion and the type of supported metal oxide. The mechanism for the activation of CO using gold on transition metal oxide support is rather complex. It includes multiple steps of chemisorption of CO on gold clusters, CO oxidation and water–gas shift reactions [62]. It is generally believed that the initial CO adsorption takes place on low-surface gold coordination number sites at corner or edge regions; their stronger interactions with the support can enhance the chemisorptivity as well as species migration between the support and the gold particle [63]. Figure 10.8 shows different modes of CO adsorbed on gold surface atoms.

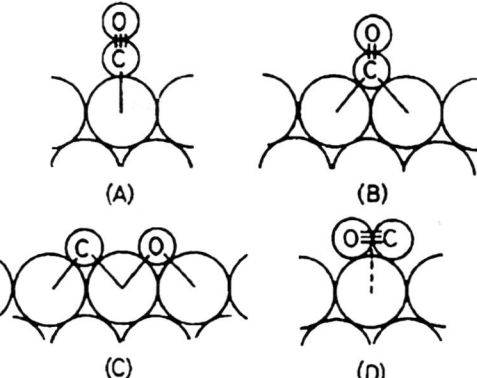

Figure 10.8. Chemisorption of carbon monoxide (CO) on a gold nanoparticle: (A) linear, (B) bridged, (C) dissociated and (D) sideways. Reproduced with the permission of World Scientific from Bond et al. [64].

In more recent progress in gold nanoparticle catalysis, Konishi and his co-workers used dodecanethiolate-protected gold as a co-catalyst for olefin oxidation in the presence of a manganese–porphyrin complex as a main catalyst species. Their result clearly showed that the presence of gold nanoparticles can catalyse regeneration of Mn(III) and Mn(V) from deactivated Mn(IV) species, which explains the enhanced activity with the combination of these two catalytic species [65]. It is interesting to point out that since Haruta's first report on gold nanoparticle catalysis, further studies of supported metal nanoparticles as catalysts have been the subject of many recent publications [66–68].

10.4.3. Engineering of nanocatalysts

Engineering of nanocatalyst materials could be considered as making nanocatalysts by design. This definition implies the ability to assemble basic components into the desired nanoarchitecture by engineering a target network so that the assembled materials could serve as catalysts for selected reactions with tailored catalytic properties. Hence, nanocatalysts can be engineered 'with a clear purpose'. Many heterogeneous catalysts often lack the abilities to differentiate between different molecules in a mixture of reactants (reactant selectivity) or to perform under severe conditions without deactivation (stability). Although this lack of discrimination can be partially overcome by tuning the operating conditions, substantial improvements are often not possible. In principle, a more logical approach is to allow a 'bottom-up' preparation of functional catalyst materials. However, the 'black magic' of catalyst preparation remains a key challenge. No recipe exists to precisely control the size and shape of metal nanoparticles, let alone the catalytic sites or their orientations thereupon. Many of most interesting catalyst materials are disordered or amorphous and their characterisation and evaluation remain an open challenge. Although the scopes of modern nanocatalyst designs are much broader with the advent of recent nanoscience/nanotechnology, it is interesting to note that the idea of making well-defined nanocatalysts by design existed 30 years ago. There has not been a lot of progress made since Schmid's pioneering ideas of pre-arranging molecules in the defined solid-state material in order to obtain reactions [69]. This clearly reflects the difficulties of the subject area but it also offers many opportunities. Where is nanoengineering of catalyst materials going? It certainly will continue to expand and seek real applications to improve activity, selectivity, stability and facilitated separation of the designed nanocatalysts. The development of new nanoparticles as nanocatalysts for a wide range of catalysis is currently proceeding at a great pace.

References

[1] C. Burda, X.B. Chen, R. Narayanan and M.A. El-Sayed, Chem. Rev. **105** (2005) 1025.
[2] C.A. Mirkin, Nanobiotechnology 2. More Concepts and Applications, Wiley-VCH, Weinheim, 2007.
[3] N.O. Fischer, A. Verma, C.M. Goodman, J.M. Simard and V.M. Rotello, J. Am. Chem. Soc. **125** (2003) 13387.
[4] R. Hong, N.O. Fischer, A. Verma, C.M. Goodman, T. Emrick and V.M. Rotello, J. Am. Chem. Soc. **126** (2004) 739.
[5] L. Wang, K.M. Wang, S. Santra, X.J. Zhao, L.R. Hilliard, J.E. Smith, J.R. Wu and W.H. Tan, Anal. Chem. **78** (2006) 646.
[6] C.N.R. Rao, A. Muller and A.K. Cheetham, The Chemistry of Nanomaterials: Synthesis, Properties and Applications, 2 volumes, Wiley-VCH, Weinheim, 2004, pp. 405–417.
[7] W.J. Parak, D. Gerion, D. Zanchet, A.S. Woerz, T. Pellegrino, C. Micheel, S.C. Williams, M. Seitz, R.E. Bruehl, Z. Bryant, C. Bustamante, C.R. Bertozzi and A.P. Alivisatos, Chem. Mater. **14** (2002) 2113.
[8] G.T. Hermanson, Bioconjugate Techniques, Academic Press, San Diego, 1996.
[9] W.C.W. Chan and S.M. Nie, Science **281** (1998) 2016.
[10] E.R. Goldman, G.P. Anderson, P.T. Tran, H. Mattoussi, P.T. Charles and J.M. Mauro, Anal. Chem. **74** (2002) 841.
[11] Y.A. Wang, J.J. Li, H.Y. Chen and X.G. Peng, J. Am. Chem. Soc. **124** (2002) 2293.
[12] B. Dubertret, P. Skourides, D.J. Norris, V. Noireaux, A.H. Brivanlou and A. Libchaber, Science **298** (2002) 1759.
[13] X.Y. Wu, H.J. Liu, J.Q. Liu, K.N. Haley, J.A. Treadway, J.P. Larson, N.F. Ge, F. Peale and M.P. Bruchez, Nat. Biotechnol. **21** (2003) 41.
[14] X. Gao, C.H. Yu, K.Y. Tam and S.C. Tsang, J. Pharm. Biomed. Anal. **38** (2005) 197.
[15] J.M. Nam, C.S. Thaxton and C.A. Mirkin, Science **301** (2003) 1884.
[16] S.C. Tsang, V. Caps, I. Paraskevas, D. Chadwick and D. Thompsett, Angew. Chem. Int. Ed. Engl. **43** (2004) 5645.
[17] E. Katz and I. Willner, Angew. Chem. Int. Ed. Engl. **43** (2004) 6042.
[18] S.C. Tsang, C.H. Yu, X. Gao and K. Tam, J. Phys. Chem. B **110** (2006) 16914.
[19] D.K. Yi, S.T. Selvan, S.S. Lee, G.C. Papaefthymiou, D. Kundaliya and J.Y. Ying, J. Am. Chem. Soc. **127** (2005) 4990.
[20] S.G. Grancharov, H. Zeng, S.H. Sun, S.X. Wang, S. O'Brien, C.B. Murray, J.R. Kirtley and G.A. Held, J. Phys. Chem. B **109** (2005) 13030.
[21] T.J. Yoon, J.S. Kim, B.G. Kim, K.N. Yu, M.H. Cho and J.K. Lee, Angew. Chem. Int. Ed. Engl. **44** (2005) 1068.
[22] L.E.M. Howard, H.L. Nguyen, S.R. Giblin, B.K. Tanner, I. Terry, A.K. Hughes and J.S.O. Evans, J. Am. Chem. Soc. **127** (2005) 10140.
[23] S.H. Sun, C.B. Murray, D. Weller, L. Folks and A. Moser, Science **287** (2000) 1989.
[24] M. Chen, J.P. Liu and S.H. Sun, J. Am. Chem. Soc. **126** (2004) 8394.

[25] C. Liu, X.W. Wu, T. Klemmer, N. Shukla, X.M. Yang, D. Weller, A.G. Roy, M. Tanase and D. Laughlin, J. Phys. Chem. B **108** (2004) 6121.
[26] C.H. Yu, C.C.H. Lo, K. Tam and S.C. Tsang, J. Phys. Chem. C **111** (2007) 7879.
[27] J.P. Liu, C.P. Luo, Y. Liu and D.J. Sellmyer, Appl. Phys. Lett. **72** (1998) 483.
[28] H. Zeng, S.H. Sun, T.S. Vedantam, J.P. Liu, Z.R. Dai and Z.L. Wang, Appl. Phys. Lett. **80** (2002) 2583.
[29] Y.K. Takahashi, M. Ohnuma and K. Hono, J. Magn. Magn. Mater. **246** (2002) 259.
[30] H. Zeng, J. Li, Z.L. Wang, J.P. Liu and S.H. Sun, Nano Lett. **4** (2004) 187.
[31] H. Zeng, J. Li, J.P. Liu, Z.L. Wang and S.H. Sun, Nature **420** (2002) 395.
[32] Z.Y. Ma, Y.P. Guan, X.Q. Liu and H.Z. Liu, Polym. Adv. Technol. **16** (2005) 554.
[33] Z.G. Peng, K. Hidajat and M.S. Uddin, J. Colloid Interface Sci. **271** (2004) 277.
[34] Z.G. Peng, K. Hidajat and M.S. Uddin, J. Colloid Interface Sci. **281** (2005) 11.
[35] I.I. Slowing, B.G. Trewyn and V.S.Y. Lin, J. Am. Chem. Soc. **129** (2007) 8845.
[36] I.I. Slowing, B.G. Trewyn, S. Giri and V.S.Y. Lin, Adv. Fun. Mater. **17** (2007) 1225.
[37] B.G. Trewyn, S. Giri, I.I. Slowing and V.S.Y. Lin, Chem. Commun. (2007) 3236.
[38] S. Giri, B.G. Trewyn, M.P. Stellmaker and V.S.Y. Lin, Angew. Chem. Int. Ed. Engl. **44** (2005) 5038.
[39] X. Gao, K.M.K. Yu, K.Y. Tam and S.C. Tsang, Chem. Commun. (2003) 2998.
[40] H.H. Yang, S.Q. Zhang, X.L. Chen, Z.X. Zhuang, J.G. Xu and X.R. Wang, Anal. Chem. **76** (2004) 1316.
[41] Z. Chen and S.G. Weber, Anal. Chem. **79** (2007) 1043.
[42] D. Ma, M. Li, A.J. Patil and S. Mann, Adv. Mater. **16** (2004) 1838.
[43] S.C. Tsang, C.H. Yu, X. Gao and K.Y. Tam, Int. J. Pharm. **327** (2006) 139.
[44] H. Kolm, J. Oberteuffer and D. Kelland, Sci. Am. **233** (1975) 46.
[45] T.M. Vickrey and J.A. Garcia-Ramirez, Sep. Sci. Technol. **15** (1980) 1297.
[46] S. Mori, Chromatographia **21** (1986) 642.
[47] J.M. Weissman, H.B. Sunkara, A.S. Tse and S.A. Asher, Science **274** (1996) 959.
[48] G.S. Pan, R. Kesavamoorthy and S.A. Asher, Phys. Rev. Lett. **78** (1997) 3860.
[49] J.H. Holtz and S.A. Asher, Nature **389** (1997) 829.
[50] W.M. Lee, S.A. Pruzinsky and P.V. Braun, Adv. Mater. **14** (2002) 271.
[51] S.G. Romanov, N.P. Johnson, A.V. Fokin, V.Y. Butko, H.M. Yates, M.E. Pemble and C.M.S. Torres, Appl. Phys. Lett. **70** (1997) 2091.
[52] M. Egen and R. Zentel, Chem. Mater. **14** (2002) 2176.
[53] M. Campbell, D.N. Sharp, M.T. Harrison, R.G. Denning and A.J. Turberfield, Nature **404** (2000) 53.
[54] I. Divliansky, T.S. Mayer, K.S. Holliday and V.H. Crespi, Appl. Phys. Lett. **82** (2003) 1667.
[55] J. Oudar, Physics and Chemistry of Surfaces, Blackie, Glasgow, 1975.
[56] E.M. McCash, Surface Chemistry, Oxford University Press, Oxford, 2001.
[57] H. Song, F. Kim, S. Connor, G.A. Somorjai and P.D. Yang, J. Phys. Chem. B **109** (2005) 188.
[58] J.W. Yoo, S.M. Lee, H.T. Kim and M.A. El-Sayed, Bull. Korean Chem. Soc. **25** (2004) 395.
[59] P. Gallezot and D. Richard, Catal. Rev. Sci. Eng. **40** (1998) 81.
[60] M. Haruta, Catal. Today **36** (1997) 153.
[61] M. Haruta, S. Tsubota, T. Kobayashi, H. Kageyama, M.J. Genet and B. Delmon, J. Catal. **144** (1993) 175.
[62] M. Haruta, Gold Bull. **37** (2004) 27.
[63] V.M. Rotello, Nanoparticles: Building Blocks for Nanotechnology, Kluwer Academic/Plenum Publishers, New York, 2004.
[64] G.C. Bond, C. Louis and D.T. Thompson, Catalysis by Gold, Imperial College Press, London, 2006, distributed by World Scientific, Singapore.
[65] Y. Murakami and K. Konishi, J. Am. Chem. Soc. **129** (2007) 14401.
[66] S.E. Collins, J.M. Cies, E. del Rio, M. Lopez-Haro, S. Trasobares, J.J. Calvino, J.M. Pintado and S. Bernal, J. Phys. Chem. C **111** (2007) 14371.

[67] L. Ilieva-Gencheva, G. Pantaleo, N. Mintcheva, I. Ivanov, A.M. Venezia and D. Andreeva, J. Nanosci. Nanotechnol. **8** (2008) 867.
[68] N. Perkas, Z. Zhong, J. Grinblat and A. Gedanken, Catal. Lett. **120** (2008) 19.
[69] G. Schmid, R. Pfeil, R. Boese, F. Bandermann, S. Meyer, G.H.M. Calis and W.A. Vandervelden, Chem. Ber. Recl. **114** (1981) 3634.

Index

ab initio methods, 181, 214, 216
absorption coefficient, 200, 220
activity, 367, 375–378
adiabatic ionisation potential, 179
adsorption, 372, 376
aerodynamic lens, 49, 51, 56–58
aerosol, 50–51
affinity, 367
alkali metals, 18–19, 24–25, 143, 152–153, 155, 157, 161, 175, 180–181, 188, 200, 208
Anderson hamiltonian, 324
anisotropy, 278–280, 285, 292–293, 301, 306, 314–318, 327–329, 335–336, 339–341, 343–347, 349–350, 353
anisotropy constants, 238–239, 262
antiferromagnetism, 254, 263–264, 266, 268–269, 279, 298–300, 302–303, 306, 309–312, 320, 336–338, 346–352
anti-Mackay packing, 41
Arrhenius relation, 232, 278
atomic magnetism, 231–232
atomic vapour, 51, 66–67
atomistic modelling, 73–75, 101, 103, 105
attachment, 368

beam depletion spectroscopies, 175, 205
bicapped octahedron, 153, 158
binary, 113, 124, 127, 131–132
biological applications, 2
(bio-)conjugation, 368–370
bio-molecules, 366–373
bio-separation/separation, 370
bisdisphedoid, 153
blocking temperature, 233, 269, 278–279, 328, 336, 345, 352
body-centred cubic (bcc), 17, 35, 40
bottom-up, 3–4, 10, 124, 135
Brownian motion, 51
Bruggeman theory, 221
bulk magnetism, 231, 233–234

calorimetric measurements, 43
capture numbers, 79–81, 88, 92
capture zone, 90–97, 107
catalysis, 2, 13
catalyst(s), 366, 372–378
chemical probe method, 244, 257
chemical stability, 370
chemisorption, 376–377

chromium, 231, 263, 350–352
circular polarisation, 283, 285
close packing, 19, 35, 39
cluster, 49–56, 58–60, 62–67
cluster beam, 49–51, 53–56
cluster growth, 73–74, 76, 78, 80, 82, 84, 86, 88, 90, 92, 94, 96, 98, 100, 102, 104, 106, 108
cluster nucleation, 73, 107–108, 110
cluster source, 49, 51, 53–54, 56, 60, 62, 66
critical micelle concentration (CMC), 117
cobalt, 236, 247, 249–251, 263, 324, 346
cobalt oxide, 346
coercivity, 279, 349, 351–352
colloid, 115, 122–123
colloid crystal(s), 373–374
colloidal particles, 2, 197–198
condensation, 120, 130–131
configuration interaction method, 144
core-shell, 113–114, 121–122, 127, 131, 133–137
core-shell particle, 50, 198–199, 205, 222
correlated superspin glass, 344
covalent, 365–368
critical island size (i), 76–79, 84–96, 107
critical temperature, 270
CTAB, 117, 119–120
cuboctahedron, 37–40

data storage, 3, 12
decahedral-fcc transition, 168
decahedron, 17, 40
denature, 368
density functional theory (DFT), 22, 24, 143, 145, 154, 156, 160, 162
dephasing time, 206, 217
deposition, 8, 10
dielectric function, 175, 199–202, 205, 208, 215, 217–218, 221–222
dimer method, 104–105, 107
dipole–dipole interaction, 238–239
dispersion, 376
divalent metals, 30, 168, 187
DNA, 2, 199
double icoashedron, 153, 155
Drude–Lorentz–Sommerfeld model, 199

easy axis, 240, 242–243, 262
effective medium theory, 222
evaporation induced self-assembly (EISA), 124
electron affinity, 176–177, 182, 184

electron beam evaporator, 66–68
electron beam lithography (EBL), 10
electronic shell model, 6–7, 18–19, 26, 30–31, 183, 187, 189–190, 192–194, 208, 210, 214–215
electrostatic attraction, 366, 369
ellipsoidal shell model, 17, 26, 28
embedded atom model (EAM), 149–150, 154, 156, 160, 162, 167
embedded clusters, 277–278, 280, 282, 284, 286, 288, 290, 292, 294, 296, 298, 300, 302, 304–306, 308, 310, 312, 314, 316, 318, 320, 322, 324, 326, 328, 330, 332, 334, 336, 338, 340, 342, 344, 346, 348, 350, 352
encapsulated, 113–114, 130–131, 133–138
entropy, 44, 46
enzyme, 368, 373
etymology, 1, 3
Euler's Law, 100
exchange bias, 278–279, 346–353
exchange interaction, 232, 234

face-centred cubic (fcc), 17, 35–36
Fano resonance, 324, 326
FePt, 370–371
ferrimagnets, 350–351
ferromagnetic surfaces, 301
ferromagnetic transition metals, 143, 158
ferromagnetism, 236, 238
focused ion beam lithography (FIB), 3, 9–10
fractal islands, 76–77, 84
functionalisation, 365–367
functionalised nanoparticles, 2

gadolinium, 254, 269–270
gas aggregation sorce, 52, 54–55
gas phase production, 2, 6, 8–9
gas-phase cluster, 50, 62
gas-phase particle, 50–51, 53, 55, 62–63, 67
generalised-gradient approximation (GGA), 146–147, 154–156, 160, 162
geometric shell model, 7, 19, 31–32, 35, 42, 193–194
Gibbs–Kelvin equation, 51
Gibbs–Thompson effect, 97
global optimisation, 143, 149, 152, 154, 156, 160–162
gold, 367, 376–377
gradient-field deflection method, 233, 251, 255, 263
graphite surfaces, 277, 318
growth, 113–116, 120, 126, 129
growth exponents, 74–75, 81
Gupta potential, 151, 154
gyromagnetic ratio, g, 236

harmonic oscillator potential, 20–21, 31
Hartree–Fock method, 144
herringbone reconstruction, 9
heterogeneous island nucleation, 89
heterogeneous nanocatalysis, 365, 374–375, 377
homoepitaxy, 77, 86, 105
homogeneous island nucleation, 90
homogeneous linewidth, 206–207
homogenous nucleation, 51
HOMO–LUMO gap, 183, 185
Hund's rule, 233–234
hydrogenation, 376
hydrolysis, 120, 130–131
hydrothermal, 114, 138
hysteresis loops, 348–351

icosahedral-decahedral transition, 168
icosahedron, 153–155, 158–159, 165
isoelectric point (IEP), 122
inhomogeneous linewidth, 206
interaction, 367, 370
interacting clusters, 277, 340
ionisation potential, 175–182, 184
γ-iron, 312, 321
iron, 238, 246–247, 251, 253, 258, 312, 321
iron oxide, 113, 125–126, 129–131, 135
iron platinum, 113–114, 121, 131, 135, 137–138
Ising model, 329
island size distribution (ISD), 73, 75, 77–78, 82–83, 85–93, 96, 99, 101
itinerant magnetism, 267

Jellium model, 17, 19, 22, 26, 178, 180–181, 189, 192, 210–213
joint probability distribution (JPD), 93–96, 100

kinetic Monte Carlo (kMC), 81, 104–107
Knudsen cell, 54, 66–68
Kohn–Sham potentials, 24
Kondo effect, 277, 279, 301, 321, 323

Landau damping, 211
Langevin function, 233, 260, 335, 341–343
laser ablation source, 49, 54–55
latent heat, 44, 46
layer-by-layer, 127, 135
liquid drop model, 4
local density approximation (LDA), 145–146
local ripening, 99
local spin density approximation (LSDA), 145–146
log-normal distribution, 52
luminescent, 126
lustre, 198
lycurgus cup, 197

Mackay icosahedron, 17, 38–39
Mackay packing, 41
magic angle, 295–296, 330–331
magic numbers, 18–19, 30–32, 35, 38–39, 41–42
magnetic, 113, 117, 121, 124–127, 130–131, 133–136
magnetic anisotropy, 231–232, 234, 238–239, 243, 261–263, 269
magnetic nanoparticles, 1, 12–13, 365, 370–374
manganese, 231, 265, 280, 302, 306, 309–312, 320, 350–351
Marks decahedron, 40
masking, 108
mass selection, 8
mass spectra, 19, 22, 29–33, 39, 41–42
matrix, 63, 66–67
Maxwell Garnett theory, 221–222
MCDAD, 277, 282–283
mean-field approximation, 74–75, 79, 81, 89, 97
medical applications, 2
melting, 7, 17, 19, 42–46
3d metal clusters, 277, 301–302
4d metal clusters, 307
5d metal clusters, 277, 307
metal precursor, 114, 116–117, 120–121, 125, 138
metal–insulator transition, 8, 161, 176, 195–196
metallic, 113, 121, 131, 139
micro-emulsion, 113, 117–120, 124–125, 127, 130, 133
micro-Hall probes, 277, 282
micro-SQUID, 277, 281–282
Mie resonance, 208–210
Mie theory, 11, 175, 201–203, 205, 207, 211, 217–219
MLDAD, 277, 282
modification, 113, 126–127
MOKE, 277, 280–281, 308, 312, 316
mole ratio, 130–131
molecular beam epitaxy, 54
molecular dynamics (MD), 74, 101–104
Møller–Plesset perturbation theory (MP4), 145, 147, 153
monodisperse, 113, 115–116, 121, 124–126, 129–131, 134–136
monomer deposition rate (F), 79, 84–85, 88, 92–94, 100
monomer diffusion rate (D), 79, 84–85, 88, 95, 109
Monte Carlo simulation, 73–75, 77–78, 81–82, 85, 87, 89, 92–93, 95–96, 101, 104, 106
Monte Carlo model, 51
morphology, 370, 374
Morse potential, 165–167
Murrell–Mottram potential, 168

nano- as prefix, 3
nano-alloy, 370, 372, 375

nanocatalyst(s), 113–116, 121, 125, 131, 136
nanocrystals, 125–126, 128
nanolens, 199
nanomaterial(s), 113–116, 121, 125, 131
nanoparticle source, 50
nanoparticle(s), 49–51, 53, 55–57, 59, 61, 113–116, 121, 124–127, 129–131, 133–135, 137, 365–374, 376–378
nanorods, 198, 217, 219
nanoscience, 1–3
nanospheres, 198, 217, 219
nanotechnology, 1–3, 114
neutral clusters, 49, 56, 59
nickel, 238, 247, 249–251, 255, 312
noble metal surfaces, 301, 303, 308
noble metals, 29, 143, 152, 155, 157, 175, 181–182, 186, 189, 191–193, 203–204, 206, 208, 214, 216
non-collinearity, 266
non-covalent, 365–367
nucleation, 113–114, 116
nudged elastic band method, 104

octahedron, 36–38, 153, 155, 158
one-dimensional chains, 279, 301, 327
optical absorption, 1, 11, 175–176, 178, 180, 182, 184, 186, 188, 190, 192, 194, 196, 198, 200, 202, 204, 206, 208, 210, 212, 214–216, 218, 220, 222
orbital moment, 231, 236–237, 241–244, 261–262, 269, 279, 283, 285, 289–290, 292, 295, 303, 305–308, 312–316, 321, 327, 329–332, 334
oxidation, 376–377
oxides, 280, 300, 350

paedophagous effect, 99
particles, 116, 122–123, 125, 130–131, 134, 137–138
Pd and Pt surfaces, 277, 302, 314
pentagonal bipyramid, 153, 155, 158–159
pentagonal pyramid, 153, 155
photoelectron spectroscopy, 11
photoemission, 11, 175–176, 184–185, 187–189, 191–196
photoexcitation, 1, 11, 175–176, 178, 180, 182, 184, 186, 188, 190, 192, 194, 196, 198, 200, 202, 204, 206, 208, 210, 212, 214, 216, 218, 220, 222
photoionisation, 175–176, 178–181, 183–184
planar structures, 154, 157–158
plasma frequency, 199–200, 214
plasmon resonance, 177, 198, 204–206, 208, 210–212, 214, 217, 219, 221
point islands, 76, 93

polarisability, 202–203, 205
polyhedra, 17, 36
polyol, 113, 120–122, 124, 126, 129, 131, 133, 137
polyol method, 10
production methods, 8
protein, 367–368, 372–373
pulsed arc cluster ion source, 49, 54–55
pulsed field mass selector, 49, 60, 62
pulsed microplasma cluster source, 49, 54, 56

quadrupole mass filter, 49, 59
quantum chemistry methods, 143–145
quantum mirage, 326–327
quantum tunnelling, 3
quenching, 236, 261

random phase approximation (RPA), 210
random Voronoi network (RVN), 90–91
rare-earths, 231, 233–234, 236–237, 254, 269
rate equations, 73, 75, 78–79, 81–83, 88–89, 93, 97, 107
reaction, 366, 368, 374–376
reactivity, 200, 231, 244, 246–247
rhodium, 241, 244, 268–269
rhombic dodecahedron, 40–41
ripening, 73, 75, 97–101, 107–108
Rudernan–Kittel–Kasuya–Yosida interaction (RKKY), 234, 269

scale-invariance, 75, 78, 85–86, 89–91, 95, 101
scaling analysis, 81–82, 84–85, 92
scanning probe microscopes, 43
s–d hybridization, 176, 194
second moment approximation, 150–151, 243
seeded supersonic nozzle source, 49, 53–54
selectivity, 366, 373, 375–378
self-assembly, 4, 8, 124, 126, 128, 199
self-learning kMC, 105
semi-empirical potentials, 143, 149, 152, 158
sensors, 2, 11, 199
shape, 114, 125–126
shell models, 17–18, 20, 22, 24, 26, 28, 30, 32, 34, 36, 38, 40, 42, 44
short-range magnetic order, 270
silica, 368–369, 371–373
single-domain particles, 232
single-molecule magnets, 280
size, 113–117, 120, 125–126, 129–131, 138
sol-gel, 113, 120, 127, 131, 133, 135
s–p hybridization, 162
spherical jellium model (phenomenological), 17–20, 22, 43
spherical jellium model (self-consistent), 17, 19, 22, 24
spherical square well potention, 20–22, 35

spill-out, 208–209, 211, 215–216
spin moment, 234, 236, 242–243, 256, 258, 260–261, 269, 286, 292–293, 295, 303, 305, 312–314, 320, 330–331, 333–334, 339–340
spin-orbit coupling, 269
spintronics, 2–3
spontaneous nucleation, 85
sputter source, 54–55
square pyramid, 153
stability, 370, 377–378
stabilizer, 120–121, 125–126, 131
step-edge, 279, 301, 328
Stern–Gerlach, 58, 62, 233, 251, 253, 259
Stokes number, 57
Stoner criterion, 235–236, 243, 267
Stoner theory, 235–236, 243, 267
strain interactions, 108
structures, 236, 244, 247–251, 256–261, 266, 268
superparamagnetism, 232–233, 252, 255, 270, 370
supersaturated vapour, 51, 54
supersaturation ratio, 51
supershells, 17, 33
surface binding, 367
surface plasmon resonance, 198, 204–206, 208, 210–212, 214, 217, 221
surface reconstructions, 107
surface/volume ratio, 1, 4
surface-enhanced Raman spectroscopy (SERS), 198, 206
surfactant, 117, 119–121, 126, 130–131, 133, 137
Sutton–Chen potential, 161

temperature accelerated dynamics (TAD), 102–105
terbium, 233, 269–270
tetrahedron, 153, 158
thermal decomposition, 120, 131
tight binding methods, 147
time of flight, 58, 61
time-dependent local density approximation (TDLDA), 210–212
top-down, 3, 9–10
transition metals, 143, 147–148, 151, 158–159, 161, 175, 192, 194, 231, 233–234, 236–239, 244, 246, 252–253, 255, 263, 269–270
trigonal bipyramid, 153–154, 158, 161
trivalent metals, 143, 152, 161, 175, 183–184, 195
truncated decahedron, 17, 40
truncated octahedron, 37
tumour therapy, 199

umbrella model, 41
uniaxial anixotropy, 240

uniform depletion approximation, 80
vanadium, 267
vapour deposition, 73, 75, 78, 81, 97, 107–108
vertical detachment energy (VDE), 176, 179, 186–188, 191
vertical ionisation potential (VIP), 179, 186–187
Voronoi tessellation, 90, 97
VSM, 277, 280

Watson–Crick base-pairing, 199
wet chemical methods, 9–10
Wien filter, 49, 60–61

Wigner–Seitz radius, 4–5, 20, 22, 25
Winsor, 117
Wood's notation, 302
Woods–Saxon potential, 20, 22, 24, 26, 31–33, 35
work function, 176–179, 182, 190
Wulff construction, 17, 35–36, 40

XMCD, 277, 279, 283–285, 289–290, 292, 295, 297–298, 300, 302–303, 308, 310–312, 315, 321, 328–329, 331–332, 339–340
XMLD, 277, 298, 300